Lipmann Symposium

Lipmann Symposium

Energy, Regulation and Biosynthesis in Molecular Biology

Editor Dietmar Richter

Walter de Gruyter · Berlin · New York 1974

Editor
Dr. *Dietmar Richter*
Professor, Head of Abt. Cell Biochemistry,
Institute of Physiological Chemistry,
University of Hamburg, Hamburg, Germany.

Photographs on pp. 699 – 701 by P. Herrlich, Z. Kučan, and M. Møller

Cover design: R. Hübler. – Printing: Druckerei Gerike, Berlin. – Binding: Lüderitz & Bauer, Berlin. –
Printed in Germany
ISBN 3 11 004976 7

Photograph by Z. Kucan

Fritz Lipmann

Dedicated to the 75th Birthday of Dr. Fritz Lipmann

Preface

Dr. Fritz Lipmann celebrated his 75th birthday on June 12th, 1974. To mark this event a symposium was arranged and took place from July 7th to 9th at the Max-Planck-Institut für Molekulare Genetik in Berlin-Dahlem, the city where, fifty years ago, the career of this great scientist began. About 80 Lipmann alumni, guests and friends of the Lipmann family foregathered to meet their teacher and mentor, to discuss the past, present and future of biochemistry, and to exchange memories of their days in the Lipmann Laboratory. They came from all over Europe and the United States, and from countries as far away as Chile, Japan, and Australia.

Because of the special nature of this versatile biochemist, the lectures of this Symposium covered many fields of biochemistry. The book begins with vivid and amusing accounts by the two "old-timers" of biochemistry, Sir Hans Krebs and Fritz Lipmann of life and conditions in the "Golden Twenties" in the laboratories of Warburg and Meyerhof at the former Kaiser-Wilhelm-Institut. The volume closes with the lecture on "Structures of Biotin Enzymes" by Dr. F. Lynen. This lecture marks the beginning of a new event in the scientific calendar, the "Fritz Lipmann Lecture", which will now be given annually, and in this connection I am especially grateful to the Boehringer Mannheim Corporation for their financial support.

As editor of this volume I am indebted to Dr. Fritz Lipmann for the time that I spent in his laboratory at the Rockefeller University in New York. As Sir Hans Krebs so aptly put it, "The desire of the followers to express a sense of loyalty, gratitude and affection" was by no means the least of the reasons why the Symposium was such a success.

It is one thing to conceive the idea of arranging such a Symposium, but its management and organization is quite another matter; it has required the assistance and support of a number of people and institutions whom I would like to thank. In particular, I would like to mention Dr. F. Lynen and Dr. H.U. Bergmeyer, both of whom made the initiation of the annual Fritz Lipmann Lecture possible; also Dr. E. Helmreich the chairman, Dr. E. Auhagen the treasurer, and Dr. H. Gibian the secretary of the Gesellschaft für Biologische Chemie. The meeting would have been much more difficult to organize without the advice and cooperation of Dr. H.G. Wittmann. The financial support given by the pharmaceutical companies Schering AG and Boehringer Mannheim Corporation, as well as the Volkswagen Stiftung, Max-Planck-Gesellschaft, Gesellschaft für Biologische Chemie and Walter de Gruyter Press is gratefully acknowledged.

The planning and organization would have been inconceivable without the help, support and continuous encouragement of my wife Heidi. For this I am very thankful.

I am grateful to the Walter de Gruyter Press for making it possible for us to publish the contributions in their entirety in a special birthday volume which is dedicated to Dr. Fritz Lipmann with best wishes.

I do not want to end this Preface without remarking on how Dr. Lipmann appeared to his former colleagues, many of whom may not have seen him for some years; everyone was pleased to see that his own personal store of high-energy phosphate bonds appears to be sufficient to keep him going for a good while yet. We wish him sincerely many more creative and energetic years.

July 1974 Dietmar Richter

Contents

List of Participants of the Lipman Symposium XIV
List of Contributors XIX

Opening Remarks

Helmreich, E.J.M.: Welcome to Professor Lipmann 1

Wittmann, H.G.: Welcome to the Max-Planck-Institut 5

Introduction

Krebs, H.A. and Lipmann, F.: Dahlem in the Late Nineteen Twenties 7

Contributions

Adler, S.P. and Stadtman, E.R.: Cascade Control of E.coli Glutamine Synthetase 28

Baddiley, J., Burnett, J.P., Hancock, I.C., and Heptinstall, J.: Structure and Biosynthesis of an Acidic Glycoprotein in an Bacterial Cell Envelope 40

Bauer, K.: Degradation of Thyrotropin Releasing Hormone (TRH). Its Inhibition by Pyroglu-His-OCH_3 and the Effect of the Inhibitor in Attempts to Study the Biosynthesis of TRH 53

Bennett, T.P.: Amphibian Cells in Culture: An Approach to Metamorphosis 63

Bessman, S.P.: The Hexokinase-Mitochondrial Binding Theory of Insulin Action 77

Bloemendal, H. and Strous, G.J.A.M.: N-Terminal Acetylation of Proteins, a Post-Initiational Event 89

Boman, H.G., Nilsson-Faye, I., and Rasmuson, T.: Why is Insect Immunity Interesting? 103

Brodie, A.F., Lee, S.-H., Prasad, R., Kalra, V.K., and Kosmakos, F.C.: Energy Transduction in Membranes of Mycobacterium phlei 115

Cabrer, B., San-Millan, M.J., Gordon, J., Vazquez, D., and Modolell, J.: Control by Peptidyl-tRNA of Elongation Factor G Interaction with the Ribosome 131

Chantrenne, H.: A Cytoplasmic Function for the Polyadenylic Sequences of Messenger RNA 144

Christman, J.K. and Acs, G.: Purification and Biological Properties of a Plasminogen Activator Characteristic of Malignantly Transformed Cells 150

Ebashi, S.: Interactions of Troponin Subunits Underlying Regulation of Muscle Contraction by Ca Ion: A Study on Hybrid Troponins 165

Edens, B., Thompson, H.A., and Moldave, K.: Studies on the Binding of Acylaminoacyl-Oligonucleotide to Rat Liver 60S Ribosomal Subunits and Its Participation in the Peptidyltransferase Reaction 179

Engström, L., Berglund, L., Bergström, G., Hjelmquist, G., and Ljungström, O.: Regulatory Phosphorylation of Purified Pig Liver Pyruvate Kinase 192

Ganoza, M.C. and Fox, J.L.: A Soluble Elongation Factor Required for Protein Synthesis with mRNA's other than Poly (U) 205

Gerlach, U. and Fegeler, W.: Variations in the Lactate Dehydrogenase Isoenzyme Pattern in Arteries 216

Gevers, W., Jones, P.A., Coetzee, G.A., and Westhuyzen, D.R. van der: Biochemical Development of the Heart in Syrian Hamsters 225

Goldberg, I.H., Kappen, L.S., and Suzuki, H.: On the Mechanism of Inhibition of Globin Chain Initiation by Pactamycin 237

Gordon, J., Howard, G.A., Stöffler, G., and Highland, J.H.: The Ribosomal Binding Site of the Antibiotic Thiostrepton 250

Haenni, A.L., Prochiantz, A., and Yot, P.: tRNA Structures in Viral RNA Genomes 264

Hess, B.: Remarks on the Acquisition of Active Quaternary Structure of Enzymes 277

Hierowski, M. and Brodersen, R.: Covalent Binding of Bilirubin to Agarose and Use of the Product for Affinity Chromatography of Serum Albumin 281

Hildebrand, J.G. and Kravitz, E.A.: Transmitter Biochemistry of Single, Identified Neurons 298

Huberman, A., Rodriguez, J.M., Franco, R., and Barahona, E.: Regulation of Apoferritin Biosynthesis in Rat Liver 308

Hülsmann, W.C. and Jansen, H.: High Lipoprotein Lipase Activity and Cardiovascular Disease 322

Kleinkauf, H. and Koischwitz, H.: Gramicidin S-Synthetase: Active Form of the Multienzyme Complex is Undissociable by Sodium Dodecylsulfate 336

Krisko, I. and Gyorkey, F.: Studies on the Biosynthesis and Structure of Renal Glomerular Basement Membrane 345

Kućan, Ž.: Ribosomes and the Target Theory 359

Lee, S.G.: Interrelation Between Tyrocidine Synthesis and Sporulation in Bacillus brevis 368

Lill, U.I., Behrendt, E.M., and Hartmann, G.R.: Intergeneric Complementation of RNA Polymerase Subunits 377

Liu, C.K., Legocki, A.B., and Weissbach, H.: Studies on Liver Elongation Factor 1 384

Maas, W.K.: Some Thoughts on the Regulation of Arginine Biosynthesis and Its Relation to Biochemistry 399

Mano, Y., Suzuki, N., Murakami, K., and Kano, K.: Mechanism of the Cell Cycle in DNA Synthesis in Sea Urchin Embryos 404

Marchis-Mouren, G., Marvaldi, J., and Cozzone, A.: Ribosome Metabolism in Starving Relaxed E.coli Cells 410

Matsuura, T. and Jones, M.E.: Subcellular Localization in the Ehrlich Ascites Cell of the Enzyme which Oxidizes Dihydroorotate to Orotate 422

Nombela, C., Nombela, N.A., and Ochoa, S.: Comparison of Polypeptide Chain Initiation Factors from Artemia salina and Rabbit Reticulocytes 435

Nose, Y., Iwashima, A., and Matsuura-Nishino, A.: Thiamine-Binding Protein of Escherichia coli 443

Ofengand, J. and Schwartz, I.: Photoaffinity Labelling of tRNA Binding Sites on E.coli Ribosomes 456

Orrego, F.: Identification of Central Transmitters with Electrically Stimulated Brain Slices 471

Park, J.H. and Meriwether, B.P.: The Acetylation of Cysteine-281 in Glyceraldehyde-3-Phosphate Dehydrogenase by an S—S Transfer Reaction 487

Parmeggiani, A., Sander, G., Marsh, R.C., Voigt, J., Nagel, K., and Chinali, G.: Regulation of Elongation Factor G-Ribosomal GTPase Activity 499

Pénit, C., Paraf, A., and Chapeville, F.: Deoxynucleotide-Polymerizing Enzymes of Murine Myelomas 511

Richter, D. and Møller, W.: Properties and Functions of Ethanol-Potassium Chloride Extractable Proteins from 80S Ribosomes and their Interchangeability with the Bacterial Proteins L7/L12 524

Roskoski, R.: Choline Acetyltransferase: Reactions of the Active Site Sulfhydryl Group 534

Schweiger, M., Herrlich, P., Rahmsdorf, H.J., Pai, S.H., Ponta, H., Hirsch-Kauffmann, M.: Control of Gene Expression by E.coli Virus T7 547

Spector, L.B.: Covalent Enzyme-Substrate Intermediates in Carboxyl Activation 564

Srere, P.A. and Singh, M.: The Role of Pantothenate in Citrate Lyase 575

Streeck, R.E., Fittler, F., and Zachau, H.G.: Cleavage of Small DNAs with Restriction Nucleases 589

Sy, J.: The Ribosomal and Nonribosomal Synthesis of Guanosine Polyphosphates 599

Szafranski, P., Filipowicz, W., Wodnar-Filipowicz, A., and Zagorska, L.: Phage f2 RNA Structure in Relation to Synthesis of Phage Proteins 610

Takeda, M. and Nishizuka, Y.: Protein Kinases from Eukaryotic Organisms 623

Tao, M., Yuh, K.-C., and Hosey, M.M.: The Interaction of Red Blood Cell Protein Factors with Cyclic AMP 636

Wakabayashi, K.: On the Linkage and Recombination of Mitochondrial Genes 647

Wiegers, U., Kramer, G., Klapproth, K., Wiegers, U., and Hilz, H.: Determination of mRNA and 28 S RNA Turnover in Proliferating HeLa Cells 658

Annual Fritz Lipmann Lecture

Lynen, F.: Structures of Biotin Enzymes 671

Photographs 699

List of Participants of the Lipmann Symposium

ACS, G.
The Mt. Sinai School of Medicine, New York, USA

AUHAGEN, E.
Treasurer of the Gesellschaft für Biologische Chemie, Wuppertal, Germany

BADDILEY, J.
The University of Newcastle, Newcastle upon Tyne, England

BAUER, K.
Technische Universität Berlin, Berlin, Germany

BENNETT, T.P.
Florida State University, Tallahasse, USA

BERGMEYER, H.-U.
Boehringer Mannheim Company, Tutzing, Germany

BESSMAN, S. P.
University of Southern California, Los Angeles, USA

BLOEMENDAL, H.
University of Nijmegen, Nijmegen, The Netherlands

BOMAN, H. G.
University of Umea, Umea, Sweden

BRODIE, A. F.
University of Southern California, Los Angeles, USA

CHANTRENNE, H.
Laboratoire de Chimie Biologique, Bruxelles, Belgium

CHAPEVILLE, F.
Institut de Biologie Moléculaire, Paris, France

CRAMER, F.
Max-Planck-Institut für Experimentelle Medizin, Göttingen, Germany

CROKAERT, R.
Université Libre de Bruxelles, Bruxelles, Belgium

EBASHI, S.
University of Tokyo, Tokyo, Japan

EHRENSTEIN, G. von
Max-Planck-Institut für Experimentelle Medizin, Göttingen, Germany

EIGEN, M.
Max-Planck-Institut für Biophysikalische Chemie, Göttingen, Germany

ELLIOTT, W. H.
University of Adelaide, Adelaide, Australia

ENGSTRÖM, L.
Institute of Medical Chemistry, Uppsala, Sweden

FISCHER, H.
Max-Planck-Institut für Immunbiologie, Freiburg, Germany

FOX, J. L.
University of Texas, Austin, USA

GERLACH, U.
Universität Münster, Münster, Germany

GIBIAN, H.
Schering AG, Berlin, Germany

GOLDBERG, I. H.
Harvard Medical School, Boston, USA

GORDON, J.
Friedrich Miescher Institut, Basel, Switzerland

HAENNI, A. L.
Institut de Biologie Moléculaire, Paris, France

HARTMANN, G.
Universität München, München, Germany

HESS, B.
Max-Planck-Institut für Ernährungsphysiologie, Dortmund, Germany

HELMREICH, E.
Chairman of the Gesellschaft für Biologische Chemie, Würzburg, Germany

HERRLICH, P.
Max-Planck-Institut für Molekulare Genetik, Berlin, Germany

HIEROWSKI, M.
University of Aarhus, Aarhus, Denmark

HILDEBRAND, J. G.
Harvard Medical School, Boston, USA

HUBERMAN, A.
Instituto Nacional de la Nutrición, Mexico City, Mexico

HÜLSMANN, W.C.
Erasmus University, Rotterdam, The Netherlands

JONES, M. E.
University of Southern California, Los Angeles, USA

KLEINKAUF, H.
Technische Universität Berlin, Berlin, Germany

KREBS, H. A.
Oxford, England

KRISKO, I.
Veterans Administration Hospital, Houston, USA

KUĆAN, Ž.
"Rugjer Bošković" Institute, Zagreb, Yugoslavia

LIPMANN, F.
The Rockefeller University, New York, USA

LYNEN, F.
Max-Planck-Institut für Biochemie, Martinsried/München, Germany

MANO, Y.
University of Tokyo, Tokyo, Japan

MARCHIS-MOUREN, G.
Institut de Chimie Biologique, Marseille, France

MØLLER, K.
Carlsberg Laboratorium, Copenhagen, Denmark

NIEMEYER, H.
Universidad de Chile, Santiago, Chile

NOSE, Y.
Kyoto Prefectural University of Medicine, Kyoto, Japan

OFENGAND, J.
Roche Institute of Molecular Biology, Nutley, USA

ORREGO, F.
Instituto Nacional de Cardiologia, Mexico City, Mexico

PARK, J.H.
Vanderbilt University Medical School, Nashville, USA

PARMEGGIANI, A.
Gesellschaft für Molekularbiologische Forschung, Stöckheim/Braunschweig, Germany

RICHTER, D.
Universität Hamburg, Hamburg, Germany

ROSKOSKI, R.
University of Iowa, Iowa City, USA

SCHUSTER, H.
Max-Planck-Institut für Molekulare Genetik, Berlin, Germany

SCHWEIGER, M.
Max-Planck-Institut für Molekulare Genetik, Berlin, Germany

SPECTOR, L.
The Rockefeller University, New York, USA

SRERE, P.A.
The University of Texas, Dallas, USA

STADTMAN, E.R.
National Institutes of Health, Bethesda, USA

SY, J.
The Rockefeller University, New York, USA

SZAFRANSKI, P.
Polish Academy of Sciences, Warsaw, Poland

TAKEDA, M.
Kobe University, Kobe, Japan

TAO, M.
University of Illinois, Chicago, USA

TRAUT, R. R.
University of California, Davis, USA

TRAUTNER, T. A.
Max-Planck-Institut für Molekulare Genetik, Berlin, Germany

URETA, T.
Universidad de Chile, Santiago, Chile

VENNESLAND, B.
Forschungsstelle Vennesland, Berlin, Germany

WAKABAYASHI, K.
University of Tokyo, Tokyo, Japan

WITT, H. T.
Technische Universität Berlin, Berlin, Germany

WITTMANN, H. G.
Max-Planck-Institut für Molekulare Genetik, Berlin, Germany

WITTMANN-LIEBOLD, B.
Max-Planck-Institut für Molekulare Genetik, Berlin, Germany

ZACHAU, H. G.
Universität München, München, Germany

List of Contributors

Acs, G. 150
Adler, S.P. 28
Baddiley, J. 40
Barahona, E. 308
Bauer, K. 53
Behrendt, E.M. 377
Bennett, T.P. 63
Berglund, L. 192
Bergström, G. 192
Bessman, S.P. 77
Bloemendal, H. 89
Boman, H.G. 103
Brodersen, R. 281
Brodie, A.F. 115
Burnett, J.P. 40
Cabrer, B. 131
Chantrenne, H. 144
Chapeville, F. 511
Chinali, G. 499
Christman, J.K. 150
Coetzee, G.A. 225
Cozzone, A. 410
Ebashi, S. 165
Edens, B. 179
Engström, L. 192
Fegeler, W. 216
Filipowicz, W. 610
Fittler, F. 589
Fox, J.L. 205
Franco, R. 308
Ganoza, M.C. 205
Gerlach, U. 216
Gevers, W. 225
Goldberg, I.H. 237
Gordon, J. 131, 250
Gyorkey, F. 345
Haenni, A.L. 264
Hancock, I.C. 40
Hartmann, G.R. 377
Helmreich, E.J.M. 1
Heptinstall, J. 40
Herrlich, P. 547
Hess, B. 277
Hierowski, M. 281
Highland, J.H. 250
Hildebrand, J.G. 298
Hilz, H. 658
Hirsch-Kauffmann, M. 547
Hjelmquist, G. 192
Hosey, M.M. 636
Howard, G.A. 250
Huberman, A. 308
Hülsmann, W.C. 322
Iwashima, A. 443
Jansen, H. 322
Jones, M.E. 422
Jones, P.A. 225
Kalra, V.K. 115
Kano, K. 404
Kappen, L.S. 237
Klapproth, K. 658
Kleinkauf, H. 336
Koischwitz, H. 336
Kosmakos, F.C. 115
Kramer, G. 658
Kravitz, E.A. 298
Krebs, H.A. 7
Krisko, I. 345
Kućan, Ž. 359
Lee, S.-H. 115
Lee, S.G. 368
Legocki, A.B. 384
Lill, U.I. 377
Lipmann, F. 7
Liu, C.K. 384
Ljungström, O. 192
Lynen, F. 671
Maas, W.K. 399
Mano, Y. 404
Marchis-Mouren, G. 410

Marsh, R.C. 499
Marvaldi, J. 410
Matsuura, T. 422
Matsuura-Nishino, A. 443
Meriwether, B.P. 487
Modolell, J. 131
Moldave, K. 179
Møller, W. 524
Murakami, K. 404
Nagel, K. 499
Nilsson-Faye, I. 103
Nishizuka, Y. 623
Nombela, C. 435
Nombela, N.A. 435
Nose, Y. 443
Ochoa, S. 435
Ofengand, J. 456
Orrego, F. 471
Pai, S.H. 547
Paraf, A. 511
Park, J.H. 487
Parmeggiani, A. 499
Pénit, C. 511
Ponta, H. 547
Prasad, R. 115
Prochiantz, A. 264
Rahmsdorf, H.J. 547
Rasmuson, T.,Jr. 103
Richter, D. 524
Rodriguez, J.M. 308
Roskoski, R. 534
Sander, G. 499
San-Millan, M.J. 131
Schwartz, I. 456
Schweiger, M. 547
Singh, M. 575
Spector, L.B. 564
Srere, P.A. 575
Stadtman, E.R. 28
Stöffler, G. 250
Streeck, R.E. 589
Strous, G.J.A.M. 89
Suzuki, H. 237
Suzuki, N. 404
Sy, J. 599
Szafranski, P. 610
Takeda, M. 623
Tao, M. 636
Thompson, H.A. 179
Vazquez, D. 131
Voigt, J. 499
Wakabayashi, K. 647
Westhuyzen, D.R.van der 225
Weissbach, H. 384
Wiegers, U. 658
Wiegers, U. 658
Wittmann, H.G. 5
Wodnar-Filipowicz, A. 610
Yot, P. 264
Yuh, K.-C. 636
Zachau, H.G. 589
Zagorska, L. 610

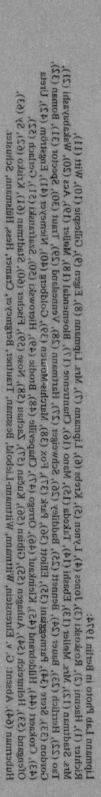

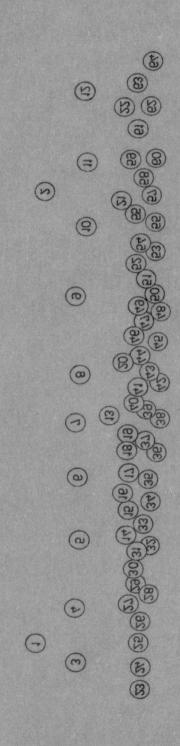

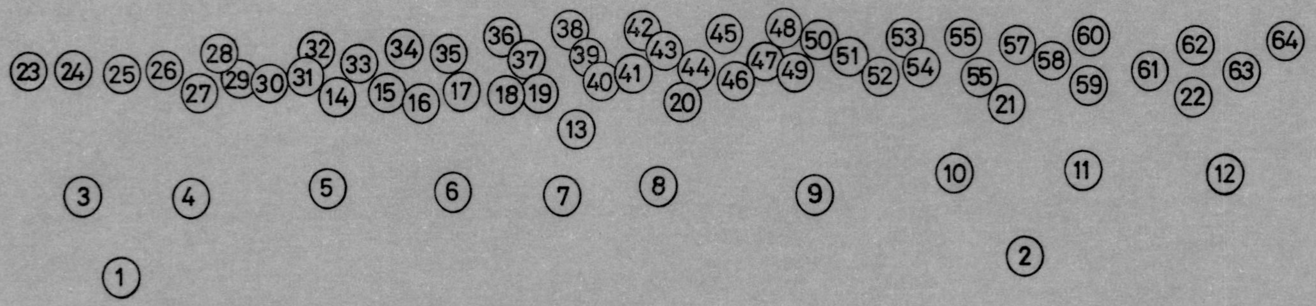

Lipmann Lab Photo in Berlin 1974:
Richter (1), Haenni (2), Roskoski (3), Jones (4), Lynen (5), Krebs (6), Lipmann (7), Mrs. Lipmann (8), Eigen (9), Gillespie (10), Witt (11), Mrs. Stadtman (12), Mrs. Møller (13), Ebashi (14), Takeda (15), Mano (16), Chantrenne (17), Bloemendal (18), Møller (19), Acs (20), Wakabayashi (21), Tao (22), Herrlich (23), Bauer (24), Bennett (25), Baddiley (26), Schweiger (27), Hartmann (28), Vennesland (29), Traut (30), Spector (31), Boman (32), Gordon (33), Srere (34), Parmeggiani (35), Elliott (36), Park (37), Fox (38), Marchis-Mouren (39), Goldberg (40), Niemeyer (41), Engström (42), Ureta (43), Crokaert (44), Hildebrand (45), Kleinkauf (46), Orrego (47), Chapeville (48), Brodie (49), Hierowski (50), Szafranski (51), Gerlach (52), Ofengand (53), Helmreich (54), Auhagen (55), Gibian (56), Kučan (57), Zachau (58), Nose (59), Fischer (60), Stadtman (61), Krisko (62), Sy (63), Huberman (64). Absent: G. v. Ehrenstein, Wittmann, Wittmann-Liebold, Bessman, Trautner, Bergmeyer, Cramer, Hess, Hülsmann, Schuster.

Welcome to Professor Lipmann

Ernst J. M. Helmreich
Vorstand der Gesellschaft für Biologische Chemie

It is indeed a memorable event to honor Professor Fritz Lipmann at the occasion of his 75th birthday here in Berlin. In 1970 Professor Lipmann wrote a personal account of his life as a Biochemist which was published by Wiley Interscience 1971 with the title: "Wanderings of a Biochemist". This delightful little book, because it is so honest and unpretentious, has impressed me as I am certain it has impressed every Biochemist who has worked at least in part during this period. Moreover, it makes clear why to honor Professor Lipmann here in Berlin, fits the occasion because it was in Berlin at the Kaiser Wilhelm Institute for Biology at Dahlem, when he joined Otto Meyerhof's Laboratory in 1927 where he was exposed to perhaps the most important intellectual influences in his life which have molded his scientific philosophy and interests. As Professor Lipmann writes: "In the Freudien sense all that I did later was subconsciously mapped out here and it started to mature between 1930 and 1940 and was more elaborately realized from then on". It is not my task to enumerate the many discoveries which Professor Lipmann and his associates have made over the years in so many different areas of Biochemistry. They are biochemical history and are taught all over the world to students. When I have selected a few examples of Professor Lipmann's contribution it is obvious that my choice was arbitrary and dictated by my own research interests. But perhaps in a time where many younger colleagues and students lack sense for historical relationships, it might be worthwhile to show on a few examples how influential Professor Lipmann's work was. This might help to arrive at a more balanced view of the events which set the pace and helped shape the Biology of our age.

Among the two examples which I have chosen one dates back to 1932/33. It was the characterization of the phosphoryl seryl ester bond in certain proteins. Professor Lipmann writes in his memoirs: "When 40 years ago I chose to probe into the binding of the phosphate in these phosphoproteins, unwittingly, it now turns out, I struck a gold mine". Phosphorylation not only of enzymes but of a variety of proteins, so different as the histone and non-histone proteins of chromatin, or membrane and contractile proteins are among the most interesting control devices which have emerged in the evolution of living systems. However, these and many other important discoveries are overshadowed by perhaps Professor Lipmann's most important contribution. That is the concept of the energy rich bond and the idea of a phosphoryl group transfer potential. There are some of us here who had the honor to be invited to contribute to the Fritz Lipmann dedicatory volume "Current Aspects of Biochemical Energetics" which was edited by Nathan O. Kaplan and Eugene P. Kennedy at the 25th Anniversary of the Publication of Fritz Lipmann's classic paper on the "Metabolic Generation and Utilization of Phosphate Bond Energy". On the cover of that book it was rightly stated that this contribution had opened the way to an understanding of the energetic basis of biosynthetic reactions and greatly clarified the relation between exergonic and endergonic processes in the living cell.

But aside from his scientific achievements there is another aspect which should not be forgotten at this occasion. We only have to look at the program of this Symposium to realize what Professor Lipmann has done for Biochemistry, especially also for European Biochemistry. The contributors to the Scientific Program of this Symposium are his students and biochemical research in Europe has greatly benefitted from them. Many of them are Heads of Departments or lead independent research groups. I am pleased to announce today that the Gesellschaft für Biologische Chemie as a token of our gratitude has installed an Annual Fritz Lipmann Lecture made possible by a generous financial contribution of Boehringer Mannheim Company. I am especially indebted to Professor H. U. Bergmeyer without its help this would not have been possible. Professor Lynen will give the first Fritz Lipmann Lecture at the conclusion of this symposium.

As a biochemist, coming from Würzburg I was thinking about something related with that University and with Biochemistry and which might be of interest to Professor Lipmann. Then I recalled that my colleague Guido Hartmann sometime ago when he was still at Würzburg University had presented me with reprints of two papers published by Robert E. Kohler from the Department of History of Science at Harvard University. The papers appeared in the Journal of the History of Biology 1971, 1972 and are entitled: "The Background to Eduard Buchner's Discovery of Cell Free Fermentation and the Reception of Eduard Buchner's Discovery of Cell Free Fermentation. As Professor Lipmann writes in his recollections: "What I delight in here is Buchner's preoccupation with the accidental". Kohler's historical study quite convincingly shows that Hans Buchner, who was an Immunologist and Professor of Medical Microbiology and Public Health in Munich was the intellectual driving force and that the discovery of zymase actually emerged from Hans Buchner's immunological studies. Eduard Buchner's experiments were actually undertaken to elucidate the role of Protoplasmic proteins for immunity. It was in the course of these studies that Eduard Buchner who later became professor of biochemistry and organic chemistry in Würzburg together with Martin Hahn developed a method for getting press juice which facilitated the study of intercellular enzymes and proved cell free fermentation in 1897. This discovery has long been recognized as the resolution of one of the most famous scientific controversies of the 19th century, namely the controversy between Louis Pasteur and Justus von Liebig over the nature of alcoholic fermentation. Therefore Guido Hartmann and I, two Biochemists from Würzburg take the pleasure to present to Professor Lipmann today together with these two reprints a reprint of a paper "Über Zellfreie Gärung" in which Eduard Buchner himself gave an account of his discovery in a lecture given 1909 and published in the Zeitschrift des "Österreichischen Ingenieur und Architekten Vereins". In this lecture Eduard Buchner said: "Der misslichste Punkt ist aber jedenfalls der, daß wir durchaus nicht in der Lage sind, die Enzyme zu isolieren. Bei Reinigungsbestrebungen ist der erste Erfolg meist der, daß auch die Wirksamkeit des betreffenden Präparates abnimmt". This sounds familiar to every enzymologist but it also shows that thanks to Professor Lipmann and the biochemists of his time we have finally solved this and many other problems because we have followed Goethe's advice with which Eduard Buchner concluded its lecture. He said: "Wenn aber einmal unsere Bemühungen tatsächlich jahrelang ohne Erfolg bleiben sollten, dann lassen Sie uns recht der Worte Goethes eingedenk sein: "Der Mensch muß bei dem Gedanken verharren, daß das Unbegreifliche begreiflich sei; er würde sonst nicht forschen!"

Before Professor Wittmann our host here in Berlin will officially open the Fritz Lipmann Symposium which was organized by Dr. Dietmar Richter, I wish to acknowledge the very generous contributions of the VW Foundation, the Schering A.G. and the Walter de Gruyter Verlag, Berlin. I am sure that all the participants are looking forward to be with friends and for having the opportunity to discuss science and reminisce about the time they spent together and with Professor Lipmann in the laboratory. I wish Professor Lipmann, his students and his friends a very good and enjoyable time while here in Berlin.

Welcome to the Max-Planck-Institut

H.G. Wittmann
Max-Planck-Institut für Molekulare Genetik
Berlin-Dahlem, Germany

Dr. Lipmann, Ladies and Gentlemen:

It gives me great pleasure to welcome you all to the Max-Planck-Institut für Molekulare Genetik in Berlin-Dahlem. It was here in Dahlem where Fritz Lipmann, our guest of honour today, began his scientific career as a biochemist almost fifty years ago. At that time, Dahlem was the scientific center in Germany. This was mainly due to the work done in the various Kaiser-Wilhelm-Institutes located here. You will hear more about this period of time in the next two lectures.

At the end of the war, many of these institutes moved from Berlin to West-Germany. During the last 25 years, the number of the institutes (renamed the Max-Planck-Institutes) increased in number to about 50 staffed by approximately 11.000 scientists, engineers and technicians. If you are interested in obtaining more details about the Max-Planck-Gesellschaft and the institutes, there is a booklet available containing information about the organization and the budget of the MPG, as well as the scientific work done in each of the institutes.

The work in our institute is mainly concerned with the biochemical and genetic aspects of nucleic acid and protein biosynthesis in bacteria and phages. There are about 200 people (scientists, technicians and graduate students) working on the following subjects: DNA replication; prophage-induction; DNA-membrane interation; mechanism of cell division and conjugation; recombination, transfection and transformation; structure and function of ribosomes, which especially includes studies on the primary structure of ribosomal proteins, on the topography of the subunits and on the functional role of the ribosomal components; protein-RNA interaction; mechanism and regulation of translation; DNA dependent protein biosynthesis in vitro. Besides these studies on prokaryotes there are two groups working on eukaryotic systems, namely on phototropism in Phycomyces and conjugation in Ciliata.

Anybody who is interested in details of the work done in our institute, he is welcome to either approach directly the group in which he is interested or, if he is unsure where to go, please come and see me so that I can make the appropriate introductions. Please regard our institute as an open house and we will try our best to make your visit in Berlin and in our institute as enjoyable as possible. Let me end by wishing all of our guests a pleasant time in Berlin and Dr. Lipmann a very happy seventy-fifth birthday.

Dahlem in the Late Nineteen Twenties

Hans A. Krebs and Fritz Lipmann

Metabolic Research Laboratory, Nuffield Department of Clinical Medicine, Radcliffe Infirmary, Oxford, England and The Rockefeller University, New York, USA

Hans A. Krebs

It was Dr. Richter who suggested that I might reminisce on the life at Dahlem in the days when Fritz Lipmann and I were working here in the laboratories of Otto Meyerhof and Otto Warburg. I was in Dahlem from 1926 to 1930, and Fritz Lipmann from 1927-1930. So I will make an attempt to convey something of the atmosphere, of the working and living conditions, of our attitudes towards our work and of our outlook in general.

The Kaiser Wilhelm Gesellschaft

Dahlem was at that time the main campus of the Kaiser Wilhelm Gesellschaft (in 1948 renamed Max-Planck-Gesellschaft). This Society was initiated in 1910 with the intention of providing outstanding scientists with first-rate research facilities. The attitude of the founders was clearly expressed by Emil Fischer when he tried to persuade - successfully - Richard Willstätter to abandon his professorship at Zürich and to join the Society. Fischer, according to Willstätter (1), described the attitude in these words: "You will be completely independent. No-one will ever trouble you. No-one will ever interfere. You may walk in the woods for a few years, if you like; you may ponder over something beautiful". On the whole this policy (based on utmost care and competence in selecting the right people) has paid magnificent dividends: Otto Warburg, Otto Meyerhof, Albert Einstein, Max von Laue, Fritz Haber, Otto Hahn, Lise Meitner, Carl Erich Correns,

Richard Goldschmidt, Michael Polanyi, Carl Neuberg and many others made the fullest use of the opportunities.

By the late 1920's, within 15 years of its foundation, and despite the upset caused by World War I, Dahlem had become one of the world centres of scientific research. Not only did it attract many of the best scientists in Germany but also young people from all over the world.

Collaborators of Warburg and Meyerhof, 1926-1930

During my stay at Dahlem the people working in the laboratory of Warburg included Erwin Negelein, Hans Gaffron, Robert Emerson, Fritz Kubowitz, Werner Cremer, Erwin Haas, Walter Christian, Walter Kempner, Akiji Fujita and several other Japanese. Meyerhof's laboratory, accommodated in the same building, a few steps away, included Karl Lohmann, Karl Meyer, Fritz Lipmann, Hermann Blaschko, Severo Ochoa, Frank Schmitt, Ralph Gerard, Dean Burk, David Nachmansohn, Louis Génévois, Ken Iwasaki. Other young biologists working in the same building were Victor Hamburger, Curt Stern and Joachim Hämmerling. Many of these left their mark on later scientific developments.

Achievements of the Laboratories of Warburg and Meyerhof in the Nineteen Twenties

The discoveries made in the middle and later part of the 1920's in Warburg's laboratory included the discovery of the aerobic glycolysis of tumours, the general occurrence of the Pasteur effect, the accurate quantitative measurements of cell respiration and cell glycolysis, the carbon monoxide inhibition of cell respiration and the light sensitivity of this inhibition which made it possible to measure the action spectrum of the oxygen transferring enzyme in respiration (now referred to as cytochrome a_3) and to identify the catalyst as an iron porphyrin, the development of spectrophotometric methods of analysis (twenty years later commercially incorporated by Beckman into his black box), the discovery of copper in blood serum and the fall of its concentration in anaemias.

Meyerhof's laboratory made decisive contributions to what is now called the Embden-Meyerhof pathway of gycolysis. It laid the groundwork leading to the discovery of hexokinase, aldolase and other enzymes. Monumental discoveries by Lohmann were those of ATP, first identified as a cofactor of glycolysis and of the "Lohmann reaction" - the interaction between ATP and creatine.

One of the secrets of these outstanding achievements in both laboratories was the creation of new methods, such as the tissue slice technique, manometry and spectrophotometry by Warburg, and Lohmann's method of distinguishing between the many different phosphate esters by measuring their rate of hydrolysis at 100° in 2 N HCl.

Now, some 45 years later, we can assess the achievements of Warburg and Meyerhof in proper perspective. Many scientific papers may seem to be very important at the time of their appearance but as the field develops it is appreciated that they were less significant than was at first thought; as time goes on the really significant contributions stand out as lasting landmarks.

An amazing feature of the teams of Warburg and Meyerhof was their smallness. Altogether there were hardly more than two or three dozen people who participated in these great developments I have listed. Meyerhof had four or five small rooms and the total number of his collaborators at any one time was not more than five. There was only one trained technician, Walter Schulz, and a part-time typist. The technician was Meyerhof's personal assistant and together with a "Diener" he looked after the general laboratory affairs such as maintenance of apparatus and ordering of materials. When I joined Warburg there was one large room for six people in all, plus several instrument rooms, plus one Diener . Warburg had no secretarial help - we all typed ourselves. There was no technician in the ordinary sense of helpers. It is true Warburg's long-standing collaborators - Negelein, Kubowitz, Christian, Haas - were originally technicians but not in the ordinary sense. They were research assistants who had been primarily trained as instrument mechanics in work shops in the factories of the Siemens Company in Berlin. They knew how to handle instruments and how to make accurate measurements, and Warburg taught them all the chemistry they needed. In 1928 Warburg obtained one more room for four extra people and when he moved into his new Institute in 1931 there were a few more places but the total number remained deliberately small. Meyerhof of course also had more space and more staff when he moved to Heidelberg in 1930.

To those who participated what we did seemed to us quite normal and natural, a matter of course. Warburg, however, was always fully conscious of the monumental nature of his contributions. He has stated this in writing more than once, for intellectual modesty was not his strongest point. He considered himself to be in direct line with the giants of biology and in particular the chemically orientated biologists - a direct successor of Pasteur, with no-one of comparable calibre between him and Pasteur (1822-1895). Meyerhof appeared much more humble, and not concerned with his own assessment of his position in the history of science. He was content to leave this to his peers and to posterity.

And not only did the genius of leadership by Warburg and Meyerhof make outstanding contribution to the subject and inspire, the small band of deeply motivated and committed young collaborators. In the process Warburg and Meyerhof also educated (I do not say "train" because "educate" means more than train; educating includes the transmission of an outlook, not merely of technicalities) a future generation of leading scientists. This happened without actually aiming at doing it, without any policy of postgraduate training programmes. It happened naturally, and I believe something of this sort will always happen naturally. Born leaders attract born followers who develop into leaders - as long as bureaucracy and the erroneous concepts of equality do not interfere.

Day-to-Day Life in the Laboratories

We all worked very hard and intensively, though the atmosphere was relaxed. In Warburg's laboratory the working hours were from 8 a.m. to 6 p.m. for six days a week. Most of the reading and most of the writing had to be done at home in the evenings, at weekends and during the summer vacation. Warburg and Meyerhof were in attendance more or less all the time and always accessible to their collaborators. There were no committee meetings and hardly any academic tourism. There was a brief luncheon interval where the younger people from different departments (especially from the laboratories of Warburg and Meyerhof) met in a common room for a simple snack consisting usually of eggs, sandwiches and milk. Coffee and tea breaks were unknown. The main vacation was rather long. Warburg closed his laboratory for 8 weeks during August and September but during this time he wrote most of his papers while on his estate in the Island of Rügen. Warburg liked to point out that the working hours were much less than they had been in his younger days. When he worked in Heidelberg in Krehl's Department of Medicine, Krehl often made a round of the laboratories on Sunday evenings and expected most of the workers to be in attendance. In Meyerhof's laboratory working hours were less rigid but hardly shorter.

Warburg's control of the laboratory was very autocratic but we never questioned the justification of his authoritarian rule because we thought he was entitled to this on account of his outstanding intellect, his achievements and his integrity, qualities which we admired enormously. On the whole his rule was benevolent but it could also be fierce. On one occasion he dismissed a research worker instantaneously after an incident in which Warburg thought he had not shown proper respect and courtesy. For Warburg autocratic control was essential in the interest of high standards of the work as well as of personal conduct. His was autocratic rule at its best. He never exploited the junior, as does autocratic rule at its worst. Democratic rule may at best make full use of the pooled resources but at worst it may create a situation where ignorance and obstruction prevails over competence and efficiency. Warburg was most generous in giving credit to his collaborators. Many pieces of research to which he had made the main contribution and which he had written were published without his name, except perhaps in an acknowledgement by the author. A review of his work (2) which established the oxygen transferring catalyst of cell respiration as an iron porphyrin ended with the passage "In concluding I wish to emphasize that the results which I have presented are largely due to the work of my collaborators, Drs. Negelein and Krebs". Yet the whole work was conceived by Warburg himself and the greater part of the critical experiments were carried out by him with his own hands.

I know of at least one specific incident where Meyerhof was

also very fair. In 1929 Lipmann had discovered (after Einar Lundsgaard had reported muscular contractions without lactate production in the presence of iodoacetate) that on anaerobic contraction muscle becomes initially alkaline even in the absence of iodoacetate, the rise of pH being due to the hydrolysis of creatine phosphate. Lipmann measured the pH change manometrically by the uptake of CO_2 which reacts with the OH^- ions formed.

This was an important finding because it helped to establish the now generally accepted concept that creatine-P, through the Lohmann reaction, can energise contraction. As it was very important to him to find a job he was anxious that he should get proper recognition and he said in a somewhat resigned spirit to one of his colleagues, Hermann Blaschko. "This will be just another paper by Meyerhof and Lipmann". Blaschko then encouraged Lipmann to ask Meyerhof whether he may not be the first author. Meyerhof's immediate reply was "But of course" (3). And Fritz was the sole author of a second paper printed directly after the joint one, in which he showed that fluoride can act similarily to iodoacetate (4).

Financial Position of Young Research Workers

Our financial position was very restricted and quite a few people in the laboratory received no salary or grant at all. Hermann Blaschko tells me that when he asked Meyerhof to be accepted in his laboratory Meyerhof eventually agreed to have him but he told him "I cannot give you any payment" whereupon Blaschko replied "I did not expect one". Fritz Lipmann was not paid during the first 2 years of his stay with Meyerhof, nor was Severo Ochoa paid. I was lucky and received a starting salary of 300 marks which rose to 400 after one year. It is difficult to equate this with present prices but it meant that we had to live frugally and count every penny. If we were careful we could afford one modest holiday a year; we could afford concerts and theatres in the cheapest seats. There were no travel grants for attending meetings. This did not mean that we were isolated because there were plenty of opportunities for learning something about new scientific developments within Berlin itself, through the colloquia at Dahlem and through the Berlin Chemical and Medical Societies. Who then, financed our maintenance? As far as I know our parents, even though they could ill afford it, for inflation had devalued the pre-war Mark by a factor of 10^{12} (a million million) by the end of 1923. Parents were willing to make sacrifices for a good training of their sons. Of course, we felt very uncomfortable to be a burden to our parents when we were between 25 and 30 years of age. Franz Knoop impressed me in 1920 by saying during his lectures to medical students that he had earned nothing until he was 37 years of age, although he had made an outstanding discovery - that of β-oxidation - when he was 30. It was understood that an academic career meant willingness to put up with very modest material standards of living. We were motivated by a keen dedication to our work and we were maintained by the hope that when we had received a thorough training - which we expected would last until we were about 30 - we would eventually get a worthwhile job - satisfying professionally as well as financially. The sacrifices needed for a long period of training meant also that only the keenest did not give up. Most of us were medical graduates and could, if we wanted to, at any time branch off into a relatively lucrative medical career - but we preferred research.

The financial and some other aspects of the scene at Dahlem were of course not unique. Erwin Chargaff - my contemporary working at that time in Vienna - recently wrote (5) "No-one who entered science within the past 30 years or so can imagine how small the scientific establishment then was. The selection process operated mainly through a form of an initial vow of poverty. Apart from industrial employment in a few scientific disciplines, such as chemistry, there were few university posts, and they were mostly ill-paid".

Nowadays there are many undergraduate students who insist on their "rights" to be fully supported by the state. We expected no rights even at the post-graduate and postdoctoral level. We were satisfied if we could work hard under reasonable conditions and learn. We did not feel entitled to expect much, let alone make demands, before we had learned a lot.

In spite of our restricted economic circumstances we were on the whole very happy because we felt that we were receiving a first rate training and were doing something worthwhile. We had no undue worries about our long-term future although this looked very uncertain in view of the economic and political difficulties in Germany. I am not aware that any of us anticipated a particularly successful career. Being close to giants of science we felt very small and Warburg himself did not do much to encourage our self-confidence. In fact when I had to leave his laboratory he told me that he considered my chances in biochemistry as slight and advised me to return to clinical medicine (which I did).

Relations between Dahlem and German Universities

Of a peculiar sort were the relations of Warburg and Meyerhof to the official representatives of German physiology and biochemistry, i. e. the German university departments. Warburg regarded himself as an outsider, and Meyerhof too, but perhaps less so. It is remarkable that German universities had not appreciated officially Meyerhof's qualities. By the time he got the Nobel Prize in 1923, at the age of 39, he held the post of an assistant in the Physiological Institute of Kiel University; he had been passed over for a junior professorial appointment (Professor Extraordinary) in favour of a man called Pütter, whose claim to distinction remained slight. After the award of the Nobel Prize to Meyerhof, Warburg succeeded in persuading the Kaiser Wilhelm Gesellschaft to offer Meyerhof a post.

This sense of being outsiders had to do with the Cinderella treatment of biochemistry by the German universities. The number of chairs and departments of biochemistry or physiological chemistry was very small in Germany. Independent departments existed in four universities only, in Frankfurt with Embden at its head, at Freiburg, Tübingen and Leipzig. In other universities biochemistry was a sub-section of physiology and the heads of these sub-sections did not have the rank of full professor. In some universities the professor of physiology was essentially a biochemist. This applied, for instance, to Heidelberg where Kossel was the Professor of Physiology and where first-rate work was done on protein chemistry.

Thus it came about that Leonor Michaelis, one of the brightest biochemists of his time could not be absorbed into the German university system and therefore left Germany in 1921 for Japan, to move later to Johns Hopkins University and the

Rockefeller Institute. While in Germany he had to earn his living as a clinical biochemist in one of the municipal hospitals in Berlin.

Hopkins, in a general address at the International Congress held in Stockholm in 1926 commented on the neglect of biochemistry in German universities and spoke about the importance of "specialised institutes of general biochemistry". He referred to the appeal by Hoppe-Seyler in Volume 1 of his Zeitschrift für Physiologische Chemie in 1877 that institutes of biochemistry should be generally set up in the universities. In 1877 the only institute of this kind was that in Strasbourg with Hoppe-Seyler at its head. Hoppe-Seyler's appeal was at once opposed by the physiologist E. Pflüger in his Journal, and 49 years later Hopkins remarked that in fact Hoppe-Seyler's appeal for the recognition of biochemistry as an independent subject had still not yet found proper response in his own country and he added "It is difficult to see how Germany can continue the lead along the path which for a long time she has almost trod alone". He emphasised that academic centres in general in Europe are in this respect behind those in America. These remarks, incidentally, were made at the suggestion of F. Knoop for the benefit of the German readership and they were reprinted in translation in the Münchener Medizinische Wochenschrift (6).

Now and Then

Today the scene of scientific research seems very different. In the 1920's pure research was still widely looked upon as a luxury or extravagance that did not deserve major support from the State. The Kaiser Wilhelm Gesellschaft, after all, was founded in 1910 as a private organisation (while the German Universities were all state-controlled).

Now, 50 years later, scientific research is regarded a necessity for the survival of a nation, and large sums are provided by Governments for training people and for pursuing research.

But although the scene has changed some fundamental principles governing successful research are still the same, and will always remain the same - the recognition of leadership in research, the value of long training, the need for hard work and for dedication, an attitude of humility.

What has gone, among other things, are the biting polemics in science in which Warburg liked to indulge, hurting and ridiculing the opponents - there are many examples of these in Warburg's books and in the Biochemische Zeitschrift (7). Gone, also, has the autocratic rule which Warburg, his teachers and his contemporaries practiced. This kind of benevolent (or mostly benevolent) dictatorship at its best as I already stated, helped to maintain high standards - but it could easily degenerate into arbitrary injustices, exploitation and mismanagement.

Today a different basis of the relations between seniors and juniors has evolved. The master may still rule, and rule firmly, but the basis of his authority is now a natural respect, a natural mixture of admiration and affection which he has earned by his work and conduct; in a good laboratory authority is no longer based on the power invested in a head of a laboratory. I think this very occasion here is an illustration of the kind of human relations between the master and his followers which nowadays exist in a really ideal team. It was the desire of the followers to express a sense of loyality, gratitude and affection which has brought this symposium into being. This assembly of so many people from so far away, motivated by mutual goodwill, I find deeply gratifying and moving, especially here today in Dahlem where, within a stone's throw in the Free University, I am told, senseless and unpleasant confrontations between teachers and taught represent a menace to academic life, a menace liable to destroy the old academic ideals, the search, in a spirit of tolerance, for knowledge and truth.

The motives which brought us together remind me of a remark which Warburg made to me during a casual chat in the laboratory. For two years I shared an island bench with him, working opposite to each other in close proximity. Although both of us were not exactly talkative while preoccupied with our experiments there were occasional conversations touching on many aspects of life. One day he remarked "The worst defeat a scientist can suffer is to die early because the fruits of his labour mature very slowly". Although these words - like many other of his casual sayings are still deeply ingrained in my memory - I did not properly appreciate their full meaning at the time. But today their meaning is clear to me. The fruits of a scientist's labour are of several kinds. Promotion to a good post and a Nobel award are some; a very important one is the long term response from students and collaborators. Much of a research scientist's time is spent in helping to shape the outlook, career and life of his juniors, and the seeds which he plants take a long time to grow. Today we see the rich harvest that has come to Fritz from helping and guiding his younger associates - a harvest not only in the form of gratitude and affection but also in the form of the intensely pleasing knowledge that there is a new generation willing to carry on the work and to uphold the standards and ideals which motivated the master.

Such kinds of thoughts, I suspect, must have been in Warburg's mind when he suddenly spoke about the importance of living to old age. Perhaps some experiences of his father (8) (a founder of a large and devoted school of physicists), had impressed him*.

*Further material on Dahlem in the 1920's and the personalities of Warburg and Meyerhof will be found in references (6), (9), (10), (11), (12) and (13).

References

(1) Willstätter, R.: Aus meinem Leben. Zweite Auflage. Weinheim Verlag Chemie, p.200 (1958).

(2) Warburg, O.: Über die chemische Konstitution des Atmungsfermentes. Naturwissenschaften 16, 345-350 (1928).

(3) Lipmann, F., Meyerhof, O.: Über die Reaktionsänderung des tätigen Muskels. Biochem. Z. 227, 84-109 (1930).

(4) Lipmann, F.: Über den Tätigkeitsstoffwechsel des fluorid-vergifteten Muskels. Biochem. Z. 227 110-115 (1930).

(5) Chargaff, E.: Building the Tower of Babble. Nature 248, 776-779 (1974).

(6) Hopkins, F.G.: Über die Notwendigkeit von Instituten für physiologische Chemie. Münch. Med. Woch. 73, 1586-1587 (1926).

(7) Krebs, H.A.: Otto Heinrich Warburg. Biographical Memoirs of Fellows of the Royal Society. 18, 629-699 (1972).
(8) Franck, J.: Emil Warburg zum Gedächtnis. Naturwissenschaften 19, 993-997 (1931).
(9) Weber, H.H.: Otto Meyerhof - Werk und Persönlichkeit, in Molecular Bioenergetics and Macromolecular Biochemistry pp. 2-13, Springer Verlag, Berlin, Heidelberg, New York, (1972).
(10) Peters, R.A. (with a contribution by H. Blaschko): Otto Meyerhof. Obituary Notices of Fellows of the Royal Society. 9, 175-200 (1954).
(11) Muralt, A. von: Otto Meyerhof. Ergebn. Physiol. 47, I-XX (1952).
(12) Nachmansohn, D., Ochoa, S., Lipmann, F.: Otto Meyerhof 1884-1951. Science 115, 363-369 (1952).
(13) Nachmansohn, D.: Biochemistry as part of my life. Ann. Rev. Biochem. 41, 1-28 (1972).

Fritz Lipmann

I am going now to describe my relationship to Dahlem when I was in Meyerhof's laboratory, and here I would also like to include Berlin. From my early youth Berlin, the great city, had been for me a magnet. I was born in a small town, Koenigsberg, then in East Prussia, and my first contact with Berlin was after absolvation of my abiturium, as we called the final examination ending gymnasium time and giving the right to enter into University. Then, my parents gave me as a gift, a week all on my own in Berlin to experience the theater and to experience the great city. I spent more time in Berlin later on studying medicine, and again I was impressed by the experience of what happens in that city.

And then I came back to Berlin for quite a long period after I had finished my medical studies. Still under the influence of medical friends, I spent the latter half of my practical year there, and the first three months with Ludwig Pick, an excellent pathologist, because it was thought that to become a physician one had to do some pathology. Then I heard about a biochemistry course which was given at the Charité, the Medical School of Berlin University, where a large number of physicians who became very good biochemists, including Hans Krebs and David Nachmansohn, were trained. That was Rona's laboratory. When I went to Ludwig Pick to tell him that I would spend the last three months of my practical year taking the biochemistry course of Peter Rona, he threw his hands up in amazement. Biochemistry then was still an unknown entity in Germany.

Rona's name is probably known to very few of you; for quite a while he was a collaborator of Leonor Michaelis. One of the reasons I wanted to mention him was that I found an amusing picture of the members of the course when I took it and I would like to show you this (Figure 1).

Figure 1

In the center is Peter Rona; in the middle is H.H.Weber as a young man - he died a few weeks ago; and in the top right corner am I, with a bow tie - some people are disappointed that I don't wear it any more as I used to do in early years.

I think it was an extraordinary course. I learned there the latest advances in the biochemistry of that time: manometry, pH measurement, electrophoresis, and so on. Actually, I stayed on with Rona and did a medical doctorate thesis which was obligatory in Germany. It did not need to be very important. Mine was on the electrophoretic behavior of iron oxide colloids, mainly concerned with the reversal of the positive charge to negative in the presence of citrate. Colloid chemistry was very modern in those days - many people used it to describe the protoplasm. It seemed enough then to call it a colloid to imply one understood something about it. We have learned better.

That was in 1921-1922. Then I decided to go back to my home town to learn chemistry since I was lucky enough to be able to do this with Hans Meerwein who was the professor of chemistry at the University. I should guess that his name is known to those of you conversant with organic chemists. He was a superior chemist, and later moved from Koenigsberg to Marburg. In three years I learned a great deal, particularly from lectures; all during those years all students had to attend his lectures and it was a tremendous pleasure. He gave them all himself; there was no substitution by assistants, and that gave us students a contact with his personality.

Then I got my Verbands-examen, which enabled me to start on a thesis. After that first step was finished I became somewhat restless. I felt it was now time for me to find a place where I could do biochemistry, for which I had been preparing myself all this time. It was not without other reasons that I chose Berlin; but I was mostly motivated by the existence in Berlin-Dahlem of the two institutions which at that time seemed to me to do work in the field I had begun to become interested in, intermediary metabolism. These were the laboratories of Carl Neuberg and Otto Meyerhof, and I debated for a time whether I should join Meyerhof or Neuberg. I eventually decided for Meyerhof because of his much more physiological leaning, and just as Hans Krebs has told you, I likewise didn't expect any salary and for the first two years I didn't get anything. I just asked him if I could work there and as I had studied chemistry and had some experience I was lucky enough to be accepted; he asked me, interestingly enough, if I had any problem to work on and I was ashamed to say that I hadn't. I had to get a problem from him.

The problems I worked on in the early days there were not very important. I did some work on fluoride inhibition of glycolysis and fermentation, which was published in Biochemische Zeitschrift. These papers I could put together and use as my chemical doctor's thesis. It reflects interestingly on the status of the heads of laboratories at the Kaiser Wilhelm Institutes that Meyerhof was unable to be my doctor "father" because he couldn't accept graduate students although he was a titular professor. However, Neuberg, whose institute was next door, could; he was a professor at the Technische Hochschule and I actually had to farm out my thesis, so to say, to Neuberg, who became my doctoral "stepfather" in a way. He always treated me very kindly.

I will now say a little about the Meyerhof laboratory. While I was sitting here just now and thinking about my choice of laboratory, I am almost surprised to discover that although most impressed by Warburg, I never even dared to think of going to work with him. One of the important aspects of Meyerhof was that he was not as stern and was much looser than Warburg. But he was Warburg's pupil and there is a paper by Warburg and Meyerhof which came from the Naples Laboratory. The Marine Laboratory in Naples was one of the meeting grounds for biochemists in the same sense that Woods Hole is or was in the U.S.A. In his earlier years Meyerhof tended to be very interested in philosophy, had joined a school of philosophers, and wrote several papers of a philosophical character. It was the influence of Warburg, I understand, that decided him to become a biochemist; and he took all the traits of Warburg, the feeling that to do good work one needs the most exact methodology and has to have full confidence in one's results.

The work in his laboratory, as you have heard already, centered about the muscle and I worked during that time largely with muscle or muscle extracts, mostly related to glycolysis. It is only in the later period after the laboratory had moved to Heidelberg that I did a fairly nice piece of work on a determination of creatine phosphate breakdown in living muscle which I measured manometrically. These manometers that I used were somewhat difficult to construct because I wanted to stimulate the muscle in the manometer and one had to seal in platinum electrodes, which was very hard to do without a leak. This work started when Meyerhof suggested that I should try to see what happens in muscle contraction at relatively high acidity, that is, in a bicarbonate solution with CO_2

in the gas phase. It was then that I found, during the early phase of a series of contractions, an alkalinization, i.e., manometrically, CO_2 absorption instead of the expected CO_2 liberation from lactic acid formation. Chemical analysis of the muscle showed that the alkali that formed early corresponded very nicely to a creatine phosphate breakdown; this had been found to yield alkali because of the strong alkalinity of the guanidinium base liberated. This was an early proof that without inhibitor under these acid conditions creatine phosphate breakdown could cause contraction.

Figure 2

It is now time to show you Meyerhof. This picture (Figure 2) is rather typical because it shows that it was not easy to approach Meyerhof. We actually talked very little and what I learned from him was largely by diffusion. But this was to influence me all through my life. This picture of us together was taken in 1941 at a conference in Madison, and you might say we look a little uneasy, which was because I had given a lecture there on the Pasteur effect and had shown disagreement with his interpretation, and he wasn't too happy about it.

Our working hours were much more relaxed than those at Warburg. I remember that we took pretty long lunch intermissions and went rather often to a little restaurant, which was next to a further-up subway station, and sat there and happily talked in the garden. When I say we, in the next picture (Figure 3) you can see some of the people who were "we".

Figure 3

You can see, from left to right, Ken Iwasaki, Karl Lohmann, Walter Schultz, and Schroeder, who was something in between a scientific assistant and a diener;then David Nachmansohn and Paul Rothschild; and again me, this time with a long tie. Ken Iwasaki, Nachmansohn, Paul Rothschild, and I were the ones who often went to that little restaurant, and not only that, we even went together to masquerade balls which were very fashionable and much attented at that time; but they were very good entertainment and as free in spirit as in present-day terms. At one of them, the socialist ball, which had nothing to do with socialists, I met Freda Hall who was to become my wife. So that was an important event during my Berlin days. Actually, at this particular ball, David Nachmansohn danced more with Freda than I did, but he was already married.

Now to return to the laboratory. I had some contact with Ralph Gerard - I think we shared a laboratory when I entered the Meyerhof Laboratory; we met again in later life and I was rather fond of him. Then I moved into another laboratory and worked very close to Ken Iwasaki who spent a good deal of time in Berlin and I am sure had a very good time there; he was not married then. I am told that Mr. Takeda, Masao's father, became a very good friend of Ken Iwasaki; they spent much time together in Berlin and are still very good friends. Ken Iwasaki is now retired from his biochemistry professorship and has a laboratory in the Takeda Company, which is one of the largest pharmaceutical companies in Japan.

The topics on which the work was done in Meyerhof's laboratory were not too varied; it was mainly concentrated on the muscle, but it also included nitrogen fixation on which Dean Burk and Ken Iwasaki worked. The status of our understanding at that time may be shown by what Karl Meyer did. He was trying, in parallel to what had been called zymase by Harden, to isolate the "glycolytic enzyme"; it wasn't quite realized then that the glycolytic enzyme was of course composed, as we now know, of numerous enzymes and that these enzymes could eventually be separated. That came not much later, but in 1928 it was really surprising that one could even aim at thinking of isolating as a unit something like a glycolytic enzyme.

Then, shortly before we moved away from Berlin, I worked for a little while with Lohmann, and I learned much from him. As you heard already, he really was an artist in determining by acid hydrolysis the different phosphorylated compounds that were in a mixture. For example, in this way he discovered the equilibration of glucose 6-phosphate with fructose 6-phosphate. The latter has a much faster hydrolysis time than glucose 6-phosphate, which is one of the most difficult to hydrolyze phosphate esters. On the other hand, fructose diphosphate is completely hydrolyzed within three hours. Lohmann used the acid hydrolysis of phosphate esters very effectively and, as I said, he discovered many new compounds.

I am very eager to show once in a while an experiment from which the wrong conclusions were drawn. Table I shows such an experiment that I did with Lohmann and which eventually became of great consequence. In this experiment fructose diphosphate was incubated in muscle extract, and I came in to try and see if this conversion of fructose diphosphate would also go without fluoride that was added in Lohmann's

Table 1: Conversion of FDP into acid - stable phosphate ester in muscle extract of winter frogs after incubation at 20°

Incubation (min)	Phosphate Bound ($mg\ P_i$)	Phosphate Acid-Hydrolyzed in 3 hr ($mg\ P_i$)	Converted Phosphate ($mg\ P_i$)	(%)
0	0.48	0.48		
20	0.50	0.39	0.12	25
60	0.50	0.25	0.27	53
120	0.49	0.19	0.33	68

Biochem.Z., 222,389,1930

earlier experiments. The table shows the change of fructose diphosphate without fluoride to what was called a difficult to hydrolyze hexose diphosphate. One can see that by the incubation in muscle extract phosphate is not released. However, the hydrolysis time of the added fructose diphosphate goes up if one uses the three hours mentioned above as a standard. One can see that with time it becomes converted, and eventually 70% of it is present as what was thought to be a different hexose diphosphate which was much more difficult to acid-hydrolyze.

To turn a little more to the history of this part of biochemistry as I did in the meeting at the Ciba conference last week where I also mentioned these experiments, Nilsson had shown a little earlier in Euler's laboratory that in the presence of fluoride, the same fructose diphosphate, with a yeast preparation, when paired with acetaldehyde, gave phosphoglyceric acid as the oxidation product parallel with reduction of acetaldehyde to ethanol. This was the first appearance of phosphoglyceric acid in the picture of fermentation and glycolvsis and nobody at that time could appreciate why this compound was formed as an oxidation product of fructose diphosphate. Nilsson came to the Meyerhof laboratory and we discussed this strange compound, phosphoglyceric acid, without seeing the light. The right idea was Embden's who essentially repeated our experiment. We had just made barium precipitates of what we considered a hexose diphosphate. He found that it was surely difficult to hydrolyze but that it was not hexose diphosphate but rather a mixture of phosphoglycerol and phosphoglyceric acid which he assumed to be formed by a dismutation reaction. Thus, we had misinterpreted the conversion of fructose diphosphate. However, in the hands of Embden it became the reason why we now talk about the Embden-Meyerhof cycle; from this reaction he then concluded the disruption of fructose diphosphate into an equilibrium mixture of two triose phosphates and mapped out the foundation of our present scheme, recognizing phosphoglyceric acid to be the oxidation product of phosphoglyceraldehyde, the biochemically dominant of the two triose phosphates formed.

That's just a sidelight on Lohmann's artistry with hydrolysis. As I said, I learned much from him to handle what I did soon afterwards. When I went to New York to work with Levene, I chose the phosphoproteins as an object of investigation. You heard in Dr. Helmreich's talk that I there isolated serine phosphate from the egg yolk phosphoprotein. Lohmann was the one who suggested that I work with Levene. He had in mind, I think, that I should work on nucleotides, but I chose the mentioned topic because Levene had isolated a very phosphate-rich protein from egg yolk which attracted my interest. The methods I used were actually borrowed from what I had learned from Lohmann. It appeared that the phosphate in the yolk protein is alkali-labile but very acid-stable, and choosing acid hydrolysis as a means of degradation of the protein, I thus isolated the serine phosphate.

Much later, I turned to what was the prominent interest in Meyerhof's laboratory, namely, bioenergetics. This came to be an underground well, so to say, that eventually opened up after I discovered acetyl phosphate and led to my writing the paper on generation and utilization of phosphate bond energy.

I would like to close by saying a little more about Berlin in those days. Before going to New York, Freda Hall and I were married, and we were amazed to see that there was such an enormous difference in the way of life between Berlin and New York, particularly among young women and young men. We read in the "Saturday Evening Post" three articles by Hergesheimer, who was then a rather fashionable novelist, in which he described Berlin as the center of Europe: people didn't go to Paris, they went, rather, to Berlin. There was the theater, there was the music, there was the dance; you could have everything. There was also a great freedom in Berlin in the late twenties, a similar freedom to that which has developed in America in recent years. It was due to the breakdown of the family ties by which young people were held because they had to depend on their families, and it had the effect that the young men and women interacted much more freely with each other. After I had moved away from Berlin and from Germany, it took a long time to forget the way of life we had experienced there.

I have been very pleased and moved by this meeting's coming about, and don't want to end without thanking the many people who have made possible this happy get-together. I first want to mention Dietmar Richter who really had the lion's share in thinking of it, writing all the letters, and arranging the whole proceedings. I wish to thank Dr. Wittmann for providing this Institute as a place for us to meet, and Dr. Helmreich for his nice words and also for the encouragement and support from Gesellschaft für Biologische Chemie. And of course I am most grateful for the support of the pharmaceutical companies of Boehringer and Schering, and of the Volkswagen Foundation, without which we could not have had it.

Thank you all very much.

Cascade Control of E. coli Glutamine Synthetase

Stuart P. Adler and Earl R. Stadtman
National Institutes of Health, Bethesda, Md. 20014, USA

One important mechanism for the regulation of glutamine synthetase activity in E. coli is the covalent attachment and removal of AMP from a specific tyrosyl residue in each of the enzymes 12 identical subunits (1-4). Adenylylation of a subunit converts it to a less active form dependent upon Mn^{++} (1). The enzyme's activity is thus controlled by the average number of adenylylated subunits per molecule which can vary from zero to 12. Both adenylylation and deadenylylation of glutamine synthetase are catalyzed by single enzyme, adenylyltransferase (ATase) (5). Adenylylation involves the transfer of AMP from ATP into an AMP-O-tyrosyl linkage (4), whereas deadenylylation involves a phosphorolysis of this bond to yield ADP (6). Although ATase catalyzes both reactions, its ability to adenylylate or deadenylylate glutamine synthetase (GS) is modulated by the regulatory protein P_{II} and metabolic effectors including α-ketoglutarate (α-KG), ATP, glutamine and Pi (4,7,8).

$$\text{ATP} + \text{GS} \xrightarrow[P_{IIA}]{\text{ATase}} \text{AMP-GS} + \text{PPi}$$

$$\downarrow\uparrow$$

$$\text{AMP-GS} + \text{Pi} \xrightarrow[\text{ATase}]{P_{IID}} \text{GS} + \text{ADP}$$

SCHEME I

The regulatory protein P_{II} exists in two interconvertible forms (7). One form, P_{IIA}, stimulates the ATase catalyzed adenylylation of glutamine synthetase, whereas the other form, P_{IID} is required for the ATase catalyzed deadenylylation (Scheme I). When P_{IIA} is incubated in the presence of UTP, ATP, α-KG, Mn^{++} or Mg^{++}, and another enzyme, uridylyltransferase (UTase), it is converted to P_{IID}; this involves the covalent attachment of UMP to the protein (Reaction 1) (9).

$$UTP + P_{II} \xrightarrow[\alpha\text{-KG, ATP, } Mg^{++}]{\text{UTase}} UMP\text{-}P_{II} + PPi$$

REACTION 1

P_{IIA} can be regenerated from P_{IID} by a uridylyl removing enzyme activity (UR enzyme) (Reaction 2).

$$P_{II}\text{-UMP} \xrightarrow[Mn^{++}]{\text{UR enzyme}} P_{II} + UMP$$

REACTION 2

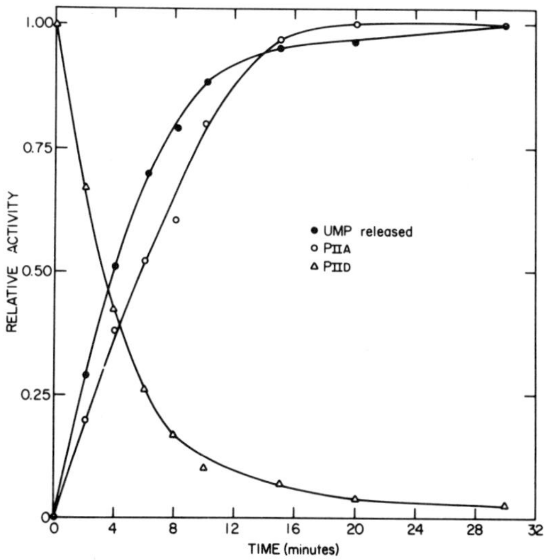

Figure 1. Reciprocal effects of UR-enzyme catalyzed deuridylylation of P_{IID}. H^3-UMP-P was incubated with partially purified preparation of UR-UTase. The P_{IIA} and P_{IID} activities as well as the release of $[^3H]$-UMP was followed. (Exp. details, see ref. 10).

Figure 1 illustrates the relationship between uridylylation and the capacity of P_{II} to stimulate adenylylation or deadenylylation of glutamine synthetase. When P_{IID} labelled with ^3H-UMP was incubated with Mn^{++} and a partially purified extract containing UR enzyme activity, there was a rapid release of covalently bound ^3H-UMP from P_{II}. This release was accompanied by a parallel increase in the ability of P_{II} to stimulate the ATase catalyzed adenylylation of glutamine synthetase, and a concomitant loss in its ability to stimulate the ATase catalyzed deadenylylation of glutamine synthetase (10). The reciprocal changes in activity associated with deuridylylation of P_{IID} clearly demonstrate the role of P_{II} uridylylation in determining the adenylylation and deadenylylation capacity of ATase.

TABLE 1

Properties of the Regulatory Protein P_{II}

Property	Value
Molecular weight, native protein	44,000
Subunit molecular weight	11,000
Number of identical subunits	4
Number of tyrosines per subunit	2
Number of iodinatable tyrosines per subunit of P_{IIA}	2
Number of iodinatable tyrosines per subunit of P_{IID}	1
Moles of UMP per mole of P_{IID} subunit	1

The P_{II} protein has been purified to homogeneity from _E. coli_ (11) and from _Pseudomonas putida_ (12). Properties are shown in Table 1. The native protein has a mol. wt. of 44,000. Based on its homogeneous behavior during disc gel electrophoresis in the presence of SDS or 8 M urea (11) and during sedimentation in 6 M guanidine·HCl (13), and also the minimum mol. wt. calculated from amino acid composition, the native protein is a tetramer of identical subunits (11). This is supported also by the facts:
(1) there are two tyrosyl residues per 11,000 mol. wt.;
(2) following iodination of the tyrosyl residues of P_{IIA} with ^{125}I, tryptic digestion yields only two radioactive peptides in

equal molar amounts. Only one of these two peptides is obtained after tryptic digestion of iodinated, fully uridylylated P_{IID}. Since substitution of a tyrosyl hydroxyl group prevents iodination of the aromatic ring, this result indicates that in P_{IID} the UMP is covalently bound in phosphodiester linkage to the hydroxyl group of one of the two tryosyl residues in each subunit. This is supported **also** by the fact that treatment of P_{IID} with phosphodiesterase results in release of the covalently bound UMP and the stoichiometric appearance of phenolate ion (pH > 11.0) as measured by ultraviolet absorption spectroscopy (14). Thus, activity of P_{II} is modulated by covalent attachment of UMP to a specific tyrosyl residue in each subunit.

Table 2 shows the effect of cations and other effectors on the enzyme activities that catalyze the uridylylation of P_{IIA} (UTase) and the deuridylylation of P_{IID} (UR - enzyme).

TABLE 2

Effect of Divalent Cations and Other Effectors on UR and UTase Activities

Effector added	Relative Activity	
	UR activity	UTase activity
Mg, ATP, α-KG[1]	55%	100%
Mn, ATP, α-KG	85	100
Mn	100	0
Mg	0	0
Mg + Mn	100	0
Mg, ATP, α-KG, GLN	-	7

[1] The concentration of effectors used was: 1 mM $MnCl_2$, 10 mM $MgCl_2$, 0.1 mM ATP, 5 mM α-ketoglutarate (α-KG), 18 mM glutamine, and 2 mM K_2Mg EDTA. These reaction mixtures also contain 2 mM K_2Mg EDTA to chelate traces of Mn^{++} that were present in the enzyme preparation. (For exp. details, see ref. 10).

Whereas Mn^{++} alone supports maximal UR activity, Mg^{++} cannot support UR activity except in the presence of ATP and α-ketoglutarate; however, activity with Mg^{++} is less than with Mn^{++} alone. In contrast, Mn^{++} and Mg^{++} support equal UTase activity in the presence of ATP and α-ketoglutarate, but neither cation supports UTase activity in the absence of these effectors. Table 2 also shows that UTase activity is inhibited by glutamine whereas other data (not shown) demonstrate no effect of glutamine on UR activity.

All attempts to separate the UR and UTase activities have failed. An example of their copurification is illustrated in Figure 2.

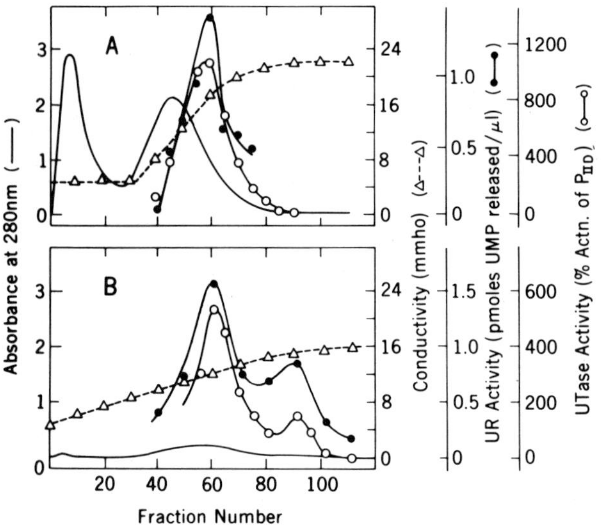

Figure 2. Chromatography of UR-UTase activities on sepharose-C_5-NH_2. (A) An extract containing UR-UTase activities was applied on a sepharose-C_5-NH_2 column (8.5 x 1.5 cm). After the unabsorbed protein was eluted, a KCl gradient up to 0.4 M was applied. (B) Fractions from experiment A containing UR-UTase activities were pooled, and reapplied to the same column as in A, only eluted with a shallower KCl gradient. See ref. **(15)**.

A partially purified extract containing UR-UTase activities was chromatographed on a hydrophobic column (Sepharose-C_6-NH_2). Even when eluted with a very shallow KCl gradient both activities cochromatograph (Figure 2B). The facts that both activities copurify through a variety of procedures, are stabilized by high ionic strength buffers (14), and in the presence of Mg^{++} require ATP and α-ketoglutarate for activity (Table 2) indicate that both activities might be catalyzed by a single enzyme or enzyme complex.

Discussion

Figure 3 summarizes current knowledge of the complex system that regulates glutamine synthetase activity in E. coli. The system consists of two opposing sets of reactions (cascades) that lead on the one hand to inactivation (adenylylation) of glutamine synthetase and on the other hand to its activation (deadenylylation). The inactivation cascade (Fig. 3A) is initiated by the action of UR enzyme which catalyzes the deuridylylation of P_{II}·UMP (i.e., P_{IID}). Deuridylylation leads to an unmodified form of P_{II}, which together with ATase promotes the adenylylation of glutamine synthetase, thus converting it from a Mg^{++}-dependent form of high catalytic potential and a pH optimum of 8.0, to a Mn^{++}-dependent form of low catalytic potential and a pH optimum of 6.9. The activation cascade (Fig. 3B) is initiated by the action of UTase, which catalyzes uridylylation of P_{II}. The P_{II}·UMP, thus formed, together with ATase promotes deadenylylation of glutamine synthetase, converting it back to the Mg^{++}-dependent form of high catalytic potential. The activities of these two opposing cascade systems are finely modulated by the concentrations of various metabolites, including UTP, ATP, α-KG, Pi, glutamine and probably other compounds as yet unidentified. The catalytic potential of glutamine synthetase is determined by its state of adenylylation (i.e., the average number of adenylylated subunits per mole of enzyme). When glutamine synthetase is incubated with ATase, P_{IIA}, P_{IID} and the effectors shown in Figure 3, it assumes a dynamic steady state of adenylylation in which the rates of adenylylation and deadenylylation are equal (16); moreover, the actual state of

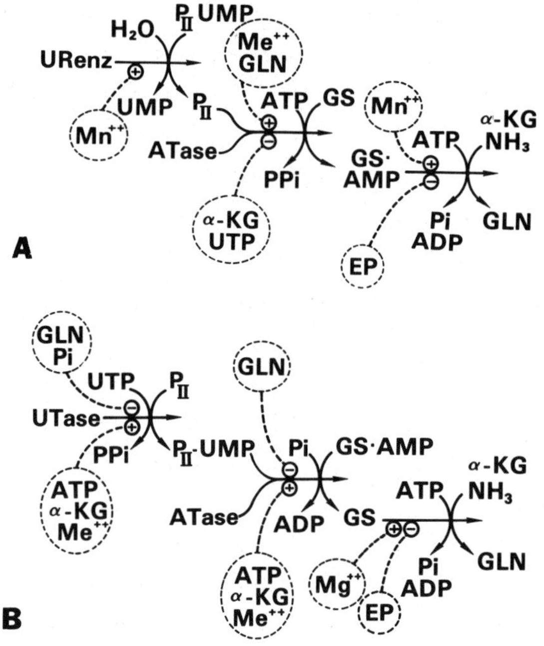

Figure 3. Cascades involved in the regulation of glutamine synthetase activity: (A) inactivation (adenylylation) of glutamine synthetase; (B) activation (deadenylylation) of glutamine synthetase. EP refers to end products of glutamine metabolism.

adenylylation ultimately obtained is determined by the relative concentrations of P_{IIA}, P_{IID} and the various effectors. Since for any given steady state the rates of adenylylation and deadenylylation of glutamine synthetase (GS) are equal, it follows from theoretical considerations that the final state of adenylylation, $\bar{n}$, is a function of the magnitude of the specific rate constants for adenylylation (k_1) and deadenylylation (k_2) and the mole fraction of P_{IIA}, $(P_{IIA})_f$, according to the equation:

$$\bar{n} = \frac{12\ (P_{IIA})_f}{k_2 + k_1\ (P_{IIA})_f - k_2\ (P_{IIA})_f}$$

Figure 4 shows how the value of $\bar{n}$ varies as a function of the $(P_{IIA})_f$ and the ratio of k_1/k_2.

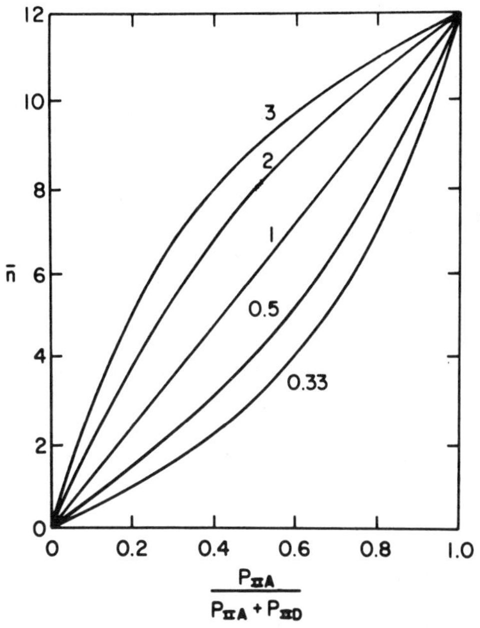

Figure 4. Dependence of the steady state level of adenylylation on the mole fraction of P_{IIA} and the relative specific rate constants for adenylylation and deadenylylation. Data are calculated values derived from equation in the text (where $(P_{IIA})_f = P_{IIA}/P_{IIA} + P_{IID}$). It is assumed that ATase is present in excess compared to P_{II} and that its activity is determined by the specific rate constants for adenylylation (k_1) and deadenylylation (k_2), and the proportions of P_{IIA} and P_{IID} present. Numbers on curves indicate the ratio, k_1/k_2.

Note that when $k_1 = k_2$, $\bar{n}$ is a linear function of the mole fraction of P_{IIA}, but that nonlinear functions are obtained if the ratio of k_1/k_2 is varied as occurs, in the response to the differential effects of metabolites on the ATase catalyzed adenylylation and deadenylylation reactions. Additional flexibility in control derives from the fact that the mole fraction of P_{IIA} can vary independently in response to changing concentrations of metabolites that affect the rates of the uridylylation and deuridylylation reactions (Fig. 3).

Cascade systems offer other advantages in the regulation of certain cellular functions. Because they consist of a series of reactions in which one catalyst acts upon another, they can amplify the response of the target enzyme to primary effectors acting on the initial enzyme in the series. This could be important if the changes in effector concentrations are small compared to the concentration of the target enzyme and in situations as described here in which the multiple protein catalysts are not present in comparable concentrations as they are in organized multienzyme complexes.

In addition, cascade systems increase the number and types of allosteric effectors that can affect the activity of the ultimate target enzyme, since each enzyme in the cascade can be independently regulated. This may be important in the regulation of enzymes such as glutamine synthetase that occupy a central position in metabolism and therefore need to receive a massive input of regulatory information from diverse cellular functions. In such cases physical and steric limitations may preclude the existence on a single enzyme of a sufficient number of allosteric sites to accommodate the required number of regulatory effectors. This may account for the fact that the direct regulation of glutamine synthetase by eight different end products of glutamine metabolism is supplemented by a cascade system in which regulation of the UR-UTase and ATase activities is mediated by six additional effectors, including divalent cations (Fig. 3). In addition 3-phosphoglycerate, fructose-6-P, P-enolpyruvate, CoA, and fructose-P_2 have been shown to inhibit the adenylylation of glutamine synthetase (17).

Finally, another advantage of cascade systems is obtained when more than one step in the cascade is subject to control by the same effector. Increased sensitivity of the system to negative control is obtained when two steps are inhibited by the same ligand. Thus, when a given concentration of a single metabolite inhibits each of two steps by 50%, the overall inhibition will be 75%. Moreover, if a metabolite stimulates two separate steps in a cascade, the net effect is to increase the apparent reaction order with respect to that effector; therefore, under appropriate conditions activation of the last enzyme in the cascade will be a sigmoidal function of the metabolite concentration. This advantage of cascade systems appears to be realized in the glutamine synthetase cascade, since as is shown in Fig. 3, two steps, (the uridylylation of P_{II} and the adenylylation of glutamine synthetase) are inhibited by glutamine and are stimulated by both α-ketoglutarate and ATP.

In the last analysis, the elaborate cascade system illustrated in Figure 3 serves as a physiological computer which is programed to sense fluctuations in the concentrations of numerous metabolites and to integrate their effects so as to modulate the activity of glutamine synthetase to meet changing metabolic demands.

References

1. Kingdon, H. S., Shapiro, B. M., Stadtman, E. R.: Regulation of glutamine synthetase, VIII. ATP: glutamine synthetase adenylyltransferase, an enzyme that catalyzes alterations in the regulatory properties of glutamine synthetase. Proc. Nat. Acad. Sci. US 58, 1703-1710 (1967).
2. Shapiro, B. M., Kingdon, H. S., Stadtman, E. R.: Regulation of glutamine synthetase, VII. Adenylylglutamine synthetase: a new form of the enzyme with altered regulatory and kinetic properties. Proc. Natl. Acad. Sci. US 58, 642-649 (1967).
3. Mecke, D., Wulff, K., Liess, K., Holzer, H.: Characterization of a glutamine synthetase inactivating enzyme from Escherichia coli. Biochem. Biophys. Res. Commun. 158, 514-525 (1966).
4. Shapiro, B. M., Stadtman, E. R.: 5'-Adenylyl-O-tyrosine: The novel phosphodiester residue of adenylylated glutamine synthetase from Escherichia coli. J. Biol. Chem. 243, 3769-3771 (1968).

5. Anderson, W. B., Hennig, S. B., Ginsburg, A., Stadtman, E. R.: Association of ATP: Glutamine synthetase adenylyltransferase activity with the P_I component of the glutamine synthetase deadenylylation system. Proc. Natl. Acad. Sci. US 67, 1417-1424 (1970).

6. Anderson, Wayne B., Stadtman, E. R.: Glutamine synthetase deadenylylation: A phosphorolytic reaction yielding ADP as nucleotide product. Biochem. Biophys. Res. Commun. 41, 704-709 (1970).

7. Brown, M. S., Segal, A., Stadtman, E. R.: Modulation of glutamine synthetase adenylylation and deadenylylation is mediated by metabolic transformation of the P_{II}-regulatory protein. Proc. Natl. Acad. Sci. US 68, 2949-2953 (1971).

8. Shapiro, B. M.: The glutamine synthetase deadenylylating enzyme system from Escherichia coli. Resolution into two components, specific nucleotide stimulation and cofactor requirements. Biochemistry 8, 659-670 (1969).

9. Mangum, John H., Magni, G., Stadtman, E. R.: Regulation of glutamine synthetase adenylylation and deadenylylation by enzymatic uridylylation and deuridylylation of the P_{II} regulatory protein. Arch. Biochem. Biophys. 158, 514-525 (1973).

10. Adler, Stuart P., Mangum, J. H., Magni, G., Stadtman, E. R.: Uridylylation of the P_{II} regulatory protein in cascade control of Escherichia coli glutamine synthetase. Third International Symposium on Metabolic Interconversion of Enzymes, Springer Verlag, New York, 221-233 (1974).

11. Adler, Stuart P., Purich, D., Stadtman, E. R.: Cascade control of E. coli glutamine synthetase: Properties of the P_{II} regulatory protein and the uridylyltransferase--uridylyl removing enzyme. Fed. Proc. 33, 1427 (1974).

12. Huang, C., Adler, S. P.: Unpublished Data.

13. Adler, S. P., Ginsburg, A.: Unpublished Data.

14. Adler, S. P.: Unpublished Observation.

15. Shaltiel, S.: Hydrophobic chromatography in the study of regulatory enzymes. Third International Symposium on Metabolic Interconversion of Enzymes, Springer Verlag, New York, 379-392 (1974).

16. Segal, A., Brown, M. S., Stadtman, E. R.: Metabolite regulation of the state of adenylylation of glutamine synthetase. Arch. Biochem. Biophys. 161, 319-327 (1974).

17. Ebner, E., Wolf, D., Gancedo, C., Elsasser, S., Holzer, H.: ATP: Glutamine synthetase adenylyltransferase from *Escherichia coli* B, purification and properties. Eur. J. Biochem. 14, 535-544 (1970).

Structure and Biosynthesis of an Acidic Glycoprotein in a Bacterial Cell Envelope

J.Baddiley, J. P. Burnett, I. C. Hancock and J. Heptinstall
Microbiological Chemistry Research Laboratory, The University, Newcastle upon Tyne, NE1 7RU, England

INTRODUCTION

Although glycoproteins are universally distributed in animals and plants, their occurrence in microorganisms is less well-documented. Some eucaryotic microorganisms are known to produce glycoproteins and in particular several yeast glycoproteins have been described, some that are enzymes[1, 2] and others that appear to be envelope or extracellular proteins[3, 4]. Mannose is the major sugar in many of these yeast glycoproteins. In contrast, there are fewer than ten reports of glycoproteins in bacteria. Everse and Kaplan[5] have described glycoprotein enzymes from Bacillus subtilis and there are reports of poorly-characterised glycoproteins in the envelopes of Escherichia coli[6] and Pseudomonas aeruginosa[7]. The bacterial glycoprotein that has been studied in most detail is that from the envelope of the marine pseudomonad BAL 31[8], but this may be a viral-specific component that results from infection with bacteriophage PM 2. N-Acetylglucosamine is the major sugar in all these bacterial glycoproteins; the only exception to this is the fucose-containing glycoprotein of a corynebacterium whose biosynthesis has been studied by Strobel and his co-workers[9].

There have been three recent reports of glycoproteins that contain phosphodiester linkages in their glycan moieties. These 'phosphoglycoproteins' have been found in Hansenula holstii[10], Cladosporium werneckii[11] and in Penicillium charlesii[12]. They represent an interesting new class of macromolecules whose function is not yet known.

We report here studies on the biosynthesis and preliminary characterisation of a glycoprotein from B. licheniformis that is rich in N-acetylglucosamine and phosphate and therefore appears to be a phosphoglycoprotein.

METHODS

Growth of Cells and Preparation of Membrane

B. licheniformis ATCC 9945 was grown to exponential phase and membranes were prepared by treatment of the cells with lysozyme in the absence of an osmotic stabiliser as previously described[13]. Membranes were suspended at about 50 mg dry wt/ml in 0.05M Tris-HCl, pH 8.0 containing 5mM ethanethiol, and were stored frozen until required.

Preparation of radioactive Glycoprotein

Large scale enzyme reactions for preparation of the phosphoglycoprotein were carried out as follows: 1.0 ml membrane suspension was incubated with 0.1 ml $MgCl_2$ (0.8 M), 0.1 ml UDP-N-acetylglucosamine (U-^{14}C; 2.98 x 10^6 cpm/μmole; 10mM) and tris buffer in a total volume of 1.5 ml at 30° for 1 h. The reaction was stopped by addition of 0.5 ml butan-1-ol and the mixture applied as a band on the origin of a preparative paper chromatogram (Whatmann 3MM). The chromatogram was developed for 18 h in solvent system A, and the material that remained at the origin was extracted into water at 60° for 3 h. More than 95% of the radioactivity from the origin was recovered in this way.

Chromatography and Electrophoresis

Paper chromatography was carried out on Whatman No. 3MM paper or No. 1 paper in the following solvents:- A propan-1-ol-ammonia (0.88 sp. gr.)-water (6 : 3 : 1 by volume). B ethylacetate-pyridine-acetic acid-water (5 : 5 : 1 : 3 by volume). Paper electrophoresis was carried out on Whatman No. 1 paper in 0.1M pyridinium acetate pH 6.5 at 40 volts/cm for $1\frac{1}{2}$ h.

Polyacrylamide gel electrophoresis in 0.1% sodium dodecylsulphate (SDS) was accomplished as described by Weber and Osborn[14]. Gels were stained for protein with Coomassie blue, and for carbohydrate material by the periodate-Schiff reagent of Segrest and Jackson[15].

Phosphate-containing compounds were detected on paper chromatograms by the method of Hanes and Isherwood[16]. Reducing sugars were stained with silver nitrate[17]. Quantitative Estimations and radioactivity measurements were carried out as previously described[18].

Preparation of polyisoprenol monophosphate

Polyisoprenols were phosphorylated as described by Popjak et al[19] and purified by column chromatography on DEAE cellulose acetate in methanolic ammonium acetate[20].

Radioactive substrates

UDP-N-(acetyl ^{14}C)-acetylglucosamine was prepared as previously described[21]. UDP-N-acetyl (U^{14}C)glucosamine was purchased from the Radiochemical Centre, Amersham, Bucks. U.K.

RESULTS

During a study of peptidoglycan biosynthesis in a membrane preparation from B. licheniformis the incorporation of radioactivity from [^{14}C]UDP-N-acetylglucosamine into polymeric material was measured in the presence and absence of UDP-N-acetylmuramyl pentapeptide (Park nucleotide), the other substrate required for peptidoglycan synthesis. Table 1 shows the results of such an experiment.

Table 1 - Incorporation of ^{14}C-UDP-N-acetylglucosamine into macromolecular material.

	With Park nucleotide	no addition
cpm incorporated into polymer	10693	7664
cpm incorporated into lipid	1105	10837

Reaction mixtures contained 0.1 ml membrane suspension, UDP-N-[acetyl-^{14}C] acetylglucosamine (0.1 μmol, 2.98 x 10^5 cpm), $MgCl_2$ (8 μmol) in a total volume of 0.13 ml. The mixture was incubated for 1 h at 30° and then the reaction mixture was applied in a 2.5 cm band to the origin of a paper chromatogram (Whatman no. 3MM paper). The paper was developed in Solvent System A for 18 h. Polymeric material remained at the origin, while lipids migrated with an R_f of 0.85.

It was found that even in the absence of Park nucleotide, when peptidoglycan synthesis is impossible, a large amount of radioactivity was incorporated into high molecular weight material, and also into lipid.

In order to ascertain whether the radioactivity remained entirely in glucosamine residues in the product, the above experiment was repeated using uniformly labelled ^{14}C-N-acetylglucosamine and the baseline area from the chromatogram, and the lipid region, were cut out and hydrolysed on the paper in 2M-HCl at 100^{0} for 3 h. The hydrolysate was chromatographed on Whatman no. 1 paper in Solvent system B for 18 h. All the radioactivity co-chromatographed with standard samples of glucosamine and N-acetylglucosamine, for both polymer and lipid.

Properties of the polymer

It was found that the material containing N-acetylglucosamine that remained at the origin of the chromatogram of reaction mixtures in Solvent A could be completely eluted from the paper into water at 60^{0} in 3 h. Material was isolated in this way from large-scale incubation mixtures as described in 'Methods'. Column chromatography of the material was carried out on Sephadex G75 in both 10 mM-tris-HCl, pH 8.0 containing 1 mM-mercaptoethanol and in the same buffer containing 0.1% sodium dodecyl sulphate (SDS). In both cases the bulk of the radioactivity was excluded from the gel and the small amount of included material was UDP-N-acetylglucosamine. The excluded material contained protein as indicated by the Lowry method and absorption at 280 nm. Analytical polyacrylamide gel electrophoresis in 0.1% SDS revealed two diffuse protein bands, one at the top of the gel and the other with R_f of about 0.75. The R_f 0.75 band contained all the radioactivity and stained positively with periodate-Schiff.

Radioactive polymer that had been extracted from large scale incubation mixtures as described was further purified by preparative SDS polyacrylamide gel electrophoresis, under the same conditions as described for analytical electrophoresis, on a 2.5 cm x 6.5 cm cylindrical gel. The fast running protein fraction (R_f 0.65 - 0.85), which contained all the radioactivity, was eluted from the gel into water, dialysed, and passed through a column of Sephadex G75, from which the excluded peak containing all the radioactivity was retained. Double-diffusion against DEAE-dextran in 1.0% agarose gel at pH 7.3 revealed two sharp precipitin lines, indicating that the purified material contained two anionic components. This partially purified labelled material was used for further chemical studies.

Acid hydrolysis in 2N HCl, 3 h at 100°, followed by paper chromatography in solvent system B and staining with silver nitrate reagent, revealed the presence of glucose and glucosamine (with a trace of N-acetylglucosamine). Paper electrophoresis of the same hydrolysate revealed a non-reducing component that stained positively for phosphate and amino groups but was unidentified. This component was also detected by paper chromatography in solvent A, in which it displayed an $R_{GlcNH_2\text{-}6\text{-}P}$ = 2.24.

Table 2 shows the amino acid composition of the glycoprotein. It contained only small amounts of the aromatic amino acids and negligible amounts of amino acids containing sulphur.

Table 2 - Amino acid composition of glycoprotein

Lysine	1.35	Proline	0.30	iLeucine	0.35
Histidine	0.17	Glycine	0.91	Leucine	0.44
Arginine	1.70	Alanine	1.61	Tyrosine	0.09
Aspartic acid	1.0	half-cystine	0	Phenylalanine	0.13
Threonine	0.35	Valine	0.44	Tryptophan	0
Serine	0.52	D.a.p.	trace	Glucosamine	0.44
Glutamic acid	1.22	Methionine	0.13		

The analysis was carried out on an 18 h hydrolysate. Aspartic acid was set arbitrarily at 1.0.

Determination of N-terminal amino acids by dansylation gave two major AAs, one of which was the dansyl derivative of glycine. The other has not been identified.

The preliminary evidence suggested that the material containing N-acetylglucosamine was an acidic glycoprotein. In order to examine this hypothesis the polymer was subjected to digestion with pronase (1 mg/ml) for 24 h at pH 7.0 in Tris-HCl containing 1mM-$CaCl_2$, 37° After 24 h a further 1 mg/ml pronase was added and incubation was continued for a further 24 h. The digestion mixture was subjected to paper electrophoresis in pyridinium acetate at pH 6.5. This yielded a small, sharply defined fraction that had a slight positive charge, was radioactive and stained with ninhydrin, together with a more diffuse, very acidic peptide fraction that contained most of the radioactivity. These peptides were eluted into water and chromatographed on a column of Sephadex G25. The acidic electrophoresis fraction yielded two peaks of radioactivity, a small one (GIA) which was excluded from the resin and a much larger one (G2A) which was included. The basic electrophoresis fraction yielded a single peak on G25 (G3B) which had approximately the same apparent molecular weight as G2A. The release of these radioactive glycopeptides from the high molecular weight material by treatment with the proteolytic enzyme confirmed that the product of the biosynthetic reaction was a glycoprotein.

Biosynthesis of the polymer

The chromatographic properties of the N-acetylglucosamine-containing lipid that was formed during biosynthesis of the polymer, and its extreme lability to acid (it had a half-life of 15 min at 100° in 0.01M-HCl in 50% MeOH) suggested that it might be a polyprenol phosphate derivative of the type known to participate in the biosynthesis of many bacterial polysaccharides[22]. In order to test this hypothesis, the effect of added C_{55}-polyprenol monophosphate on lipid and polymer synthesis was examined. Added polyprenol monophosphate stimulated lipid synthesis, but only in the presence of the nonionic surfactant Triton X-100 (Fig. 1).

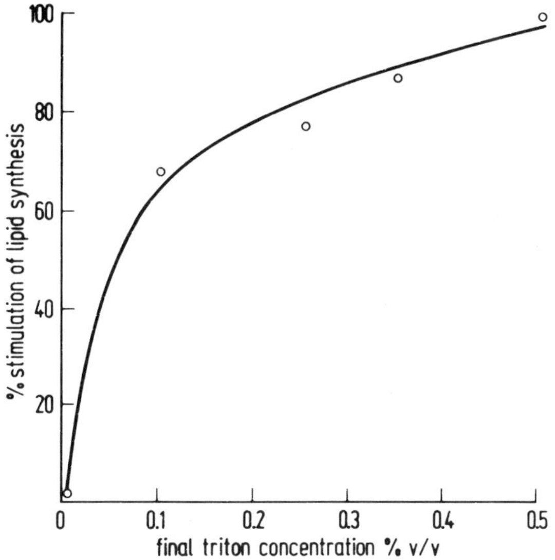

Fig. 1. The dependence of stimulation of polyprenol phosphate upon Triton X-100.

Reaction mixtures contained 0.1 ml membrane suspension, $MgCl_2$ (5 μmol), UDP-N-[^{14}C-acetyl]acetyl glucosamine (0.1 μmol, 2.98 x 10^5 cpm), and Triton X100 at the indicated concentration. The total volume was 0.13 ml. C_{55}-polyprenol phosphate (1n mol) was added to each incubation tube first in chloroform:methanol 2:1 v/v, then the solvent was evaporated off under reduced pressure before addition of the other reagents and mixing. Reaction was terminated by the addition of 0.05 ml n-butanol and lipid synthesis was measured as described for Table 1. Incubations for 1 h at 30^0.

In the presence of Triton X-100 C_{55} -polyprenol monophosphate stimulated incorporation of label from UDP-N-acetyl glucosamine into the lipid as shown in Fig. 2.

The effect of a number of different polyprenol monophosphate preparations on lipid and polymer synthesis is shown in Table 3. Only the phosphate of the all-*trans* C_{45} isoprenol, solanesol failed to stimulate both lipid and polymer synthesis.

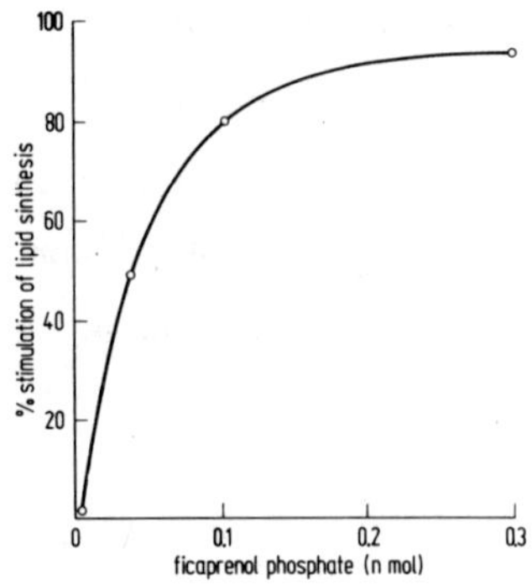

Fig. 2. Stimulation of lipid synthesis by C_{55}-polyprenol monophosphate.

Incubation mixtures were as described for Fig. 1. All incubations contained Triton X-100 at a final concentration of 0.3% v/v. Incubations were carried out at 30° for 1 h and lipid synthesis was assayed as described for Table 1.

Table 3 - Effect of polyprenol monophosphates on lipid and polymer synthesis

	no added lipid	heveaprenol phosphate	bactoprenol phosphate	solanesol phosphate
cpm in polymer	8144	13263	11652	8102
cpm in lipid	6060	12297	12235	6091

Incubation mixtures were as described for Fig. 1. Approx. 3 nmol of the indicated polyprenol phosphate was added.

This evidence suggested that the lipid was a polyprenol monophosphate or pyrophosphate N-acetylglucosamine that might be intermediate in the biosynthesis of the polymer. In order to obtain further information about the synthesis of the lipid the effects of UMP and UDP on its formation were examined. It was found that UMP strongly inhibited both polymer and lipid synthesis, while UDP inhibited the synthesis of polymer but not of lipid. This result suggested that lipid synthesis involves the transfer to polyprenol monophosphate of N-acetylglucosamine 1-phosphate from UDP-N-acetylglucosamine, with the release of UMP.

UDP-N-acetylglucosamine + lipid-P $\rightleftharpoons$ UMP + lipid-P-P-N-acetylglucosamine

UMP would inhibit this reaction by a mass-action effect. Transfer of either N-acetylglucosamine or its 1-phosphate from lipid to polymer could subsequently occur.

DISCUSSION

Our preliminary results show that in the membrane of B. licheniformis there are enzymes that catalyse the incorporation of N-acetylglucosamine, N-acetylglucosamine 1-phosphate, or both, into an endogenous glycoprotein acceptor. It is not yet known whether the product of this cell-free biosynthetic system represents the cellular end-product although work is in progress to characterise the glycoprotein of the cell.

The amino acid analysis of the glycoprotein does not indicate a large excess of acidic AA residue in the protein moiety and the polyanionic nature of the glycoprotein may therefore be attributable to the glycan part, and in particular to phosphate groups in the glycan. The high acidity of the phosphoglycoprotein and its high glycan content make measurements of its molecular weight by conventional physical techniques difficult.

The role of lipid intermediates in the biosynthesis of glycoproteins in yeasts[10] and in some mammalian systems[23, 24, 25] has received considerable attention recently, but work has been hindered by difficulties in characterising the protein acceptor and the end products in these experiments. The B. licheniformis system described here appears to provide a simpler model for lipid-mediated glycoprotein synthesis, since the product has a simpler sugar composition and is freely soluble in water, and therefore more amenable to conventional purification. The cellular location of the end product and its role in the cell remain to be determined.

SUMMARY

N-acetylglucosamine or N-acetylglucosamine-1-phosphate residues were found to be transferred from UDP-N-acetylglucosamine through a lipid intermediate into macromolecular material by a membrane-bound enzyme system from B. licheniformis. The product was water-soluble, contained protein and carbohydrate and was excluded from Sephadex G-75. After treatment with a proteolytic enzyme the material yielded glycopeptides which were included in Sephadex G-25 and were separated by paper electrophoresis. The original glycoprotein was strongly acidic, and it yielded two acidic glycopeptides and a weakly basic one, all of which contained N-acetylglucosamine. The preparation contained phosphate groups in the carbohydrate part of the glycoprotein.

REFERENCES

1. Odds, F. C. & Hierholzer (1973) J. Bacteriol. 114, 257-266.
2. Smith, W. J. & Ballon, C. E. (1974) Biochemistry 13, 355-361.
3. Sentandreau, R. & Northcote, D. H. (1969) Biochem. J. 115, 231-240.
4. Biely, P., Farkas, V. & Bauer, S. (1972) F.E.B.S. Letters 23, 153-156.
5. Everse, J. & Kaplan, N. O. (1968) J. Biol. Chem. 243, 6072-6074.

6. Okada, S. & Weinbaum, G. (1968) Biochemistry 7, 2319-2825.

7. Clarke, K., Gray, G. W. & Reaveley, D. A. (1967) Biochem. J. 105, 755-758.

8. Datta, A., Otero, R. D., Braunstein, S. N. & Franklin, R. M. (1973) Biochim. Biophys. Acta 311, 163-172.

9. Sadowski, P. L. & Strobel, G. A. (1972) J. Bacteriol. 115, 668-672.

10. Kozak, L. P. & Bretthauer, R. K. (1970) Biochemistry 9, 1115-1122.

11. Lloyd, K. O. (1972) Biochemistry 11, 3884-3890.

12. Gander, J. E., Jentoft, N. H., Drewes, L. R. & Rick, P. D. (1974) J. Biol. Chem. 249, 2063-2072.

13. Hancock, I. C. & Baddiley, J. (1972) Biochem. J. 127, 27-37.

14. Weber, K. & Osborn, M. (1969) J. Biol. Chem. 244, 4406-4412.

15. Segrest, J. P. & Jackson, R. L. (1972) Methods in Enzymol. 28, 54-62.

16. Hanes, C. S. & Isherwood, F. A. (1949) Nature (London) 164, 1107-1109.

17. Trevelyan, W. E., Procter, D. P. & Harrison, J. S. (1950) Nature (London) 166, 444-445.

18. Anderson, R. G., Hussey, H. & Baddiley, J. (1972) Biochem. J. 127, 11-25.

19. Popjak, G., Cornforth, J. W., Cornforth, R. H., Ryhage, R. & Goodman, D. S. (1962) J. Biol. Chem. 237, 56-62.

20. Lahar, M., Chin, T. H. & Lennarz, W. J. (1969) J. Biol. Chem. 244, 5890-5896.

21. Baddiley, J., Blumsom, N. L. & Douglas, L. J. (1968) Biochem. J. 110, 565-570.

22. Lennartz, W. J. & Scher, M. G. (1972) Biochim. Biophys. Acta 265, 417-441.

23. Parodi, A. J. et al (1973) Carbohydrate Res. 26, 393-400.

24. Maestri, N. & de Luca, L. (1973) Biophys. Biochem. Res. Commun. 53, 1344-1349.

25. Waechter, C., Lucas, L. & Lennartz, W. J. (1974) Biophys. Biochem. Res. Commun. 56, 343-350.

Degradation of Thyrotropin Releasing Hormone (TRH). Its Inhibition by Pyroglu-His-OCH_3 and the Effect of the Inhibitor in Attempts to study the Biosynthesis of TRH

K. Bauer
Max-Volmer-Institut, Abteilung Biochemie, Technische Universität Berlin

INTRODUCTION

Due to our interest in the mechanism of peptide synthesis in general our attention was drawn to the recently discovered group of peptides produced by the hypothalamus, the so-called releasing hormones which trigger the release of the hypophyseal hormones. We became especially interested to study first the smallest of these peptides, the thyrotropin releasing hormone (Fig.1).

Thyrotropin-releasing-hormone (TRH)

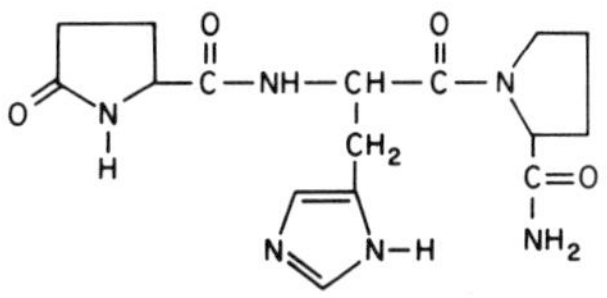

pyroGlu —— His —— Pro-NH_2

In view of the small size of the peptide the biosynthesis could be carried out enzymatically analogous to the synthesis of glutathion (1). The structural similarity of TRH with the other so far identified releasing hormone, the luteinizing hormone releasing hormone (LH-RH) with the sequence pyroGlu-His-Trp-Ser-Tyr-Gly-Leu-Arg-Pro-Gly-NH_2 makes it more likely that the synthesis is mediated by a multienzyme system as for the polypeptide antibiotics (2). However, it is also possible that TRH is synthesized by a ribosomal mechanism as part of a prohormone as demonstrated for several polypeptide hormones (3,4,5,6). The pyroglutamyl formation and the amidation at the carboxy-terminal, which is frequently found in many biologically active peptides, could be part of the cleavage mechanism or could be formed by subsequent reactions. The simple structure of TRH would make it an ideal model substance to study these questions.

RESULTS

Degradation of TRH

In the meantime it has been reported that TRH is synthesized by a particulate free extract of freeze-dried porcine hypothalamic tissue (7,8). However, when we repeated the incubations as reported using ^{14}C-proline as precursor and subjected tne methanolic extract from such an incubate to two dimensional thin layer chromatography many unidentified radiolabelled products could be detected by radioautography, but the area of the co-chromatographed marker TRH, which was visualized by the PAULY reaction, did not contain any radioactivity. Moreover with such a preparation we observed powerful degradation activity which made it impossible to find any biosynthetic activity.

A rapid inactivation to TRH by animal and human plasma was first reported by BOWERS et al. (9). The same laboratory reported recently (10) that the inactivation of TRH by serum is only due to a deamidation reaction. Our studies with TRH, ^{3}H-labelled in the proline moiety, show that this is only the first step of TRH inactivation which is followed by proteolytic cleavage yielding free proline as the major radiolabelled degradation product (Fig.2).

That pr line derives from the hydrolysis of deamido-TRH could be demonstrated by incubation of isolated deamido-TRH with serum followed by thin layer chromatography as before.

These findings confirm the results of VALE et al. (11) who suggested that there must be an additional mechanism of TRH degradation since they observed a total loss of the biological activity of TRH after incubation with serum while deamido-TRH exhibits some of the biological activity of TRH.

The high degradation activity of the tissue extract could not be explained by the serum contamination of these preparations. After incubation of TRH-^{3}H-proline with a high speed supernatant from freeze-dried porcine hypothalamic tissue, as used in the report on the biosynthesis of TRH (7,8) the degradation pattern of Fig.3 was obtained.

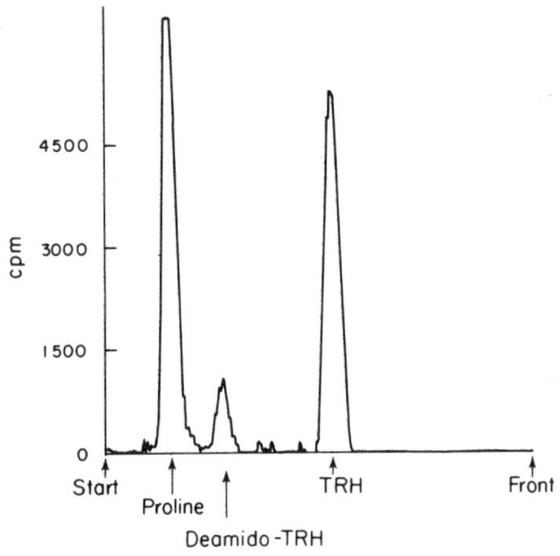

Fig.2. Degradation of TRH by Rat Serum.
10 μl of rat serum was incubated for 60 min at 37°C with 0.5 μCi TRH-^{3}H-proline (40 Ci/mM) in 20 μl buffer A (Tris-HCl pH 7.4, containing 50 mM KCl and 2 mM DTT). The reaction was stopped by the addition of methanol, the precipitate removed by centrifugation and the supernatant subjected to thin layer chromatography on silica gel (Silplate 22) using the solvent system $CHCl_3$:CH_3OH:NH_4OH (125:75:25). The separated radiolabelled split products were localized by scanning for radioactivity and identified by cochromatography with the marker substances in several other solvent systems.

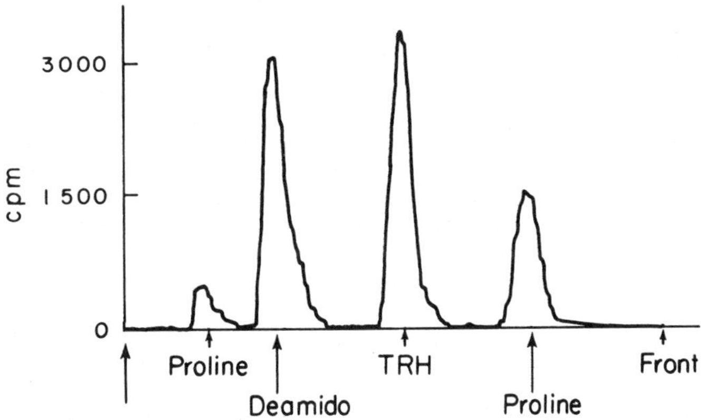

Fig.3. Degradation of TRH by Hypothalamic Tissue Extract.
The incubation was done as described in the legend of Fig.2 using a high speed supernatant prepared from a 10% homogenate of freeze-dried porcine hypothalamic fragments in buffer A. As before, the radiolabelled split products were resolved by thin layer chromatography, localized by scanning for radioactivity and identified by cochromatography with marker substances in several solvent systems.

As radiolabelled split products deamido-TRH, proline and prolineamide could be detected. The proteolytic cleavage yielding prolineamide is comparable to the reported release of the carboxy-terminal glycineamide from oxytocin by oxytocin responsive tissue(12). When ^{3}H-prolineamide was isolated and incubated with such an extract no deamidation could be observed. The free proline derives from the hydrolysis of deamido-TRH as demonstrated by the incubation of ^{3}H-deamido-TRH with this extract.

When a high speed supernatant was prepared from a homogenate of fresh rat hypothalamic or cortical tissue we could only observe a deamidation reaction while the whole homogenates possess also the peptidases yielding proline and prolineamide as radiolabelled split products (Table I).

Table I: Degradation of TRH by Various Preparations

Preparation	% of radioactivity added as TRH-^{3}H-proline			
	A	B	C	D
TRH	37	35	65	39
Deamido-TRH	33	30	35	4
Prolineamide	21	10	none	none
Proline	9	25	none	57

TRH-^{3}H-proline was incubated with the preparations listed as described in the legend of Fig.2. A: High speed supernatant of a homogenate prepared from freeze-dried hypothalamic fragments. B: Whole homogenate from fresh rat hypothalamic or cortical tissue. C: High speed supernatant from B. D: Rat serum. After thin layer chromatography the radioactivity of the radiolabelled split products were determined by scintillation counting.

Apparently these peptidases are membrane bound and become solubilized during lyophylisation. So it must be concluded that the peptidase which splits proline from deamido-TRH is different from the comparable serum enzyme. That the soluble TRH-deamidating enzyme is also different from the serum enzyme will be shown by its specific inhibition by the dipeptide ester pyroGlu-His-OCH_3 (see below).

Inhibition of TRH-Degradation

We had some hope to overcome the TRH degradative activity of the tissue preparations since it had been reported by GUILLEMIN's laboratory (11) that the methylester of the N-terminal dipeptide of TRH, pyroGlu-His-OCH_3, strongly inhibits the inactivation of the biological activity of TRH by serum. Following the effect of the inhibitor on the degradation of TRH by serum Fig.4 was obtained. At a 4 mM concentration of this material degradation was reduced by 50%.

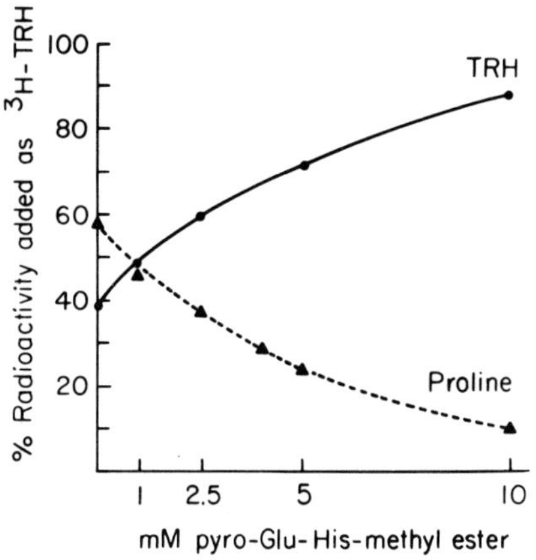

Fig.4. TRH recovery and degradation after incubation with rat serum dependent on inhibitor concentration. PyroGlu-His-OCH_3 was added to the incubation of TRH-^{3}H-proline with rat serum as in Fig.2. After thin layer chromatography the radioactivity in the TRH and proline zone was determined by scintillation counting.

When the inhibitor was added to the high speed supernatant from freeze-dried porcine hypothalamic fragments a marked difference could be shown (Fig.5). While the activity of the peptidases yielding proline and prolineamide was inhibited at inhibitor concentrations comparable with the serum enzymes the deamidation reaction was less sensitive, indicating that the TRH-deamidating enzyme of the tissue is different from the serum enzyme.

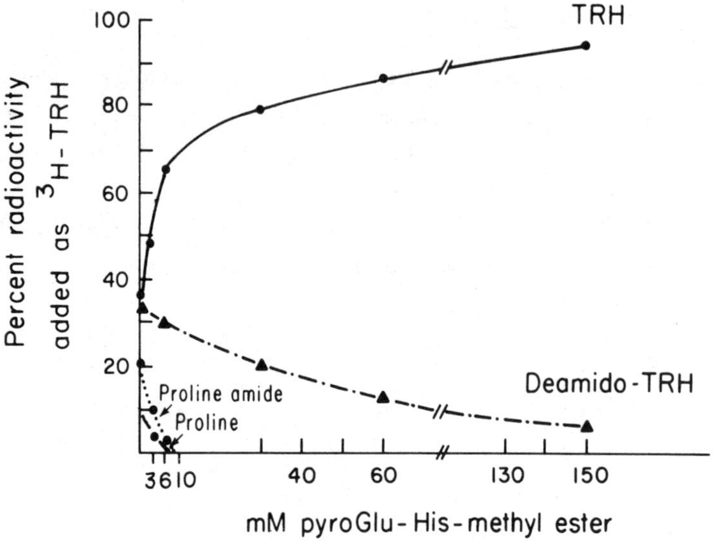

Fig.5. TRH recovery and degradation after incubation with an extract of hypothalamic tissue dependent on inhibitor concentration. The experiment was done as for Fig.4 using the high speed supernatant from a homogenate of freeze-dried porcine hypothalamic tissue.

Whether some of these TRH degradating enzymes may play a role in the regulation of the biological activity of TRH will be investigated. Preliminary results with partially purified TRH-deamidating enzyme of tissue origin indicate that this enzyme is specific for TRH.

Attempts to Study the Biosynthesis of TRH

With the information from these studies we had some hope to find biosynthetic activity for TRH by adding the dipeptide ester to an incubation of fresh rat hypothalamic tissue homogenate supplemented with amino acids, ATP, Mg^{++} and an ATP regenerating system (see legend of Table III). Since the dipeptide ester does not prevent the deamidation of TRH by hypothalamic tissue, one should expect deamido-TRH rather than TRH itself as the reaction product.

First a very effective purification procedure had to be worked out which reliably identifies deamido-TRH, present only in minute amount besides many unidentified radiolabelled metabolites. Using ^{14}C-proline as the radiolabelled precursor for the incubation and adding ^{3}H-deamido-TRH after the incubation as internal standard gave us the possibility to determine by scintillation spectrometry the recovery during the purification and so to account for the variability which allows to correlate the data obtained for quantitative evaluation. As listed in Table II the purification was achieved by five successive steps. The radiochemical purity of the material from the deamido-TRH zone of the second thin layer chromatogramm was proven by the constant $^{3}H:^{14}C$ ratio when the isolated material was subjected to thin layer chromatography in several other solvent systems, thin layer electrophoresis and chemical conversion of the isolated material into TRH by esterification followed by amidation.

Table II: Purification Scheme for the Isolation of Deamido-TRH

	Recovery$^{(+)}$ %	Purification factor$^{(++)}$
Charcoal absorption	90-95	30
CM-cellulose column	95-99	1.25
Cellulose-phosphate paper chromatography	85-90	20
TLC-Solvent System A	70-80	2.0
TLC-Solvent System B	70-80	1.1
Total	34-54	1900 fold

The incubation conditions are given in the legend of Table III. After incubation ^{3}H-deamido-TRH (90 000 dpm) and 50/ug synthetic deamido-TRH were added. After absorption on charcoal and intensive washing with 0.02 N HOAc the material was eluted by 95% pyridine, taken to dryness and applied on a short column of CM-cellulose which was eluted with water. The eluate was applied to cellulose-phosphate paper and developped twice with water. The material from the deamido-TRH zone was eluted with 2 N NH_4OH and subjected to thin layer chromatography on silica gel (Silplate 22) in the solvent system A: $CHCl_3:CH_3OH:NH_4OH$ (125:75:25). The radioactive zone corresponding to deamido-TRH was eluted with 50% methanol and used for thin layer chromatography in the solvent system B: Phenol : H_2O (75:25). After elution the radioactivity was determined by scintillation spectrometry. The recovery$^{(+)}$ is determined by the recovery of ^{3}H-deamido-TRH which was added as internal standard and the purification$^{(++)}$ is based on the ^{14}C-dpm added as ^{14}C-proline and corrected for recovery.

The incorporation obtained was unexpectedly high (30 000 dpm, which accounts for 56 pmole or 20 ng of TRH) and we began to wonder about the mechanism of incorporation. Therefore we studied first the incorporation of the other constituent amino acids. When ^{14}C-histidine was added as radiolabelled precursor to such incubation mixtures no incorporation into deamido-TRH could be observed demonstrating that the formation of deamido-TRH is not due to a de novo biosynthesis of the tripeptide.

Table III surveys the results from further studies of this reaction which is not sensitive to inhibitors of ribosomal synthesis such as puromycin, cycloheximid or RNAse and is even not energy dependent.

When we observed that the dipeptide ester is hydrolyzed by such an incubation we realized that the split is presumably due to the same peptidase that degrades deamido-TRH yielding free proline since it is well known that peptidases show some esterase activity and explaining inhibition of the degradation by a competition for the enzyme. Therefore the hydrolysis of the dipeptide ester pyroGlu-His-OCH_3 could lead to the same enzyme bound intermediate ENZYME~His-pyroGlu as the hydrolysis of deamido-TRH (pyroGlu-His-Pro-OH). By accepting proline in the reverse reaction the peptidase could incorporate proline into the dipeptide in amounts comparable to what we were finding.

Such an enzymatic process would explain that for the reaction ATP is not only not necessary but is rather inhibitory, presumably since the Mg^{++} chelating nucleotides reduce the concentration of free Mg^{++} required as cofactor of the enzyme. The specific incorporation of proline, which is not greatly reduced in the presence of all the other amino acids added in comparable molar concentration, supports this explanation. Further evidence therefore can be seen in comparing Table I with Table III. While homogenates prepared from fresh rat cerebral cortex were also active in the formation of the tripeptide, a high speed supernatant from freeze-dried hypothalamic fragments had some activity and a high speed supernatant from fresh tissue was inactive. This is in parallel with the TRH degradation activity of the different preparations in respect to the formation of proline as radiolabelled split product (see Table I).

Table III: Deamido-TRH formed on Incubation with pyroGlu-His-OCH_3 and ^{14}C-proline

Preparation	Tripeptide formed (pmole)
Hypothalamic homogenate supplemented	56
+ puromycin (100 µg/ml)	58
+ cycloheximid (300 µg/ml)	50
+ RNAse (20 µg/ml)	56
- ATP, GTP and ~P regenerating system	121
- pyroGlu-His-OCH_3	0
+ 17 amino acids	42
Homogenate of cerebral cortex	48
High speed supernatant of freeze-dried hypothalamic tissue	18
High speed supernatant of fresh tissue	0

Male rats were killed by decapitation, the tissue removed and immediately homogenized in ice cold buffer B (0.25 M sucrose, 50 mM Tris-HCl, pH 7.5, 150 mM NH_4Cl, 5 mM $MgAc_2$ and 2 mM DTT). To 300 µl of homogenate was added: 10 µmole pyroGlu-His-OCH_3, 1 µmole ATP, 0.25 µmole GTP, 2.5 µmole histidine, 2.5 µmole glutamine and 25 µCi ^{14}C-proline (250 mCi/mM). The total volume was 0.5 ml. The mixture was incubated for 45 min at 37°C and the deamido-TRH formed was isolated as described in Table II.

SUMMARY

In attempts to study the mechanism of biosynthesis of TRH very active degradation of TRH by serum and tissue preparations was observed. By addition of pyroGlu-His-OCH_3 the complete breakdown of TRH could be prevented. In the hope that this might enable us to study the biosynthesis of TRH, appropriate incubates of hypothalamic tissue homogenate with ^{14}C-proline as precursor were supplemented with the dipeptide ester. After vigorous purification radiolabelled deamido-TRH could be isolated. However, the formation of deamido-TRH is not related to the biosynthesis of TRH but is due to incorporation of proline by the reverse reaction of peptide hydrolysis.

Acknowledgment

This study was supported by a grant to Dr. F. Lipmann from the United States Public Health Service. It was done during my stay in the laboratory of Dr. Lipmann who initiated this project. I am most thankful to him as my teacher and mentor for the educational training, for all his advice, encouragement and guidance. I thank Dr.R.O. Studer, Hoffmann-La Roche A.G., Basle, for the valuable gift of peptide intermediates. Dr. R. Guillemin and his colleagues at the Salk Institute I am very much obliged for the interest and encouragement and for the support during my stay in his laboratory.

REFERENCES

(1) Mooz, E.D., Meister, A.: Tripeptide (glutathion) synthetase. Purification, properties, and mechanism of action. Biochemistry 6, 1722-1734 (1967).
(2) Lipmann, F., Gevers, W., Kleinkauf, H., Roskoski, R., Jr.: Polypeptide synthesis on protein templates: The enzymatic synthesis of gramicidin S and tyrocidines. Adv.Enzymology 35, 1-34 (1971).
(3) Steiner, D.F., Oyer, P.E.: The biosynthesis of insulin and a probable precursor by a human islet cell adenoma. Proc.Nat.Acad.Sci.US.57, 473-480 (1967).
(4) Kemper, B., Habener, J.F., Potts, jun., J.T., Rich, A.: Proparathyroid hormone: Identification of a biosynthetic precursor to parathyroid hormone. Proc.Nat.Acad.Sci.US.69, 643-647 (1972).
(5) Cohn, D.V., MacGregor, R.R., Chu, L.L.H., Kimmel, J.R., Hamilton,J.W.: Calcemic fraction-A: Biosynthetic peptide precursor of parathyroid hormone. Proc.Nat.Acad.Sci.US.69, 1521-1525 (1972).
(6) Tager, H.S., Steiner, D.F.: Isolation of glucagon-containing peptide: Primary structure of a possible fragment of proglucagon. Proc.Nat.Acad. Sci.US.70, 2321-2325 (1973).
(7) Mitnick, M.A., Reichlin, S.: Enzymatic synthesis of thyrotropin-releasing hormone: Biosynthesis by rat hypothalamic fragments in vitro. Science 172, 1241-1243 (1971).
(8) Mitnick, M.A., Reichlin, S.: Enzymatic synthesis of thyrotropin-releasing hormone (TRH) by hypothalamic 'TRH synthetase'. Endocrinology 91, 1145-1153 (1972).
(9) Bowers, C.Y., Redding, T.W., Hawley, W.D.: The effect of thyrotropin-releasing factor (TRF) in animals and man. Program of the 48th Meet. of the Endocrine Soc., Chicago, Ill., June 20-22, 1966, p.43 (abstract).
(10) Nair, R.M.G., Redding, T.W., Schally, A.V.: Site of inactivation of thyrotropin-releasing hormone by human plasma. Biochemistry 10, 3621-24, (1971).
(11) Vale, W.W., Burgus, R., Dunn, T.F., Guillemin, R.: In vitro plasma inactivation of thyrotropin-releasing factor (TRF) and related peptides. Its inhibition by various means and by the synthetic dipeptide PCA-His-OME. Hormones 2, 193-203 (1971).

Amphibian Cells in Culture: An Approach to Metamorphosis

Thomas Peter Bennett
Biological Science, Florida State University, Tallahassee, Florida 32306, USA

The liver of Rana catesbeiana undergoes metabolic differentiation during the metamorphosis of this animal from a tadpole into an adult frog. The biochemical criteria for this differentiation include the induction of synthesis of urea cycle enzymes and certain serum proteins in addition to changes in many enzyme activities which have been well documented and have been reviewed (1-4). Recently, we (5,6) have reported cytological criteria for differentiation of liver cells during thyroxine-induced and natural metamorphoses.

One may ask: "How does thyroxine control the cytological differentiation of a tadpole liver cell into a frog liver cell?" A similar question may be asked about the biochemical differentiation which occurs. In attempting to answer these questions, our first task was to establish cytological criteria by light and electron microscopy for liver cells of R. catesbeiana in vivo undergoing natural and thyroxine induced metamorphosis (5,6). These studies on liver from animals at various metamorphic stages were correlated with the morphological features of R. catesbeiana undergoing metamorphosis and are summarized in Figure 1.

The nuclei of parenchymal cells are euchromatic in early metamorphic stages and become heterochromatic about the time of forelimb development and remain so (6). They become more irregular in shape, and the number of nucleoli increases. The mitochondria increase in size, and their cristae change from a broad lamellar to a smaller, more tubular appearance during the period from hindlimb development through metamorphic climax. The rough endoplasmic reticulum (RER) proliferates throughout the cytoplasm and the cisternae become dilated during this period. The Golgi complex is associated with dense granules (indicating biosynthetic activity) during early stages of metamorphosis, appears less active during metamorphic climax, and returns to a more active appearance in froglet and adult animals.

A comparison of the fine structural changes in liver cells during natural metamorphosis with those which occur during thyroxine-induced metamorphosis (5) reveals that the alterations are qualitatively similar but often differ quantitatively in the two types of metamorphoses (see Figure 1). Further, the organelle features of liver cells in thyroxine-induced metamorphosis represent a combination of the organelle features in cells

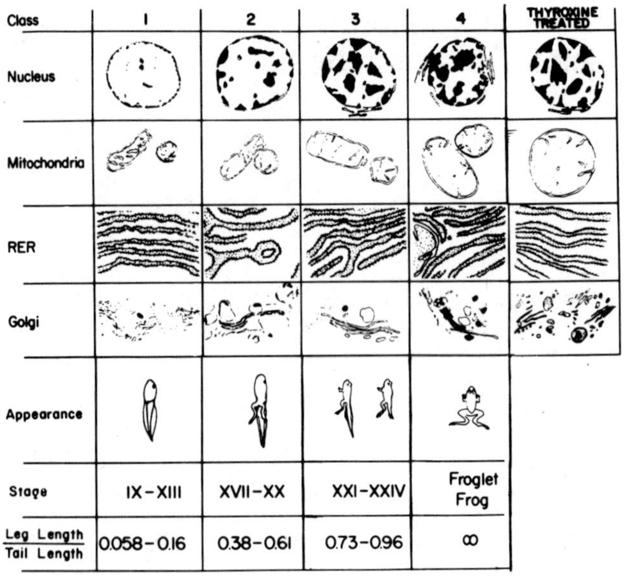

Class
1
2
3
4
THYROXINE TREATED
Nucleus
Mitochondria
RER
Golgi
Appearance
Stage
IX-XIII
XVII-XX
XXI-XXIV
Froglet Frog
Leg Length / Tail Length
0.058-0.16
0.38-0.61
0.73-0.96
∞

Figure 1. Schematic diagrams showing the salient changes in the organelles of parenchymal cells from four developmental classes of <u>Rana catesbeiana</u>. The gross appearance of respresentative members of each class is shown along with the morphological stages according to Taylor and Kollros (16) and leg length to tail length (L/T) ratios. Based on data in (5) and (6).

from animals in Classes 3 and 4. The nuclear features observed following thyroxine treatment of premetamorphic (Class 1) tadpoles resemble those of Class 3 animals; the extent of heterochromatization and the amount of perinuclear RER are similar. The mitochondria show marked increases in size during both types of metamorphoses but are considerably larger in induced metamorphosis with volumes about 2.5 times those of Class 4 mitochondria. The proliferation and cisternal features of RER in induced metamorphosis most resemble those observed in Class 3 animals. The increased number of Golgi regions in induced metamorphosis is unique; however, the appearance of individual Golgi resembles that observed in Class 4 animals. From these findings, the picture which emerges is that thyroxine-induced metamorphosis, under conditions which are frequently used in biochemical studies, results in cytological features found late in natural metamorphosis. However, in thyroxine-induced metamorphosis the extent of the changes do not precisely parallel those found in natural metamorphosis.

In preliminary reports Tata (7), Cohen (8) and Spiegle and Spiegle (9) have made some of these cytological observations on, respectively, <u>R. catesbeiana</u> and <u>R. pipiens</u>.

Because of individual differences among tadpoles in their time of response to thyroxine and tissue variability, the development of organ or cell cultures of their tissues is of considerable importance for approaching the elucidation of the mechanism(s) of thyroxine-induced changes in biochemical and cytological patterns during metamorphosis. In order to develop a model system for exploring the basis for natural and thyroxine-induced differentiation of tadpole tissues, with the ultimate objective of understanding the mechanism(s) of thyroxine action in this process, it was necessary to define the techniques for preparing and conditions for maintaining tadpole liver cells in vitro over a period of 6 days in a close approximation to their native morphological and biochemical states. This period is sufficient to observe in vivo effects of thyroxine (4,5) and should be sufficient for in vitro study of the mechanism of thyroxine action.

The extensive literature about culturing liver cells (reviewed in 10) suggested that very careful morphological studies on cells in culture must precede or be closely correlated with biochemical studies. Thus our initial studies were cytological and employed light and electron microscopy to verify the condition of cultured tadpole liver (10, 11). This was done since it is clear that, if liver cells or liver tissue are placed in sterile culture medium, stable enzyme activities may be measured for days and weeks, even though cells may be dying, lysing and releasing their contents into the sterile medium (12).

The results of our studies (10) demonstrated that the normal premetamorphic fine structure features of parenchymal cells and their organelles are well preserved for a period of six days in culture. However, on being put into culture medium and during the early period (up to about 30 hours) in culture, parenchymal cell nuclei become, according to the terminology of Fawcett (13) and Porter and Bonneville (14), heterochromatic; later they return to their so called euchromatic state. Our results (15) also suggested that in our case the "trauma," often mentioned in connection with cell culture work, that occurs when tissue adapts to in vitro conditions is reflected in defined fine structural changes (i.e., chromatin condensation) in the cultured cells. After an adjustment period these changes are reversed completely or in part.

Here we report preliminary biochemical studies on tadpole liver in organ culture. Since we are interested ultimately in an organ cultured liver system that will facilitate the analysis of thyroxine-induced changes in protein biosynthesis, we are particularly interested in establishing biochemical standards, such as wet weight, protein, RNA and DNA content, and in establishing morphological correlates to these changes when possible. The present biochemical study supports our cytological observations (10, 15) that our techniques for preparing and maintaining tadpole liver cells in vitro over a period of six days is satisfactory for keeping the cells in close approximation to their native state.

METHODS AND MATERIALS

All animals used were Class 1, that is Stage IX-XIII of Taylor and Kollros (16) R. catesbeiana tadpoles (Connecticut Valley Biological Supply Co., Southampton, Massachusetts). These tadpoles have short hind limbs but no forelimbs and are said to be in late premetamorphosis.

The culture medium used in these experiments was based on the formulation of Wolf and Quimby (17) and was purchased from the Grand Island Biological Supply Company, Grand Island, N. Y. This medium was modified to contain 100 mg glucose per liter except as indicated in individual experiments. Organ cultures were prepared as previously described (10).

The "wet weight" of tissue was determined, after it had been blotted, by weighing it on a tared glass cover slip using a Mettler (Model B1250-1) analytical balance (Mettler Corp., Princeton, N. J.). The dry weight was determined after the tissue had been dried to constant weight at 120°C and placed in a desiccator containing Drierite.

For the determination of DNA and RNA, about 2 mgs of combined pieces of liver from culture tubes were harvested, blotted dry, and accurately weighed. The tissue was homogenized in 2 ml distilled water and the homogenate was made up to 5% vol/vol with trichloroacetic acid (TCA). The mixture was centrifuged for 10 minutes at 4000 rpm in an International Clinical Centrifuge Model-CL (International Equipment Company, Boston, Mass.). The precipitate was washed two times with cold 10% TCA and once with 95% ethanol. The precipitate was then resuspended in 5% TCA and heated at 95°C for 15 minutes. The mixture was centrifuged as before, and aliquots of the supernatant were removed for the standard diphenylamine assay for DNA or for the orcinol assay for RNA (18). Protein was determined by biuret and Folin-phenol reagent methods according to the method described by Bennett (19).

C^{14}-lysine and C^{14}-valine (specific activity 336 mC/mmole and 228 mC/mmole, respectively obtained from New England Nuclear, Boston, Mass.) and H^3-6-thymidine (10,000 mC/ mmole), and H^3-5-uridine (2.0 C/mmole), obtained from Schwarz Bioresearch, Orangeburg, N. J., were used in incorporation studies. Tissue was harvested and transferred to incubation tubes containing 2 ml. Niu-Twitty solution (20), supplemented with 10 µ moles glucose. The final concentration of radioactive label in the incubation medium is indicated in the figure legends. After the incubation time indicated in the figures, the reaction was terminated by the addition of 0.10 mmoles of unlabeled compound followed by the rapid removal of the tissue. The tissue was blotted, transferred to a tared cover glass slip, and the "wet weight" of the tissue determined on the analytical balance. The tissue was then processed by a modification of Palmiter's (21) methods. It was placed in 3 ml of acetone at 0°C, in which it was insoluble, washed twice with 1 ml portions of cold 5% trichloroacetic acid (TCA) and then washed twice with 3 ml of a mixture of ethanol-ether (3:1), and, finally, the remaining tissue was dissolved in Soluene (Packard Instrument Co., Downers Grove, Ill.). One ml of Liquifluor (New England Nuclear, Boston, Mass.) was added to the sample in scintillation vials which had been previously cooled to 20°C. After 8 hours of dark-adaptation, each vial was counted using the automatic Beckman Liquid Scintillation System (Model LS-250).

Ornithine transcarbamylase (OTC) activity was determined according to the method described by Brown and Cohen (22).

Some experiments are "short term" (3 to 4 days) in duration, whereas others are "long term" (6 days). Although both long and short term microscopic experiments could be made with tissue from one tadpole liver, biochemical experiments, which require more tissue for an assay, were designed to consider either long term or short term effects. Only seldom could sufficient tissue be obtained from one animal to secure statistically valid data for some of the subtle points of short term studies as well as the more general observations of long term studies.

RESULTS

Water, Protein, DNA and RNA Content

Several biochemical parameters of cultured tissue were studied over a 3 day to 4 day period to establish the condition of liver cells after varying times in culture. Representative short term and long term experiments are summarized in Tables 1 and 2. There is essentially no change in the dry weight/wet weight ratio during increasing times in culture. Although these data do not exclude the possibility of absolute changes in the wet or dry weight of cells, they suggest that no marked changes in hydration or dehydration of the tissue fragments occur during the period in culture.

No change occurs in the amount of protein of cultured tissue over a similar period in culture (Tables 1 and 2); in contrast, a decrease in the amount of DNA and RNA occurs. In both cases the decrease is more gradual after about 40 hours in culture than during earlier periods. The overall decrease in DNA content is approximately 30 per cent; 40 per cent for RNA. The decrease in DNA and RNA content appears to have leveled off by four days and remains constant through 6 days.

Table 1

BIOCHEMICAL CONSTITUENTS IN
SHORT TERM CULTURE

Hours in Culture	Dry Weight / Wet Weight	Biochemical Parameter μgProtein	μgDNA	μgRNA
0	0.20	80 ± 10	7.5 ± 0.5	3.1 ± 0.2
20	0.18	85 ± 11	6.9 ± 0.4	2.7 ± 0.3
40	0.21	97 ± 8	6.5 ± 0.5	1.7 ± 0.5
60	0.23	85 ± 6	6.3 ± 0.6	1.8 ± 0.3
80	0.19	80 ± 9	5.7 ± 0.7	1.9 ± 0.4

Except for the ratio of dry weight to wet weight, all parameters are expressed per mg wet weight of tissue. Mean ± Standard Error.

Table 2

BIOCHEMICAL CONSTITUENTS IN
LONG TERM CULTURE

Hours in Culture	Dry Weight / Wet Weight	Biochemical Parameter μgProtein	μgDNA	μgRNA
0	0.19	94 ± 9	7.2 ± 0.6	2.9 ± 0.3
36	0.21	80 ± 7	6.4 ± 0.5	2.1 ± 0.4
72	0.20	105 ± 11	6.0 ± 0.4	1.8 ± 0.5
108	0.17	88 ± 10	5.2 ± 0.3	1.9 ± 0.3
144	0.18	98 ± 7	5.2 ± 0.4	1.8 ± 0.3

Except for the ratio of dry weight to wet weight, all parameters are expressed per mg wet weight of tissue. Mean ± Standard Error.

Ornithine Transcarbamylase (OTC) Activity

Numerous enzymatic changes are known to occur in the liver of tadpoles during metamorphosis and are thus of interest for *in vitro* culture studies. In particular, the metamorphic change from ammonotelism to ureotelism involves the introduction of the urea cycle enzymes (2) (23). Since this change in protein biosynthesis was of interest to us for future studies, the activity of one of these enzymes, ornithine transcarbamylase (OTC), was followed in time course experiments over 96 hours. Long term experiments involved assays at more extended time intervals for a total period of six days.

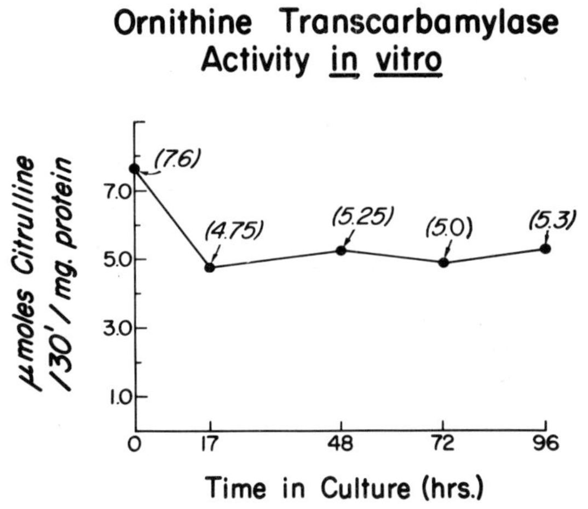

Figure 2. Ornithine transcarbamylase activity in cultured liver tissue. At the time indicated, tissue fragments were harvested and assayed as described in METHODS.

Figure 2 summarizes the fluctuations in OTC activity during the early periods in culture. There is a marked decrease in activity during the first 17 hours with subsequent activities remaining stable. This stabilized condition is shown in Table 3 for assays of cultured tissue over a period of 6 days.

Table 3

ORNITHINE TRANSCARBAMYLASE (OTC) ACTIVITY IN LONG TERM CULTURES

Hours in Culture	OTC Activity* (μmoles Citrulline/30 min mg protein)
0	7.3 ± 0.8
36	5.0 ± 0.6
72	5.4 ± 0.5
108	5.9 ± 0.7
144	5.6 ± 0.5

*Mean ± Standard Error.

Incorporation of Amino Acids, Uridine and Thymidine

In order to obtain additional information about the condition of organ cultured liver, experiments were designed to measure the overall ability of cells to transport and incorporate radioactive amino acids and nucleotides into macromolecules.

The incorporation of C^{14}-lysine and C^{14}-valine was measured by incubating pieces of tissue, after they had been cultured for either 0, 20, 44 or 68 hours, with the two radioactive amino acids under the conditions outlined in METHODS. After incubating the tissue with radioactive amino acids, etc. for the time indicated for a representative experiment shown in Figure 3, the tissue was removed and TCA precipitable radioactivity was determined according to METHODS. As shown in this figure, the average rate over three hours and level of incorporation of radioactive lysine and valine into TCA-insoluble material after three hours decreases slightly during the first 20 hours in culture and then remains at a relatively constant level during the culture period.

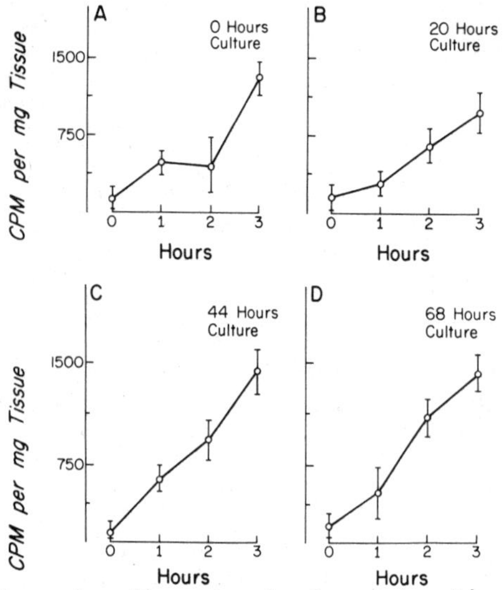

Figure 3. Incorporation of radioactive lysine and valine into TCA insoluble material. In (A) through (D) tissue was harvested after the number of hours in culture (0, 20, 44, 68) that are indicated. Tissue was incubated as described in METHODS FOR 0, 1, 2 and 3 hours before the reaction was terminated with TCA. The CPM per mg tissue were determined as described in METHODS. Each tube contained 0.5 μcuries each of C^{14}-lysine and C^{14}-valine.

The average rates of incorporation of H^3-uridine at varying times in culture for a representative experiment are shown in Figure 4. The rate of incorporation appears to decrease during the first 20 hours; the incorporation rate for tissue cultured 68 hours is decreased to about 75 per cent of the initial value.

In the case of H^3-thymidine incorporation, Figure 5, there was very little total incorporation by freshly excised tissue. A low average rate of incorporation was observed during the first 20 hours in culture. However, by 44 hours (Figure 5C) the level of incorporation increased and the rate of incorporation increased to a slightly higher level through 68 hours in culture.

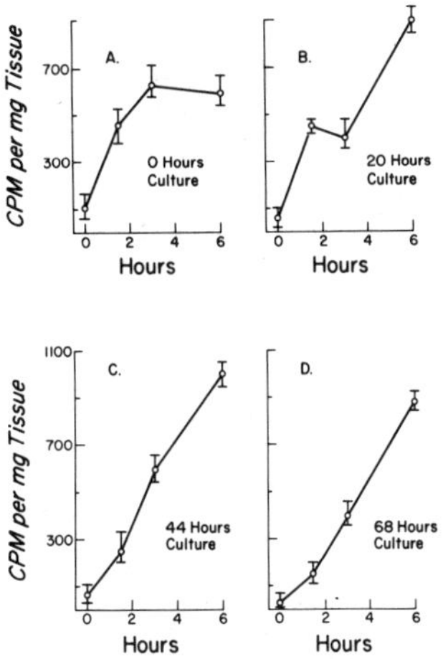

Figure 4. Incorporation of radioactive uridine into TCA insoluble material. Methods as in Figure 3, and in METHODS. Each tube contained 10 μcuries of H^3-uridine.

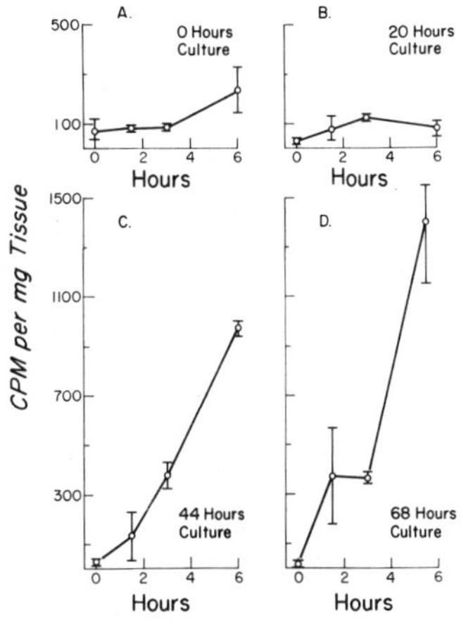

Figure 5. Incorporation of radioactive thymidine into TCA insoluble material. Methods as in Figure 3 and in METHODS. Each tube contained 10 μcuries of H^3-thymidine.

Table 4

INCORPORATION OF AMINO ACIDS, URIDINE AND THYMIDINE IN LONG TERM CULTURES

Hours in Culture	C^{14}-Lysine and C^{14}-valine (CPM/Hr/mg Tissue)	H^3-Uridine (CPM/Hr/mg Tissue)	H^3-Thymidine (CPM/Hr/mg Tissue)
0	450 ± 80	370 ± 45	155 ± 10
36	435 ± 40	300 ± 50	208 ± 35
72	510 ± 75	185 ± 15	310 ± 10
108	409 ± 60	135 ± 10	358 ± 45
144	500 ± 75	185 ± 30	369 ± 70

Average rates were calculated as in Figures 3 through 5.
Mean ± Standard Error.

As shown in Table 4, the long term cultures display a stable rate of incorporation of radioactive lysine and valine. The rate of uridine incorporation decreases with time and appears to stabilize by 72 hours in culture at a level of about 70% that of the initial value. Thymidine incorporation appears to be stabilized at an increased rate after about 72 hours in culture.

DISCUSSION

The conditions have been elaborated for the *in vitro* culture of liver fragments from premetamorphic tadpoles. Our culture methods (10,15) make use of a standard cell culture medium for amphibia (17) which has been modified to contain less glucose (10). They enable us to maintain approximately fifty to seventy-five experimental tubes, each containing two pieces of tissue from a single tadpole for experimental periods of six days. According to morphological criteria reported elsewhere (10), the parenchymal cells approximate the cytological conditions of cells immediately after excision. In addition, several biochemical criteria have now been used to establish the extent to which the molecular and metabolic condition of the cultured tissue parallels the *in vivo* situation. These criteria include measurements of biochemical constituents of the tissue and gross metabolic studies using radioisotopes.

The protein content of tadpole tissue is stable over short and long term culture periods. The values obtained are in the range of those reported by Wixom et al. (23) and Atkinson et al. (24)

for animals of a similar developmental stage. The stability of the protein content provides us with an additional parameter on which to base future in vitro measurements.

The fact that amino acid incorporation shows only a slight decrease during the early period in culture and then stabilizes during later periods in culture is consistent with the protein content data. The incorporation increase may reflect a renewal of protein synthesis after a slightly depressed period during the "trauma" associated with being excised and put into culture. As cells become established in culture, growth may occur, as reflected in increased incorporation of amino acids, but may not be measured by grosser methods, e.g. protein determination and weight measurements, which may mathematically mask such growth effects.

Ornithine transcarbamylase activity is a biochemical indicator for the onset of metamorphosis in R. catesbeiana (2,23). Although it is true that changes in protein content would affect the specific activities reported here, our data on protein content, wet and dry weights indicate that this does not happen. The demonstration of stable activity levels of this enzyme in organ cultures of R. catesbeiana liver, after an initial decrease in activity, therefore satisfies an important criterion in the development of a culture system for the in vitro study of thyroxine action. The decrease in OTC activity during early periods in culture is subject to many interpretations. Unfortunately, there are no reports of the turnover rate of OTC in the literature and one can only speculate that the decrease in activity we observed may reflect a slow-down in biosynthesis of this enzyme during the "trauma" period, often mentioned in connection with cell culture studies.

The patterns for RNA and DNA content, which show a decrease in the former after about 20 hours and a gradual decrease in the latter, are more complex than those discussed above for other biochemical features. Alone, these data would suggest that some cell death is occurring, as has been discussed in detail by Majno (12). However, the fact that a marked increase in thymidine incorporation occurs from 44 to 68 hours suggests that a more complex situation exists. One possibility is that cell division is occurring, perhaps, in a small population of parenchymal or, even, in other included (e.g. leucocytes, reticulocytes, etc.) cells. The nature of the data, however; i.e. DNA content being a gross measurement and thymidine incorporation being a sensitive measurement subject to the limitations discussed below, make it virtually impossible to estimate the proportion of cells dying to those proliferating. Further, increased incorporation of labeled thymidine does not necessarily imply an increase in DNA synthesis since incorporation depends on the intracellular concentration of thymidine as well as the actual rate of thymidine incorporation. An alternative interpretation would be that the endogenous pool of thymidine diluted the isotope during the early periods in culture and, on being depleted, permitted high rates of thymidine incorporation. Further, thymidine incorporation could reflect a sustained period of mitochondrial DNA synthesis. Although qualitative morphological studies on tadpole liver during metamorphosis (25,5,6) indicate that no liver cell division occurs, recent biochemical work in Frieden's laboratory (24) suggests that enhanced liver nuclear DNA biosynthesis does occur. This, however, is partially in conflict with the report of Cohen's group (23).

The decrease in the rate of uridine incorporation during early periods in culture is consistent with other biochemical and morphological data (15) relating to tissue trauma when adapting to culture conditions. The stabilization of the rate of uridine incorporation after a period in culture suggests: 1) that a reduced, but functional, level of RNA synthesis has been achieved or 2) alterations in uridine pool-size, etc., as mentioned above for thymidine.

In the present work, emphasis has been placed upon determining the gross metabolic features and biochemical constituents of pre-metamorphic tadpole liver in culture. The assessment of these states for tissue also rests on morphological criteria (10). The patterns for biochemical constituents during periods in culture as well as the patterns for incorporation of radioactive amino acids, uridine, and thymidine strengthen our morphological conclusions about the tissue. In addition, in some instances they provide a foundation for further biochemical studies and, in other instances, a starting point. The studies of Wixom et al. (23) and Atkinson et al. (24) provide basic information about the in vivo features of several of these parameters during metamorphosis. Although there is undoubtably some cell death occurring in our system, perhaps, as reflected in the decreases in some biochemical activities, the situation is drastically different from that which one sees in a dying cell population as reviewed by Majno (12). There are of course innumerable biochemical criteria; e.g., determination of liver specific antigens, respiratory characteristics, etc., that need to be established, as well as such cytological features as mitotic incidence. These criteria are needed to strengthen our confidence in the tadpole liver culture system as a model for investigating the mechanism of thyroxine-induced differentiation of tadpole liver.

SUMMARY

Primary explants from *Rana catesbeiana* tadpole liver have been maintained in organ culture for six days. Several biochemical parameters of the cultured liver cells have been studied during the time in culture in order to establish: 1) changes in wet weight, dry weight, as well as protein, RNA and DNA contents; 2) rates of incorporation of amino acids, uridine and thymidine; 3) changes in ornithine transcarbamylase activity. The effects of changes in the culture conditions on these biochemical characteristics are presented.

It is a pleasure to contribute this article on occasion of the Lipmann Fest since these studies were begun in the Lipmann Laboratory following my Ph.D. studies there. I thank the Gesellschaft für Biologische Chemie for their hospitality in Berlin. I acknowledge my students Henry Kriegstein and Shulamith Shafer, along with my assistants Susan Fisher and Janice Glenn, who have contributed at various times to these studies.

REFERENCES

1. Bennett, T. P., and Frieden, E.: Metamorphosis and biochemical adaptations in amphibia. In "Comparative Biochemistry" (M. Florkin and H. S. Mason, eds.), Vol. 4, pp. 483-556. Academic Press, New York (1962).
2. Cohen, P. P.: Biochemical aspects of metamorphosis: transition from ammonotelism to ureotelism. Harvey Lect. Ser. 60, 119-154 (1966).
3. Frieden, E.: Biochemistry of amphibian metamorphosis. In "Metamorphosis" (W. Etkin and R. I. Gilbert, eds.), pp. 349-398. Appleton-Century-Crofts, New York (1968).
4. Frieden, E., and Just, J.: Hormonal responses in amphibian metamorphosis. In "Mechanisms of Hormone Action" (G. Litwack, ed.). pp. 1-52. Academic Press, New York. (1970).
5. Bennett, T. P., Glenn, J. S., and Sheldon, H.: Changes in the fine structure of tadpole (Rana catesbeiana) liver during thyroxine-induced metamorphosis. Develop. Biol. 22, 232-248 (1970).
6. Bennett, T. P., and Glenn, J. S.: Fine structural changes in liver cells of Rana catesbeiana during natural metamorphosis. Develop. Biol. 22, 535-560 (1970).
7. Tata, S.R.: The formation and distribution of ribosomes during hormone-induced growth and development. Biochem. J. 104, 1-15 (1967).
8. Cohen, P. P.: Biochemical differentiation during amphibian metamorphosis. Science, 168, 533-543 (1970).
9. Spiegle, E. S. and Spiegle, M.: Some observations on the ultrastructure of the hepatocyte in the metamorphosing tadpole. Exp. Cell Res., 61, 103-112 (1970).
10. Bennett, T. P. and Kriegstein, H.: The morphology of tadpole (Rana catesbeiana) liver under varying conditions of organ culture. Anat. Rec. 176, 461-474 (1973).
11. Bennett, T. P.: Morphological and biochemical studies on tadpole liver in organ culture. Abstract 44. The American Society for Cell Biology. 11th Annual Meeting (1971).

12. Majno, G.: Death of liver tissue. In The Liver, Morphology, Biochemistry, and Physiology, (Ch. Roullier, ed.). Vol. II, pp. 267-313. Academic Press, New York (1964).
13. Fawcett, D. W.: "An Atlas of Fine Structure. The Cell, its Organelles and Inclusions." pp. 2-28. W. B. Saunders Company, Philadelphia, Pennsylvania (1966).
14. Porter, K. R. and Bonneville, M.: "Fine Structure of Cells and Tissues", 3rd Edition, pp. 20-25. Lea and Febiger, Philadelphia, Pennsylvania (1970).
15. Kriegstein, H. and Bennett, T. P.: Chromatin condensation and nucleolar segregation induced in nuclei of parenchymal cells of Rana catesbeiana tadpoles. Exptl. Cell Res., 80, 152-158 (1973).
16. Taylor, A. C. and Kollros, J. J.: Stages in the normal development of Rana pipiens larvae. Anat. Rec., 94, 7-23 (1946).
17. Wolf, K., and Quimby, M. C.: Amphibian cell culture: permanent cell line from the bullfrog (Rana catesbeiana). Science 144, 1578-1580 (1964).
18. Schneider, W. C.: Determination of nucleic acids in tissues by pentose analysis. In "Methods in Enzymology", (Colowick, S. P. and Kaplan, N. O., eds.), 3, pp. 680-684. Academic Press, New York (1957).
19. Bennett, T. P.: Membrane filtration for determining protein in the presence of interfering substances. Nature 213, 1131-1132 (1967).
20. Jacobson, A. G.: Amphibian cell culture, organ culture, and tissue dissociation. In "Methods in Developmental Biology", pp. 531-542. Thomas Y. Crowell Company, New York (1967).
21. Palmiter, R. D.: Early macromolecular synthesis in cultured mammary tissue from mid-pregnant mice. Endocrinology 85, 747-751 (1969).
22. Brown, G. W. and Cohen, P. P.: Comparative Biochemistry of Urea Synthesis. J. Biol. Chem., 234, 1769-1774 (1959).
23. Wixom, R. L., Reddy, M. K., Cohen, P. P.: A concerted response of the enzymes of urea biosynthesis during thyroxine-induced metamorphosis of Rana catesbeiana. J. Biol. Chem. 247, 3681-3692 (1972).
24. Atkinson, B. G., Atkinson, K. H., Just, J. J., and Frieden, E.: DNA synthesis in Rana catesbeiana tadpole liver during spontaneous and triod thyronine-induced metamorphosis. Develop. Biol. 29, 162-175 (1972).
25. Kaywin, L.: A cytological study of the digestive system of anuran larvae during accelerated metamorphosis. Anat. Rec. 64, 413-441 (1936).

The Hexokinase-Mitochondrial Binding Theory of Insulin Action

Samuel P. Bessman, M.D.
University of Southern California School of Medicine
Department of Pharmacology, Los Angeles, California 90033

I should like to discuss a proposal for the mechanism of action of insulin which I first presented at a session chaired by Dr. Fritz Lipmann in 1951[1] and to review it over the more than 20 years of philosophic and experimental evolution through which it has gone. The proposition simply is that insulin causes the binding of hexokinase to the mitochondrial membrane in such a way that the acceptor relation between energy generation and glucose utilization is made more efficient.

An enhancement of efficiency of oxidative phosphorylation by an insulin mediated molecular rearrangement in the cell would result in increased delivery of energy at mitochondrial sites associated with the anabolic functions of the cell. These functions include increased permeability of membranes, increased synthesis of large molecules such as protein, fat and glycogen, increased turnover of RNA, detoxication of drugs, and synthesis of small molecules such as glutathione. It is of interest to note that these physiologic results of an enhanced efficiency of mitochondrial function would seem to be those caused by insulin. In addition, a number of physiologic phenomena which are at present inexplicable may be clarified by the felicitous interaction between hexokinase and the mitochondrial membrane. For example, the tissue most sensitive to insulin action is the fat cell and this cell has its mitochondria very closely approximated to the cell membrane, pressed there by the hydrophobic fat globules contained within the cell. This would render the mitochondria more accessible to direct contact with the exogenously administered insulin.

A second phenomenon which is physiologically inexplicable at the present time on any other basis is the insulin-like result of muscular activity. Although exercise hormones have been proposed to explain the increased anabolic activity of muscle action during exercise, even in the diabetic, no such hormone has ever been demonstrated. On the other hand, the requirement for an acceptor explains both this phenomenon and even the more interesting one of the binding of ATP and ADP to the enzymes of muscle contraction. At rest, the creatine of muscle is almost entirely in the form of phosphocreatine. When the muscle becomes active, the phosphocreatine breaks down to liberate free creatine with no change in needed ATP until the exhaustion of the creatine supply[2]. Nevertheless, the oxidation in muscle increases with activity. The question then arises, "How is the muscle mitochondrion 'informed' of the need for increased oxidation when exercise has taken place?".

If the ATP of the muscle does not get converted to significant amounts of ADP, which then diffuse to the mitochondrion for the classical respiratory control phenomenon, what is the actual information transfer between the contracted fiber and the mitochondrion? The evidence to interpret this question was supplied first by Klingenberg who showed that the mitochondria of heart, skeletal muscle, and brain had a bound creatine kinase.[3]

The second information on the subject was the work of Fonyo[4], in our laboratory, who showed that creatine itself can produce the same phenomenon of respiratory control as ADP in heart sarcosomes. It now becomes clear that the signal from the contracting muscle fiber is free creatine diffusing to the mitochondrion causing the regeneration of the creatine phosphate at the mitochondrial site which restores the creatine phosphate content of the cytosol. This replenishes ATP by transphosphorylation of myosin *in situ*. The ADP formed by muscle contraction does not have to be displaced by ATP. It is interesting that this explanation fits the known fact that the binding of ADP to the contracting muscle fiber is stronger than the binding of ATP. The intermediary role of phosphocreatine in replenishing *in situ* the ATP which has broken down in contraction and in transporting new phosphate energy from the mitochondrion is thus an example of a parsimonious mechanism for producing efficiency in the presence of a limited amount of nucleotide. Otherwise, the muscle fiber would be dependent upon the diffusion process for moving the limited ATP from the mitochondrion and the ADP to the mitochondrion. This permits the rapid regeneration of muscle contraction and the substitution of creatine phosphate as an energy carrier molecule with minimal pathways of degradation for ADP and ATP for which the cytosol contains many active sites which use these molecules. Creatine is an excellent "messenger" molecule for it undergoes no reactions at all in the cytosol except transphosphorylase.

A third physiologic phenomenon which is explained by the molecular reorganization role of insulin in connecting hexokinase to mitochondria is the well-known fact that the brain is insensitive to insulin although it uses more glucose than any other tissue.

The hexokinase of brain is for the most part attached to the mitochondria as shown by Crane and Sols[5] a long time ago. If the role of insulin is to attach hexokinase to mitochondria, then the lack of sensitivity of the brain to insulin is clearly the result of the fact that the role of insulin in connecting hexokinase is no longer necessary in brain. It is interesting to note that the brain is a relatively primitive tissue and that mitochondria also have hexokinase attached to them in malignant tissues such as Hela cells.

A number of experiments of different types have been done over the past 20 years to validate certain aspects of this theory.

In the first place we would have to demonstrate that insulin does not achieve its major anabolic effects by improving the permeability of cell surfaces. This was done by Dr. Toyoda[6], in our laboratory, who measured the movement of radioactive leucine into the cell water and into the protein of surviving diaphragm sections as it was affected by insulin. Figure 1 shows the data obtained from this experiment and clearly demonstrates that under the influence of insulin there is less free leucine in a tissue space under the action of insulin. This difference in cellular free leucine from the control is accounted for by the extra incorporation of leucine into protein. This demonstration of the intracellular effect of insulin on protein synthesis was corroborated by Dr. Paul De Schepper[7] who showed that diaphragm preloaded with radioactive leucine at 4°C, at which temperature no protein synthesis occurs, would then incorporate this intracellular radioactive leucine into protein more efficiently with insulin than the control. This was true even though there were

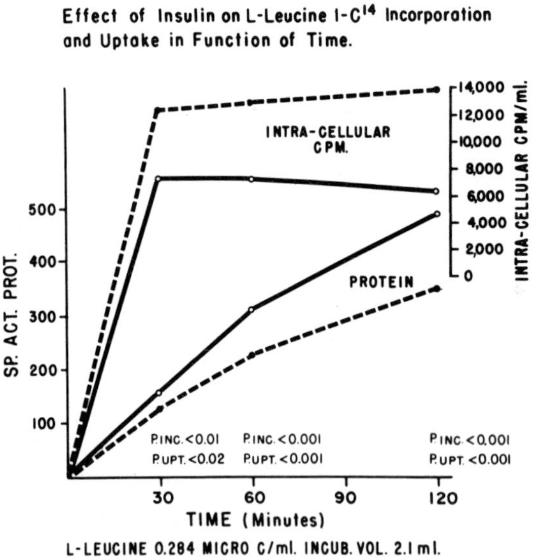

Figure 1

large amounts of non-radioactive extracellular leucine added. The preferential stimulation of protein synthesis from intracellular amino acids thus indicates that insulin does not stimulate protein synthesis by any mass action effects resulting from transfer of amino acids through the cell membrane.

The question of whether the hexokinase stimulation of anabolic activity would actually be a self-defeating one because the extra hexokinase would presumably consume all the ATP formed by the mitochondrion and therefore leave none for other anabolic reactions was investigated by Dr. Nicholas Bachur.[8] He showed that the addition of hexokinase and glucose to a mitochondrial system which furnished ATP for the acetylation of an aromatic amine dye, stimulated the acetylation of the aromatic amine, even though the Km for ATP of the amine acetylating system was higher by two orders of magnitude than the great excess of hexokinase added to the system. This experiment is very interesting from another standpoint. It suggests that different anabolic reactions have preferential places at the mitochondrial "cafeteria table".[7]

A most important question raised by the hexokinase attachment theory of insulin action is the role of glucose. If, indeed, the only role of insulin in anabolic stimulation is exercised through the hexokinase mitochondrial attachment then the presence of substrates for this enzyme would be crucial to the action of insulin. It has been shown by many workers that added glucose was not necessary for the stimulation of protein synthesis in rat diaphragm preparations. This would seem at first glance to deny the insulin-hexokinase-theory entirely. The assumption, however, that there is no free glucose in the cell is indeed erroneous. Every cell which synthesizes and breaks down glycogen also synthesizes free glucose. This free glucose comes from the amylo 1,6 glucosidase activity which breaks branches of glycogen. The presence of a large amount of glycogen in tissues is not even a prerequisite for free glucose formation because of the turnover of glycogen, which provides between 6 and 7 percent of the glucose residues as free glucose. In order to approach this problem, studies of the effect of the inhibitor 2-deoxyglucose on protein synthesis as stimulated by insulin were carried out by Dr. De Schepper.[9] It was proposed that the addition of 2-deoxyglucose would inhibit hexokinase by causing the accumulation of glucose-6-phosphate. The inhibited hexokinase would not, therefore, be able to exert an improved acceptor effect even if it was attached to the mitochondrion by insulin. Table 1 shows that addition of 2-deoxyglucose produces an inhibition of approximately 30 percent in protein synthesis from C^{14} leucine in surviving diaphragm pieces. The addition of glucose to the 2-deoxyglucose (2DG) inhibited tissue restores full protein synthesis. The addition of insulin to the 30 percent inhibited tissue had no effect even though most of the protein synthesis of the tissue remained operative. The addition of glucose plus insulin to the inhibited tissue allowed full stimulatory effect of insulin on protein synthesis. These experiments demonstrated a number of very interesting phenomena. First, they showed that intact glucose metabolism is indeed necessary for the stimulatory effect of insulin on protein synthesis.

A second point of interest is the fact that insulin apparently does not act on at least 70 percent of the capacity of the protein synthesizing process, which gives insulin a modulating role in protein synthesis but not a crucial role. This is physiologically meaningful because it is well known that protein synthesis does not cease in the insulin-deprived animal, nor does any other anabolic reaction cease. Any theory of hormone action must take into account this point: that no peptide hormone is crucial to any metabolic process. All peptide hormones exert a modulating effect only, which is superimposed upon the

Effect of 2-Deoxy-D-Glucose on basic and insulin-stimulated incorporation of L-Leucine-1-^{14}C (rat diaphragm pieces)

No. of Observa- tions	Control Value	Values after addition of				
		2-Deoxy-D-Glucose		2-Deoxy- D-glucose + Insulin	2-Deoxy-D-glucose + Glucose (1:1)	2-Deoxy- D-glucose + Glucose (1:1) + Insulin
		0.012 M	0.03 M			
6	319	201 S.E. ± 36 $p_1 < 0.05$		206 S.E. ± 19 $p_2 > 0.7$		
4	361	265 S.E. ± 29 p_1 0.05			347 S.E. ± 25; $p_1 > 0.7$ S.E. ± 11; $p_2 < 0.01$	481 S.E. ± 31 $p_3 < 0.05$
5	473		253 S.E. ± 38 $p_1 < 0.01$		446 S.E. ± 40; $p_1 > 0.5$ S.E. ± 12; $p_2 < 0.001$	594 S.E. ± 58 $p_2 > 0.05$

Table I

Note: Each row represents an independent series of observations. All values (counts per minute per milligram of protein) are means of the given number of observations. In each row, S.E. and p_1 values are calculated versus their respective controls; S.E. and corresponding p_2 values are versus the data in the 2-deoxy-D-glucose column, and S.E. and corresponding p_2 values are versus the data in the 2-deoxy-D-glucose + insulin column. The concentration of 2-deoxy-D-glucose in the last three columns is equal to that used in the corresponding line in the 2-deoxy-D-glucose columns.

* * * *

major process. The theory of hexokinase attachment by insulin permits such a modulating role by providing a means of adjusting the efficiency of a process but not the nature of the process itself. Insulin facilitates the attachment of hexokinase to mitochondria thereby making the process, the acceptor effect, more efficient, but the acceptor effect exists in the cytosol under any circumstances.

Evidence in support of the hexokinase acceptor theory has come from a number of different laboratories. McLean's group[10] reported that administration of antibodies to insulin *in vivo* to lactating rats caused mitochondrial bound hexokinase to leave the mitochondria and enter the cytosol where it could be measured as an increase in soluble hexokinase. This bound hexokinase was isozyme number two.

Further information on the binding of hexokinase to mitochondria was obtained by Borrebaek[11] who showed that incubation of fat pads with insulin and glucose caused an increase in the hexokinase bound to mitochondria. Katzen showed that the sensitivity of tissues to insulin was inversely proportional to the hexokinase isozyme[2] bound to mitochondria. Recently Dr. Gots[12], in our laboratory, completed some experiments in which he addressed himself to the question of what special role bound hexokinase played in mitochondrial metabolism.

The first question he asked was whether there was a preferential binding of hexokinase at such a site and in such a manner that endogenously generated ATP would have better access than exogenous ATP. Using liver mitochondria to which only about one percent of the cellular hexokinase is tightly bound, he showed that the apparent Km for ADP was approximately 1/3 the apparent Km for ATP when the mitochondria were supplied with substrates such as alpha-ketoglutarate. When the mitochondria had no substrate, ATP was the only nucleotide which could be utilized. The apparent Km of ADP is so much lower than that for ATP that it appears that the location of bound hexokinase must be such that ADP can enter the mitochondrion, be converted to ATP, and reach the active site of hexokinase much more efficiently than soluble ATP can diffuse to that site. This, of course, could mean that the binding site of hexokinase is inside the inner membrane where ATP would have little access.

To study this question further Dr. Gots[13] incubated liver mitochondria with ATP or ADP in the presence of a number of inhibitors of oxidative phosphorylation. He found that atractyloside, pentachlorophenol, amytal, and antimycin all inhibited phosphorylation of glucose by bound hexokinase when ADP was the nucleotide. This was to be expected because the only other way that ADP could be utilized to phosphorylate glucose would be by the myokinase reaction and apparently this reaction does not go well. What was surprising about this experiment, however, is that the same inhibitors also inhibited phosphorylation of glucose by mitochondrial bound hexokinase when ATP was the nucleotide added. (Soluble hexokinase is not inhibited by these inhibitors.) This means that ATP must be dephosphorylated and rephosphorylated in order to be available for glucose phosphorylation by bound hexokinase.

An alternative explanation could be that there is an energy coupled transfer system for ATP which is inhibited. Then we could not deduce from the previous experiments that <u>only</u> intramitochondrial generated ATP is available to the bound hexokinase. Dr. Gots then did the following experiment. He incubated mitochondria with their bound hexokinase with glucose, a-ketoglutarate, tracer amounts of inorganic P^{32} phosphate, and .001 molar ATP which was non-radioactive, and measured the specific activity of the glucose-6-phosphate formed, as well as the activity of the 7 minute phosphorous. Figure 2 shows that in the first minute the specific activity of the glucose-6-phosphate is equal to that of the tracer and the ATP is almost non-radioactive. As time goes on the ATP dilutes the tracer, the specific activity of the glucose-6-phosphate falls and that of the ATP rises somewhat. Figure 3 shows that the counts measured as glucose-6-phosphate (organic phosphate minus nucleotide P.) were indeed in glucose-6-phosphate, by column chromatography in an automatic system that we have developed.

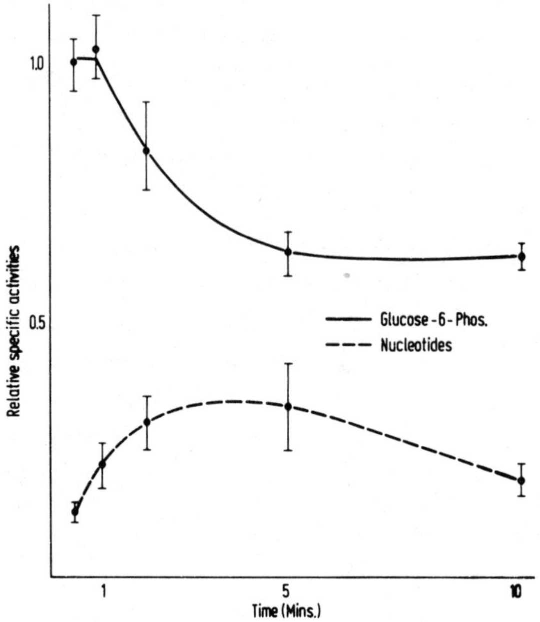

Figure 2

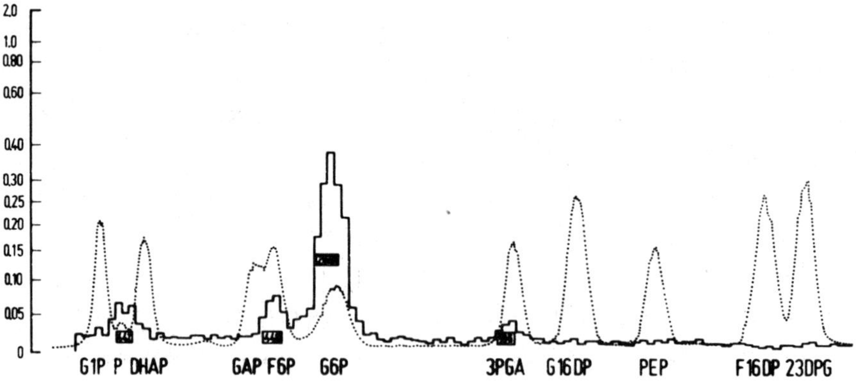

Figure 3

Thus we have shown, in the 20 years since we first presented the hypothesis, that insulin connects the hexokinase reaction as an acceptor to the mitochondrion and that a number of predictions based on it can be validated.

1. That insulin must act anabolically on something other than permeability.
2. That glucose metabolism is required for insulin action.
3. That hexokinase bound to mitochondria has a preferred location immediately adjacent to the site of oxidative phosphorylation.

Others have shown:

1. That insulin causes binding of hexokinase to mitochondria.
2. That antibodies to insulin cause unbinding.
3. That insulin activity on particular tissues is inversely proportional to the amount of bound hexokinase on the mitochondria of these tissues.

This hypothesis about insulin has furnished interesting insights into a number of enigmatic physiological phenomena.

1. The insensitivity of brain to insulin.
2. The similarity of exercise to insulin administration in the diabetic.
3. The presence of bound creatine kinase only on mitochondria of heart, brain and skeletal muscle.

In conclusion I should like to thank Dr. Lipmann for encouraging me to work on this problem and turn the hypothesis into more than an intellectual exercise.

REFERENCES

1. Bessman, S.P. A contribution to the mechanism of diabetes mellitus, in: Najjar, V. (Ed.), "Fat Metabolism", Baltimore, Johns Hopkins Press, 1954, p. 133

2. Mommaerts, W. Energetics of muscular contraction. Physiol. Rev. 49: 427 (1969)

3. Klingenberg, M. and Schollmeyer, P. Zur Reversibilitat der oxydativen Phosphorylierung. Biochem. Ztschr., 333: 335, (1960)

4. Bessman, S.P. and Fonyo, A. The Possible Role of the Mitochondrial Bound-Creatine Kinase in Regulation of Mitochondrial Respiration. Biochem. Biophys. Res. Commun., 22, 597 (1966)

5. Crane, R.K. and Sols, A. The association of hexokinase with particulate fractions of brain and other tissue homogenates. J. Biol. Chem. 203: 273 (1953)

6. Toyoda, M. and Bessman, S.P. A Non-Transport Effect of Insulin on Protein Synthesis. Fed. Proc. 19, 164 (1960)

7. Bessman, S.P. A Molecular Basis for the Mechanism of Insulin Action. Amer. J. Med. 40, 740 (1966)

8. Bachur, N. Doctoral Thesis, 1961. University of Maryland, Baltimore, Md.

9. De Schepper, P. and Bessman, S.P. Erythrose inhibition of protein synthesis. Unpublished.

10. Walters, E. and McLean, P. The effect of anti-insulin serum and alloxan-diabetes on the distribution and multiple forms of hexokinase in lactating rat mammary gland. Biochem. J. 109, 737 (1968)

11. Borrebaek, B. Mitochondrial-bound hexokinase of the rat epididymal adipose tissue and its possible relation to the action of insulin. Biochem. Med. 3, 485 (1970)

12. Gots, R.E., Gorin, F.A. and Bessman, S.P. Kinetic Enhancement of Bound Hexokinase Activity by Mitochondrial Respiration. Biochem. Biophys. Res. Commun., Vol. 49, 1249 (1972)

13. Gots, R.E. and Bessman, S.P. The Functional Compartmentation of Mitochondrial Hexokinase. Arch. Biochem. and Biophys. Acta. In Press (1973)

N-Terminal Acetylation of Proteins, a Post-Initiational Event

Hans Bloemendal and Ger J.A.M. Strous
Department of Biochemistry, University of Nijmegen,
Nijmegen, The Netherlands

Many proteins undergo modifications in their polypeptide chains during and after biosynthesis. For instance carbohydrates, phosphate, acetyl or methyl groups may become attached to suitable amino acids. The process of acetylation has been investigated in more detail in histones and acidic proteins of cell nuclei where acetyl groups have been reported to occur in internal position linked to the ε-amino group of lysine (1, 2) or to the terminal α-amino group (3). Earlier experiments in Lipmann's laboratory suggested that acetylation is independent of protein biosynthesis (4).

Meanwhile various N-terminally acetylated proteins have been detected in higher organisms, in a number of viruses and also in bacteria (see table 1). We stressed previously that the vertebrate eye lens is a unique system for the study of N-terminal acetylation (26). This organ contains three classes of structural proteins designated as α-, β- and γ crystallin. Gamma crystallin has a free N-amino terminus whereas the composing polypeptide chains of α- and β crystallin possess acetyl groups in N-terminal position.

The total primary structure of the two major polypeptides of α crystallin has recently been solved in our laboratory (5, 6). It appeared that in all α crystallin chains one methionine occurs in internal position while the other methionine residue is found at the N-terminus in an N-acetylated state.

Table 1. N-terminal sequence of a number of acetylated proteins from various sources.

α Crystallin (bovine)	Ac-Met-Asp-Ile	(5, 6)
TYMV coat protein	Ac-Met-Glu-Ile	(7)
Tropomyosin (rabbit muscle)	Ac-Met-Asp-Ile	(8)
Hemoglobin fetal F_1 (human)	Ac-Gly-His-Phe	(9)
Ovalbumin	Ac-Gly-Ser-Gly	(10)
Cytochrome c (chicken)	Ac-Gly-Asp-Ile	(11)
Cytochrome c (human)	Ac-Gly-Asp-Val	(12)
Hemoglobin (carp)	Ac-Ser-Leu-Ser	(13)
Histone F2a2 (calf thymus)	Ac-Ser-Gly-Arg	(14)
Ribosomal protein L_7 (E.coli)	Ac-Ser-Ile-Thr	(15)
Myosin (rabbit muscle)	Ac-Ser-Ser-Asp	(16)
Apoferritin (horse spleen)	Ac-Ser-Ser-Gln	(17)
Melanocyte-stimulating hormone (pig)	Ac-Ser-Tyr-Ser	(18)
TMV coat protein	Ac-Ser-Tyr-Ser	(19)
Cytochrome c (wheat)	Ac-Ala-Ser-Phe	(20)
Enolase (rabbit muscle)	Ac-Ala-Gly-Lys	(21)
Keratin (sheep)	Ac-Ala-Cys-Cys	(22)
Fibrinopeptide (bovine)	Ac-Thr-Glu-Phe	(23)
Lactate dehydrogenase (dogfish)	Ac-Thr-Ala-Leu	(24)
Actin (rabbit)	Ac-Asp-Glu-Thr	(25)

We were also able to demonstrate that the N-terminal methionine of native α crystallin is donated exclusively by initiator tRNA (27). These data together with the observation that β crystallin is N-terminally acetylated but methionine is not the terminal amino acid residue enabled us to demonstrate three facts:

1) N-terminal acetylation takes place after acetylation almost at the same moment when, in general, N-terminal methionine residues are split off;

2) acetyl-Met-$tRNA^{f}_{Met}$ cannot serve as initiator for protein synthesis;

3) free Met-$tRNA^{f}_{Met}$ is a better initiator than formyl-Met-$tRNA^{f}_{Met}$.

MATERIALS AND METHODS

The preparation and characteristics of the lens cell-free system has been described elsewhere (28). $tRNA^{f}_{Met}$ and $tRNA_{Met}$ were isolated from calf eye lenses and amino acetylated with $[^{35}S]$methionine as reported by Strous et al. (29). Enzymatic formylation of $[^{35}S]$ Met-$tRNA^{f}_{Met}$ was achieved with the aid of crude E.coli transformylase (30) while Leucovorin (2×10^{-4} M) was added as formyl donor.

Chemical formylation was carried out with p-nitrophenylformate according to Marcker and Sanger (31). A similar procedure was followed for the preparation of chemically acetylated $tRNA^{f}_{Met}$. In this case p-nitrophenylacetate was used.

In order to have 100% acetylated or formylated

$Met-tRNA^{f}_{Met}$ preparations the unblocked species were hydrolyzed with Cu^{2+} as described by Schofield and Zamecnik (32). Thereafter Cu^{2+} could be removed by gel filtration on Sephadex G-25.

Incubations were carried out at 30°C in a final volume of 0.2 ml. The incubation mixture consisted of 1 mM ATP, 0.5 mM GTP, 15 mM creatine phosphate, 10 µg creatine phosphokinase, 10 mM Tris-HCl, pH 7.4, 5 mM MgAc, 85 mM KCl, 5 mM 2-mercaptoethanol, 0.1 mM of 20 amino acids (1.2 mM Met) and about 360.000 cpm of the radioactive precursors (one of the $[^{35}S]$ Met-tRNA species). At the intervals of time indicated in the legends of figs. 1) and 2) 20 µl aliquots were added to 0.5 ml of 0.1 M KOH. After further incubation at 37°C for 10 min trichloroacetic acid was added to a final concentration of 5%. The precipitate was collected on glass fibre filters and counted in a Packard Tri-Carb liquid scintillation spectrometer with 80% efficiency. A detailed description of the separation of newly synthesized nascent peptides by strip gradient centrifugation and their thermolytic digestion products by high voltage electrophoresis has been given elsewhere (33).

RESULTS AND DISCUSSION

I. N-terminal Acetylation takes place after Initiation. We have described previously that α-, β- and γ crystallins are synthesized in the lens cell-free system as well as in lens organ culture (28). Moreover, we introduced a special gradient designated as strip gradient (33) which allows the separation of de novo synthesized α crystallin peptide chains from ribosomal subunits in the two systems just mentioned. After isolation these peptides can be fractionated according to size by gel filtration on a Sephadex G-25 column. The following molecular weight ranges were obtained:

1) 4000 dalton and higher
2) 3000-4000 dalton
3) 2000-3000 dalton
4) 1500-2000 dalton
5) 1500 dalton and lower

When these fractions are treated with thermolysin and subjected to high voltage electrophoresis the patterns shown in fig. 1 are obtained. Thermolysin cleaves acetylated as well as unblocked α crystallin polypeptides yielding N-acetyl Met-Asp and Met-Asp, respectively (26).

It can be seen that going from panel 3) to 1) there is an increasing ratio between acetyl-Met-Asp and Met-Asp. This means that N-terminal acetylation starts when the nascent peptide consists of about 25 amino acid residues and proceeds while the α crystallin chains are growing. In panels 4) and 5) there is no radioactivity located in the acetyl-Met-Asp region whatsoever. Since all radioactivity recovered in the individual fractions is derived from $[^{35}S]$ Met-$tRNA^{f}_{Met}$ and all α crystallin chains start with Met which is retained in the native molecule (34), the radioactivity behind the Met-Asp region (panels 4, 5) can only be due to methionyl peptides derived from β- and γ crystallin. Apparently the latter radioactivity vanishes in panel 3, where radioactivity due to acetyl-Met-Asp starts to appear. These observations provide evidence for the assumption that the aminopeptidase responsible for cleavage of methionine donated by the initiator tRNA becomes operative at the time when N-terminal acetylation begins.

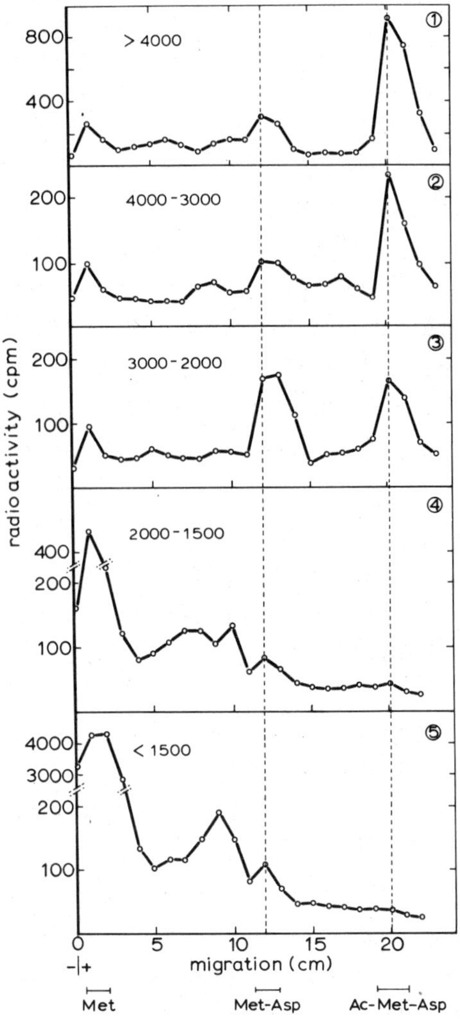

Fig. 1. Electrophoretic analysis of the N-terminal dipeptides of newly synthesized crystallin chains in vitro.
The peptides were obtained after separation of the nascent chains on Sephadex G-25 and subsequent thermolytic digestion. High voltage electrophoresis on paper was performed at pH 6.5 at 50 V/cm for 2 h in acetic acid-pyridine-water (6:200:794 by volume). 20-50 µl samples were applied.

Only in a rather limited number of proteins (like α crystallin) the N-terminal methionine resists cleavage presumably due to the nature of the N-terminal sequence.

II. Acetyl-Met tRNA is not an Initiator for Protein Biosynthesis.

For a while it was thought that various N-terminally acetylated aminoacyl tRNAs might serve as initiator in eukaryotes (35, 36, 37). The fact that the occurrence of such a tRNA species has never been demonstrated unequivocally is no proof in itself that those alternative initiators do not exist.

In order to provide experimental support for the suggestion that acetyl-Met-$tRNA^{f}_{Met}$ is not involved in the initiation mechanism of protein biosynthesis we prepared radioactive acetyl-Met-$tRNA^{f}_{Met}$, formyl-Met-$tRNA^{f}_{Met}$ and formyl-Met-$tRNA_{Met}$. The blocking groups were introduced under comparable chemical conditions. The blocked $[^{35}S]$ Met-RNAs were used as radioactive precursors in the lens cell-free incubation system. This system is fully capable to initiate, elongate and release completed crystallin polypeptide chains which are eventually assembled to "native" aggregates (28). Obviously formyl-Met-$tRNA^{f}_{Met}$ initiates α crystallin synthesis whereas the acetyl analogue is not incorporated at all into newly synthesized α crystallin chains (fig. 2). In order to verify whether or not chemically acetylated Met-$tRNA^{f}_{Met}$ lost its biological activity as far as the tRNA residue is concerned we performed a deacetylation step. It appeared that the resulting $tRNA^{f}_{Met}$ could be reacylated with methionine and used for initiation.

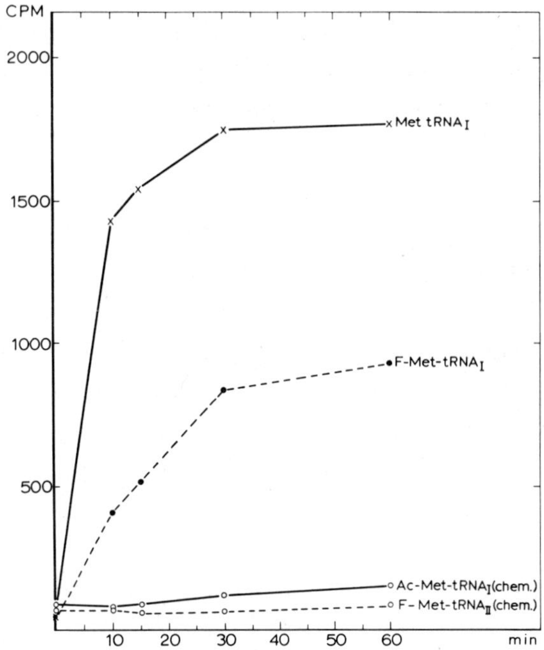

Fig. 2. Kinetics of crystallin initiation in vitro with various [^{35}S] methionyl-tRNA species. $tRNA_I = tRNA^f_{Met}$; $tRNA_{II} = tRNA_{Met}$; F = formyl; Ac = acetyl; chem = chemically blocked.

We demonstrated earlier that Met-$tRNA_{Met}$ exclusively donates its amino acid into the internal methionine position (29). As an additional control also chemically formylated Met-$tRNA_{Met}$ (here designated as F-Met-$tRNA_{II}$) has been used. Like Met-$tRNA_{Met}$ this blocked tRNA is unable to serve as initiator.

III. Met-$tRNA^f_{Met}$ is a better Initiator than Formyl-Met-$tRNA^f_{Met}$.

Besides Met-$tRNA^f_{Met}$, formyl $tRNA^f_{Met}$ has also been suggested to play a (minor) role as cytoplasmic initiator tRNA (38). Using the lens lysate as cell-free system and the formation of α crystallin as parameter for de novo protein synthesis it can easily be shown that the

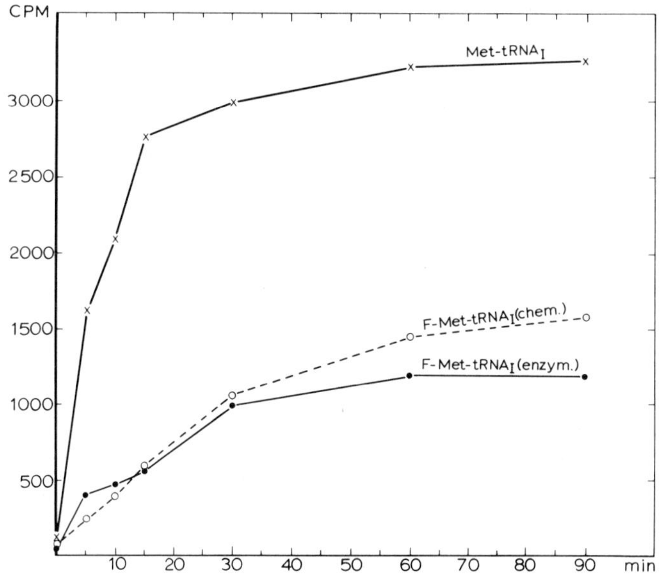

Fig. 3. Comparison between the kinetics of initiation of chemically and enzymatically blocked tRNA$^{f}_{Met}$ (for explanation of symbols see legend to fig. 2).

unblocked Met-tRNA$^{f}_{Met}$ is a more efficient initiator than formyl-Met-tRNA$^{f}_{Met}$ (compare fig. 2 upper solid curve with the dashed curve underneath). Fig. 3 shows the same result obtained from another series of incubations. From this typical experiment it may also be concluded that chemical formylation is certainly not inferior to enzymatic formylation. The incorporation of N-terminal methionine into α crystallin after 30 minutes of incubation is even higher with the chemically modified tRNA. At any rate Met-tRNA$^{f}_{Met}$ (here designated as Met-tRNA$_I$) is the more efficient initiator for α crystallin synthesis. Therefore when the initiation of protein synthesis is studied in a cell-free system the use of formyl-Met-tRNA$^{f}_{Met}$ may lead to a considerable underestimation of the initiation capacity of this system.

SUMMARY

N-terminal acetylation can successfully be studied in a lens cell-free system due to the fact that two classes of endogenous proteins are N-terminally acetylated (α- and β crystallins) while another one has a free N-terminus (γ crystallin). Moreover, only the α crystallin polypeptides are characterized by an N-terminal methionine residue donated exclusively by Met-$tRNA^{f}_{Met}$. This terminal methionine is retained in the native α crystallin aggregate.

Acetylation takes place after initiation when the nascent chains have a length which allows cleavage of N-terminal methionine in most proteins.

Acetyl-Met-$tRNA^{f}_{Met}$ is completely unable to initiate lens protein synthesis in contrast to formyl-$tRNA^{f}_{Met}$. However, the formylated species is much less effective in initiation than the unblocked initiator.

Chemical formylation of $[^{35}S]$ Met-$tRNA^{f}_{Met}$ results in an even slightly more active tRNA as compared to the initiator blocked with the aid of the E.coli transformylase.

REFERENCES

1. Gershey, E.L., Vidali, G., Allfrey, V.G.: Chemical studies of histone acetylation: The occurrence of ε-N-acetyllysine in the f2a1 histone. J. Biol. Chem. 243, 5018-5022 (1968).
2. DeLange, R.J., Fambrough, D.M., Smith, E.L., Bonner, J.: Calf and pea histone IV: The complete amino acid sequence of calf thymus histone IV: Presence of ε-N-acetyllysine. J. Biol. Chem. 244, 319-334 (1969).

3. Phillips, D.M.P.: The presence of acetyl groups in histones. Biochem. J. 87, 258-263 (1963).
4. Marchis-Mouren, G., Lipmann, F.: On the mechanism of acetylation of fetal and chicken hemoglobins. Proc. Natl. Acad. Sci. U.S.A. 53, 1147-1154 (1965).
5. Van der Ouderaa, F.J., De Jong, W.W., Bloemendal, H. The amino-acid sequence of the αA_2 chain of bovine α-crystallin. Eur. J. Biochem. 39, 207-222 (1973).
6. Van der Ouderaa, F.J., De Jong, W.W., Hilderink, A., Bloemendal, H.: The amino-acid sequence of the αB_2 chain of bovine α-crystallin. Eur. J. Biochem. (1974) in press.
7. Reinbolt, J., Peter, R., Stehelin, D., Collot, D., Duranton, H.: Étude de la structure primaire de la protéine du virus de la Mosaïque Jaune du Navet: I. Structure primaire des peptides trypsiques solubles dans l'eau. Biochim. Biophys. Acta 207, 532-547 (1970).
8. Sodek, J., Hodges, R.S., Smillie, L.B., Jurasek, L.: Amino-acid sequence of rabbit skeletal Tropomyosin and its coiled-coil strucure. Proc. Natl. Acad. Sci. U.S.A. 69, 3800-3804 (1972).
9. Schroeder, W.A., Shelton, J.R., Shelton, J.B., Cormick, J., Jones, R.T.: The amino acid sequence of the γ chain of human fetal hemoglobin. Biochem. 2, 992-1008 (1963).
10. Narita, K., Ishii, J.: N-terminal sequence in ovalbumin. J. Biochem. (Tokyo) 52, 367-373 (1962).
11. Chan, S.K., Margoliash, E.: Amino acid sequence of chicken heart cytochrome c. J. Biol. Chem. 241, 507-515 (1966).
12. Matsubara, H., Smith, E.L.: The amino acid sequence of human heart cytochrome c. J. Biol. Chem. 237, 3575-3576 (1962).
13. Hilse, K., Sorger, U., Braunitzer, G.: Polymorphism

and the N-terminal acids of carp hemoglobin. Z. Physiol. Chem. 344, 166-168 (1966).

14. Phillips, D.M.P.: N-terminal acetyl-peptides from two calf thymus histones. Biochem. J. 107, 135-138 (1968).
15. Möller, W., Groene, A., Terhorst, C., Amons, R.: 50-S ribosomal proteins - purification and partial characterization of two acidic proteins, A_1 and A_2, isolated from 50-S ribosomes of Escherichia coli. Eur. J. Biochem. 25, 13-19 (1972).
16. Offer, G.W.: Myosin: An N-acetylated protein. Biochim. Biophys. Acta 90, 193-195 (1964).
17. Suran, A.A.: N-terminal sequence of horse spleen Apoferritin. Arch. Biochem. Biophys. 113, 1-4 (1966).
18. Harris, J.I., Lerner, A.B.: Amino-acid sequence of the α-Melanocyte-stimulating hormone. Nature (London) 179, 1346-1347 (1957).
19. Funatsu, G., Fraenkel-Conrat, H.: Location of amino acid exchanges in chemically evoked mutants of Tobacco Mosaic Virus. Biochem. 3, 1356-1362 (1964).
20. Stevens, F.C., Glazer, A.N., Smith, E.L.: The amino acid sequence of wheat germ cytochrome c. J. Biol. Chem. 242, 2764-2779 (1967).
21. Winstead, J.A., Wold, F.: Studies on rabbit muscle enolase. Chemical evidence for two polypeptide chains in the active enzyme. Biochem. 3, 791-795 (1964).
22. Haylett, T., Swart, L.S.: High-sulfur proteins of reduced Merino wool III. Amino acid sequence of protein SCMKB-IIIB2. Textile Res. J. 39, 917-929 (1969).
23. Folk, J.E., Gladner, J.A.: The amino acid sequence of peptide B of co-firbin. Biochim. Biophys. Acta 44, 383-385 (1960).
24. Allison, W.S., Admiraal, J., Kaplan, N.O.: The sub-

units of dogfish M_4 lactic dehydrogenase. J. Biol. Chem. 244, 4743-4749 (1969).
25. Gaetjens, E., Barony, M.: N-acetylaspartic acid in G-actin. Biochim. Biophys. Acta 117, 176-183 (1966).
26. Strous, G.J.A.M., Van Westreenen, H., Bloemendal, H.: Synthesis of lens protein in vitro. N-terminal acetylation of α-crystallin. Eur. J. Biochem. 38, 79-85 (1973).
27. Strous, G.J.A.M., Berns, T.J.M., Van Westreenen, H., Bloemendal, H.: Synthesis of lens protein in vitro. Role of Methionyl-tRNAs in the synthesis of calf lens α-crystallin. Eur. J. Biochem. 30, 48-52 (1972)
28. Strous, G.J.A.M., Van Westreenen, H., Van der Logt, J., Bloemendal, H.: Synthesis of lens protein in vitro. The lens cell-free system. Biochim. Biophys. 353, 89-98 (1974).
29. Strous, G., Van Westreenen, J., Bloemendal, H.: Synthesis of lens protein in vitro. VI. Methionyl-tRNA from eye lens. FEBS Letters 19, 33-37 (1971).
30. Muench, K., Berg, P.: Preparation of aminoacyl ribonucleic acid synthetases from Escherichia coli; in: Procedures in Nucleic Acid Research, eds. G.L. Cantoni and D.R. Davies (Harper and Row, New York) p. 375.
31. Marcker, K.A., Sanger, F.: N-formyl methionyl-sRNA. J. Mol. Biol. 8, 835-840 (1964).
32. Schofield, P., Zamecnik, P.C.: Cupric ion catalysis in hydrolysis of aminoacyl-tRNA. Biochim. Biophys. Acta 155, 410-416 (1968).
33. Strous, G.J.A.M., Berns, A.J.M., Bloemendal, H.: N-terminal acetylation of the nascent chains of α-crystallin. Biochem. Biophys. Res. Commun. 58, 876-885 (1974).
34. Berns, A.J.M., Strous, G.J.A.M., Bloemendal, H.: Heterologous in vitro synthesis of lens α-crystallin

polypeptide. Nature New Biol. 236, 7-9 (1972).
35. Laycock, D.G., Hunt, J.A.: Synthesis of rabbit globin by a bacterial cell-free system. Nature (London) 221, 1118-1122 (1969).
36. Narita, K., Tsuchida, I., Tsunazawa, S., Ogata, K.: Formation of acetylglycylpuromycin by the incubation of Hen's oviduct minces with puromycin. Biochem. Biophys. Res. Commun. 37, 327-332 (1969).
37. Liew, C.C., Haslett, G.W., Allfrey, V.G.: N-acetylseryl-tRNA and polypeptide chain initiation during histone biosynthesis. Nature (London) 226, 414-417 (1970).
38. Yoshida, A., Lin, M.: NH_2-terminal formylmethionine- and NH_2-terminal methionine-cleaving enzymes in rabbits. J. Biol. Chem. 247, 952-957 (1972).

Why is Insect Immunity Interesting?

Hans G. Boman, Ingrid Nilsson-Faye and Torgny Rasmuson Jr.
Department of Microbiology, University of Umea, S-901 87
Umea 6, Sweden

Introduction

The industrial use of silk moths motivated the pioneering studies on insect diseases carried out first by Bassi and later by Pasteur (1). This line of work ceased with the decline of the silk industry, but it was rejuvinated when microbial control of insects was recognized as an alternative to chemical insectides (2). Despite extensive work it is still largely a mystery why certain microorganisms are insect patogens. However, it is obvious that the defense system of the insect must be of great importance for the fate of an infection.

Since about the turn of this century it has been known that insects can be vectors for microorganisms which cause severe diseases in humans, animals and plants (3). This vector property is often rather species specific and it seems possible that the immune mechanisms of the insects are of importance for maintaining the vector function.

Besides these two practical aspects there is also a more theoretical reason to study insect immunity. This depends on the fact that insects lack immunoglobulins and that homologous organ transplantations are unopposed by any defense system (4,5). Studies of invertebrate immunity may therefore provide information both on alternative recognition mechanisms and on the phylogeny of the immune systems (5).

The experimental system used

In the beginning of our work we used Drosophila (6) and together with Bertil Rasmuson we are continuing this line with the goal of making a genetic analysis of the immune mechanism. However, it soon became clear that for a biochemical characterization of the phenomena observed we needed an insect larger than Drosophila. Carroll Williams draw our attention to the fact that pupae of large silk moths can be considered as "one-test-tube-animals" and that they are a very convenient model system for the study of many biological problems. Since we have recently in detail described the experimental technique utilized (7), we will here only point out a few general aspects of the system.

In all the experiments to be described here we have used pupae of the oriental silk moth, Samia cynthia, which lives wild in some areas of the United States and southern Europe. The size of the pupae varies between 1 and 3 grams and they can contain nearly 1 ml of blood. They can be reared in the laboratory on a synthetic diet but the size then becomes smaller. Like many insects S. cynthia has a diapause, that is a state of hibernation when their metabolism is drastically reduced. This phenomenon is controlled by light and temperature and it is therefore possible to rear insects into diapause and then store them for as long as a year. When chilled diapausing pupae are brought back to room temperature their metabolism is turned back to normal rate and development of the mature insect starts. The adult moth cannot eat and its only function is to mate during the 3-5 days it lives. From the scientist's point of view it is very practical with an animal which eats all its food during one period, and which then can be stored and used at will with a minimum of attention. The diapause has been shown to be specially useful in studies of inducible reactions (8). In our case it means that we can label the components of the immune system while the background from other biosynthetic processes is depressed by the diapause.

The results to be described were obtained by essentially only two types of experiment. In the first one we have injected pupae treated in different ways with living bacteria, in most cases our own strains of E.coli. Afterwards we have at different times removed samples of blood and assayed for the number of viable bacteria. In the second type of experiment we have studied the antibacterial activity present in blood removed from pupae who previously have been immunized through a primary infection. In our standard assay we have incubated 1% of hemolymph with about 3×10^6 viable test bacteria, an ampicillin and streptomycin resistant mutant of E.coli K12 (strain D31). At different times we have withdrawn samples and spread on plates with either streptomycin or ampicillin. This semi-sterile technique eliminates possible contaminants and we have elsewhere demonstrated that we have an apparent single-hit killing with proportionality for the amount of hemolymph added (7).

Results from in vivo experiments

We have previously shown that when bacteria were injected into Drosophila two different fates were possible (6). With a patogen like Pseudomonas aeruginosa the bacteria started to grow in the insect which was killed within 24 hours. With a non-patogen like Enterobacter cloacae or Escherichia coli there was either a limited growth or no growth at all. In these cases the flies continue to live and they were seemingly unaffected by the living bacteria they carried.

When we repeated this experiment in pupae of S. cynthia we also included a pretreatment with actinomycin D. The result in Fig. 1 shows that in pupae pretreated with a physiological salt solution the E.coli injected were rapidly eliminated from the hemolymph. However, if the pupae had been pretreated with actinomycin D, there was first a limited drop and then a raise in the number of living bacteria circulating in the insect. In an indirect way this experiment implies that RNA synthesis was required for the development of the immunity in the pupa.

As in Drosophila (6) it was found possible to "vaccinate" pupae by a primary infection with a non-pathogenic bacteria (7). The defense system then induced will act against a closely related bacteria as well as against an unrelated pathogenic bacteria which otherwise would kill the insect. In such an experiment one group of pupae were given a salt injection, the other a dose of living Ent. cloacae. Two days later both groups

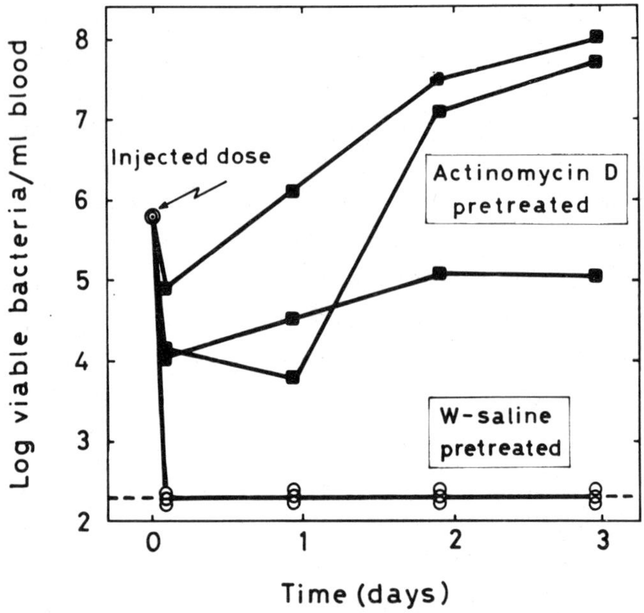

Fig. 1. Effect of actinomycin D on a primary infection. Six pupae of S. cynthia injected with living E. coli, strain D31. Samples of 5 µl of hemolymph were removed at different times and the number of viable cells of D31 was determined. The dotted line indicates the lowest detectible level of bacteria.

of pupae were challenged with an injection of living P. aeruginosa. The results in Fig. 2 show that in the vaccinated pupae Pseudomonas was rapidly eliminated from the hemolymph while in the control they grew up to a density of more than 10^9 viable cells/ml. Thus, a primary infection with Ent. cloacae protected the pupae from a secondary infection with P. aeruginosa. It should also be noted that the defense system induced eliminated P. aeruginosa faster than the living vaccine, an observation in agreement with our earlier findings in Drosophila (6).

In vitro experiments with hemolymph from immunized pupae

The fact that we use living bacteria as "substrate" has hindered the characterization of what seems to be at least in part an enzyme reaction. However, some of the properties so far known (7) are summarized in Table I. Especially the inhibition by lipopolysaccharide (LPS) has turned out to be a handy tool

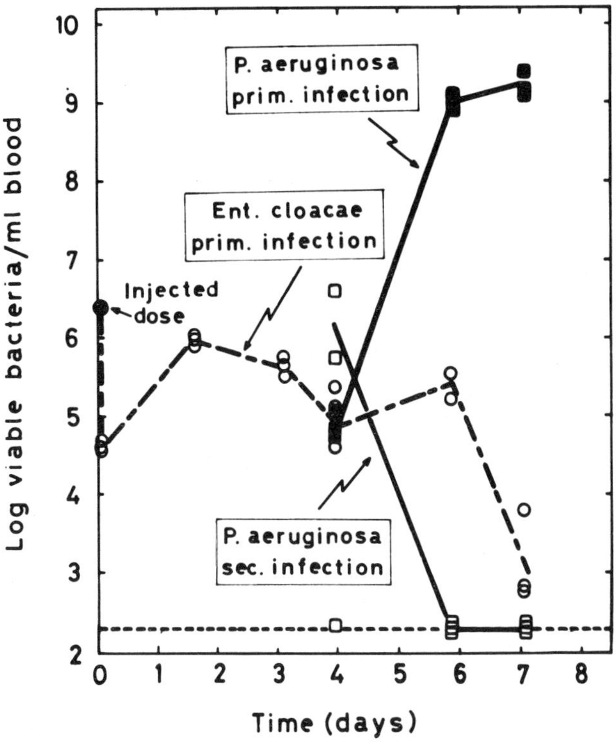

Fig. 2. "Vaccination" of pupae with viable Ent. cloacae. On day 4 immunity was tested *in vivo* by injection of viable *P. aeruginosa*. Experimental details as in Fig. 1.

for the further characterization of the system. Fig. 3 shows that LPS from strain D31 interfered with the killing of D31 while there was no effect on the killing of *Bacillus subtilis*.

Table I. Properties of Cynthia antibacterial activity

1. Activity unaffected by centrifugation.
2. Multicomponent system.
3. Activity lost by dialysis.
4. Protection by reducing agents.
5. Sensitive to pretreatment by trypsin.
6. Inhibited by 1 μg of heptose-less LPS/ml.

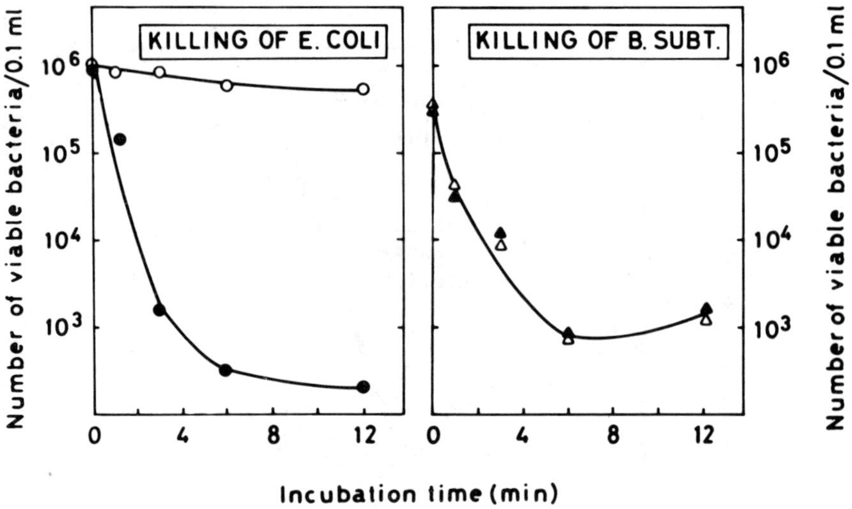

Fig. 3. Effect of LPS (50 µg/ml) on the killing of a gram positive and gram negative bacteria. With E.coli the reaction mixture contained 1%, with B. subtilis 5% hemolymph from a pupa vaccinated with strain D31. Reaction mixtures with LPS, filled symbols; reaction mixtures without LPS, unfilled symbols.

That insect blood can lyze bacteria has been observed several times and the responsible lysozyme was recently purified and characterized (9). We have therefore compared also lysis of Micrococcus lysodeikticus and E.coli, strain D31 and again found that LPS from D31 interferes only with the defense against the latter. Taken together the results in Fig.3 and 4 suggest that there are separate mechanisms for the defense against Gram positive and Gram negative bacteria. Preliminary experiments indicate that both defense systems are inducible but only the defense against Gram negative bacteria was inhibited by actinomycin D.

We have in Umea previously studied the mutational steps by which ampicillin resistance is developed in E.coli. In several cases we have found mutants with alterations in the outer membrane and one group of strains have turned out to carry muta-

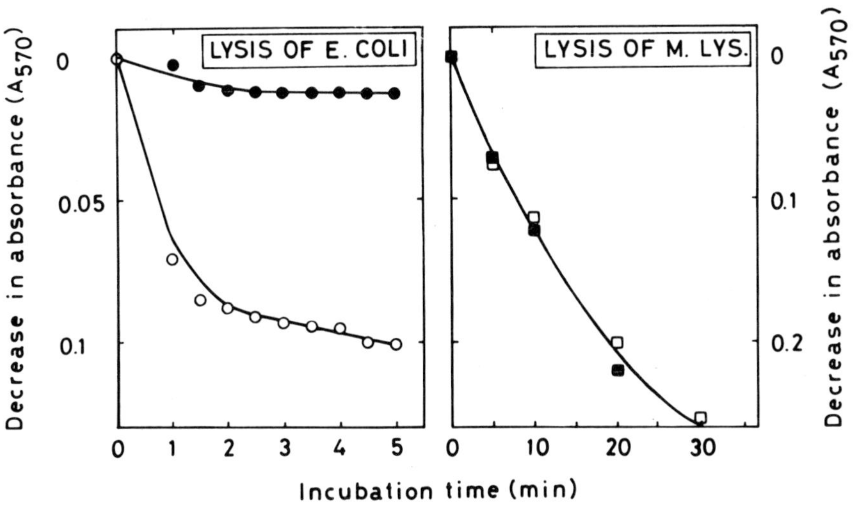

Fig. 4. Effect of LPS (50 µg/ml) on the lysis of a gram positive and gram negative bacteria. The reaction mixtures contained 1.6% of hemolymph from a pupa vaccinated with strain D31. Reaction mixtures with LPS, filled symbols; reaction mixtures without LPS, unfilled symbols.

tions affecting the biosynthesis of LPS (10). It was therefore natural for us to try such a set of mutants. As illustrated in Fig. 5 we found that the immunity system in S. cynthia could distinguish our different LPS mutants. The more carbohydrates that were lost from the LPS molecule, the more sensitive did the bacteria become to the killing action of the hemolymph.

Elsewhere we have shown that the lipopolysaccharide becomes a more potent inhibitor when increasing carbohydrate parts are lost from the molecule (7). As an extension of this result we tried lipid A liberated from LPS by mild acid hydrolysis. Despite the fact that lipid A is very poorly soluble in water there was a pronounced inhibition of the initial rate of killing (Fig. 6). Taken together the inhibition by LPS or lipid A and the increased susceptibility of LPS mutants indicate that at least one component in the hemolymph forms a strong complex with LPS. We are currently trying to isolate such a complex

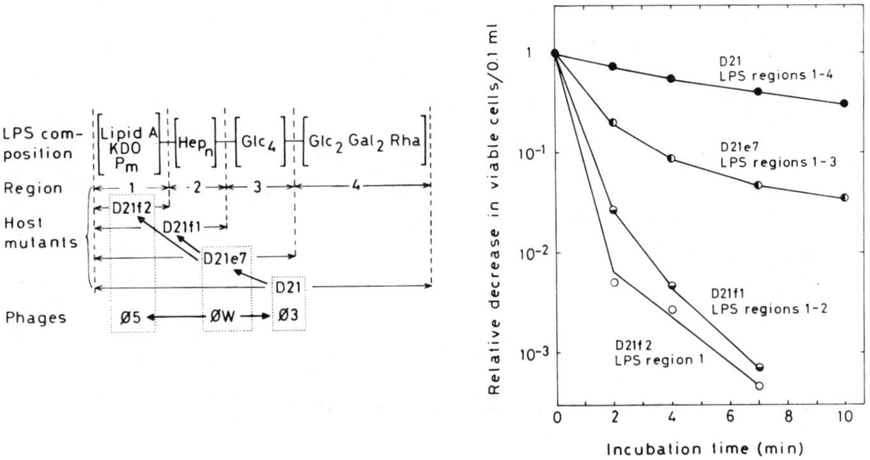

Fig. 5. Susceptibility of different LPS mutants of E.coli. The reaction mixtures contained 1% of hemolymph from a pupa vaccinated with strain D31. Left part a schematic structure showing the LPS composition of the strain used (from work by Boman and Monner).

using different isotopes and fractionation procedures.

Discussion

In a scientific context the claim that something is interesting usually means that convincing arguments exist for the scientific and practical importance of a problem. We have here presented this reasoning in our introduction. In addition, the word interesting can also cover personal motivation as well as the excitement involved in the work.

For some scientists a problem is often most stimulating

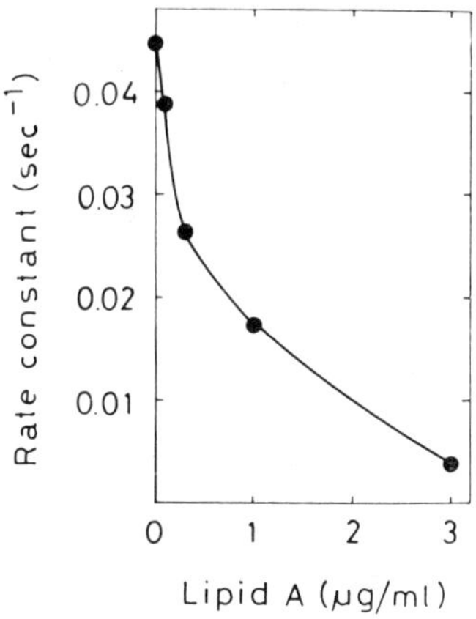

Fig. 6. Effect of lipid A on the killing of strain D31. Lipid A, obtained by mild acid hydrolysis of LPS from strain D21, was preincubated with hemolymph (1%) for 15 min before the adding of test bacteria. Rate constants were calculated from time curves assuming single-hit kinetics (7).

before it is solved. A scientific problem which offers clear questions based on solid phenomena but without obvious answers is often an interesting challenge. In the case of insect immunity we see before us the following problems:

1) In vaccinated pupae we have an antibacterial activity which is rather potent and with a mechanism of killing of the bacteria which is entirely unknown. The LPS inhibition could signalize a resemblance with the mammalian complement system. However, it may only reflect

a competitive reaction and would then indicate the presence of an active site with a strong affinity for the lipid A moiety. Other investigators have reported that LPS can induce immunity in insects (11,12). Since bacterial lysis is a terminal state and since lysis could involve the liberation of LPS in a form capable of inhibiting the killing we could in fact have a product inhibited reaction. This would superficially look senseless if not taken as evidence for a feed-back-inhibition signal to the biosynthetic machinery.

2) To understand the antibacterial activity we will have to resolve a multicomponent system where perhaps all the components are needed for the simplest yet known activity, the killing of bacteria. Together with Albert Pye we are currently approaching this problem by making a gel electrophoresis catalogue of the components of the hemolymph. We are then trying different low resolution methods for fractionation which preserve the activity and we use gel electrophoresis to rule out the participation of different components in the catalogue. Naturally, we are also trying to find conventional enzyme assays using different components isolated from the envelope of the bacteria.

3) It is important to describe the specificity of the antibacterial activity in chemical terms, especially to understand if there is a special recognition mechanism for "self" protection. Our present experience would indicate that the insects are not as cautious as mammals in recognizing "non-self" and to distinguish one intruder from another. On the other hand we have a mechanism by which the living vaccine is protected from the defense system it induces (6,7) an immunological situation resembling a symbiosis (13).

4) So far there are relatively few inducible eucaryotic systems for macromolecular synthesis. We have to search for the chemical signal(s) which start the immunity mechanism. Hopefully, we should have future possibilities to use cell or organ cultures and be able to carry out the whole immunization process in vitro.

When I (the senior author) returned from my stay at the Rockefeller Institute a British friend remarked that I had become "Lipmannized", thereby referring to my interest in protein synthesis. However, I would suggest that "Lipmannized" instead should mean the exciting spirit of searching for new scientific concepts. In this sense I have here tried to show that insect immunity is interesting to me because I once had the favour of becoming "Lipmannized". I realize that I have told you a story about a hopefully unknown fish which we have on a hook somewhere out in the sea. However, the best we now can do is to say to Dr. Lipmann: "If we can land this fish and if it is nice looking, then we will give it to you as a 75-year-birthday present".

The work was supported by grants from The Swedish Natural Science Research Council (Dnr 2453) and from The Swedish Cancer Society (Project no 157).

References

1. Porter, J.R., Bact.Rev. 37 (1973) 284.
2. Bulla, L.A. (ed), Regulation of Insect Populations by Microorganisms. Ann.N.Y.Acad.Sci. 217 (1973).
3. Romoser, W.S. The Science of Entomology. Macmillan Publishing Company, New York 1973.
4. Bernheimer, A.W., E. Caspari and A.D. Kaiser. J. Exptl. Zool. 119 (1952) 23.
5. Hildeman, W.H. and A.L. Reddy. Fed.Proc. 32 (1973) 2188.
6. Boman, H.G., I. Nilsson and B. Rasmuson. Nature 237 (1972) 232.

7. Boman, H.G., I. Nilsson-Faye, K. Paul and T. Rasmuson Jr. Infection and Immunity 10 (July 1974).
8. Wyatt, G.R. In Insect Physiology Proceedings of the Twenty-Third Biology Colloquium. The Oregon State University Press, 1962, p 23-41.
9. Powning, R.F. and W.J. Davidson. Comp.Biochem.Physiol. 45B (1973) 669.
10. Boman, H.G., K. Nordström and S. Normark. Ann.N.Y. Acad.Sci. 235 (1974) 569.
11. Chadwick, J.S. and E. Vilk. J.Invertebr.Pathol. 13 (1969) 410.
12. Schwalbe, C.P. and G.M. Boush. J.Invertebr.Pathol. 18 (1971) 85.
13. Fichtelius, K.E. Lymphology 3 (1970) 30.

Energy Transduction in Membranes of Mycobacterium phlei

Arnold F. Brodie, Soon-Ho Lee, Rajendra Prasad, Vijay K. Kalra
and Frank C. Kosmakos
University of Southern California, School of Medicine,
Department of Biochemistry, Los Angeles, CA 90033

INTRODUCTION

It is a great honor to be invited to present a paper at this Festschrift honoring Doctor Fritz Lipmann, who has contributed greatly to our understanding of bioenergetic processes. Having first been exposed to Doctor Otto Meyerhof, I (AFB) went to Doctor Lipmann's laboratory to study the mechanism of phosphate bond energy in bacteria. The role of ATP and probable existence of a mechanism for coupling phosphorylation to oxidation in bacteria was predicted by Doctor Lipmann with his studies of pyruvate oxidation by *Lactobacillus delbrueckii* (1). Since it was difficult to obtain information concerning the bioenergetic mechanism with intact mitochondria (1952) we turned our attention to bacteria. This was not done with the object of finding a new or different coupling process, but with the hope that at the mechanistic level the coupling process in bacteria would be similar to that of mammalian tissues.[1] It was hoped that the bacterial membrane system would be different enough to afford a new approach.

Two reports of a cell-free system from *E. coli* capable of oxidative phosphorylation were described during the 1952-1954 period; however, Doctor Lipmann cautioned that these systems failed to fulfill the criteria of oxidative phosphorylation and as later confirmed by Pinchot (2) the phosphorylation observed with these systems was due to glycolysis. It was from Doctor Lipmann that I learned to appreciate the nature and means of generation of phosphate bond energy and a methodological approach to answering biochemical problems. Doctor Lipmann as my professor, provided encouragement and has been my friend and mentor.

He has helped me over the past 20 years to search for an understanding of the life process in chemical and thermodynamic terms.

Oxidative phosphorylation, electron transport and active transport are membrane related phenomena which share certain common properties. Nevertheless, an understanding of the basic properties of each process and their possible relationship to one another provides some insight concerning the underlying bioenergetic mechanism(s) of these different membrane related processes. Membrane vesicles which differ in size and orientation were found to also exhibit differences in their ability to couple phosphorylation to oxidation and to carry out active transport. For example, coupled phosphorylation requires the presence of membrane bound coupling factors and is inhibited by arsenate whereas active transport of amino acids is not affected by such treatments. In addition, some essential differences between active transport and coupled phosphorylation are also observed with membrane vesicles differing in spatial orientation.

Types of membrane structures: A number of bacterial systems have been used to study membrane related phenomena; however, in most instances little is known about the nature of the membrane or its vectorial orientation, electron transport pathways or the nature of the bioenergetic processes. In contrast, the membrane vesicles from *Mycobacterium phlei* have been characterized with regard to the nature of the respiratory chains (2-5), the sequence of electron transport (3), the sites of phosphorylation (4, 6), the conformational states of the membrane during substrate oxidation (7) and active transport of metabolites (8, 9).

Various types of membrane preparations can be obtained from *M. phlei* which differ in size and vectorial orientation. Protoplast ghosts, obtained by the method of Mizuguchi and Tokunaga, are about the same size as whole cells, largely intact, and represent a membrane preparation which is oriented "right side" out (RO) (Fig. 1). Sonication of the ghosts or whole cells resulted in the formation of membrane

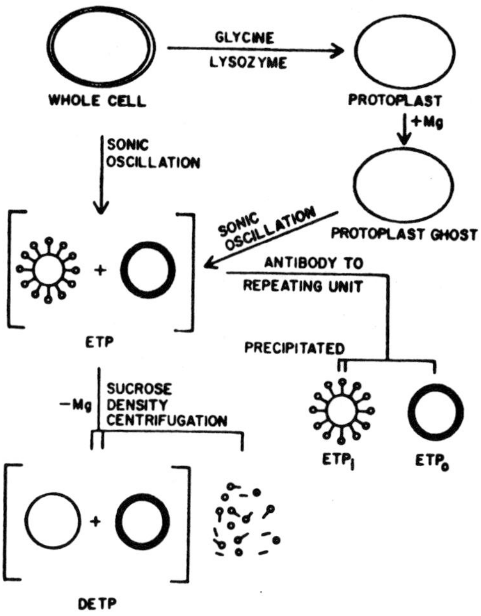

Figure 1

Types of Membrane Structures and Method of Preparation (Brodie et al., (9))

vesicles which have been referred to as the electron transport particles (ETP). These membrane vesicles range in size from 80 to 120 nm in diameter. Negative staining of the ETP revealed that the bulk of the membrane structures contain repeating units or spherical bodies (90-120 Å) attached to the membrane by a stalk (23 Å). It should be noted that although the bulk (90-95%) of the ETP are "inside out" (IO) oriented membrane vesicles they are contaminated with a small population of RO oriented membrane vesicles (12).

The membrane vesicles in the ETP fraction can be resolved into membranes depleted of membrane-bound coupling factor-latent ATPase by sucrose density gradient centrifugation in the absence of Mg^{++} ions (13, 14). The depleted electron transport vesicles (DETP) are capable of active transport and oxidation but fail to couple phosphorylation to oxidation. Morphologically the DETP appear as membrane vesicles devoid of repeating units or stalks. Restoration of oxidative phosphorylation and the reappearance of the repeating units occur upon addition of the coupling factor (BCF_4) to the DETP in the presence of Mg^{++} ions (10, 13, 14).

Nature of the respiratory chain: The nature and sequence of electron transport in the membrane vesicles from M. phlei are shown in Fig. 2. It is of interest that the same respiratory pathways are found in the protoplast ghost, ETP and DETP preparations and they contain the same amounts of quinone and cytochromes b, c, a and a_3 (nanomole/mg protein) (12). In addition, the different membrane vesicles exhibit similar oxidative activities (μatom/min/mg protein) with electron donors of the TCA cycle. In many respects, the respiratory pathways and the sequence of electron transport of the membrane vesicles from M. phlei are similar to that found in mammalian mitochondria. They differ in that the M. phlei system contains a menaquinone (MK_9(II-H)) instead of a benzoquinone, an unknown light sensitive factor on the succinoxidase pathway, and a unique phospholipid requiring enzyme, malate-vitamin K reductase, which requires FAD and converges with the NAD^+-linked pathway at the level of the menaquinone (15).

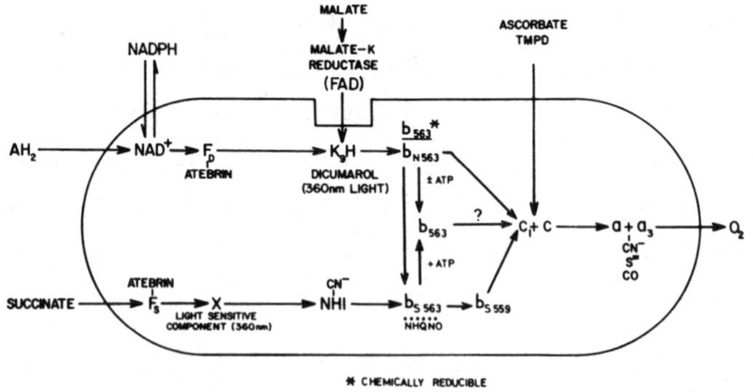

Figure 2

Electron Transport Pathways in Membrane Vesicles of M. phlei

Active transport with different types of membrane vesicles: Active transport of proline was observed with the electron transport particles, depleted ETP and cell membrane ghosts from M. phlei. Active transport with the membrane preparations exhibited a strict requirement for substrate oxidation. The uptake of proline proceeded against a concentration gradient with succinate, generated NADH, exogenous NADH or ascorbate-N,N,N',N'-tetramethyl-p-phenylenediamine (TPD) as substrates. The latter two electron donors were the most effective for the transport of proline and for oxidation but they were the least efficient for oxidative phosphorylation. The transport of proline appears to be independent of the formation of a high energy phosphorylated intermediate since it proceeds in the absence of coupling factors or phosphate and is not inhibited by high concentrations of arsenate.

Concentration gradients of proline of 35 and 170-fold with ascorbate-TPD as substrate were established with electron transport particles and membrane ghosts, respectively (9). Transport of proline with succinate was as effective as ascorbate-TPD with membrane ghost preparations, whereas with the electron transport particles, the transport of proline was 10-fold greater with ascorbate-TPD than with succinate. ATP did not support transport with either type of membrane preparation, even following treatment of the membranes with trypsin, a procedure which elicits latent ATPase activity.

The vectorial orientation of the ghost and ETP preparations was determined to be opposite of each other. Orientation was examined by the distribution of latent ATPase following sucrose density centrifugation in the absence of Mg^{++} ions. Latent ATPase may serve as a marker enzyme since it is bound to the inner surface of the cytoplasmic membrane. In addition, orientation of the membranes was also studied with uncoupling agents whose inhibitory effects on mitochondria and submitochondrial particles have been shown to be dependent on the orientation of the membrane (8, 9, 10, 12). These studies indicate that the ghost preparations are oriented RO as in the intact cell whereas the ETP preparations are oriented IO. However, the ETP contain some

RO oriented membranes and active transport of proline was inhibited only by uncoupling agents which uncouple RO oriented membranes.

Membrane vesicles (ETP) depleted of membrane-bound coupling factor-latent ATPase activity were capable of substrate oxidation but failed to couple phosphorylation to this oxidation. Nevertheless the DETP were capable of active transport at a level similar to that observed in ETP (9). Changes in pH (internal and external) in response to substrate were measured in ETP and DETP with bromthymol blue (16). Of particular interest was the finding that removal of the latent ATPase resulted in a collapse of the internal proton gradient but had no effect on the concentration of proline in the depleted membrane (17).

Active transport of proline was dependent upon substrate oxidation and the presence of Na^+ or Li^+ ions (9). The requirement for Na^+ was different from that described for Na^+ coupled transport systems (18, 19) since as the concentration of Na^+ was increased, the V_{max} of proline transport decreased but the K_m remained constant (Fig. 3). However, kinetic observations similar to those observed with the M. phlei membranes have been described in animal systems (20). With ETP or ghost preparations a change in concentration of sodium ($^{22}Na^+$) was not observed under conditions of proline active transport. The kinetic observations have been interpreted to suggest that Na^+ combines with the transport system to form a ternary complex with the metabolite that is translocated (18). Thus, Na^+ ion is necessary for translocation with the M. phlei system, but it does not affect the affinity of the transport system for proline.

Oxidative phosphorylation with different membrane preparations: A number of bacterial systems have been described which will carry out oxidative phosphorylation. In general these systems differ from intact mitochondrial systems by exhibiting lower P/O ratios. In contrast, the system from M. phlei has been shown to exhibit P/O ratio similar to that observed with intact mitochondria (21). In addition, three sites of phosphorylation have been established for the ETP preparations (6).

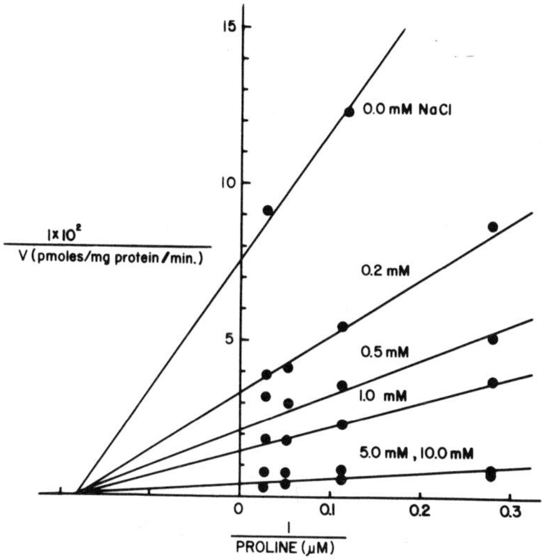

Figure 3
Lineweaver-Burke Plot of the Na^+-Dependent Transport of Proline (Hirata et al., J. Biol. Chem., in press)

In general, the low P/O ratios observed with bacterial systems were thought to be due to the harsh disruptive procedures used to prepare cell-free extracts. However, carefully prepared protoplast preparations from Micrococcus lysodeikticus yielded lower P/O ratios than that observed following disruption of the protoplast (22). Protoplast ghost from M. phlei were capable of substrate oxidation; however, their ability to couple phosphorylation to this oxidation was found to be cryptic (23). ATP formation occurred within the ghost preparations but the transfer of energy to exogenous nucleotides which was used in the assay system did not occur. Mild sonication of the ghost preparations results in the formation of ETP and in the loss of crypticity. Of particular interest was the isolation of a soluble protein component which when added to the intact ghosts resulted in the transfer of energy-rich phosphate bonds to exogenous nucleotides; the level of phosphorylation was similar to that observed with ETP (Table 1).

Table 1

Oxidative Phosphorylation in Membrane Ghost Preparations

	ΔO_2 µatoms	ΔP_i µmoles	P/O
Ghosts	4.86	0.00	0.00
Ghosts + AS II	5.72	4.02	0.70

Succinate was used as an electron donor and ADP as the phosphate acceptor.

In order to determine whether the soluble protein factor(s) caused a change in membrane permeability, the ghost preparations were preloaded with ^{14}C-ADP. The appearance of ^{14}C-ADP with exogenous ATP formed from oxidative phosphorylation was measured following addition of the soluble factor, exogenous ADP and substrate (Table 2).

Table 2

Effect of the AS II Fraction on Energy Translocation

	Membranes preloaded with ^{32}P-ADP (cpm of ^{32}P-ATP)		Membranes preloaded with ^{14}C-ADP (cpm of ^{14}C-ATP)	
	supernatant	ghosts	supernatant	ghosts
Succinate	2, 977	22, 622	1, 180	1, 240
Succinate + AS II	4, 113	3, 897	1, 280	820
AS II + ADP	43, 837	6, 900	1, 640	4, 340
Succinate + AS II + ADP	233, 798	13, 449	1, 720	10, 860

After preincubation with labeled ADP (^{32}P-ADP, 6.8 µmoles contain 5.6 x 10^6 cpm or ^{14}C-ADP, 0.1 µmoles contain 3.08 x 10^6 cpm) for 60 min at 0°C, the ghosts were washed twice and resuspended into the reaction mixture. 50 mM succinate, 12 mg of AS II and 2.5 mM ADP were added as indicated. After 20 min of incubation at 30°C, the supernatant and ghosts were separated and the ADP and ATP content of each fraction was assayed after separation on Dowex.

Differential centrifugation and analysis of the distribution of labeled ATP revealed that the ^{14}C-ATP and ^{14}C-ADP remained in the ghosts and did not appear in the external milieu. In contrast, preloading the ghost preparations with ^{32}P-ADP under similar circumstances resulted in the formation of exogenous ^{32}P-ATP upon addition of the soluble fraction. The soluble fraction has been partially purified and appears to be protein in nature. This protein fraction appears to be involved in the translocation of a high energy phosphate bond from internally formed ATP to an external phosphate acceptor.

To further substantiate these findings, ADP was coupled to Sepharose beads (Fig. 4) and was used as the exogenous source of ADP. The ADP covalently linked to the Sepharose beads is larger in size than the membrane ghosts. In the presence of the soluble fraction transfer of high energy phosphate bonds to the Sepharose-ADP occurred as assayed by the incorporation of ^{32}P or by the firefly assay method (Table 3). The transfer required the soluble protein and was inhibited by uncoupling agents such as tetraphenylboron (TPB^-), dimethyldibenzylammonia (DDA^+) and valinomycin in the presence of K^+ ions.

CH-Sepharose-ADP

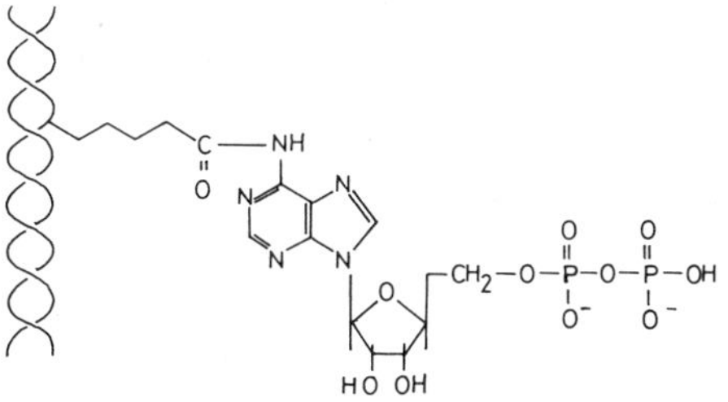

Figure 4

Phospholipase A treatment of membranes: The phospholipid composition of the ETP and DETP are essentially similar. The chief constituents being phosphatidyl ethanolamine (29%), cardiolipin (15-24%), phosphatidyl inositol (30-47%) and uncharacterized phospholipid (9%). Phospholipase A treatment of ETP resulted in 50% hydrolysis of phosphatidyl

Table 3

Energy Translocation from Ghost Preparations to Exogenous Sepharose-ADP

	Formation of Sepharose-ATP	
	Sepharose-^{32}P-ATP (cpm)	Firefly assay of Sepharose-ATP (nmoles)
Sepharose-ADP	450	65
Sepharose-ADP + AS II	8,370	2,800

Succinate was used as substrate. Ghosts (6 mg protein) were preloaded with ^{32}P-ADP as described in Table 2. The reaction mixture contained ghosts (preloaded with ^{32}P-ADP), 50 mM Hepes-KOH (pH 7.4), 10 mM $MgCl_2$, 50 mM succinate, Sepharose-ADP and the fraction AS II (12 mg protein) as indicated. Sepharose-ADP and Sepharose-ATP were separated from ghost by differential centrifugation, and Sepharose-ATP was assayed by the luciferase method.

---

ethanolamine and 5% hydrolysis of cardiolipin with no effect on the other phospholipids. Of particular interest was the finding that on removal of the membrane-bound coupling factor-latent ATPase a small increase in the hydrolysis of phosphatidyl ethanolamine was observed 15%; however, cardiolipin hydrolysis increased 42%. Conformational changes were measured with 8-anilino-1-naphthalene sulfonate (ANS) as probe. Energized fluorescences (E_f) was measured in ETP and DETP before and after phospholipase A treatment. The level of energized fluorescence decreased in both membrane preparations following phospholipase A treatment. The decrease in E_f was found to be due to a decrease in the apparent K_D and in the number of binding sites for ANS with no change in the relative quantum yield of ANS. Although more phospholipid was removed following treatment of the DETP the attachment of coupling factor to the treated membrane was not affected.

<u>Effect of Phospholipase A treatment on active transport</u>: Phospholipase A treatment of the ETP did not affect the rate or steady state level of proline transport (Table 4). In contrast, after removal of the

Table 4

Effect of Phospholipase A on Proline Transport: Reconstitution by Phospholipids

Preparations	Steady state levels of proline transport		Reconstitution of proline transport by phospholipids	
	untreated	treated	cardiolipin	phosphatidyl ethanolamine
	pmoles of proline transport			
ETP	318	315	---	---
DETP	299	119	252	179

The ETP and DETP fractions were treated as indicated for two hrs with phospholipase A and washed twice by centrifugation before use. For reconstitution, the phospholipid liposomes (0.67 μmoles) were preincubated with the membrane fractions for 15 min before measuring transport of proline.

---

membrane-bound coupling factor, phospholipase A treatment of the DETP resulted in a 60% decrease in the steady state level of proline uptake and a 20 to 35% decrease in the level of oxidation. The addition of BSA to remove the hydrolyzed fatty acid failed to influence the level of transport but increased the level of oxidation. However, partial restoration of proline transport occurred upon addition of phospholipids. The addition of cardiolipin obtained from M. phlei restored the steady state level to 85% of that observed before treatment whereas phosphatidyl ethanolamine only restored the level to 60%.

It should be noted that previous studies (9, 12) with uncoupling agents which are specific for intact mitochondria uncouple phosphorylation in ghost preparations and those specific for submitochondrial preparations uncouple phosphorylation with ETP. The finding that active transport in ghosts and ETP was inhibited only by agents which inhibit intact mitochondria and the knowledge that ETP represent a mixture of particles (most oriented "inside out") led to the assumption that only the right side out oriented particles were capable of active transport.

However, the finding that inhibition of transport by phospholipase A treatment required the removal of the membrane-bound coupling factor suggested that both types of membranes may be capable of active transport and that the inside out oriented membranes may contain patches or areas within the membrane which are oriented right side out. The various types of membranes appear to be more heterogenous than previously suspected.

Effects of Phospholipase A treatment on oxidative phosphorylation and the proton gradient: Treatment of ETP with phospholipase A resulted in an inhibition of phosphorylation; however, further treatment of the membranes with BSA followed by extensive washing resulted in the restoration of oxidative phosphorylation. The effects of phospholipase A treatment on the DETP were investigated with regard to the proton gradient and oxidative phosphorylation. As mentioned above, phospholipase A treated DETP can be reconstituted with BCF_4 (as measured by ATPase).

Membrane vesicles depleted of BCF_4 are capable of oxidation but not phosphorylation (Table 5). In addition, the removal of coupling factor-latent ATPase results in a collapse of the proton gradient. The proton gradient can be restored upon the addition of the coupling factor to the DETP. However, following phospholipase treatment of DETP the proton gradient was not restored on addition of the coupling factor or phospholipid whereas oxidative phosphorylation was restored by addition of BCF_4 alone.

In summary, it would appear that the bioenergetic mechanism(s) for active transport differs from that required for oxidative phosphorylation in membrane vesicles derived from M. phlei although both processes require substrate oxidation. Neither latent ATPase nor the proton gradient as measured by bromthymol blue are necessary for active transport of amino acids. The requirement for Na^+ ions which does not affect the affinity for proline may suggest that the driving force for active transport may be derived by an electro-chemical gradient

Table 5

Effect of Phospholipase A on Oxidative Phosphorylation and its Relationship to a Proton Gradient

Type of Membrane	ΔO_2 µatoms	ΔP_i µmoles	P/O	Decrease in absorbancy A 618-A 700 nm Δ O. D. /unit prot. (maximum change)
DETP	6.70	0.0	0.0	0.0028
DETP + BCF_4	6.78	6.2	0.91	0.0132
DETP*	1.78	0.0	0.0	0.0019
DETP* + BCF_4	4.28	4.2	0.98	0.0033

*Phospholipase A treated DETP. Phospholipase A from Crotalus Terr. Terr. was free of proteolytic activity. Oxidative phosphorylation was measured by the procedure previously described (6). Bromothymol blue response was measured with a dual wavelength spectrophotometer with succinate as the substrate (17).

established by the translocation of Na^+ ions. Caution should also be taken with regard to the nature of the orientation of the membranes and the vectorial direction of transport (9).

Although oxidative phosphorylation can be observed in the absence of a proton gradient following phospholipase A treatment, changes in the membrane, such as loss of E_f, and exposure or alteration of membrane bound proteins may account for the loss of the bromothymol blue response. In addition, evidence has been obtained which indicates a correlation between changes in conformational states of the membrane and the level of phosphorylation (7). Further insight into the mechanism of oxidative phosphorylation may be gained from an understanding of the mechanism of energy translocation with the M. phlei ghosts.

ACKNOWLEDGMENT

This work was supported by grants from the National Science Foundation (GB 32351X), the National Institutes of Health, U. S. Public Health Service (AI 05637), and the Hastings Foundation of the University of Southern California School of Medicine.

FOOTNOTE

[1] The biochemical processes of bacteria and mammalian tissues differ mainly in pathways concerned with specialized activities.

REFERENCES

1. Lipmann, F.: Coupling between pyruvic acid dehydrogenation and adenylic acid phosphorylation. Nature 143, 181, (1939).
2. Pinchot, G.B.: Mechanism of oxidative phosphorylation-observations and speculation. Persp. Biol. and Med. 8, 180-195 (1965).
3. Asano, A., Brodie, A.F.: Respiratory chains of Mycobacterium phlei. J. Biol. Chem. 239, 4280-4291 (1964).
4. Brodie, A.F., Adelson, J.: Respiratory chains and sites of phosphorylation of a bacterial system. Science 149, 265-269 (1965).
5. Brodie, A.F., Gutnick, D.: In "Electron and coupled energy transfer systems in biological systems." (King, T.E., Klingenberg, M., eds.) Marcel Dekker Inc., New York, Vol. 1 part B, pp 599-681 (1972).
6. Asano, A., Brodie, A.F.: Phosphorylation coupled to different segments of the respiratory chains of Mycobacterium phlei. J. Biol. Chem. 240, 4002-4010 (1965).
7. Aithal, H.N., Kalra, V.K., Brodie, A.F.: Temperature-induced alterations in 8-anilino-1-naphthalenesulfonate fluorescences with membranes from Mycobacterium phlei. Biochemistry 13, 171-178 (1974).
8. Hirata, H., Asano, A., Brodie, A.F.: Respiration dependent transport of proline by electron transport particles from Mycobacterium phlei. Biochem. Biophys. Res. Commun. 44, 368-374 (1971).
9. Brodie, A.F., Hirata, H., Asano, A., Cohen, N.S., Hinds, T.R., Aithal, H.N., Kalra, V.K.: In "Membrane Research." The relationship of bacterial membrane orientation to oxidative phosphorylation and active transport. (Fox, C. Fred, ed.) Academic Press, New York, pp 445-472 (1972).

10. Hirata, H., Brodie, A. F.: Membrane orientation and active transport. Biochem. Biophys. Res. Commun. 47, 633-638 (1972).

11. Mizuguchi, Y., Tokunaga, T.: Methods for isolation of deoxyribonucleic acid from mycobacteria. J. Bacteriol. 104, 1020-1021 (1970).

12. Asano, A., Cohen, N.S., Baker, R.F., Brodie, A.F.: Orientation of the cell membrane in ghosts and electron transport particles of Mycobacterium phlei. J. Biol. Chem. 248, 3386-3397 (1973).

13. Higashi, T., Bogin, E., Brodie, A. F.: Separation of a factor indispensable for coupled phosphorylation from the particulate fraction of Mycobacterium phlei. J. Biol. Chem. 244, 500-502 (1969).

14. Bogin, E., Higashi, T., Brodie, A. F.: Interchangeability of coupling factors of mammalian and bacterial origin. Biochem. Biophys. Res. Commun. 38, 478-483 (1970).

15. Asano, A., Kaneshiro, T., Brodie, A. F.: Malate-vitamin K reductase: A phospholipid requiring enzyme. J. Biol. Chem. 240, 895-905 (1965).

16. Chance, B., Mela, L.: Intramitochondrial pH changes in cation accumulation. Proc. Nat. Acad. Sci. U.S.A. 55, 1243-1251 (1966).

17. Hinds, T.R., Brodie, A.F.: Relationship of a proton gradient to the active transport of proline with membrane vesicles from Mycobacterium phlei. Proc. Nat. Acad. Sci. U.S.A., 71, 1202-1206 (1974).

18. Schultz, S.G., Curran, P.F.: Coupled transport of sodium and organic solutes. Physiol. Res. 50, 637-718 (1970).

19. Stock, J., Roseman, S.: A sodium-dependent sugar co-transport system in bacteria. Biochem. Biophys. Res. Commun. 44, 132-138 (1971).

20. Goldner, A.M., Schultz, S.G., Curran, P.F.: Sodium and sugar fluxes across the mucosal border of rabbit ileum. J. Gen. Physiol. 53, 362-383 (1969).

21. Brodie, A. F., Gray, C. T.: Phosphorylation coupled to oxidation in bacterial systems. J. Biol. Chem. 219, 853-862 (1956).

22. Ishikawa, S., Lehninger, A. L.: Reconstitution of oxidative phosphorylation in preparations from Micrococcus lysodeikticus. J. Biol. Chem. 237, 2401-2408 (1962).

23. Asano, A., Hirata, H., Brodie, A. F.: A factor(s) required for activation of oxidative phosphorylation in protoplast ghosts of Mycobacterium phlei. Biochem. Biophys. Res. Commun. 46, 1340-1346 (1972).

Control by Peptidyl-tRNA of Elongation Factor G Interaction with the Ribosome

Bartolomé Cabrer, María J.San-Millán, Julian Gordon*, David Vázquez and Juan Modolell
Instituto de Biología Celular, Velázquez 144, Madrid 6, Spain and *Friedrich Miescher-Institut, P.O.Box 273, CH-4002 Basel, Switzerland

Recent work in our and other laboratories with *Escherichia coli* and eukaryotic ribosomes has provided evidence indicating that the binding sites of aminoacyl-tRNA (the A-site) and either elongation factor (EF) G or EF 2 overlap or mutually interact on the ribosomal surface (1-8). Thus, it has been suggested that, during peptide chain elongation, EF G (EF 2) and EF Tu (EF 1) must alternate in their binding to polysomal ribosomes. The mechanism that may control such alternate interaction has been recently examined (5,8,9). It has been found that, after translocation, donor site- (P-site) bound AcPhe-tRNA inhibits interaction of EF G or EF 2 with the ribosome, but it does not prevent the EF Tu- or EF 1-dependent binding of Phe-tRNA to the A-site. Furthermore, the interaction of EF 2 or EF G with the ribosome is restored by deacylating the P-site-bound AcPhe-tRNA with puromycin.Consequently,it has been suggested (8,9) that polysomal ribosomes may only interact with EF 2 or EF G in the stage of the translational cycle when peptide bond has just been formed, peptidyl-tRNA is in the A-site, and the newly deacylated tRNA in the P-site.

Most of the available evidence for these interactions is indirect because the binding of EF G has been inferred from the parallel binding (or hydrolysis) of radioactive GTP or GDP. Also, the evidence is incomplete since the data has been obtained with the model system of poly(U),Phe-tRNA,and AcPhe-tRNA. Here, we complete the evidence with the use of purified endogenous polysomes from *E.coli*; and show, directly, the conditions for optimal binding of {^{14}C}EF G.

MATERIALS AND METHODS

Materials

To prepare *E.coli* MRE 600 polysomes, cells growing exponentially in rich medium were lysed (10), the lysate (30 ml) was layered on a 8 ml cushion of 60% sucrose in buffer A (1 M NH_4Cl, 40 mM $Mg(acetate)_2$, 10 mM Tris-HCl, pH 7.8, 2 mM EDTA and 10 mM β-mercaptoethanol) and it was centrifuged at 175,000 x g for 3 h. The supernatant was carefully removed and the pellet was left resuspending overnight at 0°C in 5 ml of buffer A. The resulting solution was clarified at 17,000 x g for 5 min, diluted to 30 ml with buffer A, and layered over a sucrose cushion as described above. Polysomes were sedimented and resuspended in a small amount of buffer B (buffer A with β-mercaptoethanol replaced by 4 mM dithiothreitol). The suspension was made 50% in glycerol and stored until used at -20°C. The polysomes thus prepared had an absolute requirement for EF G and EF T (EF Ts + EF Tu) to elongate their nascent chains and should be devoid of initiation factors (11). Endogenous GTPase activities were usually absent (see Table 3). When significant activity was present, it could be removed by a third cycle of washing and sedimentation.

Three-times washed ribosomes, EF G, {γ-^{32}P}GTP, and Ac{^{3}H}Phe-tRNA (5400 cpm/pmol, prepared from purified $tRNA_{Phe}$) were prepared as described (6,9). {^{14}C}EF G was prepared from a multiple amino acid auxotroph (strain 227 F^-, from W.Maas, New York) grown in 10 l of a minimal medium supplemented with 500 μg/ml casamino acids and containing 2.5 mCi of {^{14}C}amino acid mixture (Radiochemical Centre, Amersham). After harvesting, {^{14}C}EF G was prepared from the S100 of the cells by stepwize fractionation on DEAE-cellulose. Elongation factors were eluted with 0.3 M KCl. This was followed by back-extraction with decreasing ammonium sulphate concentrations as described earlier (12). The specific activity was 2 x 10^6 cpm/mg and 4100 cpm/µl.{^{3} H}GTP (5200 cpm/pmol), {^{3}H}GDP (2300 cpm/pmol) and {^{3}H}puromycin (1000 cpm/pmol) were from the Radiochemical Centre, Amersham.

Binding of Guanosine Nucleotide to Polysomes

EF G plus fusidic acid-dependent binding of guanosine nucleotide to polysomes was assayed in reaction mixtures (27 µl) containing: 90 mM NH_4Cl, 13 mM $Mg(acetate)_2$, 12 mM Tris-HCl, pH 7.8, 1.5 mM dithiothreitol, 2 A_{260} units of polysomes/ml, 96 µg/ml EF G, 1.9 mM fusidic acid, either 1.1 µM {^{3}H}GDP or 1.3 µM {^{3}H}GTP (preincubated immediately before use with phosphoenol pyruvate plus pyruvate kinase), and other components as specified. After 10 min at 30°C, bound nucleotide was determined by Millipore filtration. The filters were washed with buffer with the same ionic composition of the reaction mixture and 1 mM fusidic acid. Results were corrected by values obtained in reactions without EF G.

Reaction of Polysomes with {^{3}H}Puromycin

Peptidyl nascent chains of polysomes were reacted with {^{3}H}puromycin in mixtures (54 µl) prepared as for the binding of guanosine nucleotide, except that 1.4 µM GTP replaced the labeled nucleotide, polysomes were 8 A_{260} units/ml, and 4 µM {^{3}H}puromycin was present. Other details are specified in the Table.After incubation for 10 min at 30°C, peptidyl-{^{3}H}puromycin formed was determined as described (13). Results were corrected by values obtained in reaction mixtures without polysomes.

Binding of {^{14}C}EF G to Ribosomes

Prior to {^{14}C}EF G binding, ribosomes were preincubated with poly(U), purified $tRNA_{Phe}$ and, when present,0.67 µM Ac{^{3}H}Phe-tRNA, as previously described (9). Binding of {^{14}C}EF G to these ribosomes was performed in mixtures (75 or 150 µl) containing: 20 mM NH_4Cl, 50 mM KCl, 13 mM Mg(acetate)$_2$, 8 mM Tris-HCl, pH 7.8, 1.8 mM dithiothreitol, 16 to 20 A_{260} units of ribosomes/ml and {^{14}C}EF G, guanosine nucleotide, 2 mM fusidic acid and other components as specified. After incubation as indicated in the legends, {^{14}C}EF G·GDP·ribosome·fusidic acid complex formed was separated from unreacted {^{14}C}EF G and other components by filtration through small (0.65 x 8 to 13 cm) columns of Sepharose 4B or 6B. The elution buffer had the same ionic composition as the reaction mixture and contained 6 mM 2-mercaptoethanol and 1 mM fusidic acid (when also present in the reaction mixture). Unless otherwise indicated, elution was performed at room temperature. Three or four drop fractions (125 to 210 µl) were collected, 30 to 60 µl portions used to determine optical density at 260 nm and the remaining volume diluted with 2 ml of Bray's scintillation fuid (containing a thixotropic gel, Cab-o-sil, Packard), and counted.

RESULTS

Interaction of EF G with Purified Polysomes

Polysomes prepared by washing in NH_4Cl as described under "Materials and Methods" display, in glycerol density gradients, optical density profiles similar to those of unwashed polysomes, but without subunits and less than 5% of monomers (not shown). They retain most of their peptidyl-tRNA in the A-site, probably a consequence of the rapid chilling of the cells during polysome preparation that preferentially affects ribosomal translocation (12-14). This location was determined by reaction with {^{3}H}puromycin, with or without EF G and GTP. As Table 1 shows, only a small proportion of the polysomes had their peptidyl-tRNA reactive with puromycin. However, following treatment with EF G and GTP, about 50% became puromycin reactive.

Moreover, this may be a low estimate, for at the low concentration of {^{3}H}puromycin used (4 μM) part of the peptidyl residues may not be reactive (15). Table 1 also shows that the one cycle of translocation in this system is insensitive to fusidic acid, confirming earlier reports for this system (14-16).

Table 1. Reaction of {^{3}H}puromycin with nascent peptides of purified polysomes

	molecules {^{3}H}puromycin reacted per ribosome
- EF G - GTP	0.15
+ EF G + GTP	0.51
+ EF G + GTP + 1.9 mM Fusidic acid	0.45

Reaction mixtures were prepared as described in Methods with the modifications indicated above. To correct the values for the nonspecific retention of tritium to the filters (13), a control mixture lacking polysomes and EF G was run in parallel and its value was subtracted.

Table 2. EF G-dependent binding of guanosine nucleotide to pre- and posttranslocated polysomes: effect of puromycin

Addition	Nucleotide used {^{3}H}GTP	{^{3}H}GDP
	molecules bound per ribosome	
None	0.36	0.20
0.55 mM Puromycin	0.84	0.31

It was previously shown that ribosomes with AcPhe-tRNA in the A-site can translocate, but cannot stably bind EF G in the presence of fusidic acid and guanosine nucleotide, either before or after translocation (9,14). Similarly, for the polysomes with peptidyl-tRNA on the A-site described here, only a minority of the ribosomes were capable of stably binding EF G, either before (with {^{3}H}GDP) or after (with {^{3}H}GTP) translocation (Table 2). However, deacylation of the P-site-bound peptidyl-tRNA with puromycin unmasked the capacity to form the {^{3}H}GDP·EF G·ribosome·fusidic acid complex (Table 2), although this was not the case for the pretranslocated polysomes.

Table 3. EF G-dependent {γ-^{32}P}GTP hydrolysis of purified polysomes: effect of puromycin

Additions	molecules {γ-^{32}P}GTP hydrolyzed per ribosome
None	0.3
EF G	34
EF G + 0.67 mM Puromycin	134

Reaction mixtures (30 µl) were prepared essentially as for the binding of guanosine nucleotide (Methods), but Tris-HCl pH 7.8 was raised to 50 mM, polysomes were 1.1 A_{260} unit/ml, EF G was 32 µg/ml, and tritiated guanosine nucleotide was replaced by 7.4 µM {γ-^{32}P}GTP (40 cpm/pmol). After incubation at 30°C for 5 min the mixtures were analyzed for ^{32}Pi as described (14).

It has been our experience that conditions for stable interaction with EF G and fusidic acid are paralleled by the appearance of uncoupled GTPase in the absence of fusidic acid (17). This was also the case for the polysomal system described here: the results in Table 3 show that deacylation of the peptidyl-tRNA with puromycin unmasked the uncoupled GTPase.

Binding of {^{14}C}EF G to Ribosomes in Various States

Since the above results and those described previously with the AcPhe-tRNA system were inferred indirectly from the binding or the hydrolysis of the radioactively labelled nucleotide,

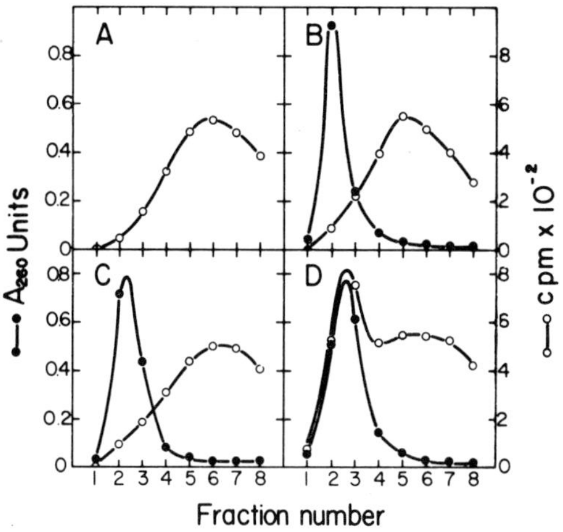

Fig.1. Requirements for stable binding of {^{14}C}EF G to ribosomes. To reaction mixtures (150 µl), with the ionic composition described in Methods and 2 µl of {^{14}C}EF G, the following components were added: A, none; B, ribosomes; C, ribosomes and 70 µM GTP; D, ribosomes, 70 µM GTP and fusidic acid. After incubation at 30°C for 5 min, the mixtures were filtered in Sepharose 6B (8 cm long) columns, and fractions collected and analyzed.

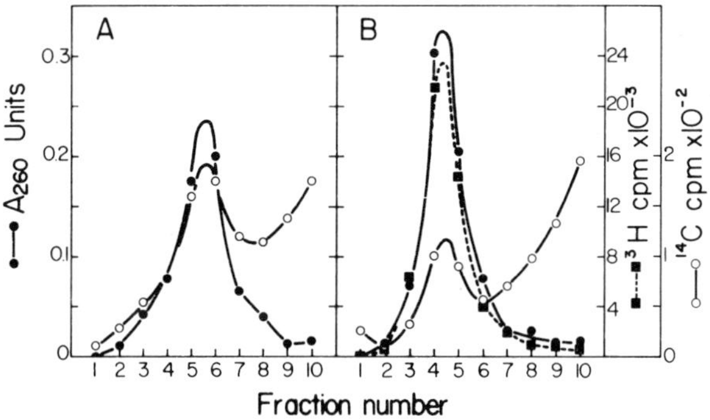

Fig. 2. Effect of A-site-bound Ac{^{3}H}Phe-tRNA on {^{14}C}EF G binding to ribosomes. Ribosomes without (*panel A*) and with (*panel B*) A-site bound Ac{^{3}H}Phe-tRNA were assayed for the binding of {^{14}C}EF G in reaction mixtures (75 µl) prepared as described in Methods. Components not specified in the text were: 3 µl {^{14}C}EF G, 1.3 µM GDP, and 0.25 mM puromycin. After incubation at 0°C for 5 min the reactions were filtered through Sepharose 6B (13 cm long) columns, which were mantained at 0°C. Fractions were analyzed as described.

we set out to confirm this directly with radioactively labelled EF G. The $\{^{14}C\}$EF G was prepared as described under "Materials and Methods", and the specificity of its binding to washed ribosomes was determined. Binding assays were carried out on Sepharose columns, as it has been our experience that EF G itself binds to nitrocellulose filters. The conditions of stable binding are shown in Fig. 1. It is seen that the presence of both GTP and fusidic acid were required for the stable binding (Fig. 1D). The omission of fusidic acid (conditions for uncoupled GTPase) eliminated this binding, even though GTP was present in the elution buffer of the column to favour interaction (Fig. 1C). In addition, further omission of GTP (Fig. 1B) or ribosomes (Fig. 1A) did not affect the elution profile of the EF G preparation. Thus, the stable binding of EF G to washed ribosomes has the same specificity as was previously inferred from the binding of the nucleotide.

Ac$\{^{3}H\}$Phe-tRNA was then bound to the A-site of the washed ribosomes. These were analysed for their capacity to stably bind $\{^{14}C\}$EF G, under conditions non-permissive for translocation (GDP) and with puromycin to remove any residual Ac$\{^{3}H\}$Phe-tRNA bound to the P-site. It can be seen (Fig. 2) that the presence of the Ac$\{^{3}H\}$Phe-tRNA blocks the binding of the $\{^{14}C\}$EF G. The inhibition was not complete probably since more ribosomes in our preparations bind EF G than AcPhe-tRNA (6,9). The Ac$\{^{3}H\}$Phe-tRNA was then translocated to the P-site by incubation in the presence of $\{^{14}C\}$EF G and GTP, and, subsequently, the binding was stabilized by fusidic acid. Again, the presence of the Ac$\{^{3}H\}$Phe-tRNA on the P-site inhibited the $\{^{14}C\}$EF G binding (Figs. 3A and B) and deacylation with puromycin partially relieved this block (Fig. 3C).

DISCUSSION

The results described here fully support the recent proposal that EF G interacts with the ribosome only transiently during each cycle of peptide bond formation (8,9,18). The proposal

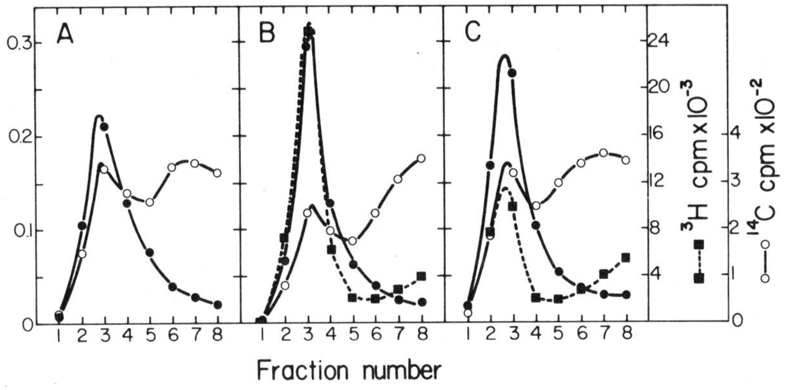

Fig. 3. Effect of P-site bound Ac{^{3}H}Phe-tRNA on {^{14}C}EF G binding to ribosomes. Ribosomes without (*panel A*) and with (*panel B and C*) bound Ac{^{3}H}Phe-tRNA were incubated at 30°C for 10 min in reaction mixtures (75 µl) prepared as described in Methods, but fusidic acid was omitted and 0.27 mM puromycin was included in the mixture corresponding to *panel C*. Other additions not specified in the text were: 3 µl {^{14}C}EF G, 1.8 µM GTP, 2.7 mM phosphoenol pyruvate and 27 µg/ml pyruvate kinase. The mixtures were chilled, 1.9 mM fusidic acid was added and incubation continued at 30°C for 5 min. Mixtures were analyzed after elution through Sepharose 4B (8 cm long) columns.

is supported by the inability of the preparation of polysomes to interact stably with EF G except when the peptidyl-tRNA on the P-site has been deacylated by puromycin, in simulation of peptide bond formation. Further, the binding of EF G measured directly is maximal only when the AcPhe-tRNA on the P-site was deacylated. Thus, in the immediately pretranslocational state, the presence of deacylated tRNA on the P-site favors EF G binding, while in the posttranslocational state, the presence of peptidyl-tRNA in the P-site blocks EF G binding. This control of EF G entry by peptidyl-tRNA could be exercised through a steric hinderance of the peptidyl chain, which when bound to the P-site prevents the entry of EF G, and,perhaps, promotes its release after translocation. Alternatively, the removal of the peptidyl chain from the P-site associated with peptide bond formation may induce a conformational change of the ribosome and/or the P-site-bound tRNA, that promotes the required configuration for EF G to interact with its site.

Earlier evidence for the cyclical interactions of the elongation factors with the ribosome during peptide chain elongation came from the fact that ribosomes complexed with EF G did not bind aminoacyl-tRNA, either in the presence or the absence of EF Tu (1-5). At this point, it is not clear whether this is the result of direct steric hindrance or indirect conformational interactions. The steric hindrance model is favored by the fact that all these interactions require ribosomal proteins L7 and L12 (19-22), and are inhibited by the antibiotic thiostrepton (23, review). However, the possibility of more indirect interaction is suggested by the finding that thiostrepton does not bind to proteins L7 and L12, but probably to L11 (Gordon, Howard, Stöffler & Highland, this volume).

SUMMARY

The interaction of elongation factor G (EF G) with ribosomes has been studied using purified endogenous *E.coli* polysomes, AcPhe-tRNA·ribosome complexes and {^{14}C}EF G. The results indicate that donor-site-bound peptidyl-tRNA prevents, after translocation, further interaction of EF G with the ribosome, and, consequently, promotes binding of aminoacyl-tRNA to the acceptor site. Since deacylation of peptidyl-tRNA by puromycin restores the ability of ribosomes to interact with EF G, it is suggested that the removal of the peptidyl chain from the donor site during peptide bond formation, may control the interaction of EF G with the ribosome.

REFERENCES

1. Cabrer, B., Vázquez, D., Modolell, J.: Inhibition by

elongation factor EF G of aminoacyl-tRNA binding to ribosomes. Proc.Nat.Acad.Sci. USA 69, 733-736 (1972).

2. Richman, N., Bodley, J.W.: Studies on translocation XII. Ribosomes cannot interact simultaneously with elongation factors EF Tu and EF G. Proc.Nat.Acad.Sci. USA 69, 686-689 (1972).
3. Miller, D.L.: Elongation factors EF Tu and EF G interact at related sites on ribosomes. Proc.Nat.Acad.Sci. USA 69, 752-755 (1972).
4. Richter, D.: Inability of *E. coli* ribosomes to interact simultaneously with the bacterial elongation factors EF Tu and EF G. Biochem.Biophys.Res.Commun. 46, 1850-1856 (1972).
5. Chinali, G., Parmeggiani, A.: Properties of elongation factor G: its interaction with the ribosomal peptidyl-site. Biochem.Biophys.Res.Commun. 54, 33-39 (1973).
6. Modolell, J., Vázquez, D.: Inhibition by aminoacyl-tRNA of elongation factor G-dependent binding of guanosine nucleotide to ribosomes. J.Biol.Chem. 248, 488-493 (1973).
7. Richter, D.: Competition between elongation factors 1 and 2, and phenylalanyl transfer ribonucleic acid for the ribosomal binding sites in a polypeptide-synthesizing system from brain. J.Biol.Chem. 248, 2853-2857 (1973).
8. Nombela, C., Ochoa, S.: Conformational control of the interaction of eukaryotic elongation factors EF-1 and EF-2 with ribosomes. Proc.Nat.Acad.Sci. USA 70, 3556-3560 (1973).
9. Modolell, J., Cabrer, B., Vázquez, D.: The interaction of elongation factor G with N-acetyl phenylalanyl transfer RNA·ribosome complexes. Proc.Nat.Acad.Sci. USA 70, 3561-3565 (1973).
10. Godson, G.N., Sinsheimer, R.L.: Lysis of *Escherichia coli* with a neutral detergent. Biochim.Biophys.Acta 149, 467-488 (1967).
11. Tai, P., Wallace, B.J., Herzog, E.L., Davis, B.D.: Properties of initiation-free polysomes of *Escherichia coli*. Biochemistry 12, 609-615 (1973).

12. Gordon, J.: Hydrolysis of guanosine 5'-triphosphate associated with binding of aminoacyl transfer RNA to ribosomes. J.Biol.Chem. 244, 5680-5686 (1969).
13. Pestka, S.: Peptidyl-puromycin synthesis on polyribosomes from *Escherichia coli*. J.Biol.Chem. 69, 624-628 (1972).
14. Modolell, J., Cabrer, B., Vázquez, D.: The stoichiometry of ribosomal translocation. J.Biol.Chem. 248, 8356-8360 (1973).
15. Pestka, S., Hintikka, H.: Studies on the formation of ribonucleic acid-ribosome complexes. XVI Effect of ribosomal translocation inhibitors on polyribosomes. J.Biol.Chem. 246, 7723-7730 (1971).
16. Burns, K., Cannon, M., Cundliffe, E.: A resolution of conflicting reports concerning the mode of action of fusidic acid. FEBS Letters 40, 219-223 (1974).
17. Conway, T., Lipmann, F.: Characterization of a ribosome-linked guanosine triphosphatase in *Escherichia coli* extracts. Proc.Nat.Acad.Sci. USA 52, 1462-1469 (1964).
18. Chinali, G., Parmeggiani, A.: Properties of the elongation factors of *Escherichia coli*. Exchange of elongation factor G during elongation of polypeptide chain. Eur.J. Biochem. 32, 463-472 (1973).
19. Hamel, E., Koka, K., Nakamoto, T.: Requirement of an *Escherichia coli* 50S ribosomal protein component for effective interaction of the ribosome with T and G factors and with guanosine triphosphate. J.Biol.Chem. 247, 805-814 (1972).
20. Kischa, K., Moller, W., Stöffler, G.: Reconstitution of a GTPase activity by a 50S ribosomal protein from *E.coli* . Nature New Biol. 233, 62-63 (1971).
21. Highland, J.H., Bodley, J.W., Gordon, J., Hasenbank, R., Stöffler, G.: Identity of the ribosomal proteins involved in the interaction with elongation factor G. Proc.Nat. Acad.Sci. USA 70, 147-150 (1973).
22. Sander, G., Marsh, R.C., Parmeggiani, A.: Isolation and

characterization of two acidic proteins from the 50S subunit required for GTPase activities of both EF G and EF T. Biochem.Biophys.Res.Commun. 47, 866-873 (1972).

23. Modolell, J., Vázquez, D.: Polypeptide chain elongation and termination. MTP International Review of Science, Biochemistry, ed. Arnstein, H.R.V. Medical and Technical Publishing Co., Oxford, in press.

A Cytoplasmic Function for the Polyadenylic Sequences of Messenger RNA

H. Chantrenne
Department of Molecular Biology, University of Brussels, Belgium

Most messenger RNAs from eukaryotic cells terminate at the 3'end with a long stretch of polyadenylic acid. Only one exception is known so far : histone mRNA. The questions raised by the existence of the poly A sequences have been discussed in several papers in which references can be found (1, 2, 3). Only a few points will be considered here, as an introduction to recent results (4) obtained in our laboratory, in collaboration with the laboratory of Dr. Littauer at Rehovot.

The mere fact that histone messenger RNA is active although it lacks poly A sequences shows that these are not required for translation per se in the eukaryotic system. They are not translated either, and their function as part of mRNA remains obscure.

Since bacterial messenger RNAs do not possess poly A sequences, it seems reasonable to search for a function of these sequences in relation with features of eukaryotic cells which distinguish them from prokaryotes, and the most obvious ones are the nuclear structures, the puzzling intranuclear RNA metabolism, the transfer of mRNA from nucleus to cytoplasm.

Thorough studies along these lines (2) established several fundamental facts about the poly A sequences. They are not transcribed from DNA, they are made by post transcriptional stepwise addition of adenylic residues; this occurs to a large extent at the 3'OH end of high molecular weight precursors of mRNA, within the nucleus.

Two drugs make it possible to dissociate message transcription and poly A addition : actinomycin D blocks transcription of the message-carrying moiety, but it does not interfere with the addition of the polyadenylic tail to already transcribed messages. On the contrary, cordycepin (3'deoxyadenosine) can be used for blocking polyadenylation without affecting transcription of the message.

Studies on the kinetics of poly A addition, on chasing of mRNA from nucleus to polyribosomes, and on the effects of cordycepin gave good evidence that the provision of new messenger to the translation system is in some way conditioned by the poly A moiety. To wit, inhibition of polyadenylation by cordycepin drastically reduces the appearance of newly made mRNA in the polyribosomes. The precise process in which poly A is involved is not ascertained; it might be related to the correct cleavage of high molecular weight precursors of messenger RNA, or to the transport of the message to the cytoplasm, or to some other unknown step.

A feature of eukaryotic messages which distinguishes them from bacterial ones is their much longer active life. Whereas the half life of bacterial messenger is only a few minutes, i.e. a small fraction of the generation time, messenger RNAs in eukaryotes are long lived. Even in exponentially growing mammalian cells, their half life is about equal to the mean generation time (5, 6), so that a sizable fraction of the messengers made in a mother cell continue to be translated in the daughter cells for a long time. In rabbit reticulocytes, hemoglobin synthesis continues for more than a day after the nucleus has been lost. In sea urchin eggs, protein synthesis starts at fertilization long before any synthesis of messenger RNA begins, and it is not prevented by actinomycin and it can be triggered by parthenogenic activation in enucleated eggs as well (7, 8). Protein synthesis in the early stages of development is therefore programmed by messengers that were made during oogenesis and were in some way stored in a stable, but not translatable form.

The possibility that poly A might be involved in the stabilization of mRNA is worth being considered. The idea was put forward (9, 10) that the length of the poly A tail might carry information as to the number of times a messenger can be translated : it was suggested that after each transcription, one adenylic residue would be cut away, and that the messenger would be destroyed when they would all be gone. A correlation is indeed found between the length of the poly A stretch and the age of the messenger : newly made messengers carry longer poly A sequences than older ones (10). However, it would seem (6) that shortening of the poly A moiety is a stochastic process with breaks occuring at random, rather than a sequential nibbling operation.

It was reported also that histone messenger RNA which is devoid of poly A tail has a shorter life than poly A containing messengers (11). The determination of active life of messengers in cells which continually generate messengers is a difficult task. Inhibitors have been commonly used for that purpose, but unspecific effects or unsuspected consequences of their action may be misleading. Kinetic methods avoiding the use of inhibitors(5) give figures of the life time somewhat at variance with those obtained with inhibitors; they confirm that messenger RNA for histone has a shorter life than the others; although not as short as was believed before. An interesting observation is that the kinetics of decay of histone mRNA is different from that of the poly A containing messengers (5, 6), but this may be related to the particular function of histones which are synthesized during a limited period of the cell cycle, as compared to the population of cell proteins which are made during most of the cell life.

Double labeling of old and newly made messengers and study of polyribosome profiles show that old mRNA with shorter poly A sequences have the same translational ability as new ones, that their capacity to initiate polypeptide synthesis remains unaffected (12). And this again leaves one with no clear function for poly A.

Attempts were made to assay the activity of messengers with and without poly A sequences, in acellular in vitro systems. The experiments reported (13) indicated that the messenger deprived of poly A was about 50 % as active as intact messenger, an observation the meaning of which is not clear, for any damage to half of the messenger population would have such a consequence.

Recent results from the Weizmann Institute (14) are more informative : they indicate that the active life of globin messenger from which the poly A sequences have been removed is shorter, in an in vitro system, than that of intact messenger. Stripped messenger is as active as normal messenger for 45 minutes, but it becomes less and less efficient for longer periods of incubation.

In vitro systems for protein synthesis can give excellent translation of added messenger; a common limitation of these system is that they do not operate perfectly for a long time. A system which proved exceptionally efficient for translating messengers is the Xenopus oocyte (15, 16). Injected hemoglobin messenger is amazingly stable in the oocyte. It was shown (16) that after one injection of messenger RNA globin synthesis continues for two weeks. This system was used in our laboratory for testing the longevity of mRNA and the possible effects of the poly A sequences.

Globin messenger RNA was prepared according to the classical procedure : isolation as a ribonucleoprotein from which the messenger was released (17), and further purification on oligo-T-cellulose (18). It was stripped from its poly A sequences by means of pure polynucleotide phosphorylase under strictly controlled conditions (14). The activity of intact mRNA and stripped mRNA was tested in Xenopus oocytes.
A convenient feature of the oocyte is that it provides an internal control of protein synthesis : it incorporates amino acids in rather large egg proteins which can be separated easily from hemoglobin on a sephadex column (15). This endogenous protein synthesis can serve as a reference for testing the synthesis caused by injected messengers.

Saturation of protein synthesis with injected Hb messenger occurs at about 0.01 γ per oocyte. At lower concentrations, the rate of hemoglobin synthesis depends on the amount of messenger

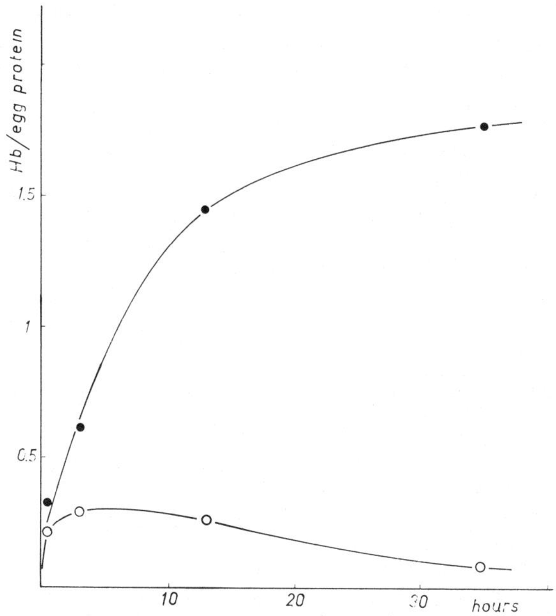

Fig. 1. Relative rate of synthesis of hemoglobin and egg protein.

120 oocytes from the same frog were injected each with 7.10^{-9} grams of mRNA. Sixty of them received native globin messenger and the others a messenger preparation from which the poly A sequence had been removed by the method of Soreq et al. (14). At different times after messenger injection, 10 oocytes of each batch were incubated in a medium containing labelled histidine, for various periods (e.g. 0 to 1 hour, 1 to 5, etc ...) The results are expressed as the ratio of counts in hemoglobin and in egg protein for different periods of incubation. Time zero is the time of injection of the messenger. For further details see (4).

injected. It is possible therefore to measure the messenger activity provided the amount injected is kept below 10^{-8}g/egg.

Under those conditions, intact and poly A stripped messengers were injected to batches of eggs taken from the same female Xenopus, and hemoglobin synthesis was compared to endogenous egg protein synthesis.

During the first hour after injection, the efficiency of both normal and stripped messengers was almost the same. This confirms that messengers without poly A tail can be translated; it shows furthermore that the preparation of stripped messenger used was not damaged in its information carrying moiety.
But the stripped messenger proved much less efficient than normal messenger in the ensuing hours.

As shown in the figure, in the eggs which have been injected with intact messenger RNA, the ratio of hemoglobin synthesis to endogenous egg protein synthesis increases regularly during the first 10 hours to become stabilized after 30 to 40 hours. On the contrary, the Hb messenger stripped of poly A sequences, after competing successfully for a few hours looses the game and is less and less translated with time.

The same RNA preparations were also assayed in an in vitro system from Krebs ascites cells. In agreement with a previous report (14) the level of activity was the same for both at the beginning of incubation and dropped faster for the stripped mRNA than for intact mRNA.

The experiments in frog eggs most clearly show that poly A is important for the maintenance of messenger integrity for a long time in the cytoplasm of the living cell.

Experiments to be reported elsewhere indicate that the loss of activity is due to the destruction of the message. These were performed in the following way. Highly radioactive DNA complementary to hemoglobin messenger RNA was made with reverse transcriptase from avian myeloblastosis virus, and used for titrating, by hybridization, the amount of hemoglobin message remaining in the oocytes as a function of time. The results which will be published soon in detail show that normal hemoglobin messenger is stable whereas messenger without poly A is destroyed.

Our conclusion is that in addition to their functions in the processing of nuclear RNA or in the export of messenger from the nucleus, poly A sequences are essential for keeping messengers active in the cytoplasm and for preventing their decay.

1.- Mendecki, J., Lee, S.Y., Brawerman, G. : Characteristics of the polyadenylic acid segment associated with messenger ribonucleic acid in mouse sarcoma ascites cells. Biochemistry, 11, 793-798 (1972).

2.- Darnell, J.E., Jelinek, W.R., Molloy, G.R. : Biogenesis of mRNA : genetic regulation in mammalian cells. Science, 181, 1215-1221 (1973).

3.- Adesnik, M., Darnell, J.E. : Biogenesis and characterization of histone messenger RNA in HeLa cells. J. Mol. Biol., 67, 397-406 (1972).

4.- Huez, G., Marbaix, G., Hubert, E., Leclercq, M., Nudel, U., Soreq, H., Salomon, R., Lebleu, B. Revel, M., Littauer, U.Z. : Role of the polyadenylic segment in the translation of globin messenger RNA in Xenopus oocytes. Proc. Nat. Acad. Sci. U.S. (in press).

5.- Greenberg, J.R. : High stability of messenger RNA in growing cultured cells. Science, 240, 102-104 (1972.)

6.- Perry, R.P., Kelley, D.E. : Messenger RNA turnover in mouse L cells. J. Mol. Biol., 79, 681-696 (1973).

7.- Brachet, J., Ficq, A., Tencer, R. : Aminoacid incorporation into proteins of nucleate and anucleate fragments of sea urchin eggs. Effect of parthenogenetic activation. Exp. Cell Res., 32, 168-170 (1963).

8.- Gross, P.R., Cousineau, G.H. : Effects of actinomycin D on macromolecule synthesis and early development in sea urchin eggs. Biochem. Biophys. Res. Comm., 10, 321-326 (1963).

9.- Sussman, M. : Model for quantitative and qualitative control of mRNA translation in eukaryotes . Nature, 225, 1245-1246 (1970).

10.- Lim, L., Canellakis, E.S. : Adenine rich polymer associated with rabbit reticulocyte messenger RNA. Nature, 227, 710-712 (1970).

11.- Craig, N., Kelley, D.E., Perry, R.P. : Lifetime of the messenger RNAs which code for ribosomal proteins in L-cells. Biochim. Biophys. Acta, 246, 493-498 (1971).

12.- Bard, E., Efron, D., Marcus, A., Perry, R.P. : Translational capacity of deadenylated messenger RNA. Cell, 1, 103-106 (1974).

13.- Williamson, R., Crossley, J., Humphries, S. : Translation of mouse globin messenger RNA from which the polyadenylic acid sequence has been removed. Biochemistry, 13, 703-707 (1974).

14.- Soreq, U., Nudel, R., Salomon, R., Littauer, U. : In vitro translation of poly(A) free rabbit globin mRNA. J. Mol. Biol. (in press).

15.- Gurdon, J.B., Lane, C.D., Woodland, H.R., Marbaix, G. : Use of frog eggs and oocytes for the study of messenger RNA and its translation in living cells. Nature, 233, 177-182 (1971).

16.- Gurdon, J.B., Lingrel, J.M., Marbaix, G. : Message stability in injected frog oocytes : long life of mammalian α and β globin messages. J. Mol. Biol., 80, 539-551 (1973).

17.- Huez, G., Burny, A., Marbaix, G., Lebleu, B. : Release of messenger RNA from rabbit reticulocyte polyribosomes at low concentration of divalent ions. Biochim. Biophys. Acta, 145, 629-636 (1967).

18.- Aviv, H., Leder, P. : Purification of biologically active globin messenger RNA by chromatography on oligothymidylic acid cellulose. Proc. Nat. Acad. Sci., 69, 1408-1412 (1972).

Purification and Biological Properties of a Plasminogen Activator Characteristic of Malignantly Transformed Cells

J.K. Christman and G.Acs
Mount Sinai School of Medicine
Fifth Avenue and 100th Street, New York, N.Y. 10029

The elucidation of tumor-specific functions is a matter of the highest priority. Their biochemical characterization would be expected to give insight into the mechanism of oncogenic transformation as well as provide a basis for selecting specific chemotherapuptic agents. However, almost all metabolic characteristics studied have proven to be shared by both neoplastic and normal cell types. It has not yet been possible to demonstrate a unique function indispensible for the transformed cell. Even the capability of tumor cells for aerobic glycolysis (1) was eventually shown to be totally absent in some tumors and to be present at widely varying levels in the remainder.

Despite the innumerable disappointments that have occurred during the search for tumor specific functions, their potential importance has kept interest in the problem alive. Recently, Reich and his co-workers, aware of the reports that neoplasia is often accompanied by changes in hydrolytic enzymes (2-6), reinvestigated the early observation of Fischer, that primary explants of viral sarcomas from chickens possessed a fibrinolytic activity absent from normal tissue explants (7). Using modern tissue culture methods and radioactive fibrin as a substrate to detect fibrinolysis, they were not only able to reconfirm Fischer's original observation but to quantitate the fibrinolytic activity and demonstrate the occurrence of this function in many other transformed lines (8, 9). Moreover, the fibrinolytic activity was shown to be dependent on two factors, one from the serum (plasminogen) and one released by the cells (plasminogen activator) (10).

There is a close correlation between appearance of the plasminogen activator and transformation of cells. Infection of chicken fibroblasts with a temperature sensitive mutant of Rous sarcoma virus at a temperature which allows viral replication but not transformation is not sufficient to cause production of this enzyme. However, a simple shift of these infected cells to the permissive temperature causes both the appearance of fibrinolytic activity and transformation (8).

In a series of experiments done in collaboration with Dr. Selma Silagi (Cornell University) (11), we have been able to demonstrate that plasminogen activator production is also coordinated with tumorogenicity. By growing a tumorogenic line of mouse melanoma cells in the presence of bromodeoxyuridine, it is possible to convert them to a non-tumorogenic state, and by removing the drug to allow reversal to the original state. We observed a close temporal link between appearance of tumorogenicity and the production of plasminogen activator by these cells. Thus, circumstantial involvement of production of this enzyme with both tumorogenicity and transformation has been established. The ultimate aim, of course, is to determine whether plasminogen activator production is obligatory for either of these processes or simply represents a cellular function derepressed coincidentally with other functions necessary for transformation.

We present here a description of the purification of the plasminogen activator, data relating to its substrate specificity and evidence for its firm association with the plasma membrane of transformed cells.

METHODS

Cell Lines and Culture Conditions

Embryonic fibroblasts were prepared by trypsinization of 10-day old golden hamster embryos. SV-40 transformed hamster cells were from a clone isolated in the laboratories of Dr. E. Reich (8). SV-40 transformants of 3T3 cells obtained from Dr. C. Basilico were isolated in our laboratory by Dr. M. Schonberg. These cells were maintained at 37° in Dulbecco modified (12) Eagle's medium supplemented with 10% fetal bovine serum. L_m cells were maintained in spinner or monolayer in Eagle's medium (13) supplemented as above.

Fibrinolytic Assay

The assay for plasminogen activator is based on release of ^{125}I-labeled fibrinopeptides from ^{125}I-labeled fibrin coated petri dishes (8). The standard 35 mm assay dish is coated with approximately 0.1 mg fibrinogen containing 60 000 cpm. The assay mixture consists of 100 uM Tris-HCl (pH 8.1), 25 μl dog serum or 4 μg dog plasminogen and the activator in a total volume of 1 ml. Aliquots of activator are chosen such that the release of radioactivity is proportional to the amount of enzyme. The activity of any preparation is reported as cpm^{125}I released in 2h into a 1 ml reaction mixture multiplied by total sample volume/volume assayed. Preparation of all assay components has been previously reported, as has the harvesting of serum-free medium containing plasminogen activator released by cells (Harvest fluid, HF (8)).

Plasma membrane preparation

Membranes were prepared according to the method of Atkinson and Summers with only slight modification (14). Cells grown in monolayer were washed three times with ice cold isotonic saline and then harvested by scraping into the same solution. The cells were further washed two times by pelleting at 1000 x g and then allowed to swell in 20 volumes of 10 mM Tris-HCl (pH 8.0), 10 mM sodium azide. After rupturing the cells with 2-3 strokes in a Dounce homogenizer, all other procedures were as described. The membrane fraction was concentrated by pelleting at 6 000 rpm for 10 minutes prior to isopycnic banding in sucrose. The resulting preparation consisted mainly of membrane vesicles contaminated with rough endoplasmic reticulum. Electron microscopic examination revealed no contamination with mitochondria or lysosomes[1].

RESULTS

Purification of the Plasminogen Activator from SV-40 Transformed Hamster Cells

The plasminogen activator of SV-40 transformed hamster cells is a serine protease, irreversibly inhibited by diisopropylfluorophosphate (15). Since it was possible to label the enzyme with radioactive diisopropylfluorophosphate, we could determine its molecular weight as 50 000 and estimate that the specific activity of this protease in the fibrinolytic assay would be $5 \cdot 10^9$ units/mg at homogeneity. As shown below, we were able to achieve this degree of purification. For a typical isolation, 8 l of serum-free medium which had been exposed to confluent monolayers of SV-40 transformed cells (HF) was centrifuged to remove cell debris, acidified to pH 3 and made to 50% of saturation with $(NH_4)_2SO_4$ (31.3 g/100 ml). The resultant pellet was suspended in one-eightieth of the initial volume in 0.05 M glycine-HCl (pH3).

Table 1. Typical Purification of Hamster Plasminogen Activator[2]

Stage	Total Protein** (mg)	Specific*** activity (units/mg protein)	Recovery (%)	Purification (fold)
1 HF	950	$4.2 \cdot 10^5$	--	--
2 50% $(NH_4)_2$ SO_4 suspension	720	$5.5 \cdot 10^5$	100	1.3
3 Supernatant 0	130	$3.2 \cdot 10^6$	102	7.5
4 0.4 m $(NH_4)_2SO_4$ eluate from SP-C25	8.8	$2.8 \cdot 10^7$	62	70
5 Amicon concentrate	1.2	$1.8 \cdot 10^8$	52	400
6 Gel eluate	0.030+	$6.0 \cdot 10^9$	45	14 000

* Based on HF from one week or 8000 cc.
** Lowry et al. (16) determination.
*** cpm/mg released per 2 h, taken from linear concentration range.
\+ Fluorescamine determination (17) (Kindly performed by Dr. S. Stein).

All of the plasminogen activator was soluble at this concentration but only 14% of the contaminating protein was dissolved (Stage 3). A further purification was accomplished with SP-C 25 Sephadex chromatography. Ninety percent of the contaminating protein was removed by elution with solutions of increasing ionic strength (to 0.15 M $(NH_4)_2SO_4$) and the plasminogen activator was eluted in high yield at 0.4M $(NH_4)_2SO_4$. The fractions containing enzyme were combined, made to 1 M urea, desalted and concentrated 25-50 fold on an Amicon PM-10 membrane. Purification at this step resulted from the loss of low molecular weight components and the removal by centrifugation of an inactive precipitate formed during concentration.

The final step employed preparative polyacrylamide gel electrophoresis. The enzyme is stable for more than 72 h in the presence of 0.1% dodecyl sulphate, permitting use of a standard analytical polyacrylamide gel

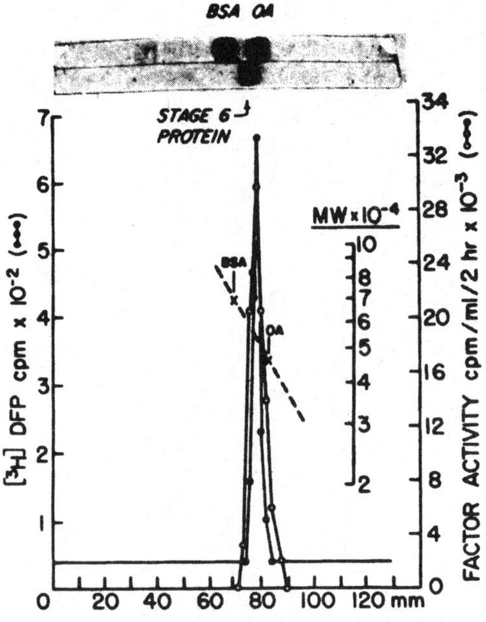

Fig. 1 Coincidence of (^{3}H)DFP inactivated cell factor, cell factor activity and Stage 6 protein on dodecylsulphate polyacrylamide gel electrophoresis. (^{3}H)DFP-labeled factor was co-electrophoresed with active Stage 6 protein. One gel was cut into 1-mm slices; each slice was split in half. One-half was assayed directly for factor activity (O-O) and the other half was digested in 0.5 ml Soluene for 2 h at 70° C. in order to determine (^{3}H)DFP counts (●-●) using a scintillation system with Liquifluor. An identical gel was fixed and stained and split longitudinally. One-half was then sliced and assayed for (^{3}H) DFP counts. The alignment of stained bands and counts was assured by the use of fine wire marker pins during the cutting. The molecular weight was accurately determined by co-electrophoresing (^{3}H)DFP-labeled factor with marker proteins and counting gel slices after staining. Gels were 12% acrylamide, 0.15% bisacrylamide, formed and run in 0.1 M sodium phosphate buffer (pH 7.2) containing 0.02 M EDTA, 2.5 M urea, and 0.1% dodecylsulphate (18). Samples were layered in 30% sucrose, 0.1% dodecylsulphate with pH and ionic strength adjusted to approximate the buffer system. Running time: 15 h at 4 mA/gel (16 mm^2). For protein staining, gels were fixed for 1 h in 12.5% trichloracetic acid, and destained in 12.5% trichloroacetic acid, BSA, bovine serum albumin; OA, ovalbumin.

system (18). The activites can be located after electrophoretic separation by incubating slices or fragments from the polyacrylamide gel in a standard assay mixture. The material electroeluted from the gel region containing plasminogen activator was found to have a specific activity of 14 000 fold higher than the starting material and to behave as a homogenous protein with molecular weight 50 000 on analytical polyacrylamide gels (Fig 1).

The stainable protein coelectrophoresed exactly with detectable plasminogen activator and with enzyme labeled by inactivation with radioactive diisopropylfluorophosphate. After removal of dodecyl sulphate by Dowex AG 1-X2 chromatography (19), the purified enzyme could be isofocused in polyacryamide gels (20). One major stainable band was found to migrate at the same pH and the factor activity, with an isoelectric point of approx. 9.5. A minor stained band was visualized within 9.25 pH units of the major band in the region of detectable activity. These data, cummulatively, are a strong indication that the enzyme has been purified to homogeneity.

We have also been able to demonstrate, utilizing preparations labeled with DFP, that brief exposure to sulphydryl reagents is sufficient to cause an apparent decrease in molecular weight by the enzyme from 50 000 to 25 000 as detected by its migration on dodecyl sulphate polyacrylamide gel electrophoresis. Since similar exposure will irreversibly inactivate the enzyme, this suggests that the plasminogen activator consists of subunits linked by disulphide bridges and that these subunits are inactive. The subunit containing the active site (DFP label) has a molecular weight of 25 000.

Substrate Specificity

The plasminogen activator is so named and owes its discovery to the fact that it is an extremely efficient activator of plasminogen. Plasmin, once formed, can be detected by most assays for non-specific proteolytic action. However, the fibrinolytic assay described above (Methods) is one of the most sensitive available and allows the indirect detection of less than 2.10^{-14} moles of activator produced by transformed hamster cells.

In an attempt to discover other functions for this enzyme or other substrates handled with as high an efficiency as plasminogen, we tested its ability to act as an activator of two other zymogens and its effect on membranes.

Chymotrypsinogen: Two assays for chymotrypsin were used: a) hydrolysis of benzoyl-L-tyrosine ethyl ester (21) with conditions such that activation of 1% of the chymotrypsinogen could be detected and b) direct chymotryptic digestion of radioactive fibrin under conditions such that activation of 0.1% of the chymotrypsinogen was detectable. Even with a molar ratio of plasminogen activator to substrate twice as high as that used for plasminogen activation and 48 h incubation, no conversion of chymotrypsinogen to chymotrypsin was observed.

Trypsinogen: Assay was by direct tryptic digestion of radioactive fibrin under conditions where activation of less than 5% of the trypsinogen in the assay mixture would be detected. With the same ratio of plasminogen activator to enzyme used for plasmin formation and with 4 h incubation at 37°, no conversion of trypsinogen to trypsin occurred.

Cell membranes: The concept that the surface of a transformed cell differs from its normal counterpart has received wide acceptance and has stimulated a large amount of research aimed at quantitating differences between the protein and carbohydrate moieties of the plasma membranes of normal and transformed cells (22-27). Since it has also been shown that mild proteolytic treatments of normal cells can affect their growth properties (28, 29), the proteins of the plasma membrane were a likely substrate for the plasminogen activator. L_m, SV-40 transformed hamster cells and SV-40 transformed 3T3 cells were labeled *in vivo* with S^{35}-methionine. The isolated plasma membranes from these cells were incubated with purified plasminogen activator at 37° for 2 h. Less than 1% of the labeled protein was converted to TCA soluble fragments by such treatment. If the membranes were subsequently disrupted with dodecyl sulphate and examined by chromatography on Sephadex G-100, no breakdown of large proteins into polypeptides could be detected. Autoradiography of preparations electrophoresed on 5% polyacrylamide-SDS slab gels (18) revealed that exposure to the enzyme did not affect the molecular weight distribution of the labeled proteins. However, if both plasminogen and the activator were added to the membranes, a general digestion of proteins occurred, which could be easily detected by either method. Concurrently, approximately 15% of the labeled protein was hydrolysed to TCA soluble fragments (Table 2). Thus, the only action of the enzyme on membranes appears to be indirect, through its activation of plasminogen.

Table 2. Digestion of Plasma Membrane Proteins Derived from Normal and Transformed Cells

	% counts released as TCA soluble
SV-40 Hamster	
+ plasminogen activator	0.2
+ plasminogen	10
+ plasminogen and activator	15
SV-40 3T3	
+ plasminogen activator	0
+ plasminogen	11
+ plasminogen and activator	17
L_m	
+ plasminogen activator	0
+ plasminogen	1
+ plasminogen and activator	16

Purified plasma membranes labeled <u>in vivo</u> with ^{35}S-methionine were incubated in 0.5 ml 0.1 M Tris-HCl at 37°. $4 \cdot 10^{-11}$ moles of plasminogen or 10^{-12} moles of purified activator were present where indicated. After 2 h the reaction was terminated by bringing the mixture to 10% with TCA. % counts released = 100· (cpm in Millipore filtrate/cpm in filtrate + cpm in retentate). Radioactivity in filtrates was determined directly in Aquasol ® and on dried filters in Liquifluor ®. Background release of radioactivity by membranes incubated with no additions has been subtracted (less than 1% total).

Plasminogen Activator Bound to the Plasma Membrane

Our plasma membrane preparations were alike in their susceptibility to digestion by plasmin, i.e.: incubation with plasminogen plus plasminogen activator solubilized approximately the same percentage of radioactive proteins. The preparations were, however, markedly different when incubated with plasminogen alone (Table 2).

Radioactive proteins in the plasma membrane preparations from SV-40 transformed hamster or 3T3 cells were hydrolyzed to TCA soluble peptides by plasminogen alone almost as efficiently as when plasminogen activator was added. L_m cell membranes were totally insensitive to incubation with plasminogen. Since L_m cells, like embryonic hamster cells, have no detectable capacity to digest fibrin when grown in monolayers on radioactive fibrin plates under conditions where both SV-40 hamster and SV-40 3T3 cells are very active, we felt that this result was a strong indication that transformed cells not only release plasminogen activator but have it tightly bound to their membranes.

As might be expected, plasma membrane preparations from SV-40 transformed cells can serve as source of plasminogen activator in the standard fibrinolytic assay while L_m preparations cannot (Table 3). Correcting for the physical loss of membranes during the purification procedure, the amount of activity in the membrane preparation from 10^7 cells is on the same order of magnitude as the amount of activity released into the medium by this number of cells in 18 h.

In view of the usual extensive washing of the transformed cells before they are lysed, this would indicate that the presence of plasminogen activator in the membrane fraction is not fortuitous. Furthermore, when L_m cells were mixed with a 10-fold excess of enzyme immediately before lysis (Methods), the resulting purified membrane preparation retained less than 1% of the starting activity.

Although our membrane preparations were not contaminated with mitochondria or lysosomes, they did contain rough endoplasmic reticulum. Thus, simply showing that the preparations contain plasminogen activator was not sufficient to prove that the enzyme is present in the plasma membrane. We have been able to demonstrate the presence of the enzyme in an indirect way by exposing plasminogen in Earle's salts (30) to thoroughly washed intact cells for brief periods of time and then assaying for plasmin with the standard assay plates. Detectable amounts of plasmin were formed with less than 15 min exposure to intact SV-40 hamster cells (Fig. 2, ●—●). There was no detectable release of enzyme to the medium during this time

Table 3. Plasminogen Activator in Isolated Plasma Membranes

	Units/ 10^7 cells
SV-40 Hamster	$5 \cdot 10^5$
SV-40 3T3	$3 \cdot 10^4$
L_m	undetectable

Membranes from 10^5-10^6 cells were incubated in the standard assay mixture with $4 \cdot 10^{-11}$ moles of plasminogen for 2 h at 37°. Unit=cpm released/ml/2h in the linear concentration range. Background release by plasminogen alone has been subtracted.

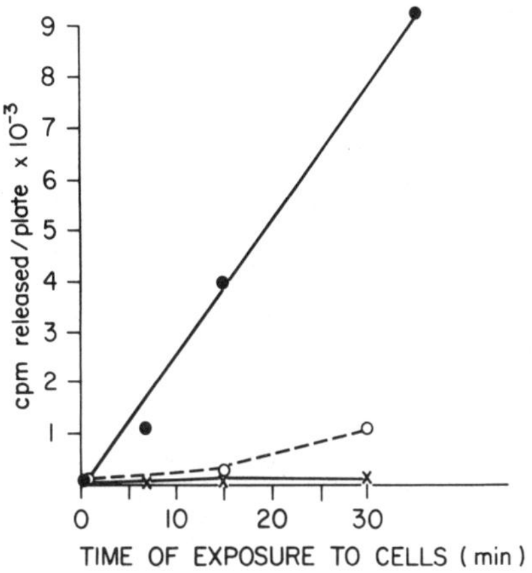

Figure 2. Activation of plasminogen by intact cells. Confluent monolayers of SV-40 hamster cells growing on 60mm petri dishes, were washed 6 times with 10 ml cold Earle's salts solution. They were then overlayered with 3 ml Earle's salts containing 4 µg/ml plasminogen and incubated at 37° for the indicated time. The solution was pipetted off and any cells or debris removed by centrifugation. One ml was then placed on a standard assay plate and incubated at 37° for 1 h to test for plasminogen activation (●——●). Control incubations to test for release of plasminogen activator into the medium were incubated with Earle's salts. Plasminogen (4 µg/ml) was added and the solutions were either incubated for 30 min at 37° before assay (o---o) or assayed immediately (x——x).

(x—x, o---o). In accordance with the observation that L_m cells have no capacity to digest fibrin, no plasmin formation occurs when plasminogen is exposed to these intact cells for as long as two h.

We interpret these results as indicating that a significant amount of plasminogen activator is bound to the cell surface of transformed cells in such a way that its active site is exposed and functional. The possibility that plasminogen in some unexplained manner aids the release of activator from these cells has not been ruled out, although we feel that this is unlikely. It is also unlikely that the enzyme released from the cells into Earle's salts is unstable since enzyme release can be detected in 1-2 h. Purified enzyme is unaffected by incubation under the same conditions.

We are in the process of comparing the membrane-bound plasminogen activator with the released enzyme and are attempting to devise additional direct proofs locating the activator in the plasma membrane.

DISCUSSION

Based on data from both Dr. Reich's group and our own, it has become clear that almost all mammalian cell types, after transformation, are capable of producing a plasminogen activator. Since it appears to make no difference whether the transforming agent is an RNA virus, a DNA virus or a chemical carcinogen, it seems reasonable to assume that the information for the production of the enzyme is contained in the host cell genome. Its appearance is then a consequence of the generalized derepression of cellular functions which occur upon transformation. The major question is whether the derepression of the information for activator production is coincidental or whether, in fact, production of this enzyme is an essential function for transformation. We hope that with our simple method for preparing large amounts of this enzyme in a pure form, antibodies to the protein can be prepared. If such antibodies prove to be specific inhibitors of the enzyme, it should be possible to obtain some information about the importance of the plasminogen activator to transformation, the maintenance of the transformed state and the growth and dissemination of tumors.

SUMMARY

A plasminogen activator characteristic of malignantly transformed cells has been purified to apparent homogeneity from SV-40 hamster cells. It is a serine protease with a molecular weight of 50 000, consisting of subunits linked by disulphide bridges. The subunit containing the active site has a molecular weight of 25 000.

Several lines of evidence are presented which indicate that the enzyme is, in all probability, firmly associated with the plasma membrane of transformed cells as well as being released into the medium.

[1] Electron micrographs were prepared by Dr. S. Silverstein of Rockefeller University.

[2] Figure 1 and Table 1 have been previously published (15).

We gratefully acknowledge the technical assistance of Mr. Cornelius Whalen, Mrs. Hannah Klett and Miss Tina Parler. Supported by USPHS (CA16890) and the American Cancer Society (NP-36K).

Bibliography

1. Warburg, O.H.: Über den Stoffwechsel der Tumoren. (1926) J. Springer, Berlin.

2. Kazakova, O.V., Orckhovich, V.N.: A neutral proteinase from the tissues of a transplantable rat sarcoma. Biokhimiya 34, 73-77 (1969).

3. Taylor, J.C., Hill, D.W., Rogolsky, M.: Detection of caseinolytic and fibrinolytic activities of BHK-21 cell strains. Exp. Cell Res. 73, 422-428 (1972).

4. Bosman, H.B.: Elevated glycosidases and proteolytic enzymes in cells transformed by RNA tumor viruses. Biochim. Biophys. Acta 264, 339-343 (1972).

5. Schnebli, H.P.: A protease-like activity associated with malignant cells. Schweiz. Med. Wochenschr. 102, 1194-1196 (1972).

6. Rubin, H.: Overgrowth stimulating factor released from Rous sarcoma cells. Science 167, 1271-1272 (1970).

7. Fischer, A.: Beitrag zur biologie der gewebezellen. Eine vergleichend biologische studie der normalen und malignen gewebezellen in vitro. Arch. Entwicklungsmech. Org. (Wilhelm Roux) 104, 210-219 (1925).

8. Unkeless, J., Tobia, A., Ossowski, L., Quigley, J.P., Rifkin, D.B., Reich, E.: An enzymatic function associated with transformation of fibroblasts by oncogenic viruses (I). J. Exp. Med. 137, 85-111 (1973).

9. Ossowski, L, Unkeless, J., Tobia, A., Quigley, J.P., Rifkin, D.B., Reich, E.: An enzymatic function associated with transformation of fibroblasts by oncogenic viruses (II). J. Exp. Med. 137, 112-126 (1973).

10. Quigley, J., Unkeless, J.: Fibrinolysin T: Purification and characterization of serum and cellular components. Fed. Proc. 32, 851 Abs. (1973).

11. Christman, J.C., Silagi, S., Acs, G.: Effect of bromodeoxyuridine on production of plasminogen activator by B_5 59 mouse melanoma. (Ms. in preparation).

12. Dulbecco, R., Freeman, G.: Plaque production by the polyoma virus. Virology 8, 396-397 (1959).

13. Eagle, H.: Amino acid metabolism in mammalian cells. Science 130, 432-437 (1959).

14. Atkinson, D.H., Summers, D.F.: Purification and properties of HeLa plasma membranes. J. Biol. Chem. 246, 5162-5175 (1971).

15. Christman, J.K., Acs, G.: Purification and characterization of a cellular fibrinolytic factor associated with oncogenic transformation: The plasminogen activator from SV-40 transformed hamster cells. Biochim. Biophys. Acta 340, 339-347 (1974).

16. Lowry, O.H., Rosenbrough, N.J., Farr, A.L., Randall, R.J.: Protein measurement with the folin-phenol reagent. J.Biol. Chem. 193, 265-275 (1951).

17. Udenfriend, S., Stein, S., Böhlen, P., Dairman, W., Leimgruber, W., Weigele, M.: Fluorescamine: A reagent for assay of amino acids, peptides, proteins and primary amines in the picomole range. Science 178, 871-872 (1972).

18. Summers, D.F., Maizel, J.V., Darnell, J.E.: Evidence for virus-specific noncapsid proteins in poliovirus-infected HeLa cells. Proc. Natl. Acad. Sci., U.S. 54, 505-513 (1965).

19. Weber, K., Kuter, D.: Reversible denaturation of enzymes by sodium dodecyl sulfate. J. Biol. Chem. 246, 4504-4509 (1971).

20. Wellner, D.: Electrofocusing in gels. Anal. Chem. 43, 59A-65A (No. 10) (1971).

21. Hummel, B.C.W.: A modified spectrophotometric determination of chymotrypsin, trypsin and thrombin. Can. J. Biochem. Physiol. 37, 1393-1399 (1959).

22. Wickus, G.G., Robbins, P.L.: Plasma membrane proteins of normal and Rous sarcoma virus-transformed chick-embryo fibroblasts. Nature 245, 65-69,(1973).

23. Greenberg, C.S., Glick, M.C.: Electrophoretic study of the polypeptides from surface membranes of mamalian cells. Biochem. 11, 3680-3685 (1972)

24. Hynes, R.O.: Alteration of cell surface proteins by viral transformation and by proteolysis. Proc. Natl. Acad. Sci., US 70, 3170-3174 (1973).

25. Perdue, J.F., Kletzien, R., Miller, K.: The isolation and characterization of plasma membranes from cultured cells. I. The chemical composition of membrane isolated from uninfected and oncogenic RNA virus-converted chick embryo fibroblasts. Biochim. Biophys. Acta 266, 505-510 (1972).

26. Wu, H.C., Meezan, E., Black, P.H., Robbins, P.W.: Comparative studies on carbohydrate-containing membrane components of normal and virus transformed mouse fibroblasts. Biochem. 8, 2509-2517 (1969).

27. Ohata, N., Pardee, A.B., McAuslan, B.R., Burger, M.M.: Sialic acid contents and controls of nor-

mal and malignant cells. Biochim. Biophys. Acta 158, 98-102 (1968).

28. Burger, M.M.: Proteolytic enzymes initiating cell division and escape from contact inhibition of growth. Nature 227, 170-171 (1970).

29. Sefton, B.M., Rubin, H.: Release from density dependent growth inhibition by proteolytic enzymes. Nature 227, 843-845 (1970).

30. Earle, W.: Production of malignancy in vitro. Natl. Cancer Inst. 4, 167-173 (1943).

Interactions of Troponin Subunits Underlying Regulation of Muscle Contraction by Ca Ion: A Study on Hybrid Troponins

S. EBASHI
Department of Pharmacology, Faculty of Medicine, and Department of Physics, Faculty of Science, University of Tokyo, Tokyo JAPAN

INTRODUCTION

The findings that the fragmented sarcoplasmic reticulum can accumulate Ca ion in the presence of ATP (1-3) and that a minute amount of Ca ion regulates the interaction of myosin and actin (1,4,5) established the physiological role of Ca ion in muscle contraction (6). The introduction of Ca ion into the field of muscle biochemistry produced a new impetus to reinvestigate the nature of myosin and actin (7), the main contractile proteins, which had been long believed to represent the whole contractile system from the physiological point of view. This led us to the discovery of a third factor in the contractile system other than myosin and actin (8). The factor, tentatively named "native tropomyosin" because of its physicochemical resemblance to tropomyosin, was later shown to be composed of tropomyosin and a new structural protein, troponin, and to be located in the grooves of actin strands with about 400 Å periodicity (see refs. 6 and 9).

There is now general agreement that troponin is composed of three subunits, the tropomyosin-binding component (troponin T or TN-T), the inhibitory component of the myosin-actin interaction (troponin I or TN-I), and the Ca-binding component (troponin C or TN-C)(10; cf. ref. 11).

Attention is now focused on the problem as to what kinds of Ca-dependent interactions of the three subunits would operate in the mechanism underlying the physiological contractile processes. Useful suggestions along this line have been provided from a spin-label study (11-15). Using sulfhydryl label, the interactions, TN-T - TN-C and TN-I - TN-C, are both shown to be Ca-dependent, the latter's Ca-dependence being in agreement with that of contraction. Furthermore, the study using a tyrosyl label indicates that in the absence of Ca ion, the interaction of TN-T and TN-C prevails over that of TN-I and TN-C, and vice verea in the presence of Ca ion (15). It thus appears that the fundamental process underlying regulation of the contractile mechanism is the Ca-dependent competition between the two interactions, TN-T - TN-C and TN-I - TN-C (11, 15).

The success in the separation of cardiac troponin into its three components enabled us to carry out the hybrid experiments using subunits of troponin derived from cardiac or skeletal muscle (16,17). Unexpectedly, the Ca-sensitizing activities of reconstituted hybrid troponins are distinctly different from hybrid to hybrid (11). A hybrid troponin showed a better activity than did natural troponin, and some hybrid troponins exhibted only very low activities, indicating the existence of unique fitness of each subunit to one another.

This article reports physiological activities of such hybrid troponins and discusses some possible interpretations as to whether and how the varied activities of such hybrid troponins conform with the above concept.

METHODS AND MATERIALS

Preparation of troponin and its subunits. Rabbit skeletal and bovine cardiac troponins were prepared by the methods reported previously (16-18). Separation of troponin into three subunits was performed by the method routinely carried out in our laboratory. Skeletal troponin, about 40 mg/ml in 6M urea containing 0.1M NaCl and 2mM $NaHCO_3$, was applied to the column of SE-Sephadex G 50 (Pharmacia). TN-I and TN-C passed through the column, leaving TN-T in the column. The mixture of TN-I and TN-T thus obtained, about 20 mg/ml, was applied to the column of DEAE-Sephadex G 25 (Pharmacia) equilibrated with 6M urea containing 0.1M NaCl and 2mM $NaHCO_3$. Most of the TN-I passed through the column, whereas TN-C was adsorbed to the gel; the latter was eluted by 6M urea containing 0.4M NaCl and 2mM $NaHCO_3$. The TN-T retained by the SE-Sephadex column was also eluted by 6M urea containing 0.4M NaCl and 2mM $NaHCO_3$. The protein concentrations of the final preparations were usually about 10 mg/ml or more.

The separation of bovine cardiac troponin was made by the same method as the above except that 0.05M NaCl was used in place of 0.1M NaCl.

Preparation of natural and hybrid troponins. Equimolar amounts of the three subunits were mixed; their molecular weights were assumed to be as follows: skeletal TN-T, TN-I and TN-C, 38,000, 22,000 and 17,000, respectively; cardiac TN-T, TN-I and TN-C, 40,000, 29,000 and 18,000, respectively (17). In some cases each component was added separately to the mixture which would subsequently be subjected to the superprecipitation experiments, but the results were essentially the same as in the premixed cases. Since the cardiac preparations were significantly contaminated by non-specific proteins (see Fig. 1 in ref. 16), the concentration each subunit determined by chemical reaction was corrected by taking the patterns of sodium dodecyl sulfate polyacrylamide gel electrophoresis into consideration.

Preparation of other proteins. Preparation of other proteins followed the methods described in a previous paper (18,19).

Estimation of Ca-sensitizing activities of troponin. The superprecipitation method was used for measuring the Ca-sensitivity of the actomyosin system induced by troponin (19). The Ca-sensitivity was defined as the ratio of the time required to reach half maximum turbidity in the presence of 10×10^{-7}M Ca ion to that required in the presence of 8×10^{-6}M Ca ion at pH 6.8 (for details, refer to Fig. 1 in ref. 19 or Fig. 6 in ref. 11). Unless otherwise specified, the test solution contained 0.04M KCl, 1mM $MgCl_2$, 0.02M Tris-maleate (pH 6.8), 0.1mM Ca-GEDTA (glycoletherdiaminetetraacetic acid, EGTA) buffer, 0.5 mg/ml troponin-tropomyosin-free myosin B, 0.35 μM tropomyosin, 0.32 μM troponin or its subunits and 0.5mM ATP; the binding constant of GEDTA for Ca at pH 6.8 was assumed as $5 \times 10^5 M^{-1}$ (20). For other methods see previous papers (18,19) or legends to tables.

RESULTS

The Ca-sensitizing activities of reconstituted troponins were listed in Table I. The activity of $T_sC_sI_s$ (T, C, or I represents TN-T, TN-C or TN-I, respectively; suffix c or s indicates the origin of the subunit, cardiac or skeletal, respectively) was almost the same as that of the original skeletal troponin, whereas that of $T_cC_cI_c$ was slightly but significantly lower than the original cardiac troponin. Therefore, the activities of hybrid troponins containing cardiac subunits must be accepted with some reservation (as will be shown below, T_c may be fully active, so that either I_c or C_c may be responsible for the lower activity). Interestingly, the hybrid combination $T_cC_sI_s$ showed far better activity than the natural combination $T_cC_cI_c$ or $T_sC_sI_s$. Other hybrid combinations showed more or less

Table I

Ca-sensitizing activities of various kinds of combinations of troponin subunits.

Combinations			Ca-sensitivities
TN-T	TN-C	TN-I	
c	c	c	15.4 ± 2.1 (6)
c	c	s	7.4 ± 1.2 (4)
c	s	c	8.6 ± 0.9 (4)
c	s	s	32.4 ± 3.4 (5)
s	c	c	3.4 ± 0.7 (4)
s	c	s	4.6 ± 0.6 (4)
s	s	c	2.6 ± 0.3 (4)
s	s	s	20.8 ± 1.7 (8)

c or s indicates the cardiac or skeletal origin of the subunit, respectively. The experimental conditions and the definition of Ca-sensitivity are described in METHODS AND MATERIALS. ± x indicates standard error of means. Figures in parentheses show the number of experiments.

lower activities that the natural combinations; it is worthy of notice that the activities of $T_S C_C I_C$ and $T_S C_C I_S$, especially the latter, were extremely low, about one fifth those of the natural combinations or lower.

As described in INTRODUCTION, there is a strong indication that the whole activity of troponin is explained by the competition between the two interactions, TN-T - TN-C and TN-C - TN-I (15). If this is so, the varied activities of these reconstituted troponins should be factorized into the two interactions.

The physiological amounts of TN-C and TN-I, i.e., 1 : 7 in their molar ratios to actin, cannot significantly sensitize the actomyosin system unless aided by TN-T, but if their amounts are increased, say, several times the physiological ratio, a considerable degree of Ca-

Table II

Ca-sensitizing activities of various kinds of combinations of TN-C and TN-I.

Combinations		Ca-sensitivities
TN-C	TN-I	
c	c	5.6 ± 0.9 (3)
c	s	3.6 ± 0.7 (3)
s	c	2.6 ± 0.6 (3)
s	s	9.2 ± 1.4 (3)

Experimental conditions were essentially the same as those for Table I except that 1.5μM of each subunit was used. For other see the legend to Table I.

sensitizing effect can be observed (see Fig. 6 in ref. 21). It is quite reasonable to assume that the interaction of TN-C and TN-I underlying the physiological processes is reflected partially in this phenomenon. Therefore we adopted this effect as a criterion of the efficacy of various combinations of TN-C and TN-I.

In Table II, the Ca-sensitizing effects of C_cI_c, C_cI_s, C_sI_s and C_sI_s in amounts five times the physiological ratio are noted. It was found that the combination C_sI_c showed particularly low activity. This may provide an explanation for the low activity of the hybrid troponin $T_sC_sI_s$.

Table III

Ca-dependence of the solubility of various kinds of combinations of TN-T and TN-C.

Combinations		pCa*
TN-T	TN-C	
c	c	4.9
c	s	4.4
s	c	7.0
s	s	6.1

* indicates the Ca concentrations to give half maximum turbidity of the solution, containing TN-T and TN-C, 100µg/ml of each, in addition to 0.02M KCl, 1mM $MgCl_2$, 0.02M Tris-maleate buffer (pH 6.8) and Ca-GEDTA buffer. The turbidity of TN-I without TN-C was taken as the maximum. For others see the legend to Table I.

As regards to interaction of TN-T and TN-C, two kinds of Ca-dependent interactions have been reported. One is the result of spin-label studies (12,15), but so far no quantitative information as to the cardiac troponin is available. The other is the Ca-dependence of the solubility of the complex of TN-T and TN-C (21): TN-T is not soluble at low ionic strengths but is solubilized by TN-C in the presence of Ca ion (as regards the actual Ca-dependence, see Fig. 8 in ref. 21). Although the physiological significance of this phenomenon is not yet well understood, it may be somehow related to the Ca-dependent affinity between TN-C and TN-T. In Table III Ca ion concentrations which give half maximum turbidities of the various combinations of TN-T and TN-C are compared.

It is interesting that the solubility of the combination T_SC_C, which forms a hybrid of very low activity with I_C, shows the highest sensitivity to Ca ion and the combination T_CC_S, of which the hybrid with I_S exhibits the highest activity, is shown to be most insensitive. This suggests that the solubilized state of the TN-T - TN-C complex represents a state in which the affinity of both subunits is too strong to be overcome by the affinity of TN-I for TN-C, and therefore the higher sensitivity of the solubility to Ca ion indicates the low degree of regulatory function of the whole troponin. If this is true, the Ca ion concentrations which give half maximum turbidities of the four combinations should be somehow related to their efficacies as a part of the regulatory system.

As shown in Table IV, the activities of four kinds of TN-T - TN-C complexes were assessed under the assumption that the troponin function could be factorized into two interactions, TN-T - TN-C and TN-C - TN-I.

The values thus estimated (c) were fairly in parallel with the Ca-dependence of the solubilities of the complexes expressed in relative values (d). At present we cannot decide whether this parallelism is a mere coincidence, or indicates the importance of this phenomenon, but it is attractive to assume that the affinity between TN-T and TN-C should balance with the affinity between TN-C and TN-I.

Table IV

Factorized Ca-sensitizing activities of viriour combinations of troponin subunits.

TN-T	TN-C	TN-I	a (T-C-I)	b (C-I)	a/b (T-C)	c (T-C)	d (T-C)
c	c	c	0.81	0.68	1.19	1.12	2.2
c	c	s	0.48	0.46	1.04*		
c	s	c	0.54	0.34	1.61	1.49	2.7
c	s	s	1.36	(1.00)	1.36*		
s	c	c	0.24	0.68	0.35*	0.52	0.1
s	c	s	0.32	0.46	0.68		
s	s	c	0.19	0.34	0.56*	0.78	(1.0)
s	s	s	(1.00)	(1.00)	(1.00)		

a and b: The Ca-sensitivities in Tables I and II were converted to relative values, taking those of $T_SC_SI_S$ and T_SI_S as unity, respectively and then linearized using the equation,

$$y=(x-1)^{2/3}$$

where x is relative Ca sensitivity and y is linearized relative activity (see the legend to Fig. 6 in ref. 11). c: Averages of the two values on the left which represent the same combinations of TN-T and TN-C. d: The pCa's in Table III were expressed in relative values, taking the values of T_SC_S as unity. * see the text.

If we look at the data in the third column from the left (a/b), there is a considerable difference between the two values ascribed to the same combination of TN-T and TN-C. It deserves attention that all four combinations containing heterozygous combination of TN-T and TN-I (with asterics) show lower activities than their corresponding combinations containing homozygous combination of TN-T and TN-I (without asterics); the difference is more marked in the cases of T_SI_C than those of T_CI_S. This may indicate the presence of an interaction of TN-T and TN-I, or a kind of unfitness between these two subunits, though it is not so prominent as that between TN-T and TN-C or TN-C and TN-I.

DISCUSSION

Since there is reasonable evidence for the concept that the activity of troponin is mainly dependent on the competition between the two interactions, TN-I and TN-C, and TN-C and TN-I (15), the results shown in Table IV may support the idea that the Ca-dependent solubility of the TN-T - TN-C complex indicates the degree of affinity between two subunits. There is no doubt that the affinity between TN-T and TN-C is an indispensable part of the mechanism of troponin function, and without this, TN-C and TN-I cannot exhibit their Ca-dependent interactions under physiological conditions. However, it is conceivable that if this affinity is too strong, it may eventually interfere with the physiological processes. Thus the appropriate balance between the affinity of TN-C for TN-T and that for TN-I seems a prerequisite for the regulatory function of troponin.

Although there is no doubt that the troponin function is chiefly based on the two interactions mentioned above, the results of this study also suggest the existence of an interaction between TN-T and TN-I. This is in accordance with the results of the spin-label study, indicating a slight but significant interaction between TN-T and TN-I (11,15)

Based on the observations made in the article as well as those in previous papers, a scheme for the troponin mechanism is presented in Fig. 1. The crucial point of this scheme is that TN-I and TN-T compete for a site on TN-C, the former predominating over the latter in the absence of Ca ion so that TN-I becomes available for tropomyosin; in the presence of Ca ion, TN-C shows more affinity for TN-I and abolishes the inhibitory action of the latter.

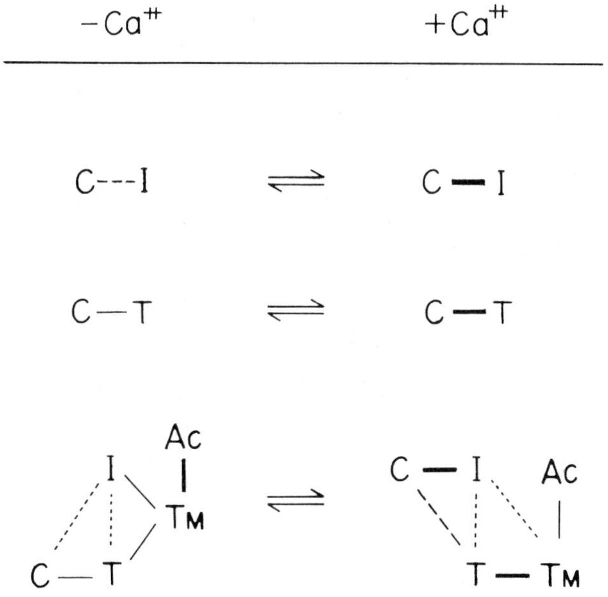

Fig. 1. Diagram illustrating possible interactions of troponin subunits by which Ca ion controls the structure of the tropomyosin(TM)-actin(AC) complex.

SUMMARY

Reconstituted natural and hybrid combinations of troponin subunits derived from either cardiac or skeletal muscle showed a wide variety of Ca-sensitizing activities. One of the hybrid troponins showed a better activity than the original troponins, and some of them very low activities. Analysis of such diverse activities of reconstituted troponin, using the Ca-sensitizing action of a large amount of the complex of TN-C and TN-I without TN-T and the Ca-dependent solubility of the complex of TN-T and TN-C as the indexes of the quality of the interaction between TN-C and TN-I and that between TN-T and TN-C, respectively, indicated that the activity of troponin depended mainly on the balance between the affinities of TN-C for TN-T and for TN-I.

ACKNOWLEDGEMENT

I wish to express my cordial and sincere thanks to Prof. F. Lipmann for his kind advice and warm encouragement since my first visit to his laboratory, where I could spend the most fruitful and enjoyable days.

My thanks are also due to Dr. F. Ebashi, Miss. T. Fujii and Miss H. Kodeki for their cooperation in this work. This project was supported in part by the research grant from Muscular Dystrophy Association of America, Inc., the Ministry of Education, Japan, the Ministry of Health and Welfare, Japan, No. 216, the Iatrochemical Foundation, Toray Science Foundation, and Mitsubishi Foundation.

REFERENCES

1. Ebashi, S.: Calcium binding and relaxation in the actomyosin system. J. Biochem. 48, 150-151 (1960).
2. Ebashi, S. and Lipmann, F.: Adenosine triphosphate-linked concentration of calcium ion in a particulate fraction of rabbit muscle. J. Cell Biol. 14, 389-400 (1962)
3. Hasselbach, W. und Makinose, M.: Die Calciumpumpe der 'Erschlaffungsgrana' des Muskels und ihre Abhängigkeit von der ATP-spaltung. Biochem. Z. 333, 518-528 (1961).
4. Weber, A.: On the role of calcium in the activity of adenosine 5′-triphosphate hybrolysis by actomyosin. J. Biol. Chem. 234, 2764-2769 (1959).
5. Ebashi, S.: Calcium binding activity of vesicular relaxing factor. J. Biochem. 50, 236-244 (1961).
6. Ebashi, S. and Endo, M.: Calcium ion and muscle contraction. Progr. Biophys. Mol. Biol. 18, 123-183 (1968).
7. Weber, A. and Winicur, S.: The role of calcium in the superprecipitation of actomyosin. J. Biol. Chem. 236, 3198-3202 (1961).
8. Ebashi, S.: Third component participating in the superprecipitation of "natural actomyosin". Nature 200, 1010-1011 (1963).
9. Ebashi, S., Endo, M. and Ohtsuki, I.: Control of muscle contraction. Quart. Rev. Biophys. 2, 351-384 (1969).
10. Greaser, M.L. and Gergely, J.: Reconstitution of troponin activity from three protein components. J. Biol. Chem. 246, 4226-4233 (1971).
11. Ebashi, S.: Regulatory mechanism of muscle contraction with special reference to the Ca-troponin-tropomyosin system. Essays in Biochemistry, 10, (1974) in press.
12. Ebashi, S., Ohnishi, S., Abe, S. and Maruyama, K.: A spin-label study on calcium-induced conformational changes of troponin components. J. Biochem. 75, 211-213 (1974)
13. Ebashi, S., Ohnishi, S., Abe, S. and Maruyama, K.: Ca-dependent interaction of troponin components as the basis of the control mechanism by Ca ion. In

Calcium binding proteins. (Drabikowski, W. and Carvalho, A.P., eds.) Polish publishers P.W.H., in press.

14. Potter, J.D., Seidel, J.C., Leavis, P.C., Lehrer, S.S. and Gergely, J.: Interaction of Ca^{2+} with troponin. ibid., in press.

15. Ohnishi, S., Abe., Maruyama, K. and Ebashi, S.: An ESR study on the calcium-induced conformational changes of troponin components. J. Biochem. in press.

16. Tsukui, R. and Ebashi, S.: Cardiac troponin. J. Biochem. 73, 1119-1121 (1973).

17. Ebashi, S., Masaki, T. and Tsukui, R.: Cardiac contractile proteins. The Myocardium, Adv. Cardiol. 12, pp. 59-69 (Karger, Basel 1974).

18. Ebashi, S., Wakabayashi, T. and Ebashi, F.: Troponin and its components. J. Biochem. 69, 441-445 (1971).

19. Ebashi, S., Kodama, A. and Ebashi, F.: Troponin, I. Preparation and physiological function. J. Biochem. 64, 465-467 (1968).

20. Ogawa, Y.: The apparent binding constant of glycoletherdiaminetetraacetic acid for calcium at neutral pH. J. Biochem. 64, 255-257 (1968).

21. Ebashi, S., Ohtsuki, I. and Mihashi, K.: Regulatory proteins of muscle with special reference to troponin. Cold Spring Harbor Symp. Quant. Biol. 37, 215-223 (1972).

Studies on the Binding of Acylaminoacyl-Oligonucleotide to Rat Liver 60S Ribosomal Subunits and Its Participation in the Peptidyltransferase Reaction

Bruce Edens, Herbert A. Thompson and Kivie Moldave

Department of Biological Chemistry, University of California, Irvine, California, 92664.

Studies in a number of laboratories demonstrated that substrates containing an oligonucleotide sequence from the 3'-terminus of tRNA, esterified with an N-blocked amino acid, could substitute for peptidyl-tRNA in the alcohol-dependent peptidyltransferase reaction with prokaryote ribosomes or the large ribosomal subunit (1). Transpeptidation with such "fragments" has been used to study the minimal structural requirements of substrates of the "donor" or "P" and "acceptor" or "A" sites (1-6); for example, CpCpA-acylaminoacyl was found to be an active substrate, and more recently, pA-fMet has been shown to participate in transpeptidation under certain conditions. Studies on the structural requirements for "A" site substrates indicate that the furanose ring must be intact and that the ester bond must be at the 3'-position (7,8). The "fragment" reaction has also been used to examine the effects of various antibiotics on the binding of substrates to the two reactive sites, or on the peptidyltransferase "catalytic center" directly (3,6,9), and to examine the role of ribosomal protein components required for peptidyltransferase (10) such as the L-11 protein (11).

The formation of acetylphenylalanyl-puromycin, from acetylphenylalanyl-tRNA and puromycin, has been demonstrated with purified rat liver 60S ribosomal subunits (12); the Vmax of the reaction was increased about 5-fold when 60S subunits were incubated with poly(U) before they were used in the alcohol-dependent peptidyltransferase reaction. Similar incubation of 60S subunits with an equivalent quantity of 40S subunits led to a severe decrease in the rate of the reaction. Recent studies (13) indicated that the effects of poly(U) and of 40S subunits were not explained by an effect on the amount of acetylphenylalanyl-tRNA that was bound to the 60S subunit. It was of interest, therefore to determine the response of the "donor" site fragment acetylphenylalanyl-oligonucleotide, which lacks most of the tRNA including the anticodon region, with respect to poly(U) and 40S subunits. This communication describes the binding of acetylphenylalanyl-oligonucleotide to 60S subunits, and the effect of various components of the translational system, on binding and subsequent transpeptidation reaction with puromycin.

METHODS

The preparation of rat liver ribosomal subunits from ribosomes (14,15) with 0.88 M KCl, charging of tRNA with [^{3}H]phenylalanine (16) and acetylation with acetic anhydride (17), digestion of acetyl[^{3}H]phenylalanyl-tRNA with ribonuclease T_1 and chromatographic isolation of acetyl[^{3}H]phenylalanyl-oligonucleotide (4,18) have been described.

Binding assay - Acetyl[^{3}H]phenylalanyl-oligonucleotide, containing 40-60 pmoles esterified [^{3}H]phenylalanine, was maintained for 10 minutes at 4° with 380 pmoles of 60S subunits, in a volume of 0.45 ml; the solutions also contained buffered salts-DTT (40 mM Tris-HCl, pH 7.5, 54 mM $MgCl_2$, 1.4 mM dithiothreitol, 33% methanol and 60 mM or 300 mM KCl) as indicated below. In some experiments, 60S subunits that had been preincubated for 5 minutes at 37°, with an amount of 40S subunits equivalent to that of 60S subunits, in buffered salts-DTT without methanol, were used. After 10 minutes at 4°, the reaction mixture was diluted with 3 ml of buffered salts-DTT solution containing the corresponding amount of KCl, and centrifuged for 4 hours at 105,000 X *g*. The sedimented pellets were dissolved in water and radioactivity was determined with a liquid scintillation counter.

Peptidyltransferase assay - The peptidyltransferase activity of preformed 60S·acetylphenylalanyl-oligonucleotide complexes, sedimented from reaction mixtures described above, was measured by resuspension of duplicate pellets in buffered salts-DTT containing 60 mM or 300 mM KCl. The resuspended pellets were incubated in the presence and absence of 0.8 mM puromycin, for 10 minutes at 20°, in a final volume of 0.45 ml. Some incubations also contained an amount of 40S subunits equivalent to that of 60S subunits. The reactions were terminated and the product extracted with ethyl acetate as described previously (2,12); the results are expressed as pmoles of labeled phenylalanine extracted, after subtraction of the values obtained without puromycin.

Peptidyltransferase assays with free acetylphenylalanyl-tRNA and puromycin were carried out in buffered salts solutions containing 4 mM $MgCl_2$ (12); when free acetylphenylalanyl-oligonucleotide was used with puromycin, the

$MgCl_2$ concentration was 54 mM. Some incubations were carried out in two steps, the first contained 60S subunits and poly(U) or 40S subunits, in buffered salts-DTT solution without methanol, for 10 minutes at 37°; the second was the peptidyltransferase assay with labeled substrate, methanol and puromycin.

RESULTS

When acetylphenylalanyl-oligonucleotide was used as substrate, peptidyltransferase activity, measured in the presence of 0.8 mM puromycin and 4 mM $MgCl_2$, was about 50% of that obtained with acetylphenylalanyl-tRNA; in control incubations without puromycin, the amount of radioactivity extracted into ethyl acetate was extremely low and nearly the same for the two substrates. The effect of $MgCl_2$ concentration on the peptidyltransferase reaction with the fragment is shown in Figure 1.

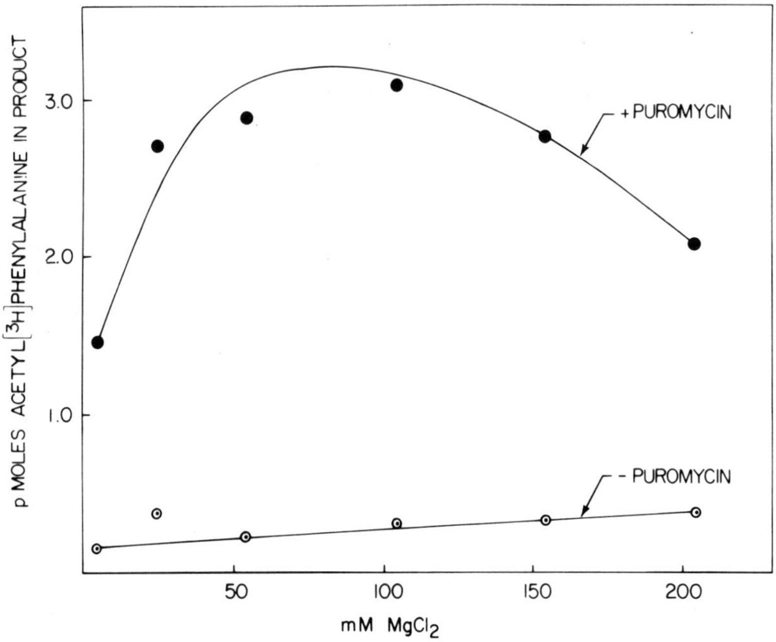

Figure 1 - Effect of magnesium ion concentration on the peptidyltransferase reaction with acetylphenylalanyl-oligonucleotide. Ribosomal 60S subunits and acetylphenylalanyl-oligonucleotide (containing 5 pmoles of esterified [^{3}H]phenylalanine) were incubated for 10 minutes at 20° as described above, with varying concentrations of $MgCl_2$.

The magnesium ion optimum was broad and ranged from 20 to 150 mM; at these concentrations the amount of product formed was approximately twice that formed at 4 mM $MgCl_2$. In contrast, the optimum for the reaction with acetylphenylalanyl-tRNA was 4 mM (12) and at 54 mM $MgCl_2$ the reaction proceeded at only one-half the rate. The broad Mg^{++}-concentration curve in Figure 1 is similar to that reported for the formation of formylmethionyl-puromycin from formylmethionyl-oligonucleotide, catalyzed by _E. coli_ 50S subunits (2).

The stimulatory effect of polynucleotides on the initial rate of the peptidyltransferase reaction with acylaminoacyl-tRNAs and puromycin, appeared to be a codon-specific effect (12); with poly(U) and acetylphenylalanyl-tRNA, the stimulation did not seem to be due to an increase in the binding of substrate to 60S subunits in alcohol (13). However, the effect on initial reaction rates and the codon-specificity could reflect a stabilization or a more appropriate alignment of the cognate acylaminoacyl-tRNA on the ribosomal site, prior to transpeptidation. It was therefore of interest to study the behavior of the acetylphenylalanyl-oligonucleotide fragment, which does not have a anti-codon region, in polynucleotide-containing reactions. Ribosomal 60S subunits were preincubated with poly(U), then assayed for peptidyltransferase activity with acetylphenylalanyl-tRNA or acetylphenylalanyl-oligonucleotide, at two magnesium ion concentrations (Table 1). As described previously, with acetylphenylalanyl-tRNA as "donor" site substrate, poly(U)-treated 60S subunits were several-fold more active than control 60S subunits in 4 mM $MgCl_2$-containing solutions; in 54 mM $MgCl_2$, poly(U) stimulated the reaction about 100%. No increase in the initial rate was seen, however, when poly(U)-treated particles were tested with acetylphenylalanyl-oligonucleotide; at 4 mM $MgCl_2$, poly(U) inhibited slightly, and at 54 mM, poly(U) caused a 40% decrease.

The effect of preincubation of 60S subunits with 40S subunits, on the peptidyltransferase reaction, is also shown in Table 1. The rate of acetylphenylalanyl-puromycin formation, from acetylphenylalanyl-tRNA and puromycin, was almost completely inhibited by preincubation with 40S subunits, at either the high or the low magnesium ion levels. The 40S effect on the reaction with the fragment was less pronounced; at 4 mM

Table 1 - The Effect of Preincubation of 60S Subunits with Poly(U) or 40S Subunits on the Peptidyltransferase Reaction With Acetylphenylalanyl-tRNA and With Acetylphenylalanyl-oligonucleotide. [a/]

First incubation components	Second incubation		
	additions	pmoles of acPhe-puromycin formed	
		4 mM Mg^{++}	54 mM Mg^{++}
60S	acPhe-tRNA	1.42	0.83
60S + poly(U)	acPhe-tRNA	5.79	1.73
60S	acPhe-oligonucleotide	0.98	2.13
60S + poly(U)	acPhe-oligonucleotide	0.79	1.30
60S + 40S	acPhe-tRNA	0.06	0.04
60S + 40S	acPhe-oligonucleotide	0.69	1.91

a/ Ribosomal 60S subunits were preincubated alone, with 50 μg of poly(U) or with 40S subunits, without methanol as described above, for 2 minutes at 37°. Then, acetylphenylalanyl-tRNA (23 pmoles of esterified phenylalanine) or acetylphenylalanyl-oligonucleotide (6 pmoles esterified radioactive phenylalanine), was added with 300 mM KCl, 33% methanol, and 0.8 mM puromycin (where present). After 10 minutes at 20°, the reactions were terminated and acetylphenylalanyl-puromycin was extracted.

$MgCl_2$, the inhibition was only 25-30% and at 54 mM, the 40S subunits had no significant effect. To further examine these effects, the formation of 60S- and 80S-substrate complexes with acetylphenylalanyl-oligonucleotide, and peptidyltransferase activity of these intermediates, were investigated. Table 2 describes the results of an experiment involving a three-step incubation procedure. The first incubation contained 60S and 40S subunits, individually or together; acetylphenylalanyl-oligonucleotide and methanol were then added and, after 10 minutes (second incubation), the particles were recovered by centrifugation. The sedimented particle-substrate complexes were resuspended as described above and assayed for peptidyltransferase activity with puromycin, in the presence and absence of added 40S subunits (third incubation). The results show that with 60S subunits alone about 60% of the particle-bound substrate recovered by centrifugation reacted to form acetylphenylalanyl-puromycin. When 40S subunits were added to the preformed 60S·acetylphenylalanyl-oligonucleotide complex, for the third incubation only, there was no effect on the amount of product formed. Experiments with 40S alone indicated that very little substrate was bound to these particles and transpeptidation did not occur. When

Table 2 - Effect of 40S Subunits on the Binding of Acetylphenylalanyl-oligonucleotide to 60S Ribosomal Subunits and its Participation in Peptidyltransferase. [a/]

First incubation components	Second incubation		Third incubation	
	additions	pmoles acPhe-oligo-nucleotide sedimented	additions	pmoles acPhe-puromycin formed
60S	acPhe-oligo + MeOH	2.73	puromycin + MeOH	1.63
60S	acPhe-oligo + MeOH	2.73	puromycin + MeOH + 40S	1.61
40S	acPhe-oligo + MeOH	0.51	puromycin + MeOH	0.01
60S + 40S	acPhe-oligo + MeOH	3.19	puromycin + MeOH	1.22

a/ Ribosomal 60S subunits were preincubated alone, or with 40S subunits without methanol, for 5 minutes at 37°. Acetylphenylalanyl-oligonucleotide (60 pmoles of esterified radioactive phenylalanine) was added and the reactants maintained with 54 mM $MgCl_2$ and 33% methanol for 10 minutes at 4°. After dilution, samples were centrifuged. One pellet of each set was dissolved and radioactivity determined. (Without 60S subunits, less than 5% of the radioactivity was sedimented.) The remaining pellets were used to determine peptidyltransferase activity.

60S plus 40S subunits were preincubated together, slightly more acetylphenylalanyl-oligonucleotide was bound, as compared to 60S particles alone, but with this substrate peptidyltransferase was inhibited about 25% only; similar experiments with acetylphenylalanyl-tRNA (13) indicated that 40S subunits inhibited the reaction over 70%.

The effect of 60 mM and 300 mM KCl concentrations on the binding of acetylphenylalanyl-oligonucleotide to 60S subunits is summarized in Table 3. As with acetylphenylalanyl-tRNA (13), acetylphenylalanyl-oligonucleotide was bound equally well in 60 or 300 mM KCl. Regardless of the KCl concentration during complex formation, the 60S-substrate intermediate reacted somewhat more rapidly in transpeptidation at 300 mM KCl than at 60 mM KCl; however, the difference between 60 and 300 mM KCl observed here was not as great as that found previously with 60S·acetylphenylalanyl-tRNA complexes, where 300 mM KCl was 3-4 fold more effective in supporting transpeptidation with puromycin. A requirement for high monovalent cation for the interaction of some aminoacyl-oligonucleotides with the "acceptor"

Table 3 - Effect of KCl Concentration on the Binding of Acetylphenylalanyl-oligonucleotide to 60S Ribosomal Subunits and its Participation in Peptidyltransferase. a/

First incubation		Second incubation	
KCl concentration	pmoles acPhe-oligonucleotide sedimented	KCl concentration	pmoles acPhe-puromycin formed
(mM)		(mM)	
60	3.0	60	1.17
60	3.0	300	1.99
300	2.8	60	1.15
300	2.8	300	1.69

a/ Ribosomal 60S subunits and 57 pmoles of oligonucleotide-bound radioactive phenylalanine were maintained, with 54 mM $MgCl_2$, 33% methanol and 60 or 300 mM KCl, for 10 minutes at 4°. The binding mixture was diluted with 3.0 ml of solutions containing 60 or 300 mM KCl, and the complexes sedimented by centrifugation. Particle-bound radioactivity was determined, and duplicate pellets, containing the 60S·acetylphenylalanyl-oligonucleotide complexes, were used to determine peptidyltransferase activity in 60 or 300 mM KCl.

site of 70S ribosomes, has also been reported by Pestka, Hishizawa and Lessard (4).

The fungal antibiotic trichodermin has been shown to inhibit peptide chain elongation and termination reactions, both of which involve peptidyltransferase activity (19-21). At 0.1 mM concentrations, trichodermin inhibited the formation of acetylphenylalanyl-puromycin from acetylphenylalanyl-tRNA and puromycin, but did not inhibit the binding of acetylphenylalanyl-tRNA to 60S subunits (13); however, trichodermin inhibited the binding of acetylphenylalanyl-oligonucleotide to 60S subunits (Table 4) by almost 70% as compared to incubations without trichodermin. The antibiotic also inhibited the transpeptidation reaction when it was added to the second incubation step, with preformed 60S·acetylphenylalanyl-oligonucleotide complex. Carrasco, Barbacid and Vazquez (22) reported that trichodermin partially inhibited binding of acetylleucyl-oligonucleotide to the "donor" site and binding of leucyl-oligonucleotide to the "acceptor" site of yeast ribosomes.

Table 4 - Effect of Trichodermin on the Binding of Acetylphenylalanyl-oligonucleotide to 60S Ribosomal Subunits and its Participation in Peptidyltransferase. a/

First incubation		Second incubation		
components	pmoles acPhe-oligo nucleotide sedimented	additions	pmoles acPhe-puromycin formed	Percent of bound substrate
60S, MeOH, acPhe-oligonucleotide	2.7	Puromycin, MeOH	1.61	60
60S, MeOH, acPhe-oligonucleotide	2.7	Puromycin, trichodermin, MeOH	0.17	6
60S, MeOH, acPhe-oligonucleotide, trichodermin	0.9	Puromycin, MeOH	0.42	47

a/ Ribosomal 60S subunits and 40 pmoles of oligonucleotide-bound [^{3}H]-phenylalanine were maintained with 54 mM $MgCl_2$ and 33% methanol, with or without trichodermin (10^{-4} M), for 10 minutes at 4°. Analyses for the amount of 60S·acetylphenylalanyl-oligonucleotide formed and for peptidyltransferase activity of the complexes were as described in Table 3.

DISCUSSION

Rat liver 60S ribosomal subunits catalyze the reaction between acetylphenylalanyl-oligonucleotide and puromycin, in the presence of methanol (33% v/v), KCl (300 mM), and $MgCl_2$ (54 mM); in the absence of puromycin, 60S·acetylphenylalanyl-oligonucleotide complexes can be isolated from the reaction mixture by centrifugation. The 60S-substrate complexes are formed equally well in 60 mM or in 300 mM KCl, but transpeptidation to puromycin is more rapid at 300 mM. In experiments not presented here, formation of 60S·acetylphenylalanyl-oligonucleotide complexes were not detected at 4 mM $MgCl_2$, even though the overall reaction with the fragment proceeds, although more slowly, at this concentration. These observations are consistent with those reported by Celma, Monro and Vazquez (3) that E. coli 50S subunits bind amino-blocked fragments in the presence of 50% ethanol, 270 mM KCl, and 13 mM MgOAc, at 0°.

Jonak and Rychlik (23) reported the magnesium-dependent binding of oligolysyl-tRNA to 50S subunits in non-alcoholic solutions; binding was stimulated by preincubation of 50S subunits with polyadenylic acid at 35°. Studies in this laboratory (12), indicated that poly(U) stimulated the alcohol-dependent formation of acylaminoacyl-puromycin, when 60S subunits were preincubated with the polynucleotide at 37° in the absence of alcohol; however, the higher rate of acetylphenylalanyl-puromycin formation was not due to an increase in the binding of substrate leading to an increase in the formation of 60S·acetylphenylalanyl-tRNA complexes (13). In studies presented here, the reaction with acetylaminoacyl-oligonucleotide, which is lacking its anti-codon region, is not stimulated by poly(U); indeed, it appears to be inhibited. These data support the suggestion that the 60S ribosomal subunit is capable of codon-anticodon recognition, which leads not to an increased binding of substrate but possibly to an increase in the stabilization or proper alignment of the particle-bound material (12, 13).

Ribosomes isolated from rat liver or formed by association of 40S and 60S subunits do not catalyze peptidyltransferase between acetylphenylalanyl-tRNA and puromycin in methanol; in contrast, the reaction with acetylphenylalanyl-oligonucleotide fragment proceeds readily with 60S or 80S particles. Thus, the observation that the amount of acetylphenylalanyl-oligonucleotide that binds to 80S particles is similar to that bound to 60S subunits is not surprising. It should be noted that the rate of transpeptidation with preformed 60S-substrate complexes was not affected when 40S subunits were added in the peptidyltransferase phase of the reaction with methanol and puromycin (13). These data support the suggestion that rat liver 80S ribosomes, either in the methanolic reaction or in poly(U)-directed aqueous systems, do not bind acetylphenylalanyl-tRNA to the donor site (16). The inhibition of acylaminoacyl-puromycin formation from acylaminoacyl-tRNA is thus a rapid and sensitive assay for the quantitative determination of 80S couples in a reaction mixture containing subunits. Preliminary experiments suggest that the equilibrium between subunits and ribosomes is frozen by the addition of 33% methanol, and that the initial rate of the subsequent peptidyltransferase reaction, measured with saturating concentrations of both substrates, is then directly proportional to the number of free 60S subunits.

Trichodermin inhibits formation of fMet-puromycin from fMet-tRNA·ribosome·AUG complexes, as well as the Release Factor-dependent release of formylmethionine (19). The formation of acetylphenylalanyl-puromycin from acetylphenylalanyl-tRNA·80S·poly(U) complexes, and formation of acetylleucyl-puromycin from acetylleucyl-oligonucleotide, with human tonsil and yeast ribosomes, is also inhibited by trichodermin; however, the binding of acetylleucyl- and leucyl-oligonucleotides to yeast ribosome "donor" and "acceptor" sites, respectively, is decreased only partly by trichodermin (22). In studies with reticulocyte ribosomes, the antibiotic has been reported to inhibit the poly(U)-directed synthesis of polyphenylalanine and peptide chain termination (21). Recent data on the interaction of [^{3}H]trichodermin with ribosomes appears to rule out more than one binding site on 60S or 80S particles. Other studies in this laboratory demonstrated that trichodermin does not affect the binding of acetylphenylalanyl-tRNA to rat liver 60S subunits in methanolic solutions (13). The present observation that binding of acetylphenylalanyl-oligonucleotide, but not acetylphenylalanyl-tRNA, is markedly inhibited by trichodermin suggests that the antibiotic binds to a site proximal to, but not directly at the "P" site, possibly the peptidyltransferase "active center"; since acetylphenylalanyl-oligonucleotide lacks much of the structure that may aid in stabilizing an interaction with the ribosome, its binding may be more easily affected by the presence of the inhibitor molecule. Thus, while acetylphenylalanyl-tRNA binds to 60S subunits in the presence of trichodermin, its 3'-terminus may be perturbed or dislocated, such that transpeptidation cannot occur. The binding data with the acylaminoacyl-oligonucleotide is not by itself inconsistent with the suggestion that trichodermin exerts its action by binding at or near the peptidyltransferase center. In this respect, it should be noted that trichodermin exhibits some properties of a competitive inhibitor of puromycin in transpeptidation (24).

SUMMARY

Rat liver 60S ribosomal subunits bind acetylphenylalanyl-tRNA in methanolic solutions containing 4 mM $MgCl_2$ but the binding of acetylphenylalanyl-oligonucleotide requires high (54 mM) concentrations of magnesium ion. Binding of both substrates occurs equally well at high (300 mM) and low (60 mM) concentrations of KCl. In 300 mM KCl-containing reaction mixtures, most of the 60S-bound acetylphenylalanyl-oligonucleotide reacts rapidly in the peptidyltransferase reaction with puromycin, but the reaction is somewhat slower in 60 mM KCl. The initial rate of the transpeptidation reaction with acetylphenylalanyl-tRNA is stimulated when 60S subunits are preincubated with poly(U) and decreased when 60S subunits are preincubated with 40S particles; with acetylphenylalanyl-oligonucleotide as the substrate, however, preincubation of 60S subunits with poly(U) significantly decreases the initial rate and 40S subunits cause only a slight decrease in the rate of transpeptidation. The antibiotic trichodermin, which inhibits peptidyltransferase with 60S-bound acetylphenylalanyl-tRNA but not the formation of 60S·acethlphenylalanyl-tRNA complex, markedly inhibits both the binding of acetylphenylalanyl-oligonucleotide and its subsequent participation in the transpeptidation phase of the reaction. The differences in the behavior of the two substrates, in various steps in peptidyltransferase, appear to reflect differences in structure in terms of the length of the aminoacylated polynucleotide chain.

Acknowledgements - This work was supported in part by research grants from the American Cancer Society (NP-88) and the U.S. Public Health Service (AM-15156). The technical assistance of Mrs. Eva Mack, Mr. Arthur Coquelin and Mr. Robert Sawchuk is gratefully acknowledged. The trichodermin was a gift of Dr. C. McLaughlin.

REFERENCES

1. Monro, R.E., Staehelin, T., Celma, M.L. and Vazquez, D.: The Peptidyl Transferase Activity of Ribosomes. Cold Spring Harbor Symp. Quant. Biol. 34, 357-368 (1969).
2. Maden, B.E.H. and Monro, R.E.: Ribosome-Catalyzed Peptidyl Transfer; Effects of Cations and pH Value. Eur. J. Biochem. 6, 309-316 (1968).
3. Celma, M.L., Monro, R.E. and Vazquez, D.: Substrate and Antibiotic Binding Sites at the Peptidyl Transferase Centre of E. Coli Ribosomes. F.E.B.S. Letters 6, 273-277 (1970).
4. Pestka, S., Hishizawa, T. and Lessard J.L.: Studies on the Formation of Transfer Ribonucleic Acid-Ribosome Complexes XIII. Aminoacyl Oligonucleotide Binding to Ribisomes: Characteristics and Requirements. J. Biol. Chem. 245, 6208-6219 (1970).
5. Mercer, J.F.B. and Symons, R.H.: Peptidyl-Donor Substrates for Ribosomal Peptidyl Transferase; Chemical Synthesis and Biological Activity of N-Acetyl Aminoacyl Di- and Tri-Nucleotides. Eur. J. Biochem. 28, 38-45 (1972).

6. Cerna, J., Rychlik, I., Krayevsky, A.A. and Gottikh, B.P.: 2'(3')-O-N-Formylmethionyl Adenosine-5'-Phosphate, A New Donor Substrate in Peptidyl Transferase - Catalyzed Reactions. F.E.B.S. Letters 37, 188-191 (1973).
7. Chladek, S., Ringer, D., and Zemlicka, J.: L-Phenylalanine Esters of Open-Chain Analog of Adenosine as Substrates for Ribosomal Peptidyl Transferase. Biochemistry 12, 5135-5138 (1973).
8. Chladek, S., Ringer, D. and Quiggle, K.: "Nonisomerizable" 2'- and 3'-O-Aminoacyl Dinucleoside Phosphates. Chemical Synthesis and Acceptor Activity in the Ribosomal Peptidyltransferase Reaction. Biochemistry 13, 2727-2735 (1974).
9. Vazquez, D.: Antibiotic Action and Protein Synthesis. Gene Expression and its Regulation (Kenney, F.T., Hamkalo, B.A., Favelukes, G., and August, J.T., eds.) Plenum Press, New York. p. 339-359 (1973).
10. Staehelin, T., Maglott, D. and Monro, R.E.: On the Catalytic Center of Peptidyl Transfer: A Part of the 50S Ribosome Structure. Cold Spring Harbor Symp. Quant. Biol. 34, 39-48 (1969).
11. Nierhaus, K.H. and Montejo, V.: A protein Involved in the Peptidyltransferase Activity of Escherichia Coli Ribosomes. Proc. Nat. Acad. Sci. U.S.A. 70, 1931-1935 (1973).
12. Thompson, H.A. and Moldave, K.: Characterization of the Peptidyltransferase Reaction Catalyzed by Rat Liver 60S Ribosomal Subunits. Biochemistry 13, 1348-1353 (1974).
13. Edens, B., Thompson, H.A. and Moldave, K.: Manuscript in preparation (1974).
14. Martin, T.E. and Wool, I.G.: Formation of Active Hybrids from Subunits of Muscle Ribosomes From Normal and Diabetic Rats. Proc. Nat. Acad. Sci. U.S.A. 60, 569-574 (1968).
15. Gasior, E. and Moldave, K.: Evidence for a Soluble Protein Factor Specific for the Interaction between Aminoacylated Transfer RNA's and the 40S Subunit of Mammalian Ribosomes. J. Mol. Biol. 66, 391-402 (1972).
16. Siler, J. and Moldave, K.: Reactions of N-acetylphenylalanyl Transfer RNA with Rat Liver Ribosomes. Biochim. Biophys. Acta 195, 130-137 (1969).

17. Haenni, A.L. and Chapeville, F.: The Behavior of Acetylphenylalanyl Soluble Ribonucleic Acid in Polyphenylalanine Synthesis. Biochim. Biophys. Acta 114, 135-148 (1966).

18. Pestka, S: Preparation of Aminoacyl-Oligonucleotides and Their Binding to Ribosomes. Methods in Enzymology (Moldave, K. and Grossman, L., eds.) Academic Press, New York XX, Part C, 502-507 (1971).

19. Tate, W.P. and Caskey, C.T.: Peptidyltransferase Inhibition by Trichodermin. J. Biol. Chem. 248, 7970-7972 (1973).

20. Barbacid, M. and Vazquez, D.: Binding of [acetyl-^{14}C] Trichodermin to the Peptidyl Transferase Centre of Eukaryotic Ribosomes, Eur. J. Biochem. 44, 437-444 (1974).

21. Wei, C. -M., Hansen, B.S., Vaughan, Jr., M.H. and McLaughlin, C.S.: Mechanism of Action of the Mycotoxin Trichodermin, a 12,13-Epoxytrichothecene. Proc. Nat. Acad. Sci. U.S.A. 71, 713-717 (1974).

22. Carrasco, L., Barbacid, M. and Vazquez, D.: The Trichodermin Group of Antibiotics, Inhibitors of Peptide Bond Formation by Eukaryotic Ribosomes. Biochem. Biophys. Acta 312, 368-376 (1973).

23. Jonak, J. and Rychlik, I.: Role of Messenger RNA in Binding of Peptidyl Transfer RNA to 30S and 50S Ribosomal Subunits. Biochem. Biophys. Acta 199, 421-434 (1970).

24. Schindler, D.: Two Classes of Inhibitors of Peptidyl Transferase Activity in Eukaryotes. Nature 249, 38-41 (1974).

Regulatory Phosphorylation of Purified Pig Liver Pyruvate Kinase

Lorentz Engström, Lars Berglund, Gunnel Bergström,
Gunilla Hjelmquist and Olle Ljungström
Institute of Medical Chemistry, Biomedical Center,
University of Uppsala, Uppsala, Sweden.

INTRODUCTION

Protein-bound phosphorylserine was first isolated by Lipmann and Levene in 1932 (ref.1). A high turnover rate of phosphate bound to serine residues of intracellular proteins was later demonstrated by Ågren _et al_. (2) indicating important metabolic functions.

The phosphorylation of serine residues at the active site of enzymes by their substrates might contribute to such a rapid phosphate turnover. However, the formation of such intermediate phosphorylenzymes has so far only been demonstrated for the phosphoglucomutase and alkaline phosphatase reactions (3).

On the other hand, regulatory phosphorylation - dephosphorylation reactions of enzymes and other proteins, catalyzed by protein kinases and phosphoprotein phosphatases, respectively, seem to be most important for the turnover of protein-bound phosphate (4).

In a recent paper, we reported the presence in rat liver cell sap of three proteins which were phosphorylated by cyclic 3',5'-AMP-stimulated protein kinase (5). One of the components was later identified as type L pyruvate kinase (ATP:pyruvate phosphotransferase; EC 2.7.1.40), the activity of which was shown to decrease due to the phosphorylation (6).

The aim of the present investigation was to obtain a liver pyruvate kinase from another species which was also regulatorily phosphorylated and which could be obtained in fairly large amounts for further studies. Therefore, type L pyruvate kinase was purified from pig liver and found to be phosphorylated by cyclic 3',5'-AMP-stimulated protein kinase with a concomitant decrease of its enzyme activity.

METHODS

Materials

γ-(^{32}P)ATP was prepared and isolated as earlier described (6). The DEAE-cellulose used was Whatman DE-52 and the phosphorylcellulose was Whatman P-11. Sepharose 6B was a product of Pharmacia. Hydroxylapatite was obtained from Biorad. Diaflo filters (Amicon Corp.) were used for ultrafiltration.

Pig liver protein kinase was prepared at 0-4°C by homogenizing the tissue in a Waring blendor with 4 volumes (v/w) of 0.25 M sucrose containing 50 mM Tris-acetic acid buffer (pH 7.5) and 1 mM EDTA. The homogenate was centrifuged at 8500 x g for 10 min and the supernatant obtained for another 60 min at 100 000 x g. 1 M acetic acid was then added to pH 5.5 (measured at 0°C). The precipitated protein was removed by centrifugation at 15 000 x g for 15 min. The pH of the supernatant was adjusted to pH 7.0 by adding 1 M potassium phosphate (pH 7.2). Solid ammonium sulfate was then added to give a final saturation of 50%. After centrifugation at 15 000 x g for 15 min the pellets were dissolved in 50 mM Tris-acetic acid buffer (pH 7.5), dialyzed and chromatographed on DEAE-cellulose with stepwise elution, as earlier described for rat liver cell sap protein (5). Histone kinase activity, assayed as before (5), was eluted with the buffer containing 0.2 M sodium acetate, and stored in small portions at -20°C. The specific activity was 0.14 units/mg of protein, defined as described earlier (5).

Analytical Methods

The radioactivity of ^{32}P-labelled phosphate was measured by determining its Cerenkov irradiation in aqueous solutions, using an Intertechnique SL 30 liquid scintillation spectrometer. Protein concentration was estimated by the biuret method (7) or by measuring the absorbance at 280 nm in a Zeiss PMQ II spectrophotometer, using an extinction coefficient of $A_{280}^{0.1\%} = 0.68$ as reported for pig liver pyruvate kinase (8).

During enzyme purification pyruvate kinase activity was estimated according to Kimberg and Yielding (9) with some modifications (6). One unit of pyruvate kinase activity is defined as that amount of enzyme which transforms one µmole of substrate per min under the conditions used. The enzyme activity was also assayed by following the decrease in absorbance at 340 nm in a reaction coupled with lactate dehydrogenase (10). The assay was performed at 30°C, and the substrate solution contained 50 mM imidazol-HCl buffer (pH 7.5), 5 mM $MgCl_2$, 100 mM KCl, 0.15 mM NADH, 1 mM ADP, 0.1 mM dithiothreitol, 0.01 or 10 µM Fru-1,6-P_2, 2 mM glucose, 1.5 units of lactate dehydrogenase and 0.5 units of yeast hexokinase. After addition of pyruvate kinase the mixture was incubated for 3 min at 30°C, and the reaction started by the addition of phosphoenolpyruvate, giving a total volume of 1.0 ml.

Phosphorylation and Polyacrylamide Gel Electrophoresis in Detergent of Purified Enzyme

22 µg of pyruvate kinase were incubated for 2 hours at 30°C with 0.09 unit of protein kinase and 0.25 mM (^{32}P)ATP in the presence of 10 mM magnesium acetate, 10 µM cyclic 3′,5′-AMP and 12.5 µM Fru-1,6-P_2. The volume was 200 µl. A blank incubation in the absence of pyruvate kinase was run in parallel. The incubations were terminated and subjected to gel electrophoresis as described earlier (6). Molecular weight estimations were performed with the

same reference proteins as before (6).

Isolation of (^{32}P)SerP from ^{32}P-Labelled Pyruvate Kinase

22 µg of purified enzyme were incubated for 30 min at 30°C with 0.01 unit of protein kinase and 0.05 mM (^{32}P)ATP in the presence of 5 mM magnesium acetate and 17 µM cyclic 3',5'-AMP. The total volume was 150 µl. The incubation was interrupted with trichloroacetic acid and the protein precipitated was hydrolyzed together with unlabelled SerP and ThrP as described before (6). The phosphoamino acids were isolated by Dowex 50 and Dowex 1 chromatographies (6).

RESULTS

Purification of Pig Liver Pyruvate Kinase

All the steps were carried out at 4°C. The centrifugations were performed in a MSE 6L Mistral centrifuge.

Homogenization. 500 g of liver were homogenized with 2 l of 0.25 M sucrose - 0.1 mM dithiothreitol - 0.1 mM EDTA. The homogenate was centrifuged at 5 000 x g for 30 min.

pH 5.2 precipitation. The pH of the extract was adjusted to pH 5.2 with 1 M acetic acid, followed by centrifugation as above. 0.5 M KOH was added to the supernatant until pH 6.8 was reached.

DEAE-cellulose chromatography. 1 vol. of 0.25 M sucrose - 0.1 mM dithiothreitol was added followed by 200 ml of DEAE-cellulose which had been equilibrated with 20 mM potassium phosphate (pH 7.0) and washed with one volume of water. The suspension was slowly stirred for 60 min. The ion exchanger was allowed to settle. It was checked that at least 90% of the enzyme activity had been adsorbed to the DEAE-cellulose. The supernatant was removed by decantation and the DEAE-cellulose suspension transferred to a column (diameter 53 mm) containing about 50 ml of DEAE-cellulose. The column was washed with 3-4 column volumes of 20 mM potassium phosphate buffer (pH 6.0) containing 30% (v/v) glycerol and 0.1 mM dithiothreitol. The enzyme was then eluted with 60 mM of the same buffer with glycerol and dithiothreitol.

Phosphocellulose chromatography. The pooled enzyme fractions from the DEAE-cellulose chromatography (about 100 ml) were diluted with one volume of 30% glycerol - 0.1 mM dithiothreitol - 0.1 mM Fru-1,6-P_2. The pH was adjusted to 5.5. The enzyme was then applied to a 100 ml (3.1 x 13.5 cm) phosphocellulose column, in equilibrium with 20 mM potassium phosphate buffer (pH 5.5). The enzyme was then eluted with 100 mM of the same buffer also containing 30% glycerol, 0.1 mM dithiothreitol and 0.1 mM Fru-1,6-P_2.

The pooled enzyme fractions were diluted with 6 volumes of 30% glycerol - 0.1 mM dithiothreitol - 0.1 mM Fru-1,6-P_2. It was then rechromatographed on a 190 ml phosphocellulose column (3.9 x 16 cm) equilibrated with 50 mM potassium phosphate buffer (pH 5.5).

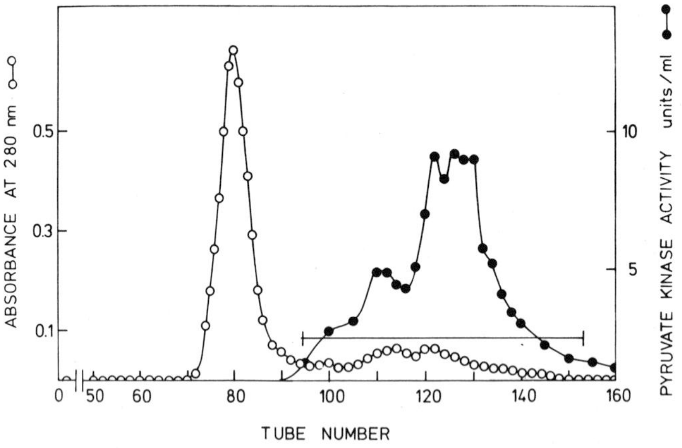

Fig.1. Phosphocellulose chromatography of pyruvate kinase. 225 mg of protein containing 2825 units of enzyme were applied to a 190 ml column and eluted with a phosphate buffer gradient as described in the text. 10 ml fractions were collected and analyzed for protein and pyruvate kinase activity. The enzyme-containing fractions were pooled as indicated.

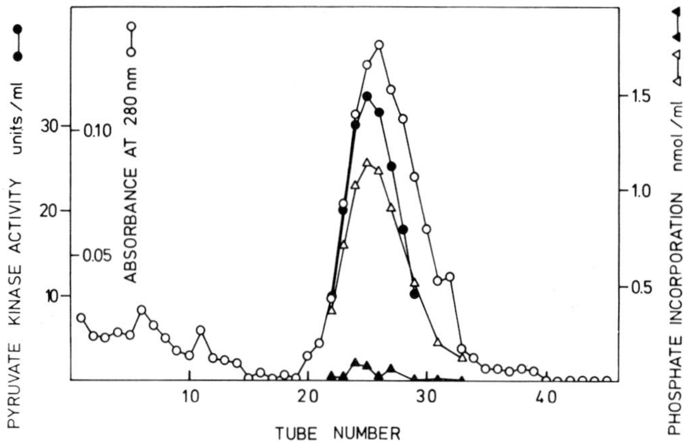

Fig.2. Sepharose 6B chromatography of pyruvate kinase. 6.2 mg of protein containing 1000 units of enzyme were chromatographed on a 170 ml Sepharose 6B column. 3.1 ml fractions were collected and analyzed for protein, pyruvate kinase activity and phosphate incorporation (△—△, in the presence of and ▲—▲, in the absence of cyclic 3',5'-AMP).

The column was eluted with a linear gradient from 1000 ml of 50 mM potassium phosphate buffer (pH 5.5) containing 30% glycerol, 0.1 mM dithiothreitol, 0.1 mM Fru-1,6-P_2 to 1000 ml of 120 mM potassium phosphate buffer with the same additions (Fig.1). The active fractions were pooled and concentrated by ultrafiltration.

Hydroxylapatite chromatography. The enzyme-containing material was dialyzed against 1 mM potassium phosphate buffer (pH 7.0) containing glycerol, dithiothreitol and Fru-1,6-P_2 as above and applied to a 10 ml (19 x 31 mm) hydroxylapatite column in equilibrium with the same buffer. The enzyme was then eluted with 10 mM potassium phosphate buffer (pH 7.0) containing 30% glycerol, 0.1 mM dithiothreitol, 0.1 mM Fru-1,6-P_2 and 50 mM KCl.

Sepharose 6B chromatography. The pooled enzyme-containing fractions from the hydroxylapatite chromatography (about 6 ml) were chromatographed on a 170 ml (2.2 x 45 cm) Sepharose 6B column in equilibrium with 100 mM potassium phosphate buffer (pH 7.0) con-

Table 1. Purification of pig liver pyruvate kinase

Purification step	Total activity (units)	Total protein (mg)	Specific activity (units/mg)	Recovery (%)
5 000 x g supernatant	9 950	53 600	0.19	100
pH 5.2 supernatant	7 150	22 100	0.32	78
DEAE-cellulose	4 800	4 130	1.16	48
First phosphocellulose	2 825	225	12.6	28
Second phosphocellulose	1 890	28.5	66	19
Hydroxylapatite	1 000	6.2	161	10
Sepharose 6B	440	2.5	176	4

taining 30% glycerol, 0.1 mM dithiothreitol and 0.1 mM Fru-1,6-P_2 (Fig.2).

Table 1 gives the result of the purification of the enzyme. The specific activity of the purified enzyme was 176 units/mg of protein. The chromatographic behaviour of the enzyme preparation on Sepharose 6B indicates a fairly high degree of purity even if some impurities can be seen in the later part of the enzyme peak.

Phosphorylation of the Purified Enzyme with Protein Kinase and (^{32}P)ATP

50 µl samples of the enzyme-containing fractions from the Sepharose 6B chromatography were incubated for 15 min at 30°C with 0.05 unit of protein kinase and 0.25 mM (^{32}P)ATP in the presence of 10 mM magnesium acetate and 10 µM cyclic 3',5'-AMP in a total volume of 250 µl. In another series cyclic 3',5'-AMP was omitted. The incubations were interrupted with trichloroacetic acid and the protein phosphorylation determined (5). A protein phosphorylation proportionate to the enzyme activity was obtained in the presence of cyclic 3',5'-AMP (Fig.2). The extent of phosphoryla-

tion was 0.4 mol of phosphate/mol of enzyme subunit, assuming a subunit molecular weight of 62 000, as discussed below. In the absence of the cyclic nucleotide there was hardly any phosphorylation at all.

In other experiments using longer incubation times and higher (^{32}P)ATP concentrations the maximal phosphate incorporation obtained was about one mol of phosphate/mol of enzyme subunit. This indicates that under optimal conditions the enzyme might be phosphorylated to an extent of one phosphoryl group per enzyme subunit.

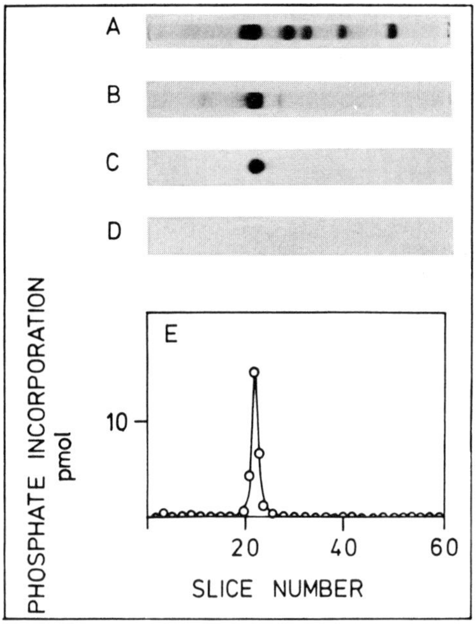

Fig.3. Polyacrylamide gel electrophoresis in detergent of pyruvate kinase.

A; Coomassie blue-stained gel of 5.6 µg of pyruvate kinase and the following reference proteins (from left to right):
4.6 µg human serum albumin, 4.6 µg ovalbumin, 18.2 µg pepsin, 4.6 µg α-chymotrypsinogen and 2.3 µg myoglobin.

B; Coomassie blue-stained gel of 11.2 µg pyruvate kinase.

C; Radioautography of gel loaded with 1.4 µg pyruvate kinase phosphorylated as described in the text.

D; As C, but pyruvate kinase omitted showing the endogenous phosphorylation of the protein kinase preparation.

E; As C, but gel sliced and phosphate incorporation measured.

Polyacrylamide Gel Electrophoresis

Polyacrylamide gel electrophoresis in detergent of the purified enzyme preparation resulted in one major band after staining with Coomassie blue (Fig.3). In addition, two faint bands were obtained showing the presence of some impurities. The molecular weight of the main component was 62 000, determined according to Shapiro et al. (11), which is in agreement with the findings of Hess and coworkers (8). When the pyruvate kinase preparation had been incubated with (^{32}P)ATP and protein kinase before gel electrophoresis followed by radioautography, only the major band was ^{32}P--labelled.

Isolation of (^{32}P)SerP from ^{32}P-Labelled Enzyme

From an acid hydrolysate of ^{32}P-labelled enzyme (^{32}P)SerP but no (^{32}P)ThrP was isolated by chromatography on Dowex 50 and Dowex 1. The molar amount of (^{32}P)phosphate bound to serine residues of the enzyme was found to be 0.3 mol/mol of enzyme subunit. The value was based on the same assumptions as earlier (6).

Effect of Phosphorylation on the Activity of Pyruvate Kinase

The effect of phosphorylation on the activity of the enzyme was examined at different phosphoenolpyruvate concentrations. It was found that the activity of the phosphorylated enzyme was markedly less than that of unphosphorylated enzyme at low substrate concentrations and at a concentration of Fru-1,6-P_2 of 0.01 μM (Fig. 4). At an increasing concentration of phosphoenolpyruvate, the inhibition became less pronounced, and at 1 mM phosphoenolpyruvate the inhibition was abolished. 0.01 mM Fru-1,6-P_2 counteracted the effect of phosphorylation. Control experiments in which pyruvate kinase was incubated with protein kinase or ATP excluded that the effect was dependent on either component alone.

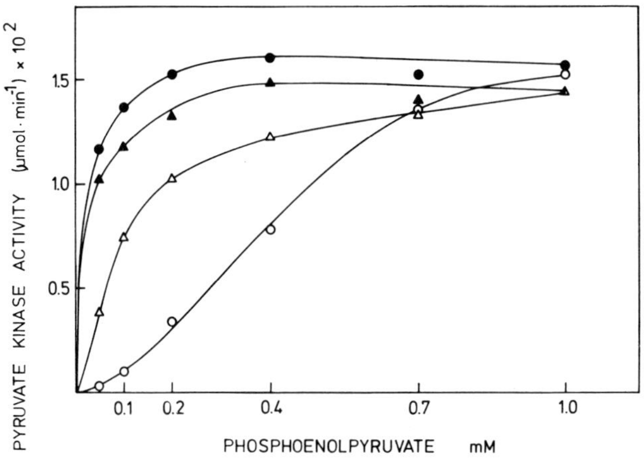

Fig.4. Effect of phosphorylation on the activity of pyruvate kinase. 11.2 µg of pyruvate kinase incubated for 15 min at 30°C with 0.01 unit of protein kinase and 0.5 mM ATP in the presence of 12.5 mM magnesium acetate and 0.05 mM cyclic 3',5'-AMP in a total volume of 100 µl. Control in the absence of protein kinase and ATP was run in parallel. The incubations were interrupted by dilution of the samples with 1.15 ml of 20 mM potassium phosphate buffer (pH 7.0) containing 30% glycerol and 0.1 mM dithiothreitol. The samples were kept at 20°C prior to analysis with the coupled assay in the presence of Fru-1,6-P_2 in the concentrations given. Phosphorylated enzyme: ○—○ , 0.01 µM Fru-1,6-P_2; ●—● , 10 µM Fru-1,6-P_2. Control: △—△ , 0.01 µM Fru-1,6-P_2; ▲—▲, 10 µM Fru-1,6-P_2.

DISCUSSION

There are two different pyruvate kinases in liver tissue (12). The amount of enzyme obtained, the elution positions on DEAE--cellulose chromatography, and the effect of Fru-1,6-P_2 on the activity of the enzyme preparation, ascertains that the L-type of the enzyme has been purified. The enzyme was purified to a fairly high degree, as judged from its chromatographic and polyacrylamide gel electrophoretic behaviour. The specific activity of the enzyme was of the same order as that of the homogenous preparation of Hess and coworkers (8). It could therefore be concluded that the main band on electrophoresis represented the enzyme, and that the enzyme was phosphorylated by cyclic 3',5'-AMP-stimulated protein kinase.

The molar phosphate incorporation has not been definitely established, but the results indicate that each subunit of the tetrameric enzyme (8) might incorporate one phosphoryl group on a serine residue.

The phosphorylation of the enzyme resulted in an inhibition of its activity at low phosphoenolpyruvate concentrations. Thus, the enzyme can be expected to be inhibited by cyclic 3',5'-AMP-dependent phosphorylation at the phosphoenolpyruvate concentrations prevailing in vivo (13). This is compatible with the stimulatory effect of cyclic 3',5'-AMP on gluconeogenesis (14) when a key glycolytic enzyme as pyruvate kinase should be inhibited.

In all probability the L type of pig liver pyruvate kinase may be added to the increasing number of enzymes whose activity is regulated by phosphorylation - dephosphorylation reactions.

SUMMARY

Pig liver pyruvate kinase (type L) (EC 2.7.1.40) was highly purified by removal of impurities by pH 5.2 precipitation followed by chromatography on DEAE-cellulose, phosphocellulose, hydroxylapatite and Sepharose 6B in phosphate buffers containing glycerol and dithiothreitol.

The enzyme was shown to be phosphorylated on incubation with (^{32}P)ATP and cyclic 3',5'-AMP-stimulated protein kinase from pig liver, as earlier described for the corresponding enzymes from rat liver. Up to 4 mol of phosphate per mol of enzyme were incorporated. From acid hydrolysates of ^{32}P-labelled enzyme (^{32}P) SerP but no (^{32}P)ThrP was isolated.

The activity of the enzyme at low phosphoenolpyruvate concentrations was decreased by phosphorylation (10-20 fold at 50 μM phosphoenolpyruvate). In all probability, the L type of pig liver pyruvate kinase belongs to the enzymes whose activity is regulated by phosphorylation - dephosphorylation reactions.

ACKNOWLEDGEMENT

This work was supported by the Swedish Medical Research Council (Project No. 50X-13).

REFERENCES

1. Lipmann, F.A. and Levene, P.A.: Serinephosphoric acid obtained on hydrolysis of vitellinic acid. J.Biol.Chem. 98, 109-114 (1932).
2. Ågren, G., deVerdier, C-H. and Glomset, J.: A study of the phosphorus-containing proteins of cells. II. The turnover rate of ^{32}P-labelled phosphoserine of the Schneider protein residues of several rat organs. Acta Chem.Scand. 8, 1570-1578 (1954).
3. Hummel, J.P. and Kalnitsky, G.: Mechanisms of certain phosphotransferase reactions: Correlation of structure and catalysis in some selected enzymes. Ann.Rev.Biochem. 33, 15-50 (1964).
4. Segal, H.L.: Enzymatic interconversion of active and inactive forms of enzymes. Science 180, 25-32 (1973).
5. Ljungström, O., Berglund, L., Hjelmquist, G., Humble, E. and Engström, L.: Cyclic 3',5'-AMP stimulated and non-stimulated phosphorylation of protein fractions from rat-liver cell sap on incubation with (γ-^{32}P)ATP. Uppsala J.Med.Sci. in the press (1974).
6. Ljungström, O., Hjelmquist, G. and Engström, L.: Phosphorylation of purified rat liver pyruvate kinase by cyclic 3',5'-AMP-stimulated protein kinase. Biochim.Biophys.Acta in the press (1974).
7. Layne, E.: Spectrophotometric and turbidimetric methods for measuring proteins. In Methods of enzymology, Vol.3, pp.447-454 (1957).
8. Kutzbach, C., Bischofberger, H., Hess, B. and Zimmermann-Telschow, H.: Pyruvate kinase from pig liver. Hoppe-Seyler's Z.Physiol.Chem. 354, 1473-1489 (1973).

9. Kimberg, D.V. and Yielding, K.L.: Pyruvate kinase. Structural and functional changes induced by diethylstilbestrol and certain steroid hormones. J.Biol.Chem. 237, 3233-3239 (1962).

10. Bücher, T. and Pfleiderer, G.: Pyruvate kinase from muscle. In Methods of enzymology, Vol.1, pp.435-440 (1955).

11. Shapiro, A.L., Vinuela, E. and Maizel, J.V.: Molecular weight estimation of polypeptide chains by electrophoresis in SDS--polyacrylamide gels. Biochem.Biophys.Res.Commun. 28, 815-820 (1967).

12. Tanaka, T., Harano, Y., Morimura, H. and Mori, R.: Evidence for the presence of two types of pyruvate kinase in rat liver. Biochem.Biophys.Res.Commun. 21, 55-60 (1965).

13. Llorente, P., Marco, R. and Sols, A.: Regulation of liver pyruvate kinase and the phosphoenolpyruvate crossroads. Eur.J.Biochem. 13, 45-54 (1970).

14. Exton, J.H. and Park, C.R.: Control of gluconeogenesis in liver. II. Effects of glucagon, catecholamines, and adenosine 3',5'-monophosphate on gluconeogenesis in the perfused rat liver. J.Biol.Chem. 243, 4189-4196 (1968).

A Soluble Elongation Factor Required for Protein Synthesis with mRNA's other than Poly (U)

M. Clelia Ganoza* and J. Lawrence Fox+
*Banting and Best Department of Medical Research, University of Toronto, Ontario, Canada and +Department of Zoology, University of Texas, Austin, Texas 78712, U.S.A.

INTRODUCTION

The introduction of the synthetic messenger RNA, poly (U), by Nirenberg in 1961 (1) provided an experimental approach to a detailed dissection of procaryotic protein synthesis. The isolation and characterization of proteins essential to polypeptide elongation occurred in the Lipmann laboratory. A transfer factor (now called EF-T) and a GTP hydrolyzing factor (now called EF-G) were elucidated (2). It was later shown that EF-T could be resolved into a heat stable fraction (EF-Ts) and a heat unstable fraction (EF-Tu) (3). The functions of these factors are now known in considerable detail and have been reviewed by several authors (4, 5, 6).

Using a more natural message, such as the mRNA isolated from f_2 bacteriophage, indicated the requirement for initiation factors. These have been isolated and largely characterized (5, 7-9). More recently termination factors have been isolated and characterized as a result of studies with natural messengers (10, 11). These studies served to emphasize the fortuitous simplicity of the poly (U) directed experimental system. While the lability of the ability of purified EF-T and EF-G to produce polypeptides from non-poly (U) messages was common laboratory folklore, in our efforts to conduct termination studies (12) with highly purified elongation factors, partially purified initiation factors (state of the art at that time) and non-poly (U) mRNAs, we were generally unable to obtain polypeptide formation unless we used elongation factor concentrations more than an order of magnitude above that needed with the poly (U) directed system. S-30's and S-100's functioned quite adequately in place of the purified elongation factors, however.

This led us to attempt to isolate an additional supernatant factor to complement EF-T (or EF-Tu and EF-Ts) and EF-G with non-poly (U) messenger RNA. This report describes the isolation and partial characterization of a supernatant, complementary factor which we call X.

METHODS AND MATERIALS

An extensive list of details is given in an earlier publication (13). Elongation factors were purified according to Gordon *et al* (14) or a gift (highly purified EF-Tu and EF-Ts) from Dr. J. Ravel (15). Ribosomes were from *E. coli* B44 and were purified by repeated 0.5 M NH_4Cl, 0.01 M $MgCl_2$-Tris pH 7.4 washing. Initiation factors were isolated as described by Iwasaki *et al* (10). Polymerizations used the procedure of Nishizuka and Lipmann (2) in 0.06 ml containing 7 mM Tris, pH 7.4, 10 to 28 mM NH_4Cl, 130 mM KCl, 10 mM $MgCl_2$, 6.9 mM DTT, 1.4 mM GTP, 40 μg of purified ribosomes and appropriate levels of elongation factors, amino acyl-tRNAs, and messenger RNA. Incubations were generally 15 minutes at 35° C.

Antisera to EF-Ts (17) and to EF-G were generous gifts of J. Gordon.

X was purified from *E. coli* or Q_{13} cells by three methods: Method 1. A crude S240 was applied to DEAE-Sephadex A-50 and eluted by a KCl gradient at roughly 0.25-0.3 M KCl. The dialyzed pooled fractions were then chromatographed on hydroxyapatite and eluted by increasing the phosphate buffer (pH 7.2) to 0.06 M. Method II. RNA was removed from the S240 by titration with protamine sulfate. The dialyzed supernatant was absorbed to alumina C γ gel and eluted by increasing the phosphate buffer to 0.2 M. Method III. The S240 was partitioned in the polyethylene glycol-dextran system of Gordon (14) followed by ammonium sulfate batchwise elution. The X activity was found in the 19.6% saturated ammonium sulfate eluant.

RESULTS

The stimulation of polypeptide synthesis by a crude supernatant fraction for various messenger RNAs is shown in Table 1. The polyphenylalanine synthesis is nearly quantitative.

Table 1

Stimulation of synthesis for mRNA's by S240

Each incubation was 0.1 ml. 61 pmoles of ^{14}C-lysyl-tRNA + 19 cold amino acyl-tRNAs or 22 pmoles of ^{14}C-phenyalanyl-tRNA, 25 μg of each initiation factor, 100 μg of S240, 5 μg of EF-G, and 10 μg of EF-T were used as appropriate.

mRNA	Factors	^{14}C-lysine or ^{14}C-phenylalanine incorporated into peptides (pmoles)	
		+mRNA	-mRNA
f_2RNA	EF-T + EF-G + IF	3.74	2.84
	EF-T + EF-G + IF + S240	22.5	4.70
Poly (A_3,U)	EF-T + EF-G	6.77	0.86
	EF-T + EF-G + IF	1.46	1.05
	S240	2.94	1.16
Poly (A)	EF-T + EF-G	0.44	0.48
	EF-T + EF-G + S240	9.15	1.80
	S240	6.30	1.48
Poly (U)	EF-T + EF-G	18.5	1.80

In the purification procedures which we used X chromatographs very close to EF-Tu. We therefore conducted a series of experiments to exclude the identity of X with the known elongation factors. The heat stability of X when compared to EF-Tu is one bit of evidence. X is inactivated at 90° C for 2 minutes, but resists 20 min at 50° C. Compare also the data in Table 2.

Table 2

Differential heat inactivation of X and EF-Tu

95 μg of protein from the Method II fraction containing both EF-Tu and X activity was used. The polymerization assay contained 20 μg EF-T and 13 μg EF-G which were saturating for the poly (U) assay.

Addition	Heat treatment	^{14}C-lysine incorporated (pmoles)	^{3}H-GDP formed (pmoles)
None	None	1.30	0.62
X + EF-Tu	None	4.17	34.7
X + EF-Tu	20 min 50° C	3.60	8.51

We demonstrated that this heat inactivated fraction with residual X activity gave linear kinetics for lysine incorporation with poly (A) over 10 minutes. At 5 minutes the lysine incorporation was linear with increasing amounts of X. This poly (A) directed stimulated synthesis was used as an assay for Method I purification. A 250 fold enrichment was obtained. EF-T and X activities were largely separated by this procedure. Contaminating EF-Ts activity could be removed by precipitation with EF-Ts antisera. Following centrifugation, residual EF-Tu activity could be inactivated by 5 min at 60° C. The effects of this are shown in Table 3.

Thus, we have shown that X can be freed of both EF-Ts and EF-Tu activities. Conversely, X will not substitute for either transfer factor.

X can be shown to be distinct from EF-G by several bits of evidence. First, a temperature sensitive mutant (18) for EF-G still possessed X activity. Secondly, a sucrose density gradient centrifugation (19) yielded a molecular weight of 50,000 for X compared to 80,000 for EF-G. Thirdly, X cannot replace EF-G inactivated by anti-G sera (Table 4). Lastly, purified EF-G does not replace X (Table 5).

Table 3

Effects of EF-Ts antisera on GDP binding and polymerization

Assays were conducted as indicated as in Methods.

Addition	^{14}C-lysine incorporated (pmoles)	^{3}H-GDP bound (pmoles)
None	0.28	1.05
EF-T + EF-G	0.45	
EF-T + EF-G + X	2.00	12.68
EF-T + EF-G + X (cleared of Ts)	1.66	1.40

Table 4

Effects of EF-G antisera on polymerization

1.7 μg EF-T, 2.5 μg EF-G, and 11.6 pmoles of ^{14}C-phenylalanyl-tRNA were used in poly (U) directed synthesis; 8.5 μg EF-T, 12.5 μg EF-G, 4 μg of X (Method III-4 fold excess) and 25 pmoles of ^{14}C-lysyl tRNA were used in poly (A) directed syntheses.

Additions	Incorporated (pmoles) ^{14}C-lysine	^{14}C-phenylalanine
None	0.38	0.26
X	0.65	
EF-T + EF-G	1.15	2.81
EF-T + EF-G + anti-EF-G	0.47	0.32
EF-T + EF-G + anti-EF-G + excess EF-G		3.19
EF-T + EF-G + X	3.81	
EF-T + EF-G + anti-EF-G + excess X	0.68	0.46

Table 5

Polymerization factor requirements

2.2 μg EF-Tu, 0.38 μg EF-Ts, 3.5 μg EF-G, and 5 μg X (Method B) were used as indicated. 38 pmoles of ^{14}C-phenylalanyl-tRNA or 22.4 pmoles of ^{14}C-lysyl-tRNA were used.

Additions	Incorporated (pmoles) ^{14}C-lysine	^{14}C-phenylalanine
None	0.26	0.19
EF-Tu	0.17	0.35
EF-Ts	0.15	0.13
EF-G	0.25	0.19
X	0.23	0.19
EF-Tu + EF-Ts	0.35	0.34
EF-Tu + EF-G	0.35	12.26
EF-Tu + EF-Ts + EF-G	0.68	23.13
EF-Tu + EF-Ts + EF-G + X	2.43	25.86

The effects of X stimulation with several other synthetic mRNAs establishes that this effect is not unique to poly (A). These results are given in Table 6.

Table 6

X polymerization stimulation with different mRNAs

12 μg of EF-T, 5 μg of EF-G, 5 μg of X (Method 3) and 27 pmoles of ^{14}C-lysyl-tRNA or 51 pmoles of ^{14}C-seryl-tRNA with 19 cold amino acyl-tRNAs were used as appropriate

mRNA	Incorporation (pmoles) EF-T + EF-G	EF-T + EF-G + X
Poly (A)	0.97	6.5
Poly (A_3, U)	0.59	3.3
Poly (A) or - Poly (A_3, U)	0.22	0.14
Poly (U, C)	2.67	14.0
Poly (U, C)	0.56	0.88

So far we have presented evidence to indicate that the known initiation factors and elongation factors required supplementation by a factor X to produce polymerization with several non-poly (U) mRNAs. We conducted a product analysis of poly (A) directed polymerization with limiting EF-Tu, EF-Ts and EF-G in the absence and presence of X. The products were deacylated and then chromatographed on carboxymethylcellulose with a 0 to 0.5 M NaCl gradient. In the absence of X dipeptide accumulates whereas the presence of X yields higher peptides. This is illustrated in Fig. 1.

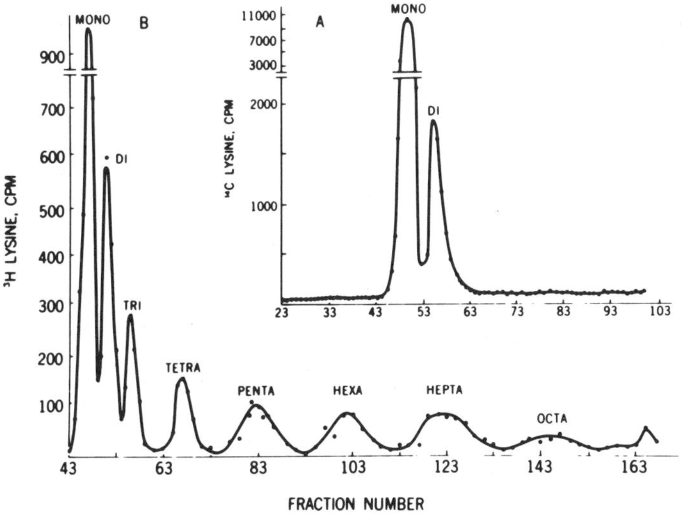

Figure 1. Carboxymethyl cellulose chromatography of reaction product in the presence and absence of X. 0.25 ml incubation. A) no X factor. 47,000 dpm ^{14}C-lysine and lysine peptide applied, 43,900 dpm recovered. B) plus 25 μg X factor. 110,000 dpm ^{14}C-lysine and lysine peptides applied, 104,000 dpm recovered.

DISCUSSION

We have presented evidence to show that highly purified EF-T and EF-G and partially purified initiation factors produce little or no polypeptide synthesis when directed by a number of non-poly (U) messenger RNAs. We have partially purified a supernatant factor which is apparently distinct from EF-Ts, EF-Tu, and EF-G by several criteria. The effect of X upon chain elongation from dipeptide to polypeptide suggests its action as part of the translocation step. We feel that the ability of X to overcome the dipeptide accumulation under poly (A) direction pretty well rules out initiation or termination functions. Since both EF-Tu and EF-G have been shown to interact at the same ribosomal site, it is not clear how EF-G functions in translocation (20-22). On the other hand X may function by accelerating peptide bond formation.

While little is presently known about the mechanism of translocation or its relationship to chain elongation, we believe that the X factor may be a useful tool for the study of these vital steps of protein synthesis.

SUMMARY

A protein (mol wt 50,000) has been isolated from Escherichia coli cells which is necessary for polypeptide synthesis with non-poly (U) messenger RNAs. It has been prepared free of EF-Ts, EF-Tu and EF-G activities and these have been shown to be unable to substitute for each other. X extends dipeptidyl-tRNA and may function in the translocation of mRNA.

ACKNOWLEDGEMENTS

We wish to thank the Medical Research Council of Canada (MCG) and the National Science Foundation (JLF) for support. The encouragement of Dr. F. Lipmann is greatly appreciated. We are grateful to Miss Anne Tirpak for technical assistance, to Dr. Julian Gordon for generous gifts of EF-Ts and EF-G antisera and especially to Dr. Joanne Ravel for generous gifts of highly purified EF-Ts and EF-Tu.

REFERENCES

1. Nirenberg, M. W. and Matthaei, J. H.: The Dependence of Cell-Free Protein Synthesis in E. coli Upon Naturally Occuring or Synthetic Polyribonucleotides. Proc. N.A.S., 47, 1588-1602 (1961).
2. Nishizuka, Y. and Lipmann, F.: Comparison of Guanosine Triphosphate Split and Polypeptide Synthesis with a Purified E. coli System. Proc. N.A.S., 55, 212-219 (1966).
3. Lucas-Lenard, J. and Lipmann, F.: Separation of Three Microbial Amino Acid Polymerization Factors. Proc. N.A.S., 55, 1562-1566 (1966).
4. Lengyel, P. and Söll, D.: Mechanisms of Protein Biosynthesis. Bacteriol. Rev., 33, 264-301 (1969).
5. Lucas-Lenard, J. and Lipmann, F.: Protein Biosynthesis. Ann. Rev. Biochem., 40, 409-448 (1971).
6. Lucas-Lenard, J. and Beres, L.: Protein Synthesis-Peptide Chain Elongation. The Enzymes, 3rd Ed., 10, In press (1974).
7. Stanley, W. M. Jr., Salas, M., Wahba, A. J. and Ochoa, S.: Translation of the Genetic Message: Factors Involved in the Initiation of Protein Synthesis. Proc. N.A.S., 56, 290-295 (1966).
8. Suttle, D. P. and Ravel, J. M.: The Effects of Initiation Factor 3 on the Formation of 30S Initiation Complexes with Synthetic and Natural Messengers. Biochem. Biophys. Res. Commun., 57, 386-593 (1974).
9. Wahba, A. J. and Miller, M. J.: Chain Initiation Factors from Escherichia coli. Methods in Enzymol., 30, 3-18 (1974).
10. Capecchi, M. R. and Klein, M. A.: Release Factors Mediating Termination of Complete Proteins. Nature, 226, 1029-1033 (1970).
11. Ganoza, M. C. and Tomkins, J. K. N.: Polypeptide Chain Termination In Vitro: Competition for Nonsense Condons Between a Purified Release Factor and Suppressor tRNA. Biochem. Biophys. Res. Commun., 40, 1455 (1970).

12. Fox, J. L. and Ganoza, M. C.: Chain Termination _In Vitro_. Studies on the Specificity of Amber and Ochre Triplets. Biochem. Biophys. Res. Commun., _32_, 1064 (1968).

13. Ganoza, M. C. and Fox, J. L.: Isolation of a Soluble Factor Needed for Protein Synthesis with Various Messenger Ribonucleic Acids Other than Poly (U). J. Biol. Chem., _249_, 1037 (1974).

14. Gordon, J., Lucas-Lenard, J. and Lipmann, F.: Isolation of Bacterial Chain Elongation Factors. Methods in Enzymol., _20_, 306-316 (1971).

15. Ravel, J. M. and Shorey, R. L.: GTP-Dependent Binding of Aminoacyl-tRNA to _Escherichia coli_. Methods in Enzymol., _20_, 306-316 (1971).

16. Iwasaki, K., Sobol, S., Wahba, A. J. and Ochoa, S.: Translation of the Genetic Message VII. Role of Initiation Factors in Formation of the Chain Initiation Complex with _Escherichia coli_ Ribosomes. Arch. Biochem. Biophys., _125_, 542-547 (1968).

17. Gordon, J., Schweiger, M., Krisko, I. and Williams, C. A.: Specificity and Evolutionary Divergence of the Antigenic Structure of the Polypeptide Chain Elongation Factors. J. Bacteriol., 100, 1-4 (1969).

18. Torchini-Valentini, G. P., Felicetti, L. and Rinaldi, G. M.: Mutants of _Escherichia coli_ Blocked in Protein Synthesis: Mutants with an Altered G Factor. Cold Spring Harbor Symp. Quant. Biol., _34_, 463-468 (1969).

19. Martin, R. G. and Ames, B. N.: A Method for Determining the Sedimentation Behavior of Enzymes: Application to Protein Mixtures. J. Biol. Chem., _236_, 1372-1379 (1961).

20. Nishizuka, Y. and Lipmann, F.: The Interrelationship Between Guanosine Triphosphate and Amino Acid Polymerization. Arch. Biochem. Biophys., _116_, 344-351 (1966).

21. Lucas-Lenard, J. and Haenni, A. L.: Release of Transfer RNA During Peptide Chain Elongation. Proc. N.A.S., 63, 93-97 (1969).

22. Richter, D.: Inability of E. coli Ribosomes to Interact Simultaneously with the Bacterial Elongation Factors EF-Tu and EF-G. Biochem. Biophys. Res. Commun., 46, 1850-1856 (1972).

Variations in the Lactate Dehydrogenase Isoenzyme Pattern in Arteries

Ulrich Gerlach and Wolfgang Fegeler
University of Münster, Medical Clinic, Germany

The contribution of a medical clinic to this symposium should discuss a biological problem which is of medical interest. So we want to mention a problem in artery metabolism during development and ageing.
The changes in enzyme activity in the arteries during development and ageing conform to two basic patterns (1-4): In the first, enzyme activity rises to a peak before falling off again with advancing age (e.g.ß-glucuronidase); in the other type, there is a decline right from the beginning (e.g.phenolsulfatase, fumarate-hydratase, sulfate activating enzymes (Lipmann:11, 16, 17), collagen synthesizing enzymes) (fig.1). The total activity of lactate dehydrogenase (LDH) in the aorta of rats conforms to the second of these two patterns.

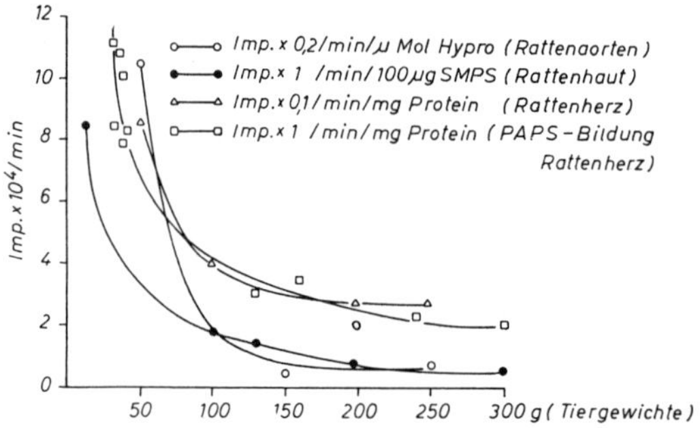

Fig. 1 Age dependency of incorporation rates for 14 C PRO into collagen, 35 SO_4 into glycosaminoglycanes, 14 C LEU into protein and the activity of sulfate activating enzyme system (10)

In this paper we want to show a connection between isoenzyme-activity of lactic dehydrogenase and metabolism in aortas.

Methods

The aortas of rats of various ages were examined. Six classes of age resp. weight were examined: 50 g, 100 g, 200 g, 250 g, and 300 g animals.After the rats had been sacrificed by decapitation the aortas were immediately excised, freed of adventitia, weighed, immersed in 1,15 % KCl in the ratio of 1: 20 and then homogenised for 60-90 sec under refrigeration. The homogenate was centrifuged for 15 min at 18.000 g and $0^{o}C$ and the clear supernatant was used for further investigation.
After determination of the total LDH activity the LDH isoenzymes were separated in agar gel. Quantitative evaluation was made using the tetrazolium method. The sum of the total weight of the 5 fractions was expressed as 100 %; the individual fractions were then calculated as percentages of this total. The readings were interpreted statistically.

Results

Comparison of Relative Values
The average values obtained from the separation of LDH isoenzymes from the aortas of rats of different ages (expressed in relative percentages) are summarized in figure 2. The graph shows that a shift in the isoenzyme pattern in the aorta of rats takes place with age.

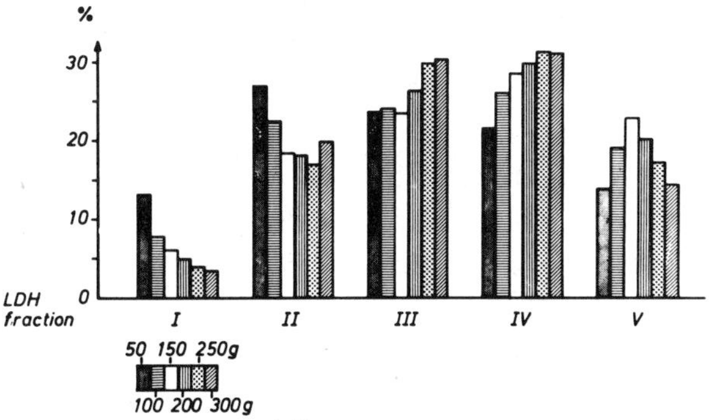

Fig. 2 LDH isoenzymes (%) in aortas of developing rats

LDH fraction 1 shows a decline. By the time a weight of 150 g is reached there is a reduction of more than a half (54.6 %), while the 300 g animals exhibit only slightly over a quarter of the initial value (26.4 %). All the differences recorded for this fraction (between 50 g and 150 g, 150 g and 300 g, and 50 g and 300 g) were found to be statistically significant.

LDH fraction 2 likewise showed a drop from 100 % in the 50 g group to 68.9 % in the 150 g group, but the LDH fraction 2 in the 300 g group rose again to 74.3 % of the initial 50 g value. The enzyme decline between the 50 g and the 150 g animals as well as between the 50 g and the 300 g animals was found to be significant. On the other hand the slight rise between the 150 g and 300 g animals was not conclusive.

LDH fraction 3 showed no difference in the readings for the 50 g and 150 g group; moreover at 99.5 % of the 50 g value the 150 g value is almost as great as that for the 50 g group (100 %). Nevertheless a significant increase does occur in LDH fraction 3 not only between the 150 g and 300 g groups but also with regard to the overall difference between the 50 g and 300 g readings, the 300 g animals exhibiting appreciably more LDH 3 activity than the animals in the 50 g group.

LDH fraction 4 likewise undergoes a significant increase with age: the 100 % of the 50 g group becomes 132 % in the 150 g group and 159 % in the 300 g group.

LDH fraction 5 reached its peak activity in the 150 g animals, the amount of LDH 5 in animals in the 50 g and 300 g groups being almost the same at 100 % and 103.5 % respectively. The differences between the 50 g and 150 g animals on the one hand and the 150 g and 300 g animals on the other are significant. There is no difference between the 50 g and 300 g groups.

Comparison of H and M Percentages

From the structure of the LDH protein it is well known that there are two subunits, H and M, whose synthesis is directed by two different gene loci. We therefore examined the percentage of H and M sub-units to determine whether they changed in keeping with the alterations in the LDH isoenzyme pattern. This calculation was based on the average values of the relative percentages. As can be seen from figure 3, a change in the relative percentage of the sub-units takes place between the 50 g and 100 g age groups, after which the pattern remains relatively constant.

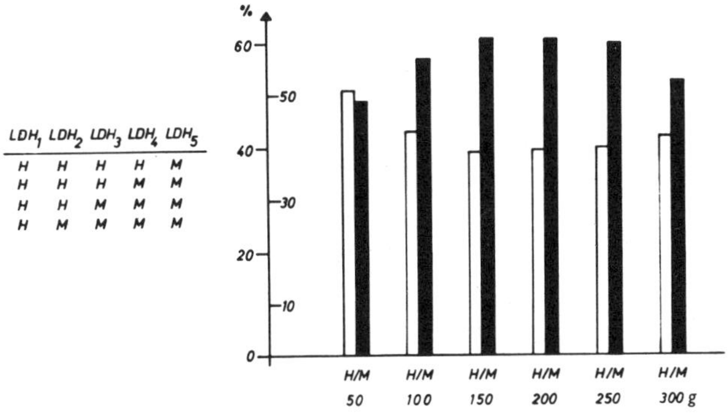

Fig. 3 Distribution in % of H and M type isoenzymes in aortas of developing rats (Total sum of sub-units = 100 %)

Discussion

The findings show that during development there is not only a decline in total activity of LDH in the aorta of rats but also a shift in the LDH isoenzyme pattern. The shift in the isoenzyme pattern described in the present study is most pronounced in the 50 g, 100 g and 150 g groups of animals and consists essentially in a fall in LDH 1 and 2 and a corresponding rise in LDH 4a and 5. The finding of Lojda and Frič (18) that the cathodic fractions of LDH increase with advancing age of the rats also agrees with these results, but we ascertained another percentage distribution of the isoenzyme pattern in the aorta of rats: We found that LDH fraction 5 had a lowest average value of 13.5 % ± 6.30 in 50 g animals and a highest average value of 22.9 % ± 9.08 in 150 g animals. In attempting to explain these findings, there are several points to be taken into consideration: The drop in total LDH in the various age groups might be connected with cell density which is reduced during ageing (Kühnau (14); Hilz (12); Gerlach (5)). The cell density in the aorta (measured in terms of DNA content) is almost twice as great in 4 weeks old rats as it is in animals aged from 4 to 15 months. If one takes into consideration here the decrease of total LDH activity in the aortas of growing rats, the congruity between the drop in total LDH and the reduction in cell density can be seen. On the other hand there is no adequate reason for assuming that alterations in the composition of the cell population are responsible for the change in enzyme activity. Moreover it is debatable whether hormonal influences, which determine growth, development and maturity during the ageing process, can also regulate the isoenzyme pattern of the aorta. In fact Lojda and Frič found alterations in the LDH isoenzyme pattern in the aortas of rats not only after castration but also in hypo- and hyperthyroidism. An increase in the cathodic fractions 4 and 5 was found after castration and in hypothyroidism, while the opposite effect was seen in hyperthyroidism. A similar shift in favour of the anodic fractions was found by Omi and Anan in human thyroid gland tissue in hyperthyroidism.

Wohlrab and Schedel describe sex differences in the isoenzyme pattern of LDH in aortas, which are probably also due to hormonal influences.

It can be assumed that there is a hormonal influence on the LDH isoenzyme pattern to the extent that the anodic fractions of LDH predominate when metabolism-enhancing hormones

are in excess while the cathodic fractions predominate in the case of hormonally depressed metabolism. The extent to which the individual LDH fractions are linked with functional metabolic centres was shown by Güttler and Clausen, who examined the intracellular distribution of the LDH isoenzyme pattern in a variety of organs of calves: The LDH fractions 4 and 5 were found predominantly in the nuclear fraction, while the LDH 1 and 2 isoenzymes were mainly localized in the mitochondrion fraction. These findings permit the advancement of the hypothesis that the changes described in the isoenzyme pattern in the aorta of our laboratory animals depend on hormonal regulation. The slowing down of general metabolic activity during development would thus be in keeping with a decrease in anodic and an increase in cathodic LDH fractions.
As mentioned before, the localisation of LDH isoenzymes in the tissue is different: According to Lojda and Frič the intima of the aorta contains more of the cathodic LDH fractions. The highest LDH activity in aorta was found in the muscle cells of the media.
With regard to the human aorta, Buddecke and Kresse reported that the age-related increase in aortic weight is chiefly conditioned by the replacement of connective tissue. During the ageing process the quantitative composition of the connective tissue also changes: a loss of elastin and an increase in hexosamines and consequently of mucopolysaccharides occur (Buddecke). The multiplication of extracellular mesenchymal substance, which occurs with advancing age, leads to a thickening and compression of what Hauss, Junge-Hülsing and Gerlach have described as the transit zone: All substrates of cellular activity (including O_2) have to be carried in and out through the transit zone, which lies between the blood stream and the parenchymal cells (in this case the cells of the vessel wall). It is obvious that when this zone is elongated and thickened, diffusion is hampered. For the cells this can mean, among other things, an impairment of

oxygen supply.
There is a connection between the LDH isoenzyme pattern and the oxygen supply to and consumption by the individual tissues, in that organs with high oxygen requirements have a greater percentage of LDH 1 and LDH 2 while those that are relatively resistant to anaerobic metabolic conditions show a higher concentration of LDH 4 and LDH 5 (Cahn, Kaplan, Levine and Zwilling, Dawson, Goodfriend and Kaplan).Because of this increase in LDH 4 and LDH 5 it can be presumed that the older vessel wall has a higher resistance to a reduced oxygen supply. Goodfriend et al. assume that a low oxygen supply increases the synthesis of M-type-isoenzyme of LDH. Also the findings of Lojda and Frič on the distribution of the LDH isoenzymes in the various layers of the aorta are in connection with this result. By way of adaptation, at the level of intracellular regulation, the cell may increase those isoenzyme fractions which predominate in tissues which are relatively resistant to a lack of O_2; in other words it may increase its LDH 4 and LDH 5 content.

Summary

93 male rats were divided into 6 age groups and an experiment was made to find out the LDH isoenzyme pattern in their aortas. It was demonstrated that the total activity of LDH depends on the development and ageing of the laboratory animals. With increasing age there is an increase in LDH 3 and 4 isoenzymes and a decrease in LDH fraction 1. The biological significance of the findings is discussed.

References

1 Buddecke,E.: Angiochemische Alterswandlungen des Aortenbindegewebes.Dtsch.Ges.Kreisl.Forsch. 24,143 (1958).

2 Buddecke,E.: Untersuchungen zur Chemie der Arterienwand, II Arteriosklerotische Veränderungen am Aortenbindegewebe des Menschen. Hoppe-Seyler 310,182(1958).

3 Buddecke,E.: Untersuchungen zur Chemie der Arterienwand, III Veränderungen und Beeinflussung des Aortenbindegewebes bei tierexperimenteller Arteriosklerose. Hoppe-Seyler 310,199(1958).

4 Buddecke,E.,and Kresse,H.: Glykosamino-Glykanohydrolasen des Arteriengewebes und ihre Aktivitätsänderungen im Alter und bei Arteriosklerose. Colloques Internationaux du Centre National de la Recherche Scientifique, Paris 1967.

5 Gerlach,U.: Altersabhängige Stoffwechselgrößen. Symposium on Experimental Gerontology Basel 23. - 25.10.1964.

6 Gerlach,U.: Abhängigkeit des Isoenzymmusters der LDH in Aorten von Entwicklung und Alter. Vortrag vor der Akademie der Wissenschaften und Literatur Mainz 1970.

7 Goodfriend,T.L., Sokol,D.M.,and Kaplan,N.O.: Control of synthesis of lactic acid dehydrogenases. 15,18(1966).

8 Güttler,F.,and Clausen,J.: Cellular Compartmentalization of Lactate Dehydrogenase Isoenzymes. Enzym.biol.clin. 8,456(1967).

9 Hauss,W.H.: Pathogenese der Koronarsklerose und des Herzinfarktes. Verh.Dt.Ges.Inn.Med. 69,554(1963).

10 Hauss,W.H., Junge-Hülsing,G.u.Gerlach,U.: Die unspezifische Mesenchymreaktion. Thieme,Stuttgart 1968.

11 Hilz,H.,and Lipmann,F.: The enzymatic activation of sulfate. Proc.nat.Acad.Sci.(Wash.) 41,880(1955).

12 Hilz,H.: Veränderungen von Zelldichte und Polysaccharidstoffwechsel im alternden Bindegewebe. Klin.Wschr. 41,332(1963).

13 Kirk,J.E.: Enzymes of the arterial wall. Academic Press, New York and London 1969.

14 Kühnau,J.: Die Biochemie des Alterns.
In: Der Mensch im Alter, 1,56(1962) Umschau, Frankfurt.

15 Lansing,A.I.: The Arterial Wall.
Williams & Wilkins Company, Baltimore 1959.

16 Lipmann,F.: Enzymatic group activation and transfer.
In: Metabolism of the nervous system. New York: Pergamon Press 1957.

17 Lipmann,F.: Biological sulfate activation and transfer.
Science 128,575(1958).

18 Lojda,Z.,and Frič,P.: Lactic dehydrogenase isoenzymes in the aortic wall.
J.Atheroscl.Res. 6,264(1966).

19 Omi,Y.,and Anan,F.K.: Lactate dehydrogenase isoenzymes of thyroid tissue in various thyroid gland diseases.
Enzym.biol.clin. 11,224(1970)

20 Wieme,R.J.: Studies on agar gel electrophoresis.
Arscia Uitgaven N.V., Brüssel 1959.

21 Wohlrab,F.,and Schedel,F.: Zur histochemischen Erfassbarkeit von Enzymaktivitätsänderungen in der menschlichen Aortenwand und ihre Beziehungen zur Atherogenese.
Dtsch.Ges.wesen 27(1972) H.31.

22 Wohlrab,F.: Persönliche Mitteilung.

Biochemical Development of the Heart in Syrian Hamsters

W. Gevers, P.A. Jones, G.A. Coetzee and D.R. van der Westhuyzen
M.R.C. Unit for Molecular and Cellular Cardiology,
University of Stellenbosch Medical School, P.O.Box 63,
Tiervlei. 7503, South Africa

INTRODUCTION

The course of most important heart diseases is profoundly influenced by the apparent inability of adult heart cells to undergo hyperplasia after injury (regeneration) and by the fact that cellular growth (hypertrophy) is the only response of the muscle cells of the organ to overload (1). An understanding of the mechanisms underlying these phenomena is thus of great importance in medicine.

We have become interested in the possibilities of studying various aspects of the development and growth of mammalian heart muscle cells in primary tissue cultures obtained from neonatal hamsters. Such systems offer considerable advantages in the ease of experimental manipulation but require careful characterization to determine whether they are suitable for the clarification of such central questions as the relationships between mitotic activity, differentiation and growth.

In the case of skeletal myocytes, there are good reasons for believing that only non-differentiated "presumptive" myoblasts can divide, while post-mitotic myoblasts undergo fusion to form syncitial myotubes and then selectively synthesize myofibrillar proteins (2, 3, 4, 5). It is not yet clear whether cellular contacts and fusion are primary events leading to suppression of mitosis and derepression of genes coding for the various proteins typical of muscle (6, 7), or whether special "quantal" mitosis, or another type of endogenous programme, lead to the state of terminal differentiation (2, 8). No analogous formulations for cardiac development have been presented to date. Heart cells do not fuse in vivo but develop elaborate plasmalemmal specializations to become a functional rather than a real syncitium (9). They can divide in young animals even when well-differentiated, but cease en masse to do so after the birth of the animal (10). There is also a tendency to polyploidy or binucleation in heart cells, which suggests that chromosomal replication and mitotic processes may not be fully coupled (11).

A relatively large number of tissue culture systems for the study of chicken, rat and hamster cardiomyocytes (amongst others) have been described (see (1)). Serum factors are important for the maintenance of myosin ATPase levels in continually dividing rat heart cells (12). Beating heart cells can be successfully cultivated in purely synthetic media (13). Mitoses have been observed in a proportion of beating, differentiated myoblasts in cultures (14), and clones of beating cells have been derived from single cells (15). None of these or other studies has involved a detailed biochemical study of the relationships between mitoses, differentiation and growth in the cultures, or has produced findings which can be correlated with the behaviour in vivo of heart cells during organismal development and growth.

METHODS

Primary Heart Cell Culture

Cultures were established, under aseptic conditions, according to HARARY and FARLEY (16), with a number of modifications. Hearts from 30-70 Syrian hamsters (2-4 days old, maintained by random inbreeding in our colony obtained from the London School of Hygiene) were cut into cubes of about 0.5 mm thick, which were washed in a trypsinizing flask with 25 ml phosphate-buffered saline (PBS) not containing Ca^{++} and Mg^{++} (17), for 10 min with constant stirring at room temperature. After decantation of the PBS, 25 ml PBS containing 0.1% trypsin (1:250, Difco Laboratories) was added, and stirring was continued at 30°C for 30 min. The disaggregated cells were collected by centrifugation at 1000 x g for 2 min and suspended in "medium" (Minimal Essential Medium buffered with Earles salts, also containing 2.2 g/l $NaHCO_3$, 10% tryptose phosphate broth, 10% calf serum inactivated at 56°C for 30 min and 100 units/ml each of Penicillin G and Streptomycin). The suspended cells were filtered through sterile No 7 nylon mesh and counted in a haemocytometer. Cells were seeded at 2-3 x 10^6 cells/5 ml "medium" per Petri dish (diameter 60 mm), and incubated at 37°C in an atmosphere of 95% air and 5% CO_2. Cells were inspected by phase contrast microscopy every day on a stage heated to 37°C. The "medium" was changed every second day.

Harvesting of attached cells was performed by treating them with 2 ml PBS, containing 0.1% trypsin, for 10 min at room temperature. They were scraped off with a stainless-steel spatula, washed three times with 10 ml "medium" and once with PBS, and counted. Cells selected for enzyme assays were suspended in 200 µl of Buffer H (10 mM tris-HCl, 1 mM EDTA, 1 mM DTT, pH 7.6) and rapidly frozen in liquid nitrogen. Cells not harvested with the aid of trypsin gave similar results but could not be easily counted. Frozen cells were allowed to swell in 0.4 ml Buffer H for 1 hr at 4°C. Homogenization was effected with 60 passes of a tight-fitting Dounce homogenizer (Blaessig Glass).

Enzyme Assays

Samples of the homogenates were centrifuged at 10 000 x g for 10 min to obtain supernatants, which were used for determination of creatine phosphokinase (18) and lactate dehydrogenase (19) activities, because they contained 80-100% of the total homogenate activities of these enzymes. The sediments contained all the cellular myosin activity, which was quantitatively extracted by homogenization in 0.1 ml 0.6 M KCl containing 1 mM ATP (pH 7.0), and frequent mixing at $0^{o}C$ for 60 min, before centrifugation in the cold at 10 000 x g for 10 min. The Ca^{++}-stimulated ATPase activity was estimated by measuring the release in 10 min at $25^{o}C$ of ^{32}P from (γ-^{32}P)-ATP, in mixtures (100 μl) containing 50 mM tris-acetate, 10 mM $CaCl_2$, 0.12 M KCl, 1 mM EDTA, 1 mM EGTA, 1 mM ATP (0.5 μCi) and enzyme. Samples were mixed with 4 volumes of 2.5% acid-washed Norit A, in 2% $HClO_4$ containing 0.05 M NaH_2PO_4, and centrifuged. Radioactivity in 25 μl of supernatant was determined by liquid scintillation counting in 5 ml Instagel (Packard Instruments). Input radioactivity was quantitated to permit expression of enzyme activities in nmoles/min at $25^{o}C$. Dependence of activity on Ca^{++} was almost total, in that omission of Ca^{++} and inclusion of 5 mM EGTA eliminated 95% of the activity, while chelation of Mg^{++} with 1-5 mM EDTA had no effect. Oligomycin sensitivity was absent.

Further samples of the primary cell homogenates were sonicated in Beckman microfuge tubes (45 sec, submerged in an Ultrasonic Cleaner, Bransonic 12), and the activities of NADH- and succinate-cytochrome c reductase systems measured spectrophotometrically at $30^{o}C$ (20).

Protein concentrations were determined colorimetrically (21).

RESULTS

About a quarter of the disaggregated neonatal myocardial cells, when seeded at 2-3 x 10^6 cells per dish, attached during the first day of culture and began to divide rapidly; the others were visible microscopically as refractile, mobile bodies and were removed during the change of medium on Day 2. Few dead

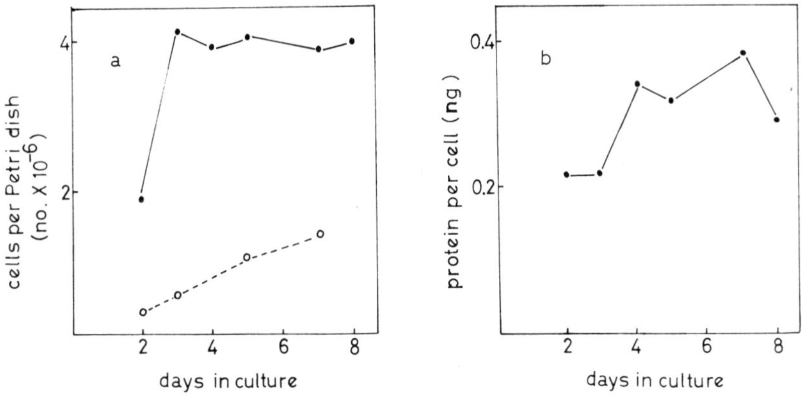

Fig.1: Growth characteristics of heart cells in culture. (a) Proliferation of cultures seeded at 2.6×10^6 cells per Petri dish, (●——●); 0.8×10^6 cells per Petri dish, (o---o). (b) Protein content per average cell of the culture, shown in (a), seeded at 2.6×10^6 cells per Petri dish.

cells were removed in this way during the medium changes on Day 4, 6 and 8, and cell turnover appeared not to be a factor that required to be taken into account. After Day 3 the cell count reached a plateau (Fig.1a). Single beating cells were observed during the first 3 days, increasing in number so that groups of "beaters" were formed and finally a synchronously beating sheet by Day 5. After this the contractile activity declined and twitches became predominant by Day 8. The cultures were demonstrably sterile throughout. The average protein content of the cells continued to increase steadily beyond the third day of culture, signifying that the cells were growing after they had stopped dividing (Fig. 1b).

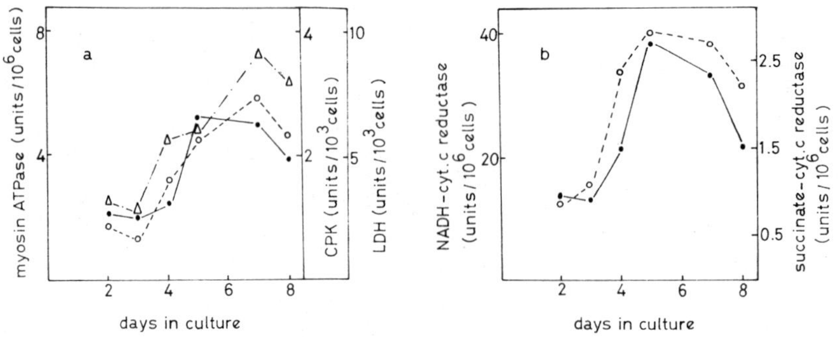

Fig.2: Enzyme activities in heart cell culture. Activities per average cell of the culture, described in Fig.1a, seeded at 2.6 x 10^6 cells per Petri dish. All enzyme units are defined as 1 nmole substrate utilized per min under the conditions given in Methods. (a) Myosin ATPase ●——●; creatine phosphokinase (CPK), o---o; lactate dehydrogenase (LDH), Δ---Δ. (b) NADH-cytochrome c reductase, ●——●; succinate-cytochrome c reductase, o---o.

Seeding the dishes at much lower cell densities led to slower growth in numbers, failure to contact-inhibit and poor sheet-beating function (Fig. 1a). This is likely to be a medium conditioning effect, although very long-lived beating cells could be obtained in this way.

Three groups of enzyme activities in the cultured cells were determined, all considered to be particularly prevalent in heart muscle, to ascertain whether the post-mitotic protein increase in the individual (average) cells could be ascribed to specialization in terms of so-called "luxury molecules" (Differentiation)

(8) or whether growth of already well-differentiated muscle cells was involved. As a guide to the behaviour of the contractile apparatus, myosin ATPase activity, wholly dependent on high concentrations of Ca^{++} ions, independent of Mg^{++} ions, and insensitive to oligomycin, was regarded as a measure specifically of the myosin content of the cells, and not of sarcoplasmic reticular, mitochondrial or other ATPases. The activity per average cell continued to increase beyond Day 3 and fell slightly again after Day 7 (Fig. 2a). The activities of creatine phosphokinase and lactate dehydrogenase increased similarly (Fig.2a), reflecting the behaviour of soluble enzymes related to the energy metabolism of the muscle cell. Although it is likely that turnover of protein occurs in such cells, and that mechanisms other than increases in the number of enzyme molecules may be responsible for some of the enhanced activities recorded, the correlation with increased protein content of the cells suggests that biosyntheses are involved.

Another group of enzymes prevalent in heart muscle cells are those in mitochondrial membrane systems. The total cellular activities of succinate- and rotenone-insensitive NADH-cytochrome c reductase activities increased 3-4 fold after the cells stopped dividing, but then declined together with the beating functions of the cells (Fig. 2b). The first mentioned activity was taken to represent the inner membranes of mitochondria and the second mentioned, the outer membranes as a first approximation.

It is noteworthy that the activity of all the enzymes measured and the total protein, expressed per average cell, did not change during the transition from Day 2 to Day 3, when the increase in cell count was greatest.

DISCUSSION

Morphological studies of heart development in various species have not yet revealed unambiguously whether development and differentiation of heart muscle cells is synchronous (9, 14). MARKWALD has recently reported that a synchrony of ultrastructural developmental status is a feature of foetal hamster hearts (22). It is thus possible that our cultures are derived by the disintegration of a population of cells, a certain percentage of which is still capable of mitotic activity and therefore, very likely, of attachment to tissue culture dishes with subsequent growth. Trypsinization may be a necessary mitogenic stimulus in this regard, to which only cells capable of mitosis can respond. On the other hand, the limited attachment that we have found, may be the result solely of irreversible proteolytically-induced damage to about three-quarters of the cells, causing failure to attach. The distinction between these two possibilities needs to be made because of the interesting implications of the first-mentioned mechanism for a study of the progressive mitotic shut-off in mammalian neonatal heart cells (10).

It is also known that trypsinization of muscle cells reversibly caused extensive disorganization of myofibrillar structure and loss of cellular proteins (23). A process which is likely to be occurring at an early stage in our cultures is therefore the single restoration of perturbed organellar structure by self-assembly, which does not involve protein synthesis but may require cellular contacts to be effective (24).

These considerations indicate that the tissue culture system may not be as satisfactory a model for cardiac differentiation as might have been desired. There may in fact be technical and theoretical barriers to the establishment of any such <u>in vitro</u>

model system. On the other hand, there is no doubt that cellular growth in our cultures does not occur for 2 or 3 days but only after termination of cell divisions, as shown by the increased total activities, per cell, of various muscle-specific enzymes or systems, coupled with the increase in total protein. By the same token, there seems to be much less evidence for specialization (enrichment with respect to muscle-specific proteins) in the cells during this time. However, aspects of the relationships between mitosis and growth, which are of great importance in the development of the normal organ, may well be amenable to detailed biochemical study in this system. Further information is required, such as the optimum age of hamsters for initiation of cultures and the accuracy of ultrastructural correlations with biochemical data, before the potential of such cultures can be fully evaluated.

SUMMARY

Cultured myocardial cells from neonatal Syrian hamsters divide rapidly for 3 days after seeding, after which the cell count remains constant, while beating cells aggregate into groups and then sheets. Cellular growth also occurs in the post-mitotic period, with respect to total protein content as well as myosin ATPase, creatine phosphokinase, lactate dehydrogenase, NADH- and succinate-cytochrome c reductase activities, all expressed per average cell.

Acknowledgements

This study was supported by the South African Medical Research Council, Atomic Energy Board and the Harry Crossley Fund. R.A. Dyer, K. Sindle and P.A.C. Weidemann are thanked for technical assistance and discussions.

REFERENCES

(1) ZAK, R.:
Cell Proliferation during Cardiac Growth.
Amer. J. Cardiol. 31, 211-219 (1973)

(2) OKAZAKI, K., HOLTZER, H.:
Myogenesis: Fusion, Myosin Synthesis and the Mitotic Cycle.
Proc. Nat. Acad. Sci. USA 56, 1484-1490 (1966)

(3) YAFFE, D.:
Cellular Aspects of Muscle Differentiation in vitro.
in: Current Topics in Developmental Biology 4, 37-77 (1969)

(4) COLEMAN, J.R., COLEMAN, A.W.:
Muscle Differentiation and Macromolecular Synthesis.
J. Cell. Physiol. 72, Suppl. 1, 19-34 (1968)

(5) DELAIN, D., MEIENHOFER, M.C., PROUX, D., SCHAPIRA, F.:
Studies on Myogenesis in vitro: Changes of Creatine Kinase, Phosphorylase and Phosphofructokinase Isoenzymes.
Differentiation 1, 349-354 (1973)

(6) WEINSTEIN, R.B., HAY, E.D.:
Deoxyribonucleic Acid Synthesis and Mitosis in Differentiated Cardiac Muscle Cells of Chick Embryos.
J. Cell. Biol. 47, 310-316 (1970)

(7) PRZYBYLA, A., STROHMAN, R.C.:
Myosin Heavy Chain Messenger RNA from Myogenic Cell Cultures.
Proc. Nat. Acad. Sci. USA 71, 662-666 (1974)

(8) HOLTZER, H., BISCHOFF, R.:
Mitosis and Myogenesis.
in: Physiology and Biochemistry of Muscle as a Food, ed. BRISKEY, E.J., CASSENS, R.G. and MARSH, B.B. (The University of Wisconsin Press, Madison, Milwaukee and London), 29-51 (1970)

(9) VIRAGH, S., CHALLICE, C.E.:
Origin and Differentiation of Cardiac Muscle Cells in the Mouse.
J. Ultrastructure Research 42, 1-24 (1973)

(10) CLAYCOMB, W.C.:
DNA Synthesis and DNA Polymerase Activity in Differentiating Cardiac Muscle.
Biochem. Biophys. Res. Comm. 54, 715-720 (1973)

(11) GROVE, D., NAIR, K.G., ZAK, R.:
Biochemical Correlates of Cardiac Hypertrophy. 3. Changes in DNA Content; the Relative Contributions of Polyploidy and Mitotic Activity.
Circ. Res. 25, 463-471 (1969)

(12) DESMOND, W., HARARY, I.: In vitro Studies of Beating Heart Cells in Culture. XV. Myosin Turnover and the Effect of Serum.
Arch. Biochem. Biophys. 151, 285-294 (1972)

(13) HALLE, W., WOLLENBERGER, A.:
Die Differenzierung isolierter Herzzellen in einem chemisch definierten Nährmedium.
Z. Zellforsch. 87, 292-314 (1968)

(14) CHACKO, S.:
DNA Synthesis, Mitosis and Differentiation in Cardiac Myogenesis.
Dev. Biol. 35, 1-18 (1973)

(15) CAHN, R.D., COON, H.G., CAHN, M.B.:
Cell Culture and Cloning Techniques.
in: Methods in Developmental Biology, ed. WILT, F., MESSEL, N. (New York, Crowell.) 493-530 (1967)

(16) HARARY, I., FARLEY, B.:
In vitro Studies on Single Beating Rat Heart Cells. I.Growth and Organization.
Exp. Cell. Res. 29, 451-465 (1963)

(17) DULBECCO, R., VOGT, M.:
Plaque Formation and Isolation of Pure Lines with Poliomyelitis virus.
J. Exp. Med. 99, 167-182 (1954)

(18) OLIVER, J.T.:
A Spectrophotometric Method for the Determination of Creatine Phosphokinase and Myokinase.
Biochem. J. 61, 116-122 (1955)

(19) BERGMEYER, H.U., BERNT, E., HESS, B.:
Lactic Dehydrogenase.
in: Methods of Enzymatic Analysis, ed. BERGMEYER, H.U. (Academic Press, New York and London), 736-743 (1963)

(20) SOTTOCASA, G.L., KUYLENSTIERNA, B., ERNSTER, L., BERGSTRAND, A.:
An Electron-Transport System Associated with the Outer Membrane of Liver Mitochondria.
J. Cell. Biol. 32, 415-438 (1967)

(21) LOWRY, O.H., ROSENBROUGH, H.T., FARR, A.L., RANDALL, R.T.:
Protein Measurement with the Folin Phenol Reagent.
J. Biol. Chem. 193, 265-275 (1951)

(22) MARKWALD, R.R.:
Distribution and Relationship of Precursor Z Material to Organizing Myofibrillar Bundles in Embryonic Rat and Hamster Ventricular Myocytes.
J. Molec. Cell. Cardiol. 5, 341-350 (1973)

(23) WOLLENBERGER, A.:
Rhythmic and Arhythmic Contractile Activity of Single Myocardial Cells Cultured in vitro.
Circ. Res. 14 and 15, Suppl. 2: 184-201 (1964)

(24) ETLINGER, J.D., FISCHMAN, D.A.:
M and Z Band Components and the Assembly of Myofibrils.
in: Cold Spring Harbor Symposia on Quantitative Biology XXXVII, 511-522 (1973)

On the Mechanism of Inhibition of Globin Chain Initiation by Pactamycin

Irving H. Goldberg, Lizzy S. Kappen and Hideo Suzuki
Department of Pharmacology, Harvard Medical School, Boston, MA. 02115, USA

INTRODUCTION

At low concentrations the anti-tumor antibiotic pactamycin selectively inhibits protein synthesis (1-8) and leads to the breakdown of rabbit reticulocyte polyribosomes and the release of completed globin chains (9,10). Globin synthesis dependent on new chain formation is especially sensitive to the antibiotic, whereas elongation and termination of nascent chains are relatively insensitive (11). These results indicate that pactamycin selectively blocks polypeptide chain initiation. The nature of the block was investigated by dissecting the steps involved in initiation and by studying the binding of pactamycin to ribosomes at various stages of their cycle in polypeptide synthesis (12-15). This paper summarizes the results of these studies.

METHODS

Methods used for the preparation of 0.5 M KCl-washed rabbit reticulocyte ribosomes, rabbit liver [^{35}S]Met-tRNA$_f$, and [^{3}H] pactamycin, and polypeptide initiation factors; for the assay of Met-tRNA$_f$ binding to ribosomes and Met-puromycin synthesis; and for the analysis of initiation oligopeptides by paper electrophoresis have been described elsewhere (6,8,12-15).

RESULTS AND DISCUSSION

Pactamycin does not inhibit the binding of [^{35}S]Met-tRNA$_f$ to 40S ribosomal subunits but prevents the formation of 80S initiation complexes and the synthesis of Met-puromycin at 2 mM Mg^{2+} (Table 1).

Table 1: Effect of Pactamycin on Met-$tRNA_f$ Binding to Reticulocyte Ribosomes and Met-Puromycin Formation.

Experiment 1		Experiment 2	
$[^{35}S]$Met-$tRNA_f$ Binding		Met-Puromycin Formation	
Pactamycin (μM)	cpm Bound (% Inhibition)	Pactamycin (μM)	Met-Puromycin (cpm) (% Inhibition)
0	7823	0	3433
2	7791 (0)	1	766 (78)
10	7190 (8)	1 (at 7 min)	2952 (14)
		2	665 (81)
		2 (at 7 min)	2495 (27)
		10	267 (92)

Experimental conditions have been described elsewhere (12).

The synthesis of a functional initiation complex, as judged by its capacity to form $[^{35}S]$Met-puromycin was impaired by pactamycin only if the antibiotic was added before or during complex formation; if added after binding was completed, the puromycin reaction proceeded normally (12,16). Furthermore, the sensitivity of Met-puromycin formation to pactamycin was dependent on the concentration of Mg^{2+}. The higher the Mg^{2+} concentration, the less was the inhibition by the antibiotic (12).

The formation of an active 80S initiation complex proceeds via a 40S ribosomal subunit-mRNA-Met-$tRNA_f$ intermediate to which a 60S ribosomal subunit joins (17,18). The overall process requires initiation factors and GTP. The analysis of binding reactions similar to those described in Table 1 by sucrose density gradient centrifugation showed that at 2 mM Mg^{2+} the formation of the puromycin-sensitive 80S initiation complex was inhibited by pactamycin and that this was associated with a pronounced increase in the amount of $[^{35}S]$Met-$tRNA_f$ bound to the smaller initiation complex, which sediments at about 50S (Figure 1). This effect was much

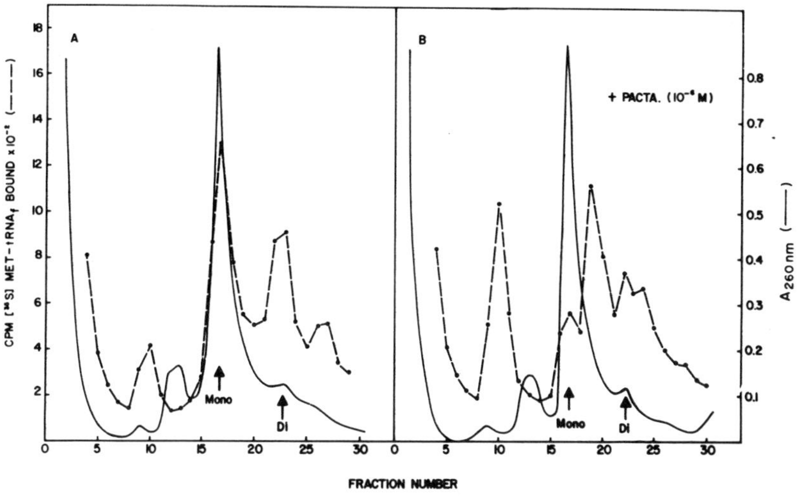

Figure 1. Sucrose gradient analysis of $[^{35}S]$Met-$tRNA_f$ bound to salt-washed reticulocyte ribosomes, and the effect of pactamycin. Reaction conditions are the same as for Table 1 and centrifugation is described elsewhere (12).

less at 5 mM Mg^{2+} (12). It should be noted, as shown in Figure 1, that in addition to the accumulation of radioactivity in the 50S region of the gradient, new peaks of radioactivity appear between the monomer and dimer and between the dimer and trimer regions of the gradient. Evidence has been obtained that these peaks represent mono- or di-somes bearing 40S ribosomal subunits ("1 1/2 mer and 2 1/2 mer") without the corresponding 60S subunit (12). Such ribosomal structures have been described before under various experimental conditions (17,19) and are thought to represent an intermediate stage in ribosomal protein synthesis. These experiments show that at low Mg^{2+} concentrations, where cellular polypeptide synthesis is presumed to take place, pactamycin prevents the formation of a functional 80S initiation complex (or its equivalent on a polyribosomal structure) either because the 60S subunit does not join to form the complex or because the 80S structure is unstable and dissociates. Levin, Kyner and Acs (20) have obtained similar results in a different eukaryotic system. These experiments were carried out with initiation factors and salt-washed ribosomes. When similar experiments were performed in whole reticulocyte lysates, however, the block in binding of $[^{35}S]$ Met-$tRNA_f$ to 80S ribosomes or polysomes was usually less pronounced. These results suggested the possibility that the fractionated system, but not the lysate, was deficient in a factor necessary for the stable joining of the 60S ribosomal subunit and for the formation of a puromycin reactive initiator-tRNA in the presence of pactamycin. In fact, we have recently isolated a protein factor from a preparation of crude initiation factors that when added in excess obliterates the pactamycin inhibition of the puromycin reaction (14). This factor is eluted from a DEAE-cellulose column between 0.22 and 0.25 M KCl and is referred to as F-0.25. As shown in Table 2, the puromycin reaction depends on

Table 2: Reversal by F-0.25 of Pactamycin Inhibition of Met-Puromycin Formation.

	$[^{35}S]$Met-Puromycin Formation				
	Control	Pactamycin			
F-0.25 Added		0.2 μM		2 μM	
(μg)	cpm	cpm	Inhib. %	cpm	Inhib. %
0	415	---	---	---	---
1.6	1397	492	65	353	75
3.2	2356	929	61	736	69
6.4	4201	1746	58	1389	67
18.9	7127	5200	27	4245	40
37.7	8705	7750	11	6517	25

Met-puromycin formation was carried out with salt-washed ribosomes and initiation factors required for Met-$tRNA_f$ binding to the small ribosomal subunit as described elsewhere (14).

the presence of F-0.25 in addition to the other initiation factors required for Met-$tRNA_f$ binding to the small ribosomal subunit.

The amount of Met-puromycin synthesized was almost linear with the amount of F-0.25 added to the reaction mixture. In this system, using isolated initiation factors, the ability of F-0.25 to overcome the inhibition of pactamycin was clearly observed. Since the reaction was linear for at least up to 20 min., the value obtained at 10 min. incubation (v) was plotted according to Lineweaver and Burk. Such an analysis shows that the interaction of pactamycin and F-0.25 is of a competitive type under the conditions examined. Addition to excess of none of the other initiation factors was able to reverse the pactamycin inhibition of the puromycin reaction.

The ability of F-0.25 to act as a joining factor was demonstrated by sucrose density gradient centrifugation analysis of the binding of [^{35}S]Met-$tRNA_f$ to washed ribosomes bearing endogenous mRNA in the presence of partially purified initiation factors (14). With a mixture of initiation factors excluding F-0.25, almost all [^{35}S]Met-$tRNA_f$ was associated with the small ribosomal subunits and only a very small amount of radioactivity was observed on the 80S ribosomes. Addition of increasing amounts of F-0.25 was accompanied by a progressive shift of radioactivity from the smaller ribosomal subunits to the 80S region of the gradient. These data suggest strongly that F-0.25 promotes the joining of the 60S subunit to the 40S initiation complex to form the 80S initiation complex. Pactamycin at a concentration of 2 μM resulted in the accumulation of radioactivity on the small subunit initiation complex and in the inhibition of 80S initiation complex formation, just as had been found in Figure 1. At a near saturating level of F-0.25 (6 μg), pactamycin blocked 80S initiation complex formation about 50% and as with Met-puromycin formation, increasing amounts of F-0.25 reduced the degree of inhibition significantly, so that it was 35% with 19 μg F-0.25 and only

14% with 38 μg F-0.25.

The precise nature of the interaction between pactamycin and joining factor remains to be clarified. Since pactamycin binds to the ribosome (6,8,15), it is possible that the antibiotic competes with the joining factor for binding to a common or overlapping site on the ribosome. From what follows below, this appears to be unlikely. It is also possible that pactamycin does not directly prevent the joining factor from promoting 80S complex formation but blocks the subsequent release of the factor from the complex, producing a stable initiation complex of a type similar to that formed with fusidic acid at the translocation step (21). In this manner, the catalytic action of joining factor would be inhibited and the formation of 80S initiation complexes would result in the sequestration of stoichiometric quantities of factor. This formulation could explain why the pactamycin inhibition of the joining reaction is always less than complete and inversely related to the amount of factor present.

That the block in the 60S subunit joining step <u>per se</u> may not account for the inhibition of polypeptide synthesis by pactamycin is shown in experiments where the joining is permitted to take place (at 5 mM Mg^{2+} or where excess F-0.25 is present) but polypeptide synthesis remains inhibited. These experiments were designed to determine whether the initiation complex formed in lysates or on the 80S ribosome at the higher Mg^{2+} concentration in the presence of pactamycin was able to partake in oligopeptide formation. To test for this, [^{35}S]Met-$tRNA_f$ was incubated in the presence of antibiotic with lysates or salt-washed ribosomes and all the other components necessary for extended polypeptide synthesis, and the ribosome bound radioactivity was analyzed to determine if it had been converted into peptides. As shown in Figure 2, when pactamycin was included there was an accumulation of the initial dipeptide, methionyl-valine (13). Different patterns were seen with other antibiotics. For instance, the

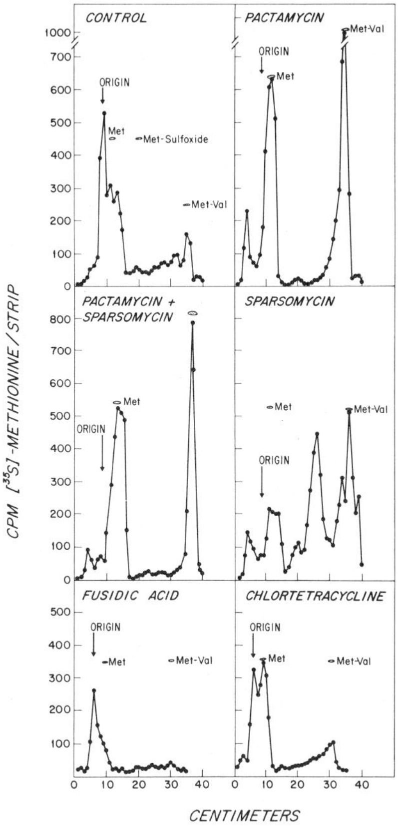

Figure 2. Electrophoretic analysis of peptides synthesized in the presence of pactamycin and other inhibitors. 0.2 ml reaction mixtures were incubated and centrifuged through a sucrose gradient as described (13). The inhibitors (PM, 2 x 10^{-6}M; sparsomycin, 1 x 10^{-5}M; fusidic acid, 2 x 10^{-3}M; chlortetracycline, 1 x 10^{-3}M) when present were added at the beginning. Mono- and poly-somes were isolated and the pellet was incubated with 4 µmoles of NaOH at 37°C for 15 minutes. 80 µl sample was applied on Whatman No. 3 MM paper and electrophoretic analysis was performed ad described (13). The positions of the markers met and met-val are shown in the figure

ribosome bound radioactivity found with sparsomycin is in di-, tri- and oligo-peptides, as expected of an agent that inhibits peptide bond formation independent of the chain length. Similar results with pactamycin have also been obtained in a different system (22).

These data indicate that pactamycin prevents the formation of tripeptide while permitting dipeptide to accumulate. The first peptide bond can form to a considerable extent in the presence of the antibiotic but there is a block at some step after this. Since the dipeptidyl moiety can be released by puromycin, its translocation to the P-site of the ribosome must have occurred before the block, probably at the A-site, was manifest. It is tempting to speculate that the possible interference with A-site function is connected with the observed effect of pactamycin at the ribosomal joining step. It is possible, for instance, that the A-site is obstructed or distorted by a joining factor complex that is unable to dissociate and release the factor for its recycling. Why this block should be expressed at the second time the A-site is used, and not the first (i.e., when Val-tRNA is bound), is not clear.

The above results show that some step(s) early in polypeptide formation is especially sensitive to inhibition by pactamycin. It is possible, however, that an important factor determining selectivity for new chain starts is not only the step being blocked but also at what stage in polypeptide synthesis on ribosomes the antibiotic interacts with its target. If, for instance, pactamycin can interact with (or bind to) the protein synthesizing system only at the time of initiation complex formation or at an even earlier step, but not at all once elongation has begun, then the overall effect is clearly one on initiation even though the step being inhibited is one involved in chain elongation. Thus, it was important to know the precise nature of the interaction of the antibiotic and the polypeptide synthesizing system.

It was known that pactamycin irreversibly inactivates ribosomes but not the soluble factors involved in protein synthesis (23). We found that tritiated pactamycin binds rapidly to the smaller (30S) ribosomal subunit from bacteria (6,15) and also interacts with the smaller (40S) ribosomal subunit from mammalian cells (8).

At 0°C, pactamycin binds to the rabbit reticulocyte 40S ribosomal subunit, as well as to the 80S ribosome, but not to polyribosomes. The binding to 80S ribosomes is presumably by way of the smaller subunit, since there is no binding to the isolated larger subunit. Further, binding to the free smaller subunit has the highest specific activity. Single ribosomes derived from polyribosomes by RNase treatment appear not to bind pactamycin at 0°C. On the other hand, 80S ribosomes produced by NaF treatment of reticulocytes by blocking initiation, bind pactamycin readily. For the most part, they are run-off ribosomes and thus free of mRNA and likely other components involved in polypeptide synthesis. These data plus those using f2RNA as mRNA with E. coli ribosomes (15) indicate that ribosomes that bear even a small fragment of mRNA are not able to bind pactamycin at 0°C. In contrast to the negligible binding of pactamycin to mRNA-bearing monosomes and polyribosomes at 0°C, adding ^{3}H-pactamycin at 37°C results in binding of the antibiotic to polyribosomes. Unlike the rapid binding at 0°C to ribosomes lacking mRNA, binding of pactamycin to polyribosomes is slow, accumulative and sensitive to inhibitors of elongation.

These data suggest that pactamycin binds preferentially to free smaller ribosomal subunits, possibly even after initiation complex formation (and possibly to run-off single ribosomes) before monosome formation. The specificity of pactamycin for initiation may thus rest on the fact that the antibiotic binds to free smaller ribosomal subunits or run-off ribosomes, but not to ribosomes (in polysomes) already engaged in chain elongation. The binding of pactamycin to polysomes found at temperatures where protein synthesis takes place would result from the association of a new initiation complex bearing pactamycin to the initiation site on the mRNA recently vacated by ribosomal progression down the mRNA. That polypeptide synthesis is required to vacate this site is indicated by the fact that elongation inhibitors decrease the amount of pactamycin that becomes bound to the polysomes. In this way, the step in the protein synthesis sequence blocked by pactamycin need not be one peculiar to initiation but may be one involved in elongation, such as translocation or acceptor site binding. Thus, only the last ribosome inactivated by bound antibiotic would be unable to move down the mRNA. Such an effect would be viewed grossly (as measured by polysome degradation and globin chain completion) as one on "initiation" even though it is actually "elongation" that is being affected.

SUMMARY

(1) Pactamycin does not inhibit the binding of [^{35}S]Met-$tRNA_f$ to 40S ribosomal subunits but prevents the formation of 80S initiation complexes and the synthesis of Met-puromycin at 2 mM Mg^{2+}. [^{35}S]Met-$tRNA_f$ accumulates on 40S complexes and oligosomes bearing 40S complexes (e.g., so-called one-and-one half mer, etc.). These effects are much less at 5 mM Mg^{2+}. (2) A factor can be obtained by DEAE-cellulose chromatography of the 0.5 M KCl wash of ribosomes which is required for the joining of 60S subunits to 40S initiation complexes to form 80S initiation complexes and which, when present in excess, can overcome the pactamycin effects described in (1) above in a competitive manner. (3) In systems capable of forming polypeptide, pactamycin results in the accumulation of the initial dipeptide, Met-Val, on the P-site of ribosomes under conditions where the "joining reaction" is not blocked indicating that the antibiotic blocks the formation of tripeptide and higher peptides. (4) Pactamycin binds to the 40S ribosomal subunit at 0°C and does not bind to polyribosomes or to monosomes bearing mRNA at 0°C but does at 37°C.

These data have been interpreted in terms of possible models for the mechanism of action of pactamycin on polypeptide chain initiation.

FOOTNOTE

This work was supported by U.S. Public Health Service Research Grant GM 12573 from the National Institutes of Health.

REFERENCES

1) Colombo, B., Felicetti, L. and Baglioni, C.: Inhibition of protein synthesis in reticulocytes by antibiotics. Biochim. Biophys. Acta 119, 109-119 (1966).

2) Young, C.W.: Inhibitory effects of acetoxycycloheximide, puromycin and pactamycin upon synthesis of protein and DNA in asynchronous populations of HeLa cells. Mol. Pharmacol. 2, 50-55 (1966).

3) Bhuyan, B.K.: Pactamycin, an antibiotic that inhibits protein synthesis. Biochem. Pharmacol. 16, 1411-1420 (1967).

4) Cohen, L.B. and Goldberg, I.H.: Inhibition of peptidyl-sRNA binding to ribosomes by pactamycin. Biochem. Biophys. Res. Commun. 29, 617-622 (1967).

5) Cohen, L.B., Herner, A.E. and Goldberg, I.H.: Inhibition by pactamycin of the initiation of protein synthesis. Binding of N-acetylphenylalanyl transfer ribonucleic acid and poly-uridylic acid to ribosomes. Biochemistry 8, 1312-1326 (1969).

6) Cohen, L.B., Goldberg, I.H. and Herner, A.E.: Inhibition by pactamycin of the initiation of protein synthesis. Effect on the 30S ribosomal subunit. Biochemistry 8, 1327-1335 (1969).

7) Cundliffe, E. and McQuilen, K.: Bacterial protein synthesis: The effects of antibiotics. J. Mol. Biol. 30, 137-146 (1967).

8) Macdonald, J.S. and Goldberg, I.H.: An effect of pactamycin on the initiation of protein synthesis in reticulocytes. Biochem. Biophys. Res. Commun. 41, 1-8 (1970).

9) Stewart-Blair, M.L., Yanowitz, I.S. and Goldberg, I.H.: Inhibition of synthesis of new globin chains in reticulocyte lysates by pactamycin. Biochemistry 10, 4198-4206 (1971).

10) Lodish, H.F., Housman, D. and Jacobsen, M.: Initiation of hemoglobin synthesis. Specific inhibition by antibiotics and bacteriophage ribonucleic acid. Biochemistry 10, 2348-2356 (1971).

11) Ayuso, M. and Goldberg, I.H.: Pactamycin inhibition of eucaryotic polypeptide synthesis dependent on added initiation factors. Biochim. Biophys. Acta 294, 118-122 (1973).

12) Kappen, L.S., Suzuki, H. and Goldberg, I.H.: Inhibition of reticulocyte peptide-chain initiation by pactamycin: accumulation of inactive ribosomal initiation complexes. Proc. Nat. Acad. Sci. USA 70, 22-26 (1973).

13) Kappen, L.S. and Goldberg, I.H.: Inhibition of globin chain initiation in reticulocyte lysates by pactamycin: accumulation of methionyl-valine. Biochem. Biophys. Res. Commun. 54, 1083-1091 (1973).

14) Suzuki, H. and Goldberg, I.H.: Reversal of pactamycin inhibition of methionyl-puromycin synthesis and 80S initiation complex formation by a ribosomal joining factor. Proc. Nat. Acad. Sci. USA, in press (1974).

15) Stewart, M.L. and Goldberg, I.H.: Pactamycin binding to E. coli ribosomes: interference by formation of the protein synthesizing complex with f2 viral RNA. Biochim. Biophys. Acta 294, 123-137 (1973).

16) Seal, S.N. and Marcus, A.: Reactivity of ribosomally bound methionyl-tRNA with puromycin and the locus of pactamycin inhibition of chain initiation. Biochem. Biophys. Res. Commun. 46, 1895-1902 (1972).

17) Crystal, R.G., Shafritz, D.A., Prichard, P.M. and Anderson, W.F.: Initial dipeptide formation in hemoglobin biosynthesis. Proc. Nat. Acad. Sci. USA 68, 1810-1814 (1971).

18) Weeks, D.P., Verma, D.P.S., Seal, S.N. and Marcus, A.: Role of ribosomal subunits in eukaryotic protein chain initiation. Nature 236, 167-168 (1972).

19) Hoerz, W. and McCarty, K.S.: Evidence for a proposed initiation complex for protein synthesis in reticulocyte polyribosome profiles. Proc. Nat. Acad. Sci. USA 63, 1206-1213 (1969).

20) Levin, D.H., Kyner, D. and Acs, G.: Protein synthesis initiation in eukaryotes: Characterization of ribosomal factors from mouse fibroblasts. J. Biol. Chem. 248, 6416-6425 (1973).

21) Bodley, J.W., Zieve, F.J., Lin, L. and Zieve, S.T.: Studies on translocation. III. Conditions necessary for the formation and detection of a stable ribosome-G factor-guanosine diphosphate complex in the presence of fusidic acid. J. Biol. Chem. 245, 5656-5667 (1970).

22) Cheung, C.P., Stewart, M.L. and Gupta, N.K.: Protein synthesis in rabbit reticulocytes: Evidence for the synthesis of initial dipeptides in the presence of pactamycin. Biochem. Biophys. Res. Commun. 54, 1092-1101 (1973).

23) Felicetti, L., Colombo, B. and Baglioni, C.: Inhibition of protein synthesis in reticulocytes by antibiotics. II. The site of action of cycloheximide, streptovitacin A and pactamycin. Biochim. Biophys. Acta 119, 120-129 (1966).

The Ribosomal Binding Site of the Antibiotic Thiostrepton

J. Gordon, G.A. Howard, G. Stöffler and J.H. Highland
Friedrich Miescher-Institute, Basel, Switzerland and
Max-Planck-Institut für Molekulare Genetik, Berlin-Dahlem,
BRD

INTRODUCTION

The antibiotic thiostrepton has aroused interest recently as it appears to be a specific inhibitor which acts on the overlap between the sites of interaction of the elongation factor EF-G and the A site on the 50S subunit of the *E. coli* ribosome (reviewed in refs. 1 and 2 and discussed by Cabrer, San-Millan, Gordon, Vazquez and Modolell elsewhere in this volume). The ribosomal proteins L7 and L12 have been implicated in this area since their presence is required for EF-G and A site interactions (3-8). However, a report by Sopori and Lengyel (9) showed evidence that CsCl cores of the 50S subunit, which do not contain proteins L7 and L12 (10), still had full thiostrepton binding activity. We therefore set out to define in more detail this binding. The first step was to determine whether the binding measured directly with radioactive thiostrepton corresponded to that which was biochemically significant (below and in detail in ref. 11). We then determined which ribosomal protein was responsible for the binding of thiostrepton (below and in detail in ref. 12). Finally, we investigated the relationships between this protein and proteins L7 and L12 (below and in detail in ref. 13).

METHODS

$[^{35}S]$thiostrepton was prepared by growth of the producing organism, *Streptomyces azureus*, in a medium containing $[^{35}S]$sulfate. Isolation of the antibiotic by extraction with ethylene chloride, is described in detail elsewhere (11). The binding of the $[^{35}S]$thiostrepton was determined on Sepharose 6B columns (11). Cores were prepared with the procedure of Hamel *et al.* (3) with ethanol, NH_4Cl, and various temperatures. For clarity, we have modified the notation of Hamel *et al.* (3) in such a way that the subscript indicates the temperature of the treatment as follows. P_0 indicates a core that has been extracted with ethanol-NH_4Cl at 0°, P_{37} indicates a core that has been prepared by extraction of the 50S subunit at 37°, and P_{0-37} indicates a core that has been prepared by treatment of the P_0 core at 37°. Cores were also prepared from the 50S subunits by treatment with LiCl according to the method of Homann and Nierhaus (14), and the proteins split off from these were fractionated on G100 Sephadex columns by a procedure modified from Nierhaus and Montejo (10). Ribosomal proteins were identified by two dimensional gel electrophoresis, using the method of Kaltschmidt and Wittmann (15), as miniaturized by Howard and Traut (16).

RESULTS

1. Characterization of the Thiostrepton Binding

The binding of radioactive thiostrepton to ribosomes was investigated in detail in order to ensure that the measured binding was that which was biochemically significant. The results of this are summarized in Table 1. The results given show that the binding was specific for the 50S bacterial subunit, and the binding was blocked by pre-bound EF-G. This parallels the finding of Highland *et al.* (17) that the presence of pre-bound EF-G prevents the inactivation of ribo-

Table 1 Characteristics of the Binding of Thiostrepton to Ribosomes

Reaction Mixtures	Thiostrepton Bound (Molecules/Ribosome)
50S subunit	1.29
30S subunit	0.12
50S + EF-G	0.58
Rabbit retic., 80S	0.10
Chick liver, 80S	0.15

Reaction mixtures were prepared with the components indicated, and were analyzed on Sepharose 6B columns, except the 30S subunit, where the data was derived from sucrose gradient analysis of the reaction mixture. EF-G was pre-bound to the ribosomes in the presence of fusidic acid and GTP. Details to be published (11).

somes by thiostrepton. The extent of inhibition of thiostrepton binding was the same as the extent of binding of EF-G to this ribosome preparation. Thus, all the characteristics of the binding conform with what would be expected from the biochemical activity. The binding comes close to one mole per mole of ribosomes from titration studies (11).

2. Identification of the Binding Protein

Having established the characteristics of thiostrepton binding, we went onto identify in detail the binding site. We adopted three approaches to this end. The first was to remove selectively as few proteins as possible from the ribosome, and show if the selective removal of an individual protein eliminated the thiostrepton binding capacity of the resultant cores. The second approach was to produce as simple a core fraction as possible, and show the restoration of the thiostrepton capacity by re-binding of an individual protein. The third approach was to use specific antibody techniques to determine which antibody against an individual ribosomal protein could be used to selectively eliminate the thiostrepton binding capacity of the core. While any approach alone may yield equivocal results, all three approaches confirmed the same answer, and so support each other. The core fractions prepared by the method of Hamel et al. are known to be devoid of proteins L7 and L12 (3). Additional proteins which can be removed at higher temperatures, are both important for restoration of the capacity to interact with EF-G (3) and to restore the binding of L7 and L12 (18). These proteins have not previously been identified. Before determining whether any of these proteins were responsible for the binding of thiostrepton, we analyzed the proteins removed by this treatment. The results are summarized in Table 2. In addition, the thiostrepton binding capacity of the core

Table 2 Maximal Particle with Reduced Thiostrepton Binding Activity

Particle Designation	Proteins Removed	Thiostrepton Bound (% of 50S)
P_0	L7,L12	100
P_{37}	L7,L12,L10	92
P_{0-37}	L7,L12,L10,L11	23

The various particles were as designated under Methods, and their protein contents and thiostrepton binding activities analyzed. Details to be published (17).

particles was determined. The results in the table show that one step of treatment at 37° removed protein L10 in addition to proteins L7, L12. This did not affect the thiostrepton binding. However, sequential treatments at 0° and 37° resulted in the loss of L11, as well as a considerable reduction in the thiostrepton binding. This implicates protein L11 as the protein responsible for the binding of thiostrepton.

To confirm this, we prepared more extensively reduced cores by treatment with LiCl (14). In addition, we restored the binding activity by re-addition of the split protein fractions. The results of these studies are given in Table 3. It can be seen that treatment with 0.5 M LiCl removed about 50% of the thiostrepton binding capacity, while treatment with 1 M LiCl removed it quantitatively. Further, re-addition of the 1 M LiCl split fraction to its respective core fully

Table 3 Minimal Particle with Full Thiostrepton Binding Activity

Core	Split	Thiostrepton Bound (% of 50S)
1/2 M LiCl	—	48
1 M LiCl	—	0
1 M LiCl	1 M LiCl	100
4 M LiCl	1 M LiCl	100
RNA	1 M LiCl	0
4 M LiCl	G100 - 1 M LiCl	100

The various cores and splits described were prepared and mixed together in equivalent amounts. The G100 fraction was the peak of activity in this assay following G100 fractionation of the 1 M LiCl split fraction. To be published in detail (12).

restored its thiostrepton binding capacity. In order to approach a minimal system, both core and split fractions were fractionated further. The 1 M LiCl core could be fractionated further with 4 M LiCl without loss in ability to complement the 1 M LiCl split protein fraction in thiostrepton binding activity. However, removal of the remaining 7 proteins from this core (RNA in Table 3) eliminated the activity. The proteins in the 1 M LiCl split were reduced in number by fractionation on Sephadex G100. This fraction could fully restore the thiostrepton binding capacity of the 4 M LiCl core. This fraction was mainly L11, with traces of proteins L1, L8 and L9. This again implicates L11 as the thiostrepton binding protein.

The protein responsible for thiostrepton binding was finally confirmed as L11 by specific antibody studies. With intact ribosomes we were unable to obtain any antibody inhibition of thiostrepton binding by any of the specific antibodies against individual ribosomal proteins. We therefore treated the 1 M LiCl split protein fraction with the antibodies. Table 4 summarizes the results obtained.

It can be seen that of all the specific antibodies directed against the proteins in the 1 M LiCl split fraction, only that directed against L11 inhibited the thiostrepton binding. Similarly, when the reconstituted minimum core fraction was treated with antibodies against its protein constituents, only antibody to L11 inhibited. Thus, these results, too, implicate L11 as the protein responsible for the binding of thiostrepton.

Table 4 Antibody Inhibition of Thiostrepton Binding

Treatment	Thiostrepton Bound (% of untreated)
Proteins in 1 M LiCl split treated:	
L11	21
L1, L2, L5-L10, L12, L14-L16, L18, L25-L30, L33	90-100
Proteins of reconstituted minimal core treated:	
L11	30
L3, L4, L13, L17, L21, L22, L23	96-100

Either the 1 M LiCl split fraction was treated with specific antibodies to the proteins in it, as indicated, or the reconstituted minimal particle: 4 M core, Sephadex G100 fractionated 1 M LiCl split as described in Table 2. Each was treated with antibodies directed against the individual protein constituents as indicated. Details to be published (17).

3. Relationship of Protein L11 to L17 and L12

Since proteins L7 and L12 have been implicated in the set of reactions which thiostrepton inhibits (1-8), it seemed of interest to determine if we could detect an interaction between these proteins and L11. Since the ethanol-NH_4Cl treatment removes this group of proteins (Table 1, ref. 12,13), we have the opportunity to test whether L11 is required for the re-binding of L7 and L12. By suitable mixtures of the three kinds of core and split fractions described in Table 2, it was possible to construct complete ribosomal protein mixtures with single protein omissions. We then determined which proteins were re-bound in these single omission mixtures, and the resultant reconstitutions are summarized in Table 5. The results in the table show that when all the removed proteins were added back, they all rebound either when the P_O core, the P_{37} core or the P_{O-37} cores were mixed with their respective protein complements. However, when protein L10 was absent from the mixture, L7 and L12 failed to rebind. When L11 was absent from the mixture, L10 as well as L7 and L12 failed to rebind. Thus, we conclude that L11 was required for the rebinding of L10, which in turn was required for the re-binding of L7 and L12. Hence, we can define an interaction, albeit indirect, between L11 and L7 and L12.

Table 5 Reconstitutibility of Various Ethanol Core and Split Fractions

Core	Split	Proteins in Reconstitution	Proteins Re-Bound
P_O	P_O	complete	L7,L12
P_{37}	P_{37}	complete	L7,L12,L10
P_{37}	P_O	L10 absent	none
P_{O-37}	P_O+P_{O-37}	complete	L7,L12,L10,L11
P_{O-37}	P_{37}	L11 absent	none

The core and split fractions were mixed, the reconstituted particle re-isolated, and the protein content of the particles analyzed. Details to be published (13).

DISCUSSION

The binding site for the antibiotic thiostrepton has been defined as protein L11 by a number of criteria. Thiostrepton inhibition was the first evidence for an overlap between the sites of interaction of EF-G and that of the aminoacyl tRNA·EF-Tu·GTP complex. Subsequently EF-G and aminoacyl tRNA binding were found to be mutually exclusive on the ribosome, and were thus suggested to have overlapping sites of interaction (1,2). Since proteins L7, L12 were found to be involved in both of these interactions (3-8), it might be concluded that the overlap was on the proteins L7, L12. The finding that L11 is involved in the binding of thiostrepton suggests that L11, as well as L7 and L12, may be a common part of the sites of interaction of EF-G and aminoacyl tRNA. However, none of the data rules out an indirect interaction.

The inter-relationship between L11 and L7, L12 is supported by other data. First, our finding that L10 is required for the re-binding of L7, L12 is consistent with the finding that anti-L10 prevents the re-binding of L7, L12 (21) and that cores lacking L10 are unable to reconstitute EF-G dependent GTPase activity (22). Furthermore, there is evidence that L10 and L11 are neighbors, since these also have been claimed as components of a pair cross-linkable by bi-functional reagents (23). Another indication of a role of L11 in this region of the ribosome comes from the observation that a photo-affinity label derived from GTP attaches to L11, amongst others, when bound with EF-G and fusidic acid (24).

Evidence has been presented in favor of protein L11 being part of the peptidyl transferase catalytic center (10,19), although thiostrepton itself is not an inhibitor of the peptidyl-transferase activity (1). The evidence for L11 being part of the peptidyl transferase center came from the finding of restoration of peptidyl transferase by re-addition of protein L11 to LiCl cores (10), and from inhibition of the hydrolytic step of chain termination by antibody to L11 (19). This reaction is considered to be mediated by the peptidyl transferase (20). However, the termination hydrolysis was inhibited by the action of the antibody on the intact ribosome (19). On the other hand, with the same antibody preparation, we only observed inhibition of thiostrepton binding by the treatment of separated proteins, or of a reconstituted incomplete particle. This suggests that the antigenic determinants in the peptidyl transferase region of the molecule are distinct from the determinants in the thiostrepton binding site, even though both may be on protein L11.

SUMMARY

The antibiotic thiostrepton has been prepared by growth of the producer organism in a medium containing [^{35}S]sulfate. The antibiotic bound stoichiometrically and specifically to the large subunit of E. coli ribosomes. It did not bind to eukaryotic ribosomes. The binding was blocked by the presence of pre-bound EF-G to the same extent as the EF-G was bound to the ribosomes. The ribosomal protein responsible for the binding of thiostrepton was identified as L11 by the following criteria: selective removal of L11 by treatment of the 50S subunit with ethanol, NH_4Cl and temperature also removed the thiostrepton binding capacity; selective re-addition of a preparation containing mainly protein L11 restored the thiostrepton binding capacity of ribosomal cores containing only 7 other ribosomal proteins; only antibodies to protein L11 eliminated the thiostrepton binding capacity either by separate treatment of a protein fraction, or by treatment of the 4 M LiCl core after reconstitution. A link between protein L11 and L7, L12 was deduced from reconstitution experiments. Various fractions obtained with the ethanol, NH_4Cl, temperature treatment permit the construction of mixtures consisting of a complete set of ribosomal proteins, but with the omission of L11 or L10. From these reconstitution mixtures it was deduced that L11 is required for the rebinding of L10 and L10 is required for the rebinding of L7, L12.

REFERENCES

1. Pestka, S., Bodley, J.W.: The thiostrepton group of antibiotics. In "Antibiotics", Gottlieb & Shaw, Eds., Springer-Verlag, Berlin, 1974. In press.

2. Modolell, J., Vazquez, D.: Polypeptide chain elongation and termination, chapter 4 in "MTP International Review of Science; Biochemistry". Vol. 7, H.R.V. Arnstein, Ed., Medical and Technical Publishing Co., Oxford, 1974. In press.

3. Hamel, E., Koka, K., Nakomoto, T.: Requirement of an Escherichia coli 50S ribosomal protein component for effective interaction of the ribosome with T and G factors and with guanosine triphosphate. J. Biol. Chem. 247, 805-814 (1972).

4. Kischa, K., Möller, W., Stöffler, G.: Reconstitution of a GTPase activity by a 50S ribosomal protein from E. coli. Nature New Biol. 233, 62-63 (1971).

5. Highland, J.H., Bodley, J.W., Gordon, J., Hasenbank, R., Stöffler, G.: Identity of the ribosomal proteins involved in the interaction with the elongation factor G. Proc. Nat. Acad. Sci. USA 70, 147-150 (1973).

6. Highland, J.H., Ochsner, E., Gordon, J., Bodley, J.W., Hasenbank, R., Stöffler, G.: Coordinate inhibition of elongation factor G function and ribosomal subunit association by antibodies to several ribosomal proteins. Proc. Nat. Acad. Sci. USA 71, 627-630 (1974).

7. Highland, J.H., Ochsner, E., Gordon, J. Hasenbank, R., Stöffler, G.: Inhibition of phenylalanyl-tRNA binding and EF-Tu dependent GTP hydrolysis by antibodies specific for several ribosomal proteins. J. Mol. Biol. (1974). In press.

8. Sander, G., Marsh, R.C., Parmeggiani, A.: Isolation and characterization of two acidic proteins from the 50S subunit required for GTPase activities of both EF-G and EF-T. Biochem. Biophys. Res. Commun. 47, 866-873 (1972).

9. Sopori, M.L., Lengyel, P.: Components of the 50S ribosomal subunit involved in GTP cleavage. Biochem. Biophys. Res. Commun. 46, 238-244 (1972).

10. Nierhaus, K.H., Montejo, V.: A protein involved in the peptdidyltransferase activity of Escherichia coli ribosomes. Proc. Nat. Acad. Sci. USA 70, 1931-1935 (1973).

11. Gordon, J., Highland, J.H.: Binding of thiostrepton to the ribosomes of E. coli: Characterization and stoichiometry of the binding. In preparation.

12. Highland, J.H., Howard, G.A., Ochsner, E., Stöffler, G., Hasenbank, R., Gordon, J.: Identification of the ribosomal protein responsible for the binding of thiostrepton to E. coli ribosomes. J. Biol. Chem. (1974). In press.

13. Highland, J.H., Howard, G.A.: Assembly of ribosomal proteins L7, L10, L11, L12 on the 50S subunit of E. coli. J. Biol. Chem. (1974). In press.

14. Homann, H.E., Nierhaus, K.H.: Protein compositions of biosynthetic precursors and artificial subparticles from ribosomal subunits in Escherichia coli K12. Eur. J. Biochem. 20, 249-257 (1971).

15. Kaltschmidt, E., Wittmann, H.G.: Ribosomal proteins. VII. Two-dimensional polyacrylamide gel electrophoresis for fingerprinting of ribosomal proteins. Anal. Biochem. 36, 401-412 (1970).

16. Howard, G.A., Traut, R.R.: Separation and radioautography of microgram quantities of ribosomal proteins by two-dimensional polyacrylamide gel electrophoresis. FEBS Letters 29, 177-180 (1973).

17. Highland, J.H., Lin, L., Bodley, J.W.: Protection of ribosomes from thiostrepton inactivation by the binding of G factor and guanosine diphosphate. Biochemistry 10, 4404-4409 (1971).

18. Brot, N., Marcel, R., Yamasaki, E., Weissbach, H.: Further studies on the role of 50S ribosomal proteins in protein synthesis. J. Biol. Chem. 248, 6952-6956 (1973).

19. Tate, W.P., Caskey, C.T., Stöffler, G.: Inhibition of peptide chain termination by antibodies specific for ribosomal proteins. In preparation.

20. Caskey, C.T., Beaudet, A.L., Scolnick, E.M., Rosman, M.: Hydrolysis of fMet-tRNA by peptidyl transferase. Proc. Nat. Acad. Sci. USA 68, 3163-3167 (1971).

21. Stöffler, G., Hasenbank, R., Bodley, J., Highland, J.H.: Inhibition of protein L7/L12 binding to 50S ribosomal cores by antibodies specific for proteins L6, L10 and L18. J. Mol. Biol. (1974). In press.

22. Schrier, P.I., Maassen, J.A., Möller, W.: Involvment of 50S ribosomal proteins L6 and L10 in the ribosome dependent GTPase activity of elongation factor G. Biochem. Biophys. Res. Commun. 53, 90-98 (1973).

23. Clegg, C., Hayes, D.: Identification of neighbouring proteins in the ribosomes of *Escherichia coli*. A topographical study with the cross-linking reagent dimethyl suberimidate. Eur. J. Biochem. 42, 21-28 (1974).

24. Maassen, J.A., Möller, W.: Identification by photo-affinity labelling of the proteins in E. coli ribosomes involved in EF-G dependent GDP binding. Proc. Nat. Acad. Sci. USA 71, 1277-1280 (1974).

tRNA Structures in Viral RNA Genomes*

A.L. Haenni, A. Prochiantz and P. Yot
Laboratoire de Biochimie du Développement, Institut de Biologie Moléculaire du C.N.R.S. et de l'Université Paris VII, 2 Place Jussieu, 75005 Paris, FRANCE.

INTRODUCTION

As opposed to the long-lived postulate whereby a given macromolecule performs one specific function in the living cell, it is becoming increasingly apparent that at least certain macromolecules are pleitropic, operating at various levels in cell development.

With respect to tRNAs, it has been discovered that beside their fundamental function in the translation of the genetic code, they can be involved in other very different functions. For example, they participate in the synthesis of the bacterial cell wall donating certain amino acids, and as such they are confronted with cellular components that are completely different from those they encounter during messenger-dependent protein synthesis. Furthermore, certain tRNA molecules can play a regulatory role in gene expression, as in the case of the histidine operon. More recently, a tRNA has appeared as a possible primer for RNA-dependent DNA synthesis.

In this paper, we wish to report on yet another intriguing situation involving tRNAs, namely the existence of a tRNA structure as an intrinsic part of several viral RNAs. We present a summary

*It is our pleasure to dedicate this article to Dr. Fritz Lipmann with all our admiration and gratitude.

of the existing evidence for the presence of tRNA structures in viral RNAs, and then show that to the already extensive list of tRNA-specific enzymes capable of recognizing certain plant viral RNAs, can be added the *E. coli* elongation factor EF-T (=EF-Tu and EF-Ts). Finally, we discuss the possible physiological role of these structures in the development of viruses.

Of the several plant viral RNAs screened, Turnip Yellow Mosaic Virus (TYMV) RNA has so far brought the most conclusive evidence for the existence of a tRNA structure. Since moreover it is the model with which we are the most familiar, we will primarily use it in the description of the experiments to follow.

RESULTS

Esterification of TYMV RNA Preparations by Valine

We have shown (Table 1) that valine is esterified to the 3' end of TYMV RNA in the presence of ATP and an *E. coli* cell-free extract devoid of tRNAs (1, 2). The incubation conditions and the kinetics of valyl-RNA formation were the same as those reported for the esterification of valine to tRNA. Although the ratio of valine bound per RNA molecule of 2×10^6 daltons fluctuated between 0.3 and 0.8 depending on the batch of RNA, it never exceeded one. No other amino acid could be bound to the RNA and no other nucleoside-triphosphate could replace ATP.

Table 1. Requirements for the binding of valine to TYMV RNA

Conditions	^{14}C-Valine bound (pmoles)
Complete system	37
- Proteins	<1
- ATP	<1
- TYMV RNA	<1

The incubation contained 50 pmoles of TYMV RNA. From Pinck *et al.* (1).

In order to determine the nature of the linkage formed between the amino acid and the RNA of TYMV, we resorted to four methods:

1. The stability of TYMV valyl-RNA was measured at pH 8.6 and compared to that of valyl-tRNA: in both cases the half-life time was 60 min at 37°C, suggesting an ester linkage between valine and the viral RNA.

2. After chemical acetylation of valyl-RNA of TYMV in conditions in which the α-amino group of aminoacyl-tRNA is acetylated, it could be shown that acetylvalyl-RNA was formed, indicating that the amino group of valine in valyl-RNA is free.

3. Using this acetylvalyl-RNA, it was observed that the peptidyl-tRNA hydrolase releases acetylvaline as it does with acetyl-valyl-tRNA.

4. When TYMV valyl-RNA was submitted to digestion by pancreatic RNase or snake venom phosphodiesterase, the chromatographic analysis showed that the products formed were respectively valyl-adenosine and valyl-AMP.

The conclusion of these experiments was that in TYMV RNA valine is attached to the 3' position of a terminal adenosine by an energy rich ester bond.

Valine is Esterified to High Molecular Weight RNA

At this stage it was imperative to show that the valine acceptor was not a tRNA molecule that co-purified with TYMV RNA. All attempts at dissociating such a putative complex using treatments with sodium dodecylsulfate, urea, EDTA or dimethylsulfoxyde and formaldehyde were unsuccessful: after all these treatments valine co-chromatographed with high molecular weight RNA. However, although rather unlikely, it could not *a priori* be excluded that in the cell a $tRNA^{Val}$ had become covalently linked

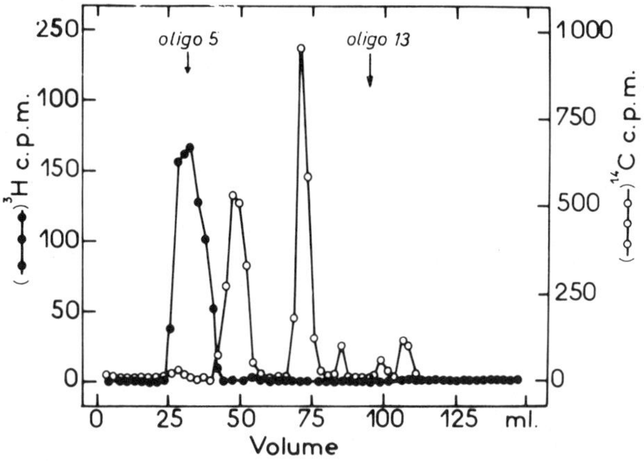

Figure 1. DEAE-cellulose chromatography of T_1 RNase digests of TYMV ^{3}H-valyl-RNA and cabbage ^{14}C-valyl-tRNA. The positions of oligo 5 and oligo 10 were defined by treating *E. coli* ^{14}C-valyl-tRNA with T_1 RNase and analyzing the digest in similar conditions. From Yot *et al.* (2).

to the 3' end of the TYMV genome by a hypothetical RNA-ligase.

The following experiment was carried out to check this hypothesis. tRNA was extracted from the uninfected host (Chinese cabbage leaves), charged with ^{14}C-valine, submitted to T_1 RNase digestion together with ^{3}H-valyl-RNA of TYMV, and the valyl-oligonucleotides formed were analyzed using a DEAE-cellulose column in the presence of urea. Figure 1 shows that the valyl-oligonucleotide from valyl-RNA of TYMV is different from those pertaining from the host valyl-tRNA: the former yields a pentanucleotide and the latter two valyl-oligonucleotides, one of about seven and the other of about nine nucleotides in length (2).

Our general conclusion from these experiments was that indeed valine forms an ester bond with the adenosine at the 3' end of the viral genome itself and not of a contaminating tRNA.

TYMV RNA as Substrate of the tRNA Nucleotidyltransferase

In order to define whether valyl-tRNA synthetase was responsible for the charging of valine onto TYMV RNA, the *E. coli* extract was replaced by pure valyl-tRNA synthetase in the charging reaction.

Table 2. Acceptor activity of *E. coli* tRNA and TYMV RNA for ^{14}C-valine

Enzymes added	*E. coli* tRNA (cmp/μg RNA)	TYMV RNA (cmp/μg RNA)
DEAE-enzyme preparation	425	56
Valyl-tRNA synthetase	406	5
tRNA nucleotidyltransferase, and valyl-tRNA synthetase	375	53

All enzymes were from *E. coli*. From Yot *et al.* (2).

As seen in Table 2, no esterification of valine to the viral RNA was observed in these conditions. However the addition of tRNA nucleotidyltransferase restored the charging of TYMV RNA. Furthermore, when ^{3}H-ATP was used in the presence of this enzyme, AMP became covalently bound to TYMV RNA (3). Moreover phosphodiesterase-treated TYMV RNA incorporated two CMP per AMP residue (4).

These results indicate that the RNA as it is extracted from the virion is devoid of a 3'-terminal adenosine residue and suggest that, as with tRNA, the complete -CCA sequence is required for charging to occur.

tRNA-Like Structure at the 3' End of TYMV RNA

The fact that three enzymes normally thought to be specific of tRNAs, namely tRNA-nucleotidyltransferase, valyl-tRNA synthetase and peptidyl-tRNA hydrolase, recognize the 3' region of TYMV RNA led us to conclude that this part of the molecule must possess a tRNA-like structure. It thus suggested that other enzymes specific for tRNAs, or even the components of the protein synthesizing machinery might possibly also recognize that part of the molecule.

1. Recognition by an endonuclease. In order to define if RNase P, an endonuclease involved in the maturation of tRNA from its 5S

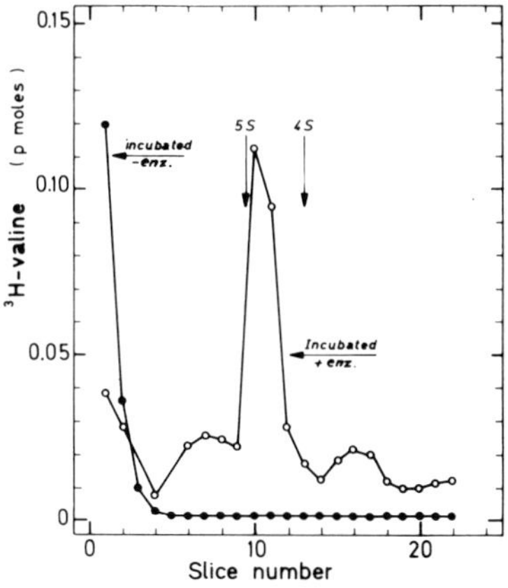

Figure 2. Effect of endonuclease on TYMV ^{3}H-valyl-RNA. Conditions of incubation and of gel electrophoresis were as indicated in Prochiantz and Haenni (5).

precursor, is also active on TYMV RNA, an enzyme fraction behaving similarly to RNase P was prepared from *E. coli* ribosomes and incubated with TYMV RNA charged or not with valine (5). In both cases, there resulted the formation of a 4.5S fragment corresponding to the 3' region of TYMV RNA as judged by gel electrophoresis (see Figure 2) or by Sephadex G-100 filtration. This implies that there exists a great similarity between the structures in the 5S tRNA precursor and in the 3' region of TYMV RNA. Under appropriate conditions, the isolated fragment could also be acceptor of valine indicating that other parts of the viral genome do not participate in the tRNA-like structure.

2. Recognition of valyl-RNA by the ribosomal system. Using an *E. coli* protein synthesizing system, valyl-RNA of TYMV, a mixture of tRNAs and all the amino acids except valine, it was found that

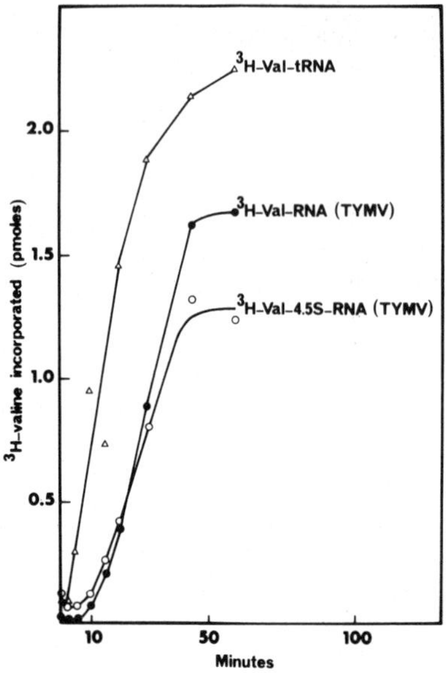

Figure 3. Kinetics of incorporation.The conditions of incubation were as already described (6) except that in all cases 2 nmoles of TYMV RNA were added as mRNA and that where indicated *E. coli* ^{3}H-valyl-tRNA (26 nmoles of tRNA charged with 20 pmoles of ^{3}H-valine), or TYMV ^{3}H-valyl-RNA (170 pmoles of RNA charged with 20 pmoles of ^{3}H-valine) preincubated or not with endonuclease were included. Total incubation mixtures were 5 ml; at various times, 0.5 ml aliquots were removed for hot TCA determinations.

some of the valine was recovered in the hot TCA precipitate (6). We verified that this corresponded to the direct transfer of valine from TYMV RNA, and not to incorporation after a transacylation process.

As compared to the incorporation of valine from valyl-tRNA which proceeded after a short lag-time of 2 to 3 min, a considerably longer lag-time was observed with valyl-RNA of TYMV as valine donor. In order to examine whether this prolonged lag-time might reflect the liberation of the valyl-4.5S-RNA fragment that would preceed incorporation of valine, the 4.5S-RNA fragment charged with valine was used directly in the incubation mixture. In

Figure 3 we compare the incorporation of valine from various valine donors into the hot TCA precipitate, having added TYMV RNA as mRNA in all cases. Valyl-RNA and valyl-4.5S-RNA are about half as active as valyl-tRNA in donating their amino acid for polypeptide synthesis. The lag-time observed with valyl-4.5S-RNA was not reduced when compared to that seen with valyl-RNA. Therefore, the liberation of a valyl-RNA fragment which most likely preceeds incorporation of valine into peptide chains appears not to be the limiting factor responsible for the lag-time. This lag-time might correspond to a further modification of the valyl-4.5S-RNA prior to incorporation.

3. Recognition by the elongation factor EF-T. Since valine from valyl-RNA is used by the ribosomal system, it appeared likely

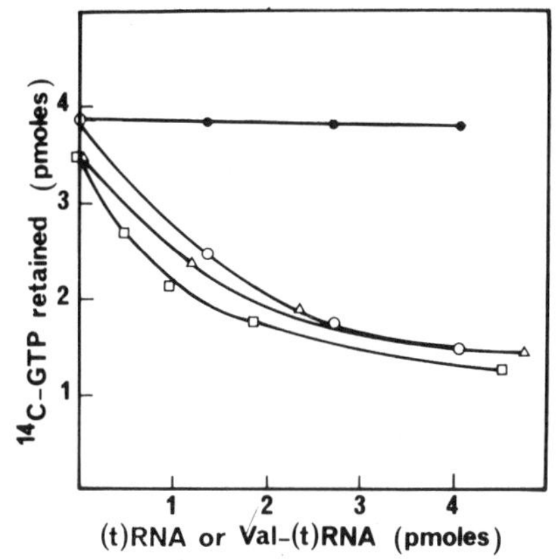

Figure 4. Interaction of TYMV valyl-RNA with EF-T and GTP. Various amounts of acylated or non acylated (t)RNAs were incubated for 10 min at 0°C in a total volume of 200 µl containing 2 µmoles $MgCl_2$, 10 µmoles NH_4Cl, 2 µmoles Tris-HCl pH 7.5, 50 µmoles ^{14}C-GTP (260 mCi/mmole), 15 µg EF-Ts and 0.21 µg EF-Tu. The reaction was stopped by dilution with 2 ml of a cold buffer solution containing 10 mM $MgCl_2$, 50 mM NH_4Cl, 10 mM Tris-HCl pH 7.5. The mixture was then filtered over a Millipore filter and the radioactivity retained determined. ●—● Uncharged (t)RNA; △—△ Val-tRNA (*E. coli*); ○—○ Val-RNA (TYMV); □—□ Val-RNA (TYMV) preincubated with endonuclease.

that it might be used during elongation of peptide chains after interaction with EF-Tu. Indeed, Litvak *et al.* (7) have shown that valyl-RNA of TYMV interacts with the wheat germ elongation factor EF-1 in the presence of GTP.

Using highly purified *E. coli* EF-Tu and EF-Ts (kindly supplied by Dr. A. Parmeggiani), the formation of a ternary complex was examined using the Millipore binding assay. Figure 4 demonstrates that both valyl-RNA and valyl-4.5S-RNA of TYMV form a ternary complex with the elongation factors and GTP, and that the stoichiometry of valyl-RNA to GTP in the complex is comparable to that observed with valyl-tRNA. As with tRNAs, the uncharged RNA did not lead to complex formation. These results however provide no information about the respective stabilities of the different complexes.

DISCUSSION

Since the 3' region of TYMV RNA possesses all the known functions of tRNAs, we conclude that it must contain a structure similar to that of tRNAs as suggested in the model presented in

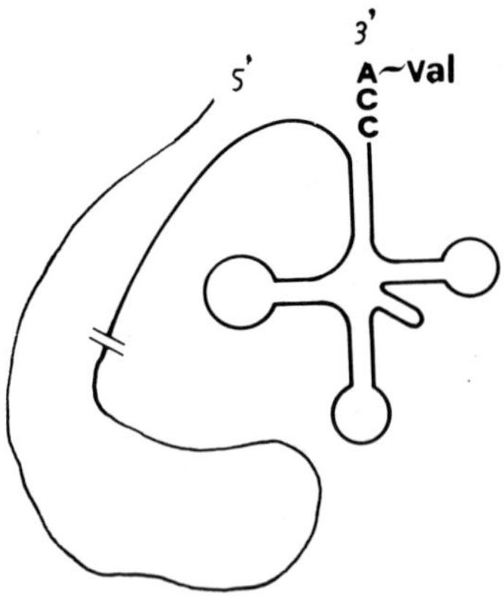

Figure 5. Possible model for TYMV valyl-RNA.

Figure 5. The presence of such a structure, which at least *in vitro* is not masked by the 98 or 99% of the remaining part of the viral RNA, must be of primordial importance for virus development.

TYMV RNA is not the only viral RNA capable of accepting an amino acid. The RNAs of other Tymoviruses, such as the Okra Mosaic Virus and the Eggplant Mosaic Virus, also accept valine (8); the RNAs of the Bromoviruses as well as of Cucumber Mosaic Virus accept tyrosine (9 and R.J. Kohl and T.C. Hall, personal communication);Tobacco Mosaic Virus RNA binds histidine (10). With respect to animal systems, only mengovirus RNA has so far been shown to accept an amino acid, namely histidine (11).

These experiments might indicate that other viral RNAs contain a tRNA-like structure in their 3' region. It should however be pointed out that the RNA of several animal viruses are terminated at their 3' end by a poly(A) sequence. This nevertheless does not exclude the possibility that the poly(A) sequence follows a tRNA-like structure in the genome: it is known that tRNAs are efficiently used as primers by the poly(A) synthetase.

Very little can be said about the presence of similar structures in RNA phages, although it is well established that these RNAs possess the 3' terminal -CCA sequence common to all tRNAs. Recently, using fragmented R17, MS2 and Qβ RNA, we found that the tRNA nucleotidyltransferase recognizes some of these fragments in conditions in which the enzyme is highly specific for tRNAs (A. Prochiantz *et al.*, submitted for publication). However, up to the present time,it has been impossible to charge a phage RNA with an amino acid and it therefore appears that in RNA phage genomes only vestiges of tRNA-like structures might be present.

The physiological role of tRNA-like structures in viral genomes remains obscur. It is likely that these structures represent signals at some stage of viral development. At least two possible types of signals can be envisaged:

1. The tRNA-like structure could be involved in the recognition of the genome by the RNA replicase. If the replicating system of these viruses were to contain the elongation factors as subunits, as in the case of Qβ replicase, then the tRNA structure could represent the binding site for the replicase, and as such could be involved in the initiation of the replicating process.

2. The tRNA-like structure could play a role in the translation process either by allowing preferential synthesis of virus specific protein(s) as in the case of tRNAs specific of certain DNA bacteriophages, or by interfering with host valyl-tRNA, thereby inhibiting host specific protein synthesis.

The fact that several viral RNAs contain a tRNA-like structure raises the question of the origin of these structures. In the case of TYMV, for example, it would be interesting to know the primary sequences of all the cytoplasmic and chloroplastic tRNAs isoacceptor of valine, and compare them to that of the 3' region of TYMV RNA. If significant similarities were to be found with one of these isoacceptors, interesting conclusions could undoubtedly be drawn as to the possible origin of these viruses.

SUMMARY

It is shown that TYMV RNA of 2×10^6 daltons is esterified at its 3' end by valine in the presence of valyl-tRNA synthetase. Furthermore, the 3' region of this RNA is recognized by all the tRNA-specific enzymes tested: tRNA nucleotidyltransferase, peptidyl-tRNA hydrolase, an endonuclease activity analogous to RNase P, and EF-T. An RNA fragment of 4.5S corresponding to the 3' region of the genome has been isolated. It contains all the tRNA properties of total valyl-RNA, including the capacity to donate its valine for the synthesis of peptide chains. The possible physiological role of these tRNA-like structures in the viral genome is discussed.

ACKNOWLEDGEMENTS

We are grateful to Dr. F. Chapeville in whose laboratory this work was carried out for his interest and his constant encouragement. We also wish to thank Dr. A. Parmeggiani for having kindly received one of us (A.L. H.) in his laboratory, and C. Bénicourt, O. Bernard and J.P. Dumas for help in some of the experiments and for stimulating discussions.

This work was supported by research grants from NATO (n°769) and from the ATP Differenciation Cellulaire (n°4910), Centre National de la Recherche Scientifique.

REFERENCES

1. Pinck, M., Yot, P., Chapeville, F., Duranton, H.M.: Enzymatic binding of valine to the 3' end of TYMV RNA. Nature 226, 954-956 (1970).

2. Yot, P., Pinck, M., Haenni, A.L., Duranton, H.M., Chapeville, F.: Valine-specific tRNA-like structure in Turnip Yellow Mosaic Virus RNA. Proc. Nat. Acad. Sci.US. 67, 1345-1352 (1970).

3. Litvak, S., Carré, D.S., Chapeville, F.: TYMV RNA as a substrate of the tRNA nucleotidyltransferase. FEBS Letters 11, 316-319 (1970).

4. Litvak, S., Tarrago-Litvak, L. Chapeville, F.: TYMV RNA as a substrate of the tRNA nucleotidyltransferase II.Incorporation of CMP and determination of a short nucleotide sequence at the 3' end of the RNA. J. Virol. 11, 238-242 (1973).

5. Prochiantz, A., Haenni, A.L.: TYMV RNA as a substrate of the tRNA maturation endonuclease. Nature New Biology 241, 168-170 (1973).

6. Haenni, A.L., Prochiantz, A., Bernard, O., Chapeville, F.: TYMV valyl-RNA as an amino acid donor in protein biosynthesis. Nature New Biology 241, 166-168 (1973).

7. Litvak, S., Tarrago, A., Tarrago-Litvak, L., Allende, J.E.: Elongation factor-viral genome interaction dependent on the aminoacylation of TYMV and TYMV RNAs. Nature New Biology 241, 88-90 (1973).

8. Pinck, M., Chan, S., Genevaux, M., Hirth, L., Duranton, H.M.: Valine specific tRNA-like structure in RNAs of two viruses of Turnip Yellow Mosaic Virus group. Biochimie 54, 1093-1094 (1972).

9. Hall, T.C., Shih, D.S., Kaesberg, P.: Enzyme-mediated binding of tyrosine to Brome-Mosaic-Virus ribonucleic acid. Biochem. J. 129, 169-176 (1972).

10. Öberg, B., Philipson, L.: Binding of histidine to Tobacco Mosaic Virus RNA. Biochem. Biophys. Res. Commun. 48, 927-932 (1972).

11. Salomon, R., Littauer, V.Z.: Enzymatic acylation of histidine to mengovirus RNA. Nature 249, 32-34 (1974).

Remarks on the Acquisition of Active Quaternary Structure of Enzymes

Benno Hess
Max-Planck-Institut für Ernährungsphysiologie,
Dortmund / Germany

When a peptide chain is born by ribosomal synthesis its first function is to fold itself to form a three-dimensional highly specific structure. In case of oligomeric enzymes the folding process will be succeeded by an assembly of folded monomeric units leading to the formation of quaternary structures. Whereas conventional points of view indicate that native proteins are the thermodynamicly most stable structures, evidence and arguments have recently been presented showing the operation of kinetic constraints in the folding and selfassembly of native proteins. Indeed, the folding of relatively small peptide chains might proceed straight forward in the range of msec in an aqueous milieu supported by suitable mono- or divalent kations and anions, but the folding of large chains with molecular weights up to 50.000 and more is a complex process. Since a random search mechanism of folding involves a power function depending on the chain length, the folding time would be indefinitely long, if no kinetic constraints would operate (for summary see 1, 2, 3). Thus, a process of selfassembly can be visualized as:

1. A folding process in which a "disordered" monomeric polypeptide chain yields a specific ordered native conformation involving a nucleation step

and subsequently a growth event. The formation of a nucleation centre implies a limited number of amino acid residues. During the growth helical and non-helical regions, pleated sheets and interchain regions are generated.

2. In case of oligomeric enzymes an association of "rightly" folded monomers towards higher state of aggregation follows.

The function of a specific renaturation primer can be illustrated in case of the yeast enzyme pyruvate kinase. The enzyme of S. carlsbergensis is a tetrameric protein composed of four identical subunits (4). Each subunit contains one mole of L-valine noncovalently bound. The enzyme readily dissociates into monomeric units. So far, no renaturation was observed. Recently, it was found that L-valine is a specific primer of the renaturation process (5). The amino acid induces renaturation with a $K_{0.5}$ of 17 μM and a pseudo first order rate constant of 0.019 min^{-1} at 25° with respect to the monomeric species, indicating that L-valine influences the folding of the monomeric form from a disordered state to its native conformation being followed by a spontaneous reassociation with formation of the tetrameric enzyme. This action of the amino acid is highly specific and only shared, however to a lesser degree of yield, by the analogues γ-hydroxyvaline and norvaline. All natural amino acids as well as D-valine are inactive in this function. An analysis of the structural requirements shows that both charge groups as well as the hydrophobic cluster of the L-configuration specificly interact with the peptide chain in the induction process. This novel function of an amino acid in the renaturation process of pyruvate kinase of yeast might well be fulfilled in case of other proteins by low molecular ligands such as co-enzymes (6,7) or divalent ions (5).

The high complexity of the process of folding of large polypeptides, such as occuring in case of the glycolytic enzymes, rises the question of a possible similarity of the folding pathway depending on a given structural size and enzymic function. Indeed, the discovery of similarities of the general topology of the tertiary chain structure in enzymes of the class of dehydrogenases as well as of a phosphotransferase might support this view in addition to the concept of evolutionary relationship (8, 9). Also it should be considered that the establishment of active centres and binding sites seems to be a small problem for the folding of the primary amino acid sequence. In comparison to the surface of the active site the rest of a supporting globular monomeric unit with a molecular weight of 50.000 comprises approximately the 50-fold of the surface of the active site. We suggest that similarities of tertiary structure topology of enzymes might result from the evolution of common folding pathways, which are kineticly and energeticly favoured implying a minimum of energy demand as well as a time requirement which is compatible with the high turnover of protein synthesis and degradation in cellular life.

References:

1. Wetlaufer, D.B., Ristow, S.: Acquisition of three-dimensional structure of proteins. Annu. Rev. Biochem. 42, 135-138 (1973).

2. Wetlaufer, D.B.: Nucleation, rapid folding, and globular intrachain regions in proteins. Proc. Nat. Acad. Sci. USA 70, 697-701 (1973).

3. Hess, B., Bornmann, L.: Renaturation mechanism of pyruvate kinase of Saccharomyces carlsbergensis in: Metabolic Interconvertible Enzymes, Springer Verlag Berlin, Heidelberg, New York pp. 361-367 (1974).

4. Hess, B., Sossinka, J.: Pyruvate kinase of yeast. Properties and crystals. Naturwissenschaften 61, 122-124 (1974).

5. Bornmann, L., Hess, B., Zimmermann-Telschow, H.: Mechanism of renaturation of pyruvate kinase of Saccharomyces carlsbergensis: Activation by L-valine and magnesium and manganese ions. Proc. Nat. Acad. Sci. USA 71, 1525-1529 (1974).

6. Deal, W.C. : Metabolic control and structure of glycolytic enzymes. IV. Nicotinamide-adenine dinucleotide dependent in vitro reversal of dissociation and possible in vivo control of yeast glyceraldehyde 3-phosphate dehydrogenase synthesis. Biochem. 8, 2795-2805 (1969).

7. Fischer, E., Krebs, E.G.: Relationship of structure to function of muscle phosphorylase. Fed. Proc. 25, 1511-1520 (1966).

8. Schulz, G.E., Schirmer, A.H.: Topological comparison of adenylate kinase with other proteins. Nature (1974) in press.

9. Buehner, M., Ford, G.C., Moras, D., Olsen, K.W., Rossmann, M.G.: D-glyceraldehyde-3-phosphate dehydrogenase: three-dimensional structure and evolutionary significance. Proc. Nat. Acad. Sci. USA 70, 3052-3054 (1973).

Covalent Binding of Bilirubin to Agarose and Use of the Product for Affinity Chromatography of Serum Albumin

Marian Hierowski and Rolf Brodersen
Institute of Medical Biochemistry, University of Aarhus (Denmark)

INTRODUCTION

Albumin in the blood plasma reversibly binds several biological substances. The affinity is especially high for bilirubin[1] and the long chain fatty acids[2]. This fact suggested the use of bilirubin or a fatty acid as a ligand for the study of the binding properties of albumin by affinity chromatography. A suitable fatty acid-agarose was prepared by Peters, Taniuchi and Anfinsen[3]. Since fatty acids and bilirubin are bound to different high-affinity sites on the albumin molecule[4], a bilirubin-agarose would offer a new approach to the study of binding sites on albumin and would facilitate the isolation of intracellular acceptors for bilirubin. Attempts made to couple bilirubin covalently to agarose were unsuccessful because of sensitivity to light and to oxidation[3]. The present paper reports the synthesis of different bilirubin-amino--alkylamino-agarose gels and their binding properties.

MATERIALS

The following agarose preparations and derivatives were obtained from Pharmacia, Sweden: Sepharose 4B (agarose), CNBr-activated Sepharose 4B (batch No. 6743), AH-Sepharose 4B (aminohexylamino agarose with a content of 6-10 μmoles free amino group per ml swollen gel). The latter was washed before use with water to remove lactose. Ethylenediamine and 3,3'-iminobispropylamine were purchased from Aldrich-Europe, Belgium. Hexamethylenediamine and decamethylenediamine were products of Merck-Schuchardt, and 1,4-butanediamine was from Fluka AG. Bilirubin, trinitrobenzenesulfonic acid, and 1-ethyl-3(3-dimethylaminopropyl)-carbodiimide HCL were Sigma products. N,N-dimethylformamide, containing 1 per cent water, was from Fluka AG. Lyophilized agarase, B grade, was a product of Calbiochem. Crystalline human hemoglobin was obtained from Nutritional Biochemicals Corporation and human blood serum from Behringwerke AG (Normal Range Control Serum). Crystalline rat serum albumin fraction V, Pentex, was purchased from Miles Serevac Ltd., England. The parameters of bilirubin binding of this material have been estimated by the peroxidase method[1], $n_1 = 1$, $k_1 = 5x10^7\ M^{-1}$, and by spectrophotometry, n_2 about 2, k_2 about $5x10^5\ M^{-1}$ (Brodersen, R. and Juhl, H., unpublished). These figures are similar to those obtained for human serum albumin, using the same methods.

METHODS

Preparative procedures

Aminoalkylamino derivatives of agarose, which served as intermediates for coupling of bilirubin to the gel matrix, were prepared according to Cuatrecasas and Anfinsen[5,6], except that the cyanogen bromide activated Sepharose 4B was stirred overnight with the appropriate amine at pH 10.0 in the dark at 2°C. Derivatives so prepared contained 8-12 μmoles free amino group per ml swollen agarose (250 mg dry weight).

Coupling of bilirubin to aminoalkylamino agarose was accomplished by the following modification of the carbodiimide procedure [7-10]. 0,34 mmoles bilirubin was dissolved in 200 ml N,N-dimethylformamide. 4 g 1-ethyl -3(3-dimethyl-aminopropyl)-carbodiimide hydrochloride, dissolved in 200 ml water, was added and pH of the mixture was adjusted to 4.75 with 1 M HCL. After a reaction time of 5 min this solution was added drop by drop in the course of 30 to 60 min to a suspension of aminoalkylamino - agarose derivative, 10 g dry wt, preswollen with 80 ml water. The pH of the alkaline suspension dropped to 9.0 and was maintained at this value by slow addition of 1 M KOH. The mixture was then stirred gently for 16 hours. The substituted agarose was washed on a Büchner funnel with continous suction, first quickly with 99 % dimethylformamide and then extensively with 50 % aqueous dimethylformamide until bilirubin was absent from the effluent, as tested by spectrophotometry at 450 nm. Completeness of washing is important, since traces of non-covalently bound bilirubin later on, when albumin is added, might bind to some albumin and prevent retention of an equimolar amount of the protein.

Blocking of unreacted amino groups with acetic anhydride[3] was attempted but resulted in decomposition of the bound bilirubin, as seen from changes of the light absorption spectrum.

All operations involving bilirubin were carried out at 2°C and under the light of a dark-room bulb.

The bilirubin-aminoalkylamino-agarose derivatives were stored in a mixture of dimethylformamide and water (1:2 by vol) in the dark at 2°C and were stable for several weeks.

Analytical Procedures

Determination of NH_2-groups in aminoalkylamino derivatives of agarose was carried out with trinitrobenzenesulfonic acid according to Failla & Santi[11].

Determination of bilirubin in bilirubin-aminoalkylamino-agarose was done after solubilization of the gel and oxidation to biliverdin as follows. A suspension, containing the equivalent of 1 ml swollen agarose in 5.0 ml 50 % acetic acid, was heated to 75°C for 2 hours and measured by spectrophometry at λmax = 370 nm (ε = 49000), or, with similar results, at λmax = 680 nm (ε = 22000).

Light absorption spectra of the pigment present in bilirubin-aminoalkylamino-agarose gels were obtained after digestion of the agarose with agarase. 0.05 ml of the swollen agarose derivative, suspended in 2.5 ml of 0,15 M KCL, 0.02 M sodium phosphate buffer, pH 6.0, was digested with agarase, 0.1 mg, activity 540 units per mg, 3250 units per mg protein. The time, 2 hours, of enzymatic digestion at 25°C needed for total solubilization of the gel was found by spectrophometric measurements of increasing absorbancy of the supernatant at λ_{max} = 450 nm.

Electrophoresis were carried out on cellulose acetate strips in an apparatus from LKB, Sweden, after concentrating by ultrafiltration in Centriflo cones (Amicon Corp., Lexington, Mass.). Quantitation was done after staining with lissamine green SF[11,12], cutting of the strips, elution, and spectrophotometry at 635 nm.

Assay of Chromatographic Proporties

Bilirubin-aminoalkylamino-agarose preparations were packed in polypropylene tubes, 6 x 40 mm, bed volume

1 ml. One to 2 ml of a protein solution containing 1 to 2.5 mg per ml, or serum diluted 1:50, was passed through these columns in the dark at 2°C in a buffer, 0.01 M Tris, pH 7.4, 0.15 M NaCl. After washing with 10 ml of the same buffer the columns were eluted with 5 ml of the eluent. For re-use the columns were washed and equilibrated with Tris-NaCl buffer. Proteins were determined by absorbance at 280 nm (hemoglobin at 407), or by the method of Lowry[14]. Rat serum albumin or human blood serum with known protein concentrations was used for standard determinations. Sodium salicylate, used as an eluent in experiments with rat serum albumin, interfered with these procedures. The bromcresol green method[15,16] for determination of albumin was found suitable in these cases.

The maximal capacity for uptake of rat serum albumin was measured by saturation of the column with a solution of the protein, 2 mg/ml (about 5 ml was needed), washing with 10 ml Tris-NaCl buffer, and subtraction of the amount of albumin in the combined effluents from the amount applied.

Uptake of proteins from blood serum was similarly studied after passing 5 ml human standard serum, diluted 1:50 with the buffer, through the column and washing as above. Albumin was determined by the bromcresol green method, total protein according to Lowry et al., and the globulins by electrophoresis.

RESULTS

Immobilization of Bilirubin on Agarose

Bilirubin was successfully tied to agarose through an aminoalkylamino link. Five different links were used with varying "arm"-length, from two to ten carbon atoms in the alkyl group (Table IV). Coupling of bilirubin to the aminoalkylamino-agarose derivatives was accomplished by reaction with ethyl-(dimethylaminopropyl)-carbodiimide as a reagent and was conducted in a mixture of dimethylformamide and water in two steps, at pH 4.75 and subsequently at pH 9.0 Results of varying the pH during the second step are seen in Table I.

Table I
Effect of pH during the second step of coupling of bilirubin to aminohexylamino-agarose

Buffer	pH	Bilirubin coupled µmoles per ml bed volume
0.1 M sodium citrate	6.0	0.16
0.1 M sodium phosphate	7.5	0.42
0.1 M sodium borate	8.5	0.96
0.1 M sodium borate	9.5	1.02
0.1 M sodium carbonate	10.5	0.88
0.1 M sodium carbonate	11.5	0.24

The reaction mechanism is treated in the Discussion.

The amount of bilirubin linked to the gel was less than the amount of free amino groups present. In aminohexylamino-agarose the content of $-NH_2$ was 6.5 µmoles per ml swollen gel, and 5.4 µmoles/ml remained after coupling with bilirubin. Determination of bilirubin in the product gave 1.1 µmoles/ml, equal to the decrease of free amino groups. This seems to indicate that one molecule of bilirubin was tied to the gel through one amino group.

The spectrum of the bilirubin-aminohexylamino-agarose after digestion with agarase showed a maximum around 450 nm (Fig. 1.)

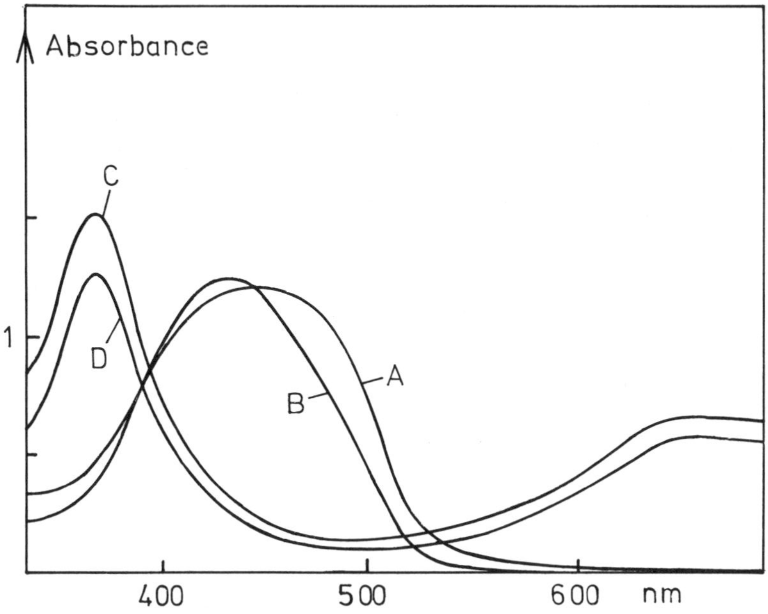

Fig. 1. Light absorption spectra. A, bilirubin-aminohexylamino-agarose after digestion with agarase. B, 26 µM bilirubin in alkaline solution. C, bilirubin-aminohexylamino-agarose after 2 hours in 50 % acetic acid at 75°. D, 26 µM biliverdin in 50 % acetic acid.

The maximum is broader than that of bilirubin in alkaline solution, but shows some similarity to the spectrum of conjugated bilirubin[17]. The spectrum did not indicate presence of biliverdin, and no signs of oxidation were observed under the conditions of the experiments. Working under red light is essential to obtain this result.

After solubization of the gel and oxidation in the presence of air with 50 % acetic acid at 75°C for 2 hours a spectrum of biliverdin was obtained (Fig. 1), identical with that seen after similar treatment of a bilirubin solution.

Addition of diazotized sulfanilic acid in acidic 50 % methanol did not result in development of increased light absorption at 540 nm, not even after prolonged incubation. The pigment present in the gel thus was diazo-negative. The identity of the pigment is treated in the Discussion.

Binding of Albumin and Other Proteins to the Gels at pH 7.4. Columns of the various agarose preparations were loaded with 4 mg rat serum albumin, as described in the Methods and washed with 10 ml buffer. The amount of protein retained by the gel after washing was determined and is seen in Table II.

Table II

Retention of rat serum albumin after application of 4 mg of the protein to 1-ml columns and washing with 10 ml buffer, pH 7.4.

	Percentage of albumin retained on		
	Agarose	Aminohexylamino-agarose	Bilirubin aminohexylamino-agarose
Albumin	0	8	100
Bilirubin albumin (1:1)	5	12	78[x)]

x) Elution was still in progress by the end of washing with 10 ml buffer.

Little or no protein was retained by the unsubstituted agarose and its aminohexylamino derivative. Retention was complete on the bilirubin-aminoalkylamino-agarose gels, and binding of albumin to this material in fact had a rather high affinity since no protein was eluted with 10 ml buffer. Albumin to which one molecule of bilirubin was prebound per molecule of the protein was retained with a lower affinity, as seen from the fact that some albumin was still being eluted even at the end of washing with 10 ml buffer.

Elution of the serum albumin, bound to bilirubin-aminoalkylamino-agarose, could be accomplished by 8 M urea, which causes unfolding of the protein, or by sodium salicylate or sulfisoxazole, substances which compete with bilirubin for the high-affinity site on albumin[18,19] Table III.

Table III

Elution of rat serum albumin (4 mg applied) from bilirubin-aminohexylamino-agarose, bed volume 1 ml.

Eluent, 10 ml vol.		Percentage of albumin eluted
0.001 M Sulfisoxazole	pH 7.4	25
0.01 m Sulfisoxazole	pH 7.4	100
0.001 M Sodium salicylate	pH 7.4	10
0.01 M Sodium Salicylate	pH 7.4	98
4.0 M Urea		0
6.0 M Urea		20
8.00 M Urea		100

The maximal binding capacity of the gels for rat serum albumin increased with the length of the aminoalkylamino "arm", as seen in Table IV.

Table IV

Retention of serum protein by various bilirubin-amino-alkylamino-agarose gels

Aminoalkylamino group	Maximal binding capacity rat serum albumin	Retention of human serum globulins from 5 ml serum, diluted 1:50 (2.9 mg globulin)
	mg per ml bed volume	
Ethyl	1,5	0.0
Butyl	2.9	0.0
Hexyl	5.1	0.2
3,3'-Iminobispropylamine	3.2	0.3
Decyl	6.8	0.4

With hexyl and longer alkyl links small amounts of globulins were bound besides the albumin, when diluted blood serum was passed through the gels, and this amount increased with the length of the alkyl group.

Electrophoretic studies (Table V) showed that a 1-ml column of bilirubin-aminohexylamino-agarose retained all the albumin and traces of α_1- and α_2-globulins from human serum.
The globulins could be eluted with 4 M urea which did not elute albumin. Pure albumin could then be obtained by elution with 8 M urea or 0.01 sodium salicylate.

Binding of hemoglobin, alone and together with albumin was studied as follows. Hemoglobin was partially retained by the bilirubin-aminohexylamino-agarose column, 61 % of 3 mg was bound to the gel. When a solution of rat serum albumin and hemoglobin, 3 mg each, was applied, 31 % of the hemoglobin and all the albumin was retained, while 10 % only was retained when 3 mg hemoglobin was applied to a column which was presaturated with rat serum albumin. No albumin was lost from the saturated gel when hemoglobin was introduced. Vice versa, a column presaturated with hemoglobin, 2.3 mg in 1 ml bed volume, lost 40 % of this when 5.1 mg rat

Table V

Separation of human serum proteins on bilirubin-aminohexylamino-agarose

	Percentage of total amount of protein			
	Present in serum	Retained on column	Eluted by 4 M urea	Subsequently eluted by 8 M urea
Albumin	60	60	0	60
α_1-globulin	3	2	2	0
α_2-globulin	10	1	1	0
β-globulin	12	0	0	0
γ-globulin	15	0	0	0
Total	100	63	3	60

serum albumin was applied. All the albumin was bound. The 1.4 mg hemoglobin left in the gel thus did not reduce the capacity for albumin. These results indicate that considerable amounts of hemoglobin could be retained by the gel, partly bound to the same sites as albumin, partly to other loci. The affinity for binding of albumin was much higher than for hemoglobin, as seen from displacement of the latter protein by application of the first, whereas bound albumin could not be displaced by hemoglobin.

DISCUSSION

Coupling of bilirubin to aminoalkylamino-agarose was conducted in two steps. First, the pigment was mixed with the carbodiimide reagent at pH 4.75 in order to obtain an O-acyl urea derivative by reaction with carboxyl groups of bilirubin. The mixture was then added to a suspension of the aminoalkylamino-agarose. The optimal pH-value during the second step was found to be 9.5, in agreement with the expected reaction with the unprotonated amino group. The product is unstable at higher pH.

Peters, Taniuchi and Anfinsen[3] coupled long-chain fatty acids to agarose under similar circumstances, but in one step at pH 10. We could not obtain coupling with bilirubin, unless a lower pH-value was used during the first part of the reaction.

The identity of the pigment present in the product seems to be ascertained by the light absorption spectrum obtained after digestion with agarase, which is similar to that of bilirubin, and by the fact that a biliverdin spectrum was found after oxidation. The latter demonstrates the presence, after oxidation, of the biliverdin chromophore, extending over all four pyrrole rings and the two vinyl groups. The spectrum after digestion with agarase is however not identical to that of bilirubin. The differences might be ascribed to changes of intramolecular hydrogen bonding between the chromophores and carboxyl groups[20,21] caused by the presence of aminoalkylamino-galactose residues which remain attached to bilirubin carboxyl after enzymatic hydrolysis of the agarose. Conjugated bilirubins, which have hexuronic acid moieties in these positions, show similar spectra[17]. Alternatively it is possible that the point of attachment of bilirubin to the gel matrix is not a carboxyl group but is located in one of the chromophores. The fact that no reaction occurred which diazotized sulfanilic acid in acidic 50 % methanol could be explained hereby. The electrophilic diazonium reagent attacks bilirubin at the central pyrrole rings, at the carbon atoms connected to the methylene bridge. A high electron density is present in these points. The distribution of electric charge would be shifted, if the bilirubin is bound to the gel matrix through the chromophores. Such bonding would explain the deviation of the spectrum and the lack of diazo reactivity. It can at present not be stated wether the bilirubin is bound through carboxyl or otherwise, and the nature of the first step in the coupling procedure, taking place at pH 4.75, remains to be elucidated.

Serum albumin, from rat or man, is bound with high affinity to the bilirubin-aminoalkylamino-agarose. This is seen from the fact that it could not be washed out with large volumes of buffer. Other proteins in serum or hemoglobin are bound with less affinity, as is also serum albumin with one molecule of bilirubin prebound to the high-affinity site. Albumin was eluted with substances, salicylate and sulfisoxazole, which compete with bilirubin for binding to this site. Altogether this seems to point to specific binding of serum albumin to the gel through the high-affinity site. Additional amounts of albumin may be bound through weaker bonds, especially in the gels with long aminoalkylamino "arms".

The maximal capacity for specific binding of albumin to bilirubin-aminohexylamino-agarose, 5.2 mg (0.08 µmoles) per ml bed volume, is considerably lower than the content of bilirubin, about 1 µmole/ml. Only about one tenth of the bilirubin groups in the gel are able to bind albumin. This might point towards heterogeneity of the type of bond between bilirubin and the gel matrix.

The gel was successfully used for affinity chromatography of serum albumin, which hereby could be isolated from other proteins in blood serum. It would further appear suitable for studies of binding of bilirubin to serum albumin, as well as for isolation of intracellular receptor proteins.

SUMMARY

A method is described for covalent coupling of bilirubin to agarose through an aminohexylamino group, using a carbodiimide reagent. The product contained 1.1 μmoles bilirubin per ml of the packed gel (250 mg dry weight) and could reversibly bind 0.08 μmoles rat serum albumin. Elution was effected by salicylate and sulfisoxazole, substances competing for the high-affinity bilirubin site on albumin, and by 8 M urea. Albumin could be separated from human serum by affinity chromatography on this material. Lengthening of the aminoalkylamino "arm" increased the capacity for albumin, but also increased binding of other serum proteins.

ACKNOWLEDGEMENTS

The authors wish to thank Dr. Theodore Peters, jr., the Mary Imogene Bassett Hospital, Cooperstown, N.Y., for helpful communication during the work. H. Røjgaard Petersen, M.Sc. has given valuable advice in cellulose acetate electrophoresis. The skilful technical work of Jetta Bach Haulrik is acknowledged.

REFERENCES

1 Jacobsen, J.: Binding of bilirubin to human serum albumin - determination of the dissociation constants, Fed. Eur.Biochem. Soc.Lett. 5, 112-114 (1969)

2 Spector, A.A., John, K., and Fletcher, J.E.: Binding of longchain fatty acids to bovine serum albumin, J.Lipid.Res. 10, 56-66 (1969).

3 Peters, T. Jr., Taniuchi, H., and Anfinsen, C.B.Jr.: Affinity chromatography of serum albumin with fatty acids immobilized on agarose, J.Biol.Chem.248, 2447-2457 (1973).

4 Thiessen, H., Jacobsen, J., and Brodersen, R.:Displacement of albumin-bound bilirubin by fatty acids, Acta Pædiat.Scand. 61, 285-288 (1972).

5 Cuatrecasas, P., and Anfinsen, C.B.: Selective enzyme purification by affinity chromatography, Proc. Natl.Acad.Sci. U.S. 61, 636-643 (1968).

6 Cuatrecasas, P., and Anfinsen, C.B.: Affinity chromatography, Methods Enzymol. 22, 345-385 (1971).

7 Hoare, D.G., and Koshland, D.E.Jr.: A procedure for the selective modification of carboxyl groups in proteins, J.Amer.Chem.Soc. 88, 2057-2058 (1966).

8 Hoare, D.G., and Koshland, D.E.Jr.: A method for the quantitative modification and estimation of carboxylic acid groups in proteins, J.Biol.Chem. 242, 2447-2453 (1967).

9 Riehm, J.P., and Sheraga, H.A.: Structural studies of ribonuclease. XVII A reactive carboxyl group in ribonuclease, Biochemistry 4, 772-782 (1965).

10 Wilchek, M., Frensdorff, A., and Sela, M.: Modification of the caboxyl groups of ribonuclease by attachment of glycine or alanylglycine, Biochemistry 6, 247-252 (1967).

11 Brakenridge, C.J.: Optimal staining conditions for the quantitative analysis of human serum protein fractions by cellulose acetate electrophoresis, Anal.Chem. 32, 1353-1356 (1960).

12 Brakenridge, C.J.: Variable dye uptake in the quantitative analysis of abnormal globulins by cellulose acetate electrophoresis, Anal.Chem. 32, 1357-1359 (1960).

13 Failla, F., and Santi, D.V.: A simple method for quantitating ligands covalently bound to agarose beads, Anal.Biochem. 52, 363-368 (1973).

14 Lowry, O.H., Rosebrough, N.J., Farr, A.L. and Randall, R.J.: Protein measurement with the folin phenol reagent, J.Biol.Chem. 193, 265-275 (1951).

15 Doumas, B., Watson, W.A. and Biggs, H.G.: Albumin standards and the measurement of serum albumin with bromcresol green, Clin.Chim.Acta 31, 87-96 (1971).

16 McPherson, I.G., and Everard, D.W.: Serum albumin estimation: Modification of the bromcresol green method, Clin.Chim.Acta 37, 117-121 (1972).

17 Jacobsen, J.: Dimerisation of bilirubin diglucuronide and formation of a complex of bilirubin and the diglucuronide, Scand.J.Clin.Lab. Invest. 26, 395-398 (1970).

18 Silvermann, W.A., Andersen, D.H., Blanc, W.A., and Crozier, D.N.: A difference of mortality rate and incidence of kernicterus among premature infants allotted to two prophylactic antibacterial regimens, Pediatrics 18, 614-625 (1965).

19 Odell, G.B.: The dissociation of bilirubin from albumin and its clinical implications, J. Pediatrics 55, 268-279 (1959).

20 Brodersen, R., Flodgaard, H., and Krogh Hansen, J.: Intramolecular hydrogen bonding in bilirubin, Acta

Che.Scand. 21, 2284-2285 (1967).

21 Kuenzle, C.C., Weibel, H.H., Pelloni, R.R., and Hemmerich, P.: Structure and conformation of bilirubin. Opposing views that invoke tautomeric equilibria, hydrogen bonding and a betaine may be reconciled by a single resonance hybrid, Biochem.J. 133, 364-368 (1973).

Transmitter Biochemistry of Single, Identified Neurons

John G. Hildebrand and Edward A. Kravitz
Department of Neurobiology, Harvard Medical School,
Boston, MA. 02115, USA

Chemical synaptic transmission between a neuron and its follower cells depends upon release of a specific neurotransmitter substance from the terminals of the presynaptic neuron, evoked by depolarization of the terminals by an invading action potential or by electrotonic currents. Thus a neuron must synthesize and accumulate, at least partly in a releasable pool, the appropriate transmitter substance. (For references and brief reviews of these and related topics, see 1.) In order to understand thoroughly this presynaptic production and storage of neurotransmitters, one must: identify the transmitter employed by the neuron in question; characterize the substrates, enzymes, and other molecules involved in transmitter metabolism and storage; and elucidate the genetic and other regulatory mechanisms governing these components.

Work in our laboratory has aimed at gaining this understanding of transmitter accumulation using the nervous system of the lobster, *Homarus americanus*. The neuroanatomy of Crustacea, such as *Homarus*, favors investigation of the physiology and biochemistry of individual neurons. One can isolate somata or axons of single, physiologically identified neurons from a series of animals, thereby obtaining sufficient material for biochemical studies using micro-techniques. We have focused mainly on three types of nerve cells in the lobster: the excitatory (E) and inhibitory (I) neurons whose cell bodies are in the ganglia of the central nervous system and whose axons innervate exoskeletal muscles; and sensory (S) neurons whose cell bodies lie in the periphery, associated with various sensory receptor organs, and whose axons enter the central ganglia and synapse on ganglionic neurons.

The identities of the neurotransmitters released by these three types of neurons are known with different degrees of certainty. A variety of physiological and pharmacological observations from this and other laboratories, as well as biochemical findings to be described below, established that γ-aminobutyrate (GABA) is the inhibitory neuromuscular transmitter in Crustacea (for reviews, see 1-3). The excitatory neuromuscular transmitter in the lobster is probably glutamate, but the existing evidence is less compelling than that for GABA (4). Finally, increasing evidence supports the suggestion of Florey and Biederman (5) that acetylcholine (ACh) functions as a sensory transmitter in Crustacea (reviewed in 6-8).

We review here the recent work in our laboratory on the cellular distribution of the components of the metabolic pathways for GABA and ACh. These studies support the likely proposition that the distribution of the transmitter-synthesizing enzymes determines the transmitter chemistry of individual neurons.

METHODS

All procedures used in this work have been described in detail in the various publications cited under References. What follows is only a brief summary of our experimental approach.

Single, identified E-, I-, and S-axons were dissected from live *Homarus americanus* (3,7,8). E- and I-axons innervating the dactyl musculature of the walking leg were exposed in the meropodite segment and identified by electrical stimulation. Lengths of individual axons, usually 4 to 6 cm, were isolated from other neural and connective tissues. S-axons of the abdominal muscle stretch receptor organ were identified by location of their source at the receptor cell bodies and freed from other tissues, yielding 2 to 4 cm of a purely sensory axon pair.

Isolated axons were homogenized in appropriate buffer mixtures in glass micro homogenizers, and enzymes and substrates of interest were quantified in aliquots of the resulting extracts. The micro-assay procedures are summarized in Table 1.

Table 1

Summary of Micro-Assay Methods

Component	Assay Principle	Ref.
GABA Pathway:		
Glutamic decarboxylase (GAD)	glutamate-U-^{14}C ⟶ ^{14}CO + ^{14}C-GABA	3
GABA-glutamate transaminase (GT)	^{14}C-GABA —α-ketoglutarate⟶ ^{14}C-succinic semialdehyde —dehydrogenase + NAD in extract⟶ ^{14}C-succinate	3
GABA	GABA + NADP^{+} —bacterial enzyme system⟶ succinate + NADPH (NADPH measured fluorometrically)	9,10
Glutamate	glutamate —bacterial decarboxylase⟶ GABA (then as for GABA above)	9
ACh Pathway:		
Choline acetyl-transferase (ChAc)	choline + ^{14}C-acetyl-CoA ⟶ ^{14}C-ACh + CoASh	7,8
Acetylcholin-esterase (AChE)	acetylthiocholine ⟶ thiocholine + acetate (thiocholine measured spectro-photometrically using Ellman's reagent)	8,11
Choline	choline + ATP-γ-^{32}P —choline kinase⟶ ^{32}P-phosphorylcholine + ADP	8
ACh	isolated ACh —alkali⟶ choline + acetate (then as for choline above)	8

RESULTS

The enzymes and key substrates of the GABA and ACh metabolic pathways were quantified in extracts of isolated E-, I-, and S-axons. The results are summarized in Tables 2 and 3.

Table 2
Components of the GABA Pathway
in E-, I-, and S-Axons

Components	I	E	S
Glutamate	70 mM	70 mM	
GAD	90 $\frac{pmol}{hr \cdot cm}$	0	0
CO_2			
GABA	100 mM	<1 mM	<1 mM
GT	30 $\frac{pmol}{hr \cdot cm}$	25 $\frac{pmol}{hr \cdot cm}$	10 $\frac{pmol}{hr \cdot cm}$
α-keto-glutarate	3 mM	3 mM	
Succinic semialdehyde			
NAD^+, H_2O → dehydrogenase → NADH $+H^+$	not limiting	not limiting	not limiting
Succinate			

These data are abstracted from Refs. 3,8,9,13.

Table 3
Components of the ACh Pathway
in E-, I-, and S-Axons

Components	E + I	S
Choline	1.9 $\frac{\text{pmol}}{\mu\text{g protein}}$	1.9 $\frac{\text{pmol}}{\mu\text{g protein}}$
+	(*ca*·0.5 mM)	(*ca*·0.5 mM)
Acetyl-CoA		
↓ ChAc	0.032 $\frac{\text{pmol}}{\text{hr}\cdot\text{cm}}$	17 $\frac{\text{pmol}}{\text{hr}\cdot\text{cm}}$
ACh	0.012 $\frac{\text{pmol}}{\mu\text{g protein}}$	1.9 $\frac{\text{pmol}}{\mu\text{g protein}}$
↓ H_2O	(*ca*·0.003 mM)	(*ca*·0.5 mM)
AChE	48 $\frac{\text{nmol}}{\text{hr}\cdot\text{cm}}$	27 $\frac{\text{nmol}}{\text{hr}\cdot\text{cm}}$
Acetate + Choline		

These data are abstracted from Ref. 8.

DISCUSSION

The results summarized in Table 2 showed that the inhibitory transmitter, GABA, and its biosynthetic enzyme, GAD, are selectively accumulated in lobster I-axons. In contrast is the distribution of the GABA-degrading enzymes, GT and succinic semialdehyde dehydrogenase, which are present in all three types of axons. These findings suggested that regulation of transmitter accumulation in neurons occurs at the level of the key biosynthetic enzyme (7). It appeared that the presence of a particular transmitter-synthesizing enzyme, such as GAD, might determine the transmitter chemistry of the neuron.

Further work indicated that control of GABA accumulation in I-neurons probably involves product feedback inhibition of GAD. The concentrations of enzymes in extracts of isolated axons (Table 2) are such that I-axons can synthesize more GABA than they can destroy. But lobster GAD is inhibited by GABA (12). In the presence of 100 mM GABA (the concentration found in I-axons), GAD activity decreases to about the level of GT activity (25-30 pmols/(hr·cm)), suggesting that the level of GABA maintained in I-axons represents a steady state of synthesis and destruction.

To test the generality of the notion that transmitter-synthesizing, but not degrading, enzymes are selectively distributed, we turned to a second transmitter system: ACh, a likely sensory transmitter in the lobster. Our observations on the distribution of the components of the ACh pathway in the three types of axons, summarized in Table 3, agreed well with the previous results for GABA. S-axons showed a ChAc activity more than 500 times greater than that in E- and I-axons, but AChE activity was high in all three types of fibers. ACh itself was about 160 times more concentrated in S-axons than in E- and I-axons, while the choline level was equal in all fibers. The apparent concentration of ACh in S-axons, assuming uniform distribution in the axoplasm, was ca. 0.5 mM or about 200 times lower than that of GABA in I-fibers. We have no explanation for this difference in the levels to which neurotransmitters accumulate in the two kinds of neurons. Nor do we understand how the ACh level is regulated in S-neurons. ChAc activity drops only 50 per cent in the presence of 500 mM ACh (8), so that it seems unlikely that product feedback control governs

ACh levels, and the activity of the degradative enzyme (AChE) is orders of magnitude greater than that of the synthetic enzyme (ChAc). Until we know more about the physical location of ACh, ChAc, and AChE in S-neurons, we cannot speculate further on biosynthetic control.

The work summarized here, as well as that reported by others (for example, 14-15), suggests that transmitter-degrading enzymes may be very widely distributed, produced and retained by many (or all) types of neurons. On the other hand, each transmitter-synthesizing enzyme may accumulate only in those neurons destined to employ its product as transmitter. The development of the capacity to synthesize and retain the biosynthetic enzyme for the appropriate transmitter thus must be a crucial event in the differentiation of a neuron.

NOTE

[1]This paper is gratefully and respectfully dedicated to Fritz Lipmann on the occasion of his 75th birthday.

ACKNOWLEDGEMENTS

This work has been supported by NIH grants Nos. 5P01 NS 02253 and 5R01 NS 07848 and by an Established Investigatorship of the American Heart Association to J.G.H. We thank E. Livingston and S. Wilson for assistance in preparation of this manuscript.

REFERENCES

1. Hall, Z.W., Hildebrand, J.G., Kravitz, E.A.: Chemistry of Synaptic Transmission. Newton, Mass.: Chiron Press, Inc. (1974)
2. Kravitz, E.A.: Acetylcholine, γ-aminobutyric acid, and glutamic acid: Physiological and chemical studies related to their roles as neurotransmitter agents. In: The Neurosciences, ed. Quarton, et al., New York: Rockefeller University Press, 433-444 (1967)
3. Hall, Z.W., Bownds, M.D., Kravitz, E.A.: The metabolism of γ-aminobutyric acid (GABA) in the lobster nervous system - enzymes in single excitatory and inhibitory axons. J. Cell Biol. 46, 290-299 (1970)
4. Kravitz, E.A., Slater, C.R., Takahashi, K., Bownds, M.D., Grossfeld, R.M.: Excitatory transmission in invertebrates - Glutamate as a potential neuromuscular transmitter compound. In: Excitatory Synaptic Mechanisms, ed. Andersen and Jansen, Oslo: Universitetsforlaget, 85-93 (1970)
5. Florey, E., Biederman, M.A.: Studies on the distribution of Factor I and acetylcholine in crustacean peripheral nerve. J. Gen. Physiol. 43, 509-522 (1960)
6. Hildebrand, J.G., Barker, D.L., Herbert, E., Kravitz, E.A.: Screening for neurotransmitters: A rapid radiochemical procedure. J. Neurobiol. 2, 231-246 (1971)
7. Barker, D.L., Herbert, E., Hildebrand, J.G., Kravitz, E.A.: Acetylcholine and lobster sensory neurones. J. Physiol. 226, 205-229 (1972).
8. Hildebrand, J.G., Townsel, J.G., Kravitz, E.A.: Distribution of acetylcholine, choline, choline acetyltransferase, and acetylcholinesterase in regions and single, identified axons of the lobster nervous sys-

tem. J. Neurochem., in press (1974)
9. Kravitz, E.A., Molinoff, P.B., Hall, Z.W.: A comparison of the enzymes and substrates of gamma-aminobutyric acid metabolism in lobster excitatory and inhibitory axons. Proc. Nat. Acad. Sci. 54, 778-782 (1965)
10. Jakoby, W.B., Scott, E.M.: Aldehyde oxidation III. succinic semialdehyde dehydrogenase. J. Biol. Chem. 234, 937-940 (1959)
11. Ellman, G.L., Courtney, K.D., Andres, V., Featherstone, R.M.: A new and rapid colorimetric determination of acetylcholinesterase activity. Biochem. Pharmacol. 7, 88-95 (1961)
12. Molinoff, P.B., Kravitz, E.A.: The metabolism of γ-aminobutryic acid (GABA) in the lobster nervous system - glutamic decarboxylase. J. Neurochem. 15, 391-409 (1968)
13. Kravitz, E.A., Potter, D.D.: A further study of the distribution of γ-aminobutyric acid between excitatory and inhibitory axons of the lobster. J. Neurochem. 12, 323-328 (1965)
14. Giller, E. Jr., Schwartz, J.H.: Choline acetyltransferase in identified neurons of abdominal ganglion of Aplysia californica. J. Neurophysiol. 34, 93-107 (1971)
15. Giller, E. Jr., Schwartz, J.H.: Acetylcholinesterase in identified neurons of abdominal ganglion of Aplysia californica. J. Neurophysiol. 34, 108-115 (1971)

Regulation of Apoferritin Biosynthesis in Rat Liver

A. Huberman, J.M. Rodriguez, Rebecca Franco and E. Barahona, Instituto Nacional de la Nutrición, Departamento de Bioquimica México 22, D.F.

INTRODUCTION

Ferritin is a soluble protein which is found in the cytoplasm of practically all mammalian cells. The organs richest in ferritin content are the spleen, the liver and the bone marrow (1). Its distribution in nature is very wide as it is found in invertebrates,fish, plants and in the fungus Phycomyces. The most studied ferritin is horse spleen ferritin which is relatively easy to isolate due to its stability to heating and its crystallization with cadmium sulphate (2).

The ferritin molecule is made of two parts. The protein, apoferritin which forms a virus-like empty sphere with an internal and external radii of 37 Å and 61 Å respectively. The molecular weight of apoferritin is approximately 450,000 dalton. It is made up of 24 subunits of 18,500 daltons eaxh, which confers on the whole molecule an octahedral symmetry (3). Inside this protein sphere is found the other part of the molecule, an inorganic micelle of ferric hydroxyphosphate. The content of iron of each ferritin molecule can vary between 0 (apoferritin) and 4000-5000 atoms.

The only known function of ferritin is to serve as an intracellular iron store and its role in intermediate metabolism possibly is to sequester this metal, eliminating it from circulation. In this way it would neutralize the toxic effects of free iron while being able to liberate it in accordance with the metabolic demands of the organism. It can also have a regulating role in iron absorption because any excess of ingested iron could be trapped by intestinal ferritin and eliminated with the shedding of mucosal cells.

Hours after parenteral administration of iron there is an accumulation of ferritin in the liver which was shown to be due to synthesis de novo (4). This induction of ferritin synthesis is abolished by cycloheximide (5, 6). On the other hand, the administration of actinomycin D will not inhibit iron induction of ferritin synthesis (7). These data indicate that iron is not exerting its effect at the transcriptional level but most probably is acting at the translational level.

The purpose of our work has been to find out the mechanism by which iron regulates the synthesis of ferritin.

METHODOLOGY

The induction of ferritin synthesis was accomplished through the intraperitoneal injection of 25 mg of Fe/100 g of body weight of a pharmaceutical preparation of iron-dextran (HI-DEX, Mead Johnson of Mexico), except when the dose effect was explored. The animals used were female Wistar albino rats of 200 g body weight. Iron quantification was performed with our own method which is specific for nonheme iron (8).For the isolation of ferritin, the method of Drysdale and Munro was followed (9), but gel filtration was done at room temperature. Other methods are decribed in the legends to the figures.

RESULTS

After one injection of iron-dextran, the content of ferritin (in mg/g of liver) increased from a control value of 1.1 to 2.6 after 24 h, while the nonheme iron content (in µg of Fe/g of liver) increased from a control value of 104 to 1700 after 72 h. That is, while there is a doubling of the ferritin content of the liver, which is stable after many days, the iron content continues increasing up to three days after stimulation, indicating an increase in the iron saturation of each ferritin molecule.
These two parameters (ferritin synthesis and iron saturation) were analyzed in the experiments depicted in Fig. 1. Iron stimulation was done with increasing doses in different groups.

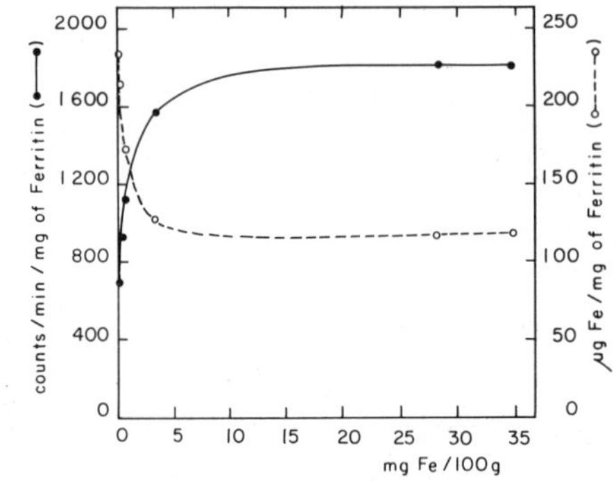

Fig. 1. Effect of increasing doses of iron-dextran on the specific activity and iron content of liver ferritin. Each pair of rats received a different dose of iron-dextran diluted with saline so that the final volume was equal in all cases. After 4 h, each one received a pulse of 5 µCi of ^{14}C-L-leucine (262 mCi/mmole) per 100 g body weight through a tail vein. Two hours later they were sacrificed, ferritin was isolated from the livers and specific activity and iron content measured. Solid circles: iron content.

It is seen that the stimulation of ferritin synthesis is a function of the dose until a plateau is reached at about 10 mg of Fe/100 g. At the same time, the iron content per mg of ferritin decreases continually until it stabilizes at the same value as above. This is interpreted as a dilution of the incoming iron in a growing population of ferritin molecules. Only after 6 h, does the iron saturation of ferritin begin to increase steadily.

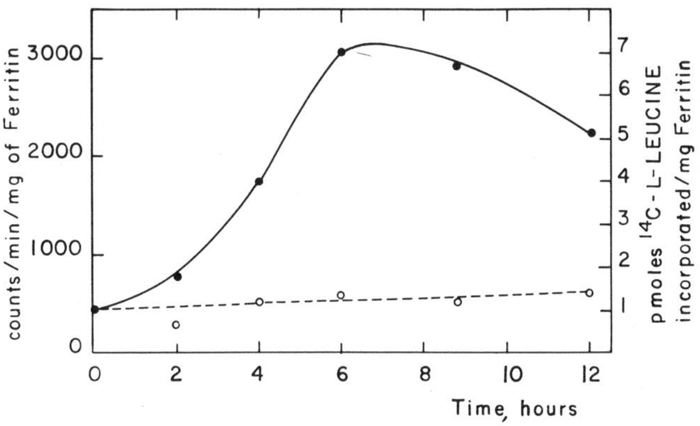

Fig. 2. Induction of ferritin synthesis after the intraperitoneal injection of iron-dextran. Each rat received a dose of 25 mg Fe/100 g body weight. At 0, 2, 4, 6, and 10 hours, each pair of animals is injected with 5 µCi of ^{14}C-L-leucine (262 mCi/mmole) per 100 g body weight, via a tail vein. After 2 h they were sacrificed, liver ferritin was isolated and its specific activity determined. In a fragment of the same liver, the specific activity of total soluble protein was determined. Solid circles: ferritin; open circles: total soluble protein.

Fig. 2 illustrates a time course where the dose is fixed. The specific activity increases very rapidly after two hours and reaches a maximum after 6 h. As the animals are receiving a true pulse of radioactivity, the decrease in specific activity after the plateau is reached (6-8 h) means that the initial stimulus is being lost. That the stimulus is specific

for ferritin can be seen in the discontinous line which corresponds to the specific activity of total soluble protein of the same liver.
The previous time course did not allow an analysis of the induction before 2 h, so a different series of experiments were done in which time 0 was almost simultaneous for iron and radioactivity injections. Fig. 3 shows that there is a time lag of 15 min between the iron injection and the increase in specific activity of ferritin.

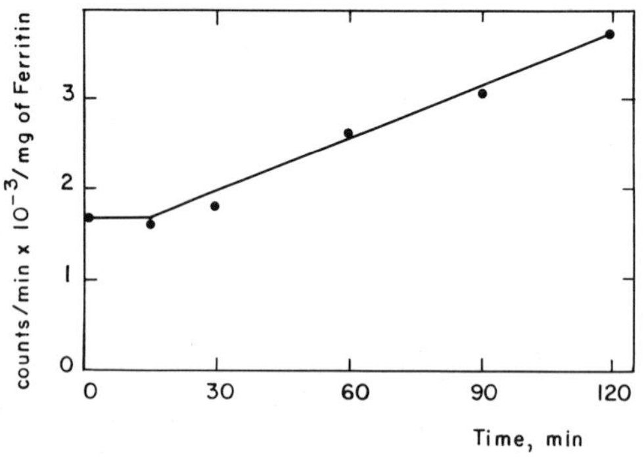

Fig. 3. Incorporation of a radioactive amino acid into ferritin at short times after the intraperitoneal injection of iron-dextran. Each rat received a dose of 25 mg Fe/100 g body weight followed immediately by 5 µCi of ^{14}C-L-leucine (306 mCi/mmole) per 100 g body weight, via a tail vein. After 15, 30, 60, 90, and 120 minutes, pairs of rats were sacrificed, ferritin isolated and its specific activity determined.

The primary structure of ferritin has been studied only in horse spleen ferritin (10, 11). Rat liver ferritin, which is the one used for regulation studies in mammals, has not been studied from this point of view. As a first step, we undertook to study the amino end of rat liver ferritin. We have found that the amino end is blocked, being resistant to Sanger's reagent after denaturation. By means of combined hydrazino-

lysis and dinitrophenylation we have found that the blocking group is an acetyl.
Pronase digestion of rat liver apoferritin allowed us to isolate a ninhydrin-negative peptide by ionic interchange chromatography on Dowex 50. This peptide was purified on a column of Dowex 2.
The analysis of this peptide showed that its sequence is N-acetyl-Ser-Ser-Gln (12), which coincides partially with the tryptic peptide found by Suran (10): N-ac-Ser-Ser-Gln-Ile-Arg. These data, together with the findings of Liew, Haslett and Allfrey (13) induced us to think perhaps there was a tRNA specific for the initiation of ferritin and that it would appear as a result of the iron induction.
tRNA was isolated from control and stimulated rat livers. After deacylating, it was charged with ^{14}C-L-serine under optimal conditions. The results were (in pmoles of serine per µg of tRNA) 3.16 $\pm$ 0.24 for the control tRNA, and 3.18 $\pm$ 0.23 for the stimulated tRNA (mean value of three determinations $\pm$ standard deviation). This means that no previously inexistent tRNA is appearing, nor is the level of a previously existent tRNA increasing but one cannot discard the possibility that an isoacceptor is modified after iron stimulation so that it can now accept an acetyl group.
To explore this latter possibility, rats were stimulated with iron and at 4 h the liver was taken out and connected to a perfusion machine. To the perfusion fluid, 50 µCi of ^{14}C-L-serine and 800 µCi of ^{3}H-acetate (sodium) were added. After 15 min of perfusion, the tRNA was isolated. The same procedure was followed with controls, but these were injected with saline. Both batches of tRNA were digested with bovine

pancreatic ribonuclease, snake venom (Crotalus atrox) and alkaline phosphatase, during 6 h at 37°, with the purpose of liberating the amino-acid (serine or N-acetyl-serine) and at the same time eliminate phosphoserine. After acidifying to eliminate the enzymes, the incubation medium was concentrated and submitted to high voltage electrophoresis, as explained in Fig. 4. It is seen in (A) that the control tRNA only shows serine in the origin, but in (B) besides this peak corresponding to serine in the origin, there is another peak of mixed radioactivity (^{14}C + ^{3}H) at 19 cm towards the anode, which coincides with a N-acetyl-serine standard.

This means that in the stimulated animal, a new species of tRNA is appearing which corresponds to N-acetyl-seryl-tRNA.

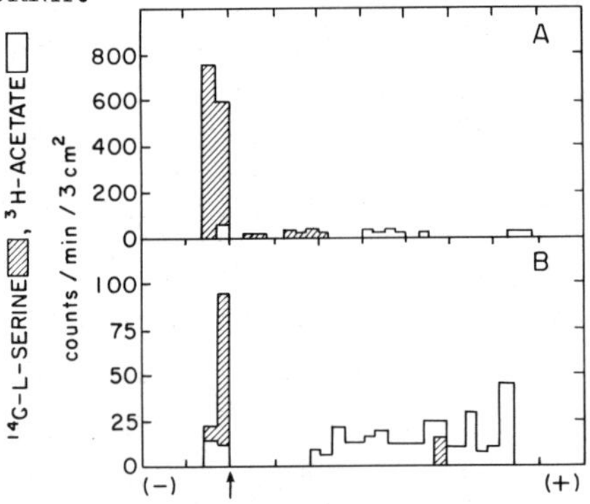

Fig. 4. Electrophoretic analysis of ^{14}C-seryl-tRNA from rat liver. Paper electrophoresis on Whatmann 3MM, buffer: pyridine:acetic acid:water (136:14:3000, v/v/v), pH 6.0, 50 V/cm (3800 volt.), 45 min. Control rats received saline and stimulated rats received 25 mg Fe/100 g body weight. At 4 h the livers were excised and connected to a perfusion machine. To the perfusion fluid were added 50 µCi of ^{14}C-L-serine (140 mCi/mmole) and 800 µCi of ^{3}H-acetate (sodium) (100 mCi/mmole). After 15 min, perfusion was stopped, tRNA was isolated and digested with bovine pancreatic ribonuclease, snake venom (Crotalus atrox), and microbial alkaline phoshatase. (A): control; (B): stimulated with iron.

In order to test if the changes were taking place in the tRNA or in the supernatant, a transacetylation in vitro experiment was done as explained in Fig. 5 and 6.

In Fig. 5 (control tRNA) there is a small peak of mixed radioactivity in the origin, corresponding to the position of N-acetyl-serine at pH 2.0. The proportion of total serine counts which appear at this position is approximately 10%.

Fig. 6 (stimulated tRNA) the peak with mixed radioactivity in the position of N-acetyl-serine is more conspicuous now corresponding to about 40% of the total serine counts.

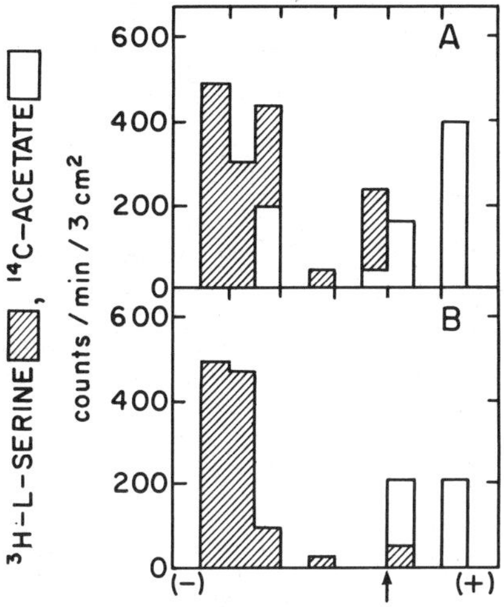

Fig. 5. Analysis of the product of the in vitro transacetylation reaction. Paper electrophoresis in Whatman 3MM, with 0.5M formic acid, pH 2.0, 450 volts, 75 min. The tRNA isolated from control rats was charged with ^{3}H-L-serine (1.23 Ci/mmole) using rat liver pH 5.0 fraction. The ^{3}H-seryl-tRNA was recovered and incubated with a 250,000 xg rat liver supernatant from control rats (A) or from stimulated rats (B), in the presence of ^{14}C-acetyl-coenzyme A (48.6 mC/mmole). The tRNA was isolated, submitted to alkaline hydrolysis, concentrated and electrophoresed.

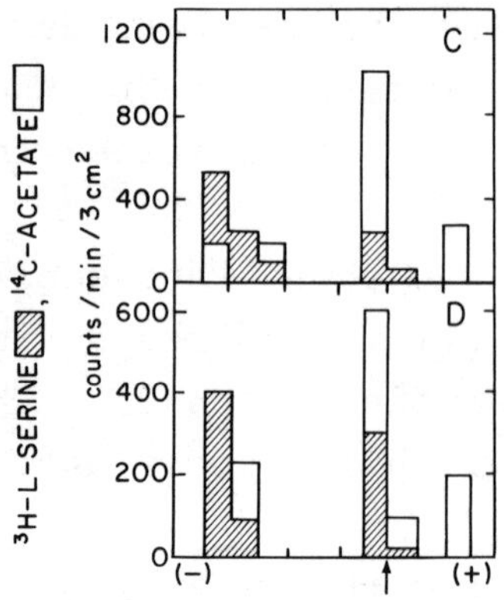

Fig. 6. Analysis of the product of the in vitro transacetylation reaction. The same conditions as in Fig. 5, but in this case, the tRNA used was from livers stimulated with iron.

In both figures (5 and 6) it is noticeable that there is no difference whether one uses control or stimulated supernatant but there is a definite difference whether one uses control or stimulated tRNA. In other words, there is in the supernatant a transacetylase which seems to have as a substrate a seryl-$tRNA^{Ser}$ isoacceptor but that the recognition process takes place only after iron stimulation, perhaps due to a modification.

DISCUSSION

The biosynthesis of ferritin can be induced by means of the acute administration of iron. This induction is post-transcriptional because it cannot be abolished by actinomycin D, and as a corollary, its specific messenger RNA should be of a long half life. Its translation must be subjected to a regulatory mechanism which involves iron. Vassart et al. (14) have concluded from a simulated computer experiment that the regulation of translation should be most effective at the initiation level. The fact that we have found that apoferritin has its amino end blocked with N-acetyl-serine, and that after the stimulation we can show the presence of a N-acetyl-seryl-tRNA point in the direction of initiation as the center of regulation of apoferritin biosynthesis. On the other hand, these facts, by themselves are not a categorical proof that this particular tRNA is acting as an initiator.

Most investigators agree that in general, initiation of protein biosynthesis in eukaryotes is accomplished with one of the two methionine isoacceptors, met-$tRNA^{fMet}$, but it is interesting that a growing number of proteins are found which have its amino end acetylated.

In Table I are listed the known eukaryotic proteins which bear an N-acetyl-serine in its amino end.

TABLE 1

PROTEINS WHICH BEAR A N-ACETYL-SERYL MOIETY ON ITS AMINO END

- Tobacco mosaic virus (cover protein)
- Histone (calf thymus), fractions f2a1 and f2a2
- Glyceraldehyde 3-phosphate dehydrogenase (lobster)
- Apoferritin (horse spleen)
- Apoferritin (rat liver)
- Hemoglobin alpha (carp)
- Myosin (rabbit muscle)
- Alpha-melanocyte stimulating hormone (pig)
- Carbonic anhydrase (human and bovine erythrocyte)
- Mitochondrial structural protein (beef heart)
- Phosphoglycerate kinase (human erythrocyte)
- Factor XIII (bovine plasma)

The list would be more extensive if we would take into account all the proteins which are acetylated in

its amino end, but even so it is noteworthy the limited number of amino acids which can be acetylated: glycine, alanine, threonine, methionine and aspartic acid (besides serine). This gives N-terminal acetylation a special significance even though we don't know its mechanism nor its finality.
Whether this N-terminal acetylation is linked to initiation or is posteior to it has been a source of contradicting data. Marchis-Mouren and Lipmann, studying the acetylation of human F_1 hemoglobin concluded that the acetylation was independent of protein synthesis (15). In the case of alpha-crystallin (16, 17) Bloemendal shows that acetylation of the N-terminal methionine occurs after initiation. They suggest that all proteins are initiated by the same mechanism, that is, with methionyl-$tRNA^{fMet}$ and that a ribosomal aminopeptidase cleaves the terminal methionine before the chain is completed. Then, acetylation would take place on the contiguous amino acid. In the specific case of alpha-crystallin, methionine would not be eliminated due to steric hindrance of the contiguous sequence N-ac-Met-Asp-Ile. The only two other proteins which have N-ac-Met on its amino end (turnip yellow mosaic virus coat protein and rabbit muscle tropomyosin) have as contiguous amino acid, Glu and Asp respectively. Nevertheless, this argument is weakened by the fact that rabbit muscle actin has N-ac-Asp on its amino end.
On the other hand, Liew, Haslett and Allfrey (13) have shown that f2a histones, which have in common a N-ac-Ser end, the induction of histone biosynthesis after partial hepatectomy brings about the appearance of N-acetyl-Seryl-tRNA. They have also found nascent peptides liberated as peptidyl-puromycin which had N-acetyl-Ser on its amino end and they have concluded

that N-ac-Seryl-tRNA is acting as an initiator in histone biosynthesis.
Narita (18) has come to the same conclusion with respect to N-acetyl-glycine, which is the N-terminal amino acid of ovalbumin.
Menninger (19) has described a peptidyl-tRNA hydrolase in rat liver which hydrolyzes various N-acyl-aminoacyl-tRNA's but which does not attack aminoacyl-tRNA. This enzyme hydrolyzes f-met-$tRNA^{fMet}$ from E. coli (which is not hydrolyzed by the E. coli hydrolase). On the contrary, the rat liver hydrolase does not attack N-ac-Seryl-tRNA (which is hydrolyzed by the E. coli enzyme). This confers on N-ac-Seryl-tRNA a special metabolic role in the hepatocyte cytoplasm.
In conclusion, we think that there is a relation between ferritin biosynthesis in rat liver and N-acetyl-Seryl-tRNA. Its presence there might indicate that in eukaryotes there are more than one mode of initiation.

SUMMARY

The induction of ferritin biosynthesis in rat liver has been shown to be related to the appearance of a N-acetyl-Seryl-tRNA in the cytoplasm. It has also been shown that rat liver apoferritin has an N-terminal sequence which starts with N-ac-Ser and that this sequence coincides with the initial tripeptide of horse spleen apoterritin. Transacetylation experiments have demonstrated that the tRNA of iron-stimulated livers is capable of accepting the acetyl moiety of acetyl-coenzyme A.

ACKNOWLEDGEMENTS

This work was done with a grant from Mead Johnson de Mexico, that also provided the iron-dextran preparation, Hi-Dex. Dr. J. M. Rodriguez was a Fellow of the National Council of Science and Technology (CONACYT) of Mexico.

REFERENCES

1. Granick, S.: Ferritin. IV. Occurrence and immunological properties of ferritin. J. Biol. Chem. 149, 157-167 (1943).

2. Granick, S.: Ferritin. I. Physical and chemical properties of horse spleen ferritin. J. Biol. Chem. 146, 451-461 (1942).

3. Hoy, T. G., Harrison, P. M., Hoare, R. J.: Quaternary structure of apoferritin: the rotation function at 9 Å resolution. J. Mol. Biol. 84, 515-522 (1974).

4. Fineberg, R. A., Greenberg, D. M.: Ferritin biosynthesis. II. Acceleration of synthesis by the administration of iron. J. Biol. Chem. 214, 97-106 (1955).

5. Chu, L. L. H., Fineberg, R. A.: On the mechanism of iron-induced synthesis of apoferritin in HeLa cells J. Biol. Chem. 244, 3847-3854 (1969).

6. Millar, J. A., Cumming, R. L. C., Smith, J. A., Goldberg, A.: Effect of actinomycin D, cycloheximide, and acute blood loss on ferritin synthesis in rat liver. Biochem. J. 119, 643-649 (1970).

7. Drysdale, J. W., Munro, H. N.: Failure of actinomycin D to prevent induction of liver apoferritin after iron administration. Biochim. Biophys. Acta 103, 185-188 (1965).

8. Rodriguez, J. M., Huberman, A.: Improved method for the determination of nonheme iron in animal tissues. I. Methodology. Rev. Invest. Clin. (Mex.) 26, 169-173 (1974).

9. Drysdale, J. W., Munro, H. N.: Small-scale isolation of ferritin for the assay of incorporation of ^{14}C-labelled amino acids. Biochem. J. 95, 851-858 (1965).

10. Suran, A. A.: N-terminal sequence of horse spleen apoferritin. Arch. Biochem. Biophys. 113, 1-4 (1966).

11. Mainwaring, W. I. P., Hofmann, T.: Horse spleen apoferritin: N-terminal and C-terminal residues. Arch. Biochem. Biophys. 125, 975-980 (1968).

12. Huberman, A., Rodriguez, J. M., Franco, R., Barahona, E.: Mechanism of induction of apoferritin biosynthesis in rat liver. Federation Proceed. (Abstract) 33:1404 (1974).

13. Liew, C. C., Haslett, G. W., Allfrey, V. G.: N-acetyl-seryl-tRNA and polypeptide chain initiation during histone biosynthesis. Nature 226, 414-417 (1970).

14. Vassart, G., Dumont, J. E., Cantraine, F. R. L.: Translational control of protein synthesis: a simulation study. Biochim. Biophys. Acta, 247, 471-485 (1971).

15. Marchis-Mouren, G., Lipmann, F.: On the mechanism of acetylation of fetal and chicken hemoglobins. Biochemistry 53, 1147-1154 (1965).

16. Strous, G. J. A. M., van Westreenen, H., Bloemendal, H.: Synthesis of lens proteins on vitro. N-terminal acetylation of alpha-crystallin. Eur. J. Biochem. 38, 79-85 (1973).

17. Strous, G. J. A. M., Berns, A. J. M., Bloemendal, H.: N-terminal acetylation of the nascent chains of alpha-crystallin. Biochem. Biophys. Res. Commun. 58, 876-884 (1974).

18. Narita, K.: The N-terminal acetyl group of proteins and its possible implications for protein biosynthesis. In: Proteins. Structure and function. M. Funatsu, ed. Vol. 2, p. 227-259. Halsted Press, Tokyo. 1972.

19. Menninger, J. R., Deery, S., Draper, D., Walker, Ch.: The metabolic role of peptidyl-tRNA hydrolase. II. Peptidyl-tRNA hydrolase activity on washed rat liver ribosomes.Biochim. Biophys. Acta 335, 185-195 (1974).

High Lipoprotein Lipase Activity and Cardiovascular Disease

W.C. Hülsmann and H. Jansen
Dept. Biochemistry I, Erasmus University,
P.O.Box 1738, Rotterdam, The Netherlands

Lipoprotein Lipases of Heart

Lipolytic activity released into the circulation by heparin has been used as an indirect measure of lipoprotein lipase, an enzyme involved in the removal of plasma triglycerides and possibly localized at the endothelial surface of various organs. Korn (1) reported that the increase of Clearing Factor (lipoprotein lipase) in the plasma after heparin injection correlated with a depletion of the enzyme from heart tissue. Payza *et al*. (2) reported that under those conditions the enzyme was also depleted from kidney and lung. Unfortunately these studies were carried out with Ediol as the artificial lipolytic substrate, which in addition to triglyceride contains monoglyceride (3). However, Borensztajn and Robinson (4), who tested Clearing Factor lipase activity with chylomicron triglyceride substrate were able to confirm that heparin released Clearing Factor lipase from the perfused rat heart. C.T. Bartels in our laboratory (unpublished) confirmed this using Intralipid, labeled with radioactive trioleate, as the lipolytic substrate. These studies, in agreement with those of Borensztajn and Robinson (4), also showed that in vitro perfusion of rat hearts with heparin and serum-containing perfusion media, resulted in the release of about half of the cardiac lipase activity to the perfusion medium. The activity remaining in the heart after 15 min heparin perfusion did not decrease further after additional 15 min perfusion. This suggests the presence of two pools of lipoprotein lipase in the heart: one readily releasable by heparin and one probably present within the cardiocytes. That the former is involved in the utilization of triglycerides from the serum was shown by Borensztajn and Robinson (4), who showed that preperfusion of the heart with heparin resulted in a loss of 80% of the rate of $^{14}CO_2$ production from labeled triglyceride-containing chylomicrons, subsequently perfused through the hearts. The enzyme remaining in the heart after heparin perfusion, may be involved in the utilization of triglycerides, stored in the cardiocytes. The enzyme has a pH optimum between 8 and 9 and is enriched in the microsomal and soluble fractions of homogenates as judged by De Duve plots of fractions obtained by differential centrifugation.

Myocardial Damage by Long-Chain Fatty Acids

The activity of lipoprotein lipase increases by rapeseed oil feeding when compared with olive oil feeding (5). Rapeseed oil, rich in erucic acid (cis-13-docosenoic acid), causes triglyceride storage in heart and skeletal muscles (6), especially between 3 and 6 days feeding. After 6 days feeding the fat stores diminish and after 4 weeks feeding almost normal values are obtained. The fatty acids of the accumulated triglycerides in the heart reflect the dietary fatty acids (7), suggesting an inhibited oxidation of all the dietary fatty acids in the heart. However, an increased uptake of long-chain fatty acids during the initial period of rapeseed oil feeding should also be considered (comp. ref. 8) together with a decreased rate of fatty acid oxidation (9,10). The increased cardiac lipoprotein lipase activity (5), present in the period when the amounts of triglyceride stored in the heart decrease, may be partly responsible for the fibrotic changes that occur when rapeseed oil or certain marine oils are given for prolonged periods of time (comp. ref. 11). Higher fatty acids at a relatively high concentration may cause damage to structures (e.g. mitochondrial swelling, lysosomal rupture etc.) and function (membrane ATP-ases). In the liver, the bulk of the triglycerides can be disposed of as very low density lipoproteins (VLDL). In heart this is not possible and triglycerides can only be disposed of after lipase activity. A part of the fatty acids produced is exported and another part is utilized by the heart. Ischemia therefore may interfere with both removal systems, since the fall of the coronary flow rate will interfere with fatty acid export and low oxygen tensions will interfere with fatty acid oxidation. A correlation between arrhythmias and high extracellular levels of free fatty acids has been disputed (12-14). It is doubtful whether such a correlation exists, since the frequency of arrhythmias and heart disease tends to be lower in periods of prolonged fasting (e.g. World War II), when free fatty acid levels in the blood are increased. Increased triglyceride storage within the heart, which may be accompanied by high cardiac lipase activity (5), on the other hand may be more harmful, since in the cardiocytes serum albumin, which can complex free fatty acids, is absent.

Occurrence of Atherosclerosis in Conditions of High Lipoprotein Lipase Activity

A high activity of extracellular cardiac lipoprotein lipase, involved in the uptake of triglycerides from blood as was mentioned above, may also be harmful. Borensztajn and Robinson (comp. the excellent review article of Robinson (15)), in confirmation with Alousi and Mallov (16), have found that fasting for a relatively short period causes an increase of cardiac clearing factor. This also holds for the diabetic state (17) where insulin levels are also low, and contrary to the high muscle activity, adipose tissue activity is low (15). A correlation between uncontrolled diabetes and coronary atherosclerosis has often been described and may be related to the high clearing factor activity of the coronary system and the simultaneous existence of high levels of VLDL in the blood (15). Since these triglyceride-rich lipoproteins are the substrate for clearing factor, it is conceivable that in these states, especially near the vascular endothelium of (heart)muscle, the concentration of the reaction products, low density lipoproteins (LDL) and free fatty acids (FFA), is also high.

Yudkin (18) put forward the hypothesis that the incidence of ischemic heart disease may be augmented by the excessive dietary intake of sucrose. Indeed, when starch in the diet of rats was replaced by sucrose, the lipid level in the aortas was found to increase (19). Under these dietary conditions lipoprotein lipase from adipose tissue seems to be decreased (20). We (21) were able to show that fructose feeding to rats leads to a lowering of postheparin serum triglyceridase and palmitoyl-CoA hydrolase activities. Heparin-releasable lipoprotein lipase from heart contributes very little to postheparin serum lipase (22, 23). Since dietary (and hormonal) influences generally have opposite effects on lipoprotein lipase from adipose tissue and heart (15), it is tempting to speculate that while fructose feeding decreases the adipose tissue enzyme it increases the heart enzyme. Therefore increased rates of lipolysis are likely to occur in the coronary vessels. The liver enzyme, which in contrast to immunologically different extrahepatic enzyme(s), hydrolyses palmitoyl-CoA as well as triglycerides (22,23) and probably also phospholipids, could be involved in the removal of beta-lipoproteins. Its low activity during fructose feeding might therefore also play a role in atherosclerosis, in which the subendothelial zone has been shown to contain lipid droplets, strongly resembling LDL (24).

Also pointing in the direction that recurrent high activities of lipoprotein lipase activity predispose for atherosclerosis, is the recently described (25) phenomenon of increased incidence of atherosclerosis in patients subjected to intermittent hemodialysis. Chronic renal failure is known to be accompanied by hyperlipemia (26,27). Especially triglyceride levels of the blood may be very high. During and after dialysis clearing of the blood is very rapid, hence in chronic renal failure peaks of high lipoprotein lipase activity frequently occur (28).

In type II hyperbetalipoproteinemia the incidence of coronary heart disease is also very high (ref. 29; the incidence is also quoted in ref. 25). Here the main biochemical feature is the increased serum level of LDL, often accompanied by clearcut increases of VLDL (29). Since LDL is the product of VLDL hydrolysis by lipoprotein lipase, this enzyme must be active in this disease. Indeed, in type I hyperlipemia, in which lipoprotein lipase is deficient, LDL levels are generally low and clinically no predisposition for cardiovascular disease occurs (29). It is likely that in type II hyperlipoproteinemia the rate of removal of LDL by the liver is also diminished (30).

Vascular Changes by High Concentration of Long-chain Fatty Acids

Fatty acids like octanoate and palmitate increase the coronary flow rate as has been observed by Challoner and Steinberg (31) in hearts perfused according to the Langendorff technique. We (32) were able to confirm this and showed that the higher fatty acids on a molar basis are more efficient than the more water-soluble lower fatty acids. Local lipoprotein lipase activity then may give rise to local long-chain fatty acid production and hence may cause vasodilation. Local fatty acid intoxication may be present in fat embolism, a systemic disease in which the permeability of the microcirculation is increased. In a model study De Ruiter (33) was able to show that relatively high concentrations of oleate promote the rate of extravasation, capillary stasis and the ultimate formation of circumscript loci where erythrocytes or carbon black particles (added to the perfusion medium as contrast-particles, to help the microscopic inspection of the perfused mesenterium) were seen to penetrate. It is therefore tempting to speculate that the extravasation of intravascular lipid, as seen in atherosclerosis, also proceeds in this manner. Majno and Palade (34) showed that the vasodilators histamine and serotonin also caused widening of gaps between endothelial cells. These gaps are probably important routes for the entry of LDL from the plasma to the arterial wall. A striking correlation between the concentration of LDL in the intima and the cholesterol level in serum of patients was found by Smith and Slater (35), which favours the idea of transudation between endothelial cells rather than transport by pinocytosis. Duncan et al. (36) already proposed that the vascular wall may be stretched and that interendothelial junctions might open. Stein and Stein (37) made in vitro retrograd perfusions through the thoracic aorta and coronary system with a perfusion medium containing labeled lipoproteins, isolated by ultracentrifugation. They concluded that LDL and high density lipoproteins are transported through the endothelial surface in contrast to the larger VLDL. The lipoprotein lipase reaction, acting on VLDL to produce FFA and LDL may be essential for the accumulation of lipid under the endothelial surface. In the first place vasodilator (and porosity promoting; see below) FFA is produced and secondly smaller particles, relatively enriched in cholesterol, which can pass the endothelial barrier.

That long-chain fatty acids, even when bound to albumin so that their free concentration is in the micromolar range, reversibly cause vasodilation, has been shown by us (32). Vasodilation, by opening the microcirculation promotes metabolism. Hence it is possible that fatty acids as vasodilators may stimulate their own utilization in the tissues. At excessively high concentrations, however, long-chain fatty acids may promote intravascular stasis and may cause vascular leakage. This has been demonstrated by De Ruiter (33) as has been mentioned above. The mechanism of this phenomenon might be based on the occurrence of a calcium-containing cementing substance between endothelial cells (38), which could "soften" by removal of Ca^{2+} (in this case by fatty acids that form insoluble calcium-soaps). The presence of a calcium-containing intercellular substance has often been disputed, but evidence in favour of such a rutenium-red staining mucopolysaccharide substance accumulates (39). The leaks caused by fatty acids may occur at the intercellular junctions (34,38), or even preferentially where three cells meet (40). The local excessive lipolysis, the basis of the hypothesis formulated by De Ruiter (33) in 1966 as the cause of vascular leakage in traumatic fat embolism, might also be the basis of atherosclerosis, where vascular porosity apparently exists since lipid particles and thrombocytes are able to escape from the bloodstream to the subendothelial zone. In this connection it may be of interest to note that recently Maca and Hoak (41) were able to demonstrate in rabbits that a strong lipolytic response on subcutaneously injected adrenocorticotropic hormone evoked detachment of aorta endothelial cells, followed by platelet attachment to subendothelial structures. Recently Zilversmit (42) presented a literature survey from which he also concludes that lipoprotein lipase activity might be an important link in the atherogenic process. In this paper Zilversmit (42) draws attention to the possible benificial effect of heparin in the treatment of atherosclerosis, being a substance that releases lipoprotein lipase from the potential site of atherosclerosis, while causing intraluminal clearing of plasma triglycerides.

SUMMARY

A high activity of lipoprotein lipase in combination with high concentrations of triglyceride-rich lipoproteins, may lead to cardiovascular disease. A high activity of intracellular cardiac lipoprotein lipase, possibly involved in the removal of stored triglycerides, may lead to structural changes (such as fibrosis) and functional changes (such as arrhythmias). A high activity of extracellular cardiac lipoprotein lipase, involved in the uptake of triglycerides from the blood, may lead to increased vascular porosity, resulting in a deposition of intravascular components, such as the relatively cholesterol-rich and small low density lipoproteins, in the subendothelial zone of the vessels.

ACKNOWLEDGEMENTS

One of us (W.C.H.) gratefully recalls the stimulating discussions with Dr. H. de Ruiter on lipolysis and vascular permeability changed some ten years ago, which were the basis of the present paper.

REFERENCES

1. Korn, E.D.: Clearing Factor, a heparin-activated lipoprotein lipase. I. Isolation and characterization of the enzyme from normal rat heart. J. Biol. Chem. 215, 1-14 (1955).
2. Payza, A.N., Eiber, H.B., Walters, S.: Studies with clearing factor V. State of tissue lipases after injection of heparin. Proc. Soc. Exp. Biol. & Med. 125, 188-192 (1967).
3. Biale, Y., Shafrir, E.: Lipolytic activity toward tri- and monoglycerides in postheparin plasma. Clin. Chim. Acta 23, 413-419 (1969).
4. Borensztajn, J., Robinson, D.S.: The effect of fasting in the utilization of chylomicron triglyceride fatty acids in relation to clearing factor lipase (lipoprotein lipase) releasable by heparin in the perfused rat heart. J. Lipid. Res. 11, 111-117 (1970).
5. Jansen, H., Hülsmann, W.C., Struyk, C.B., Houtsmuller, U.M.T.: Influence of rapeseed oil feeding on the lipoprotein lipase activity of rat heart. Submitted for publication (1974).
6. Abdellatif, A.M.M., Vles, R.O.: Pathological effects of dietary rapeseed oil in rats. Nutr. Metab. 12, 285-295 (1970).
7. Beare-Rogers, J.L., Nera, E.A., Craig, B.M. : Accumulation of cardiac fatty acids in rats fed synthesized oils containing C_{22} fatty acids. Lipids 7, 46-50 (1972).
8. Gumpen, S.A., Norum, K.R.: The relative amounts of long-chain acylcarnitines, short-chain acylcarnitines and carnitine in heart, liver and brown adipose tissue from rats fed on rapeseed oil.

Biochim. Biophys. Acta 310, 48-55 (1973).

9. Christophersen, B.O., Bremer, J.: Inhibitory effect of erucylcarnitine on the oxidation of palmitate by rat heart mitochondria. FEBS Lett. 23, 230-232 (1972).

10. Swarttouw, M.A.: The oxidation of erucic acid by rat heart mitochondria. Biochim. Biophys. Acta 337, 13-21 (1974).

11. Abdellatif, A.M.M., Vles, R.O.: Short-term and long-term pathological effects of glyceryl trierucate and of increasing levels of dietary rapeseed oil in rats. Nutr. Metabol. 15, 219-231 (1973).

12. Opie, L.H.: Effect of fatty acids on contractility and rhythm of the heart. Nature (Lond.) 227, 1055-1056 (1970).

13. Opie, L.H., Thomas, M., Owen, P., Norris, R.M., Holland, A.J., van Noorden, S.: Failure of high concentrations of circulating free fatty acids to provoke arrhythmias in experimental myocardial infarction. Lancet i, 818-822 (1971).

14. Oliver, M.F.: Metabolic response during impending myocardial infarction. II. Clinical implications. Circulation 45, 491-500 (1972).

15. Robinson, D.S.: The function of plasma triglycerides in fatty acid transport. In Florkin, M. & Stotz, E.H. (Eds) Comprehensive Biochemistry Vol. 18, 51-116 (1970).

16. Alousi, A.A., Mallov, S.: Effects of hyperthyroidism, epinephrine and diet on heart lipoprotein lipase activity. Amer. J. Physiol. 206, 603-609 (1964).

17. Kessler, J.J.: Effect of diabetes and insulin on the activity of mycardial and adipose tissue lipoprotein lipase of rats. J. Clin. Invest. 42,

362-367 (1963).

18. Yudkin, J.: Sucrose and cardiovascular disease. Proc. Nutr. Soc. 31, 331-337 (1972).
19. Khan, I.H., Yudkin, J.: The lipid content of the aortas of rats given sucrose. Proc. Nutr. Soc. 31, 10A (1972).
20. Bruckdorfer, K.R., Kang, S.S., Yudkin, J.: The hyperlipaemic effect of sucrose in male and female rats. Proc. Nutr. Soc. 31, 11A (1972).
21. Jansen, H., Hülsmann, W.C.: Long-chain acyl-CoA hydrolase activity in serum: identity with clearing factor lipase. Biochim. Biophys. Acta 296, 241-248 (1973).
22. Jansen, H., van Zuylen-van Wiggen, A., Hülsmann, W.C.: Lipoprotein lipase from heart and liver: an immunological study. Biochem. Biophys. Res. Commun. 55, 30-37 (1973).
23. Jansen, H., Hülsmann, W.C.: Liver and extrahepatic contributions to postheparin serum lipase activity of the rat. Submitted for publication (1974).
24. Page, I.H.: Atherosclerosis. An introduction. Circulation 10, 1-27 (1954).
25. Lindner, A., Charra, B., Sherrard, D.J., Scribner, B.H.: Accelerated atherosclerosis in prolonged maintenance hemodialysis. N. Engl. J. Med. 290, 697-701 (1974).
26. Roodvoets, A.P., van Neerbos, B.R., Hooghwinkel, G.J.M., Hulsmans, H.A.M., Beukers, H.: Hyperlipemia in patients on regular dialysis treatment. Proc. Eur. Dial. Transpl. Assoc. 4, 257-263 (1967).
27. Bagdade, J.D., Porte Jr., D., Bierman, E.L.: Hypertriglyceridemia; a metabolic consequence of chronic renal failure. N. Engl. J. Med. 279, 181-

185 (1968).

28. Tsaltas, T.T., Friedman, E.A.: Plasma lipid studies of uremic patients during hemodialysis. Amer. J. Clin. Nutr. 21, 430-435 (1968).

29. Fredrickson, D.S., Levy, R.I.: Familial hyperlipoproteinemia. In Metabolic Basis of Inherited Disease (Stanbury, J.B., Wyngaarden, J.B., Fredrickson, D.S., Eds.) McGraw Hill Book Cy, p. 545-614 (1972).

30. Langer, T., Strober, W., Levy, R.I.: Familial type II hyperlipoproteinemia; a defect of beta apoprotein catabolism. J. Clin. Invest. 48, 49A (1969).

31. Challoner, D.R., Steinberg, D.: Effect of free fatty acid on the oxygen consumption of perfused rat heart. Amer. J. Physiol. 210, 280-286 (1966).

32. Hülsmann, W.C., Kurpershoek-Davidov, R.: Problems in the study of gluconeogenesis in the isolated perfused rat liver. Hoppe Seyler's Z. f. Physiol. Chemie 354, 1205 (1973).

33. de Ruiter, H.: Traumatic fat embolism. M.D. thesis Amsterdam, Drukkerij Noord Holland N.V. Hoorn, the Netherlands (1966).

34. Majno, G.,Palade, G.E.: The effect of histamine and serotonin on vascular permeability. An electron microscope study. J. Biophys. Biochem. Cytol. 11, 571-605 (1961).

35. Smith, E.B., Slater, R.S.: Lipids and low density lipoproteins in intima in relation to its morphological characteristics. In Atherogenesis Initiating Factors, Ciba Foundation Symposium 12, 39-52 (1973).

36. Duncan, L.E., Birck, K., Lynch, A.: The effect of pressure and stretching on the passage of labeled

albumin into canine aorta wall. J. Atherosclerosis Res. 5, 69-79 (1965).

37. Stein, Y., Stein, O.: Lipid synthesis and degradation and lipoprotein transport in mammalian aorta. In Atherogenesis Initiating Factors, Ciba Foundation Symposium 12, 165-179 (1973).

38. Chambers, R., Zweifach, B.W.: Intercellular cement and capillary permeability. Physiol. Rev. 27, 436-463 (1947).

39. Luft, J.H.: Ultrastructural basis of capillary permeability. In The inflammatory process (Zweifach, B.W., Grant, L., McClusky, R.T., Eds.) Academic press New York and London, p. 121-159 (1965).

40. Krogh, A.: Anatomy and physiology of capillaries, Yale University Press, New Haven, Conn., U.S.A., 2nd ed. (1929).

41. Maca, R.D., Hoak, J.C.: Endothelial injury and platelet aggregation associated with acute lipid mobilization. Lab. Invest. 30, 589-595 (1974).

42. Zilversmit, D.B.: A proposal linking atherogenesis to the interaction of endothelial lipoprotein lipase with triglyceride-rich lipoproteins. Circulation Res. 23, 633-638 (1973).

Gramicidin S-Synthetase: Active Form of the Multienzyme Complex is Undissociable by Sodium Dodecylsulfate

H. Kleinkauf and H. Koischwitz
Max-Volmer-Institut, Abteilung Biochemie, Technische Universität Berlin

INTRODUCTION

The formation of gramicidin S (GS), tyrocidines (TY), linear gramicidins, and edeines (1-3), by various strains of Bacillus brevis has been shown to proceed on multienzyme complexes. The constituent amino acids are activated on protein sites comparable to those of aminoacyl-tRNA synthetases. The noncovalently bound aminoacyladenylates are transferred to thiol groups instead of hydroxyl groups of tRNA. Evidence for direct location of the thiol at the activation site is given by affinity labelling with L-proline-chloromethylketone resulting in the loss of the L-proline dependent ATP-pyrophosphate exchange reaction (4).

Peptide bond formation occurs from thioesters in four types of reactions (Fig.1):

1. Initiation Reaction: Thiolester-bound D-phenylalanine is transferred from the racemase (E.C.5.1.1.11) to the thiolester-bound L-proline on the multienzyme complex (gramicidin S-synthetase = GSS, or tyrocidine intermediate = TYS I). No transfer is observed with chemically prepared D-phenylalanine-thiolesters, indicating catalytic involvement of racemase sites (5).
2. Elongation Reaction: In this reaction a thiolester-bound peptide is transferred to a thiolester-bound amino acid. This transfer involves transthiolation of peptides to the 4'-phosphopantetheine (PAN). All active thiol groups must be in reach of this 'peptide carrier', which limits the size of PAN-containing multienzyme complexes.
3. Transfer Reaction: Here a peptide is transferred to an activated amino acid on another multienzyme complex. It is not known if this transfer proceeds over an intermediate step similar to reaction 2.

4. Termination Reaction: The simplest termination reaction is the diketopiperazine formation of the enzyme-bound intermediate D-phenylalanyl-L-proline. The decapeptide-cyclization of tyrocidine may be similar. During GS formation, however, two peptides react head-to-tail in a reaction that is not yet fully understood (6).

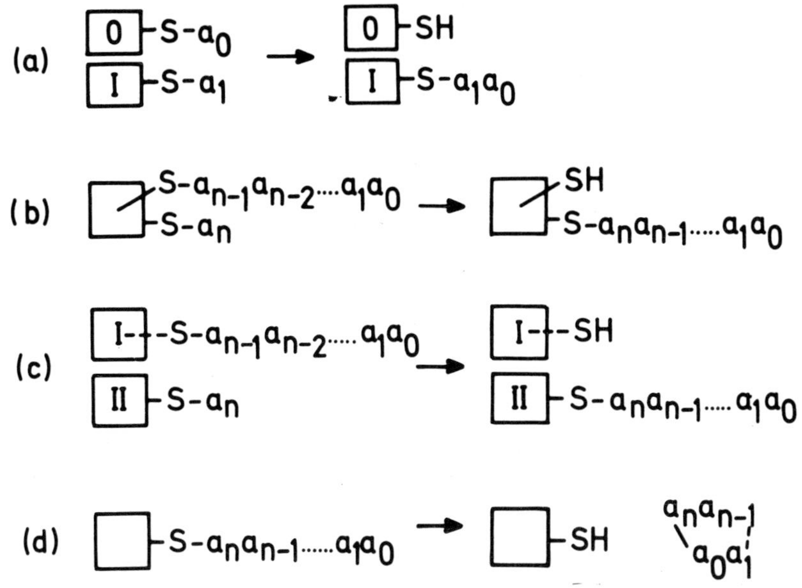

Fig.1. Types of reaction involved in enzyme-mediated sequential peptide bond formation. Squares indicate individual enzymes; subscripts show amino acid sequence; the term '0' stands for initiating step.

There appears to be a direct relationship between the number of catalytic functions and size of PAN-containing enzyme complexes. As first observed by LIPMANN's group there was an almost constant ratio of molecular weight and number of amino acid activation sites in GSS, TYS I, and TYS II (5). The limit of size and catalytic abilities comes from the requirement that all active thiols or catalytic centers have to be available to PAN. Evaluation of similar models that have been discussed by LYNEN for yeast fatty acid synthetase leads to the conclusion that length and charged groups of bound intermediates are responsible for specific binding and control of sequence.

Observations of subunit structure of aminoacyl-tRNA synthetases (7),and characterization of PAN-containing proteins (8) led to speculations on the subunit structure of peptide-synthesizing complexes. LIPMANN's group postulated similar sized subunits of about 70,000 daltons for amino acid activation together with small PAN-proteins (9).

The search for subunits, however, became complicated by factors in crude extracts able to dissociate TYS I, TYS II, and GSS. Purified synthetases could not be dissociated except for a small proportion in the case of TY (9) or a large proportion (at least 50%) of GSS not containing all catalytic functions of the complex.

Thus peptide synthesizing enzyme complexes appear to be single chain polypeptides with several different active centers.

METHODS

Preparation of Enzyme: Growth of cells and purification steps were essentially as described by GEVERS et al. (10), and KLEINKAUF et al. (11), after DEAE-cellulose chromatographic fractions active in the L-ornithine dependent ATP-pyrophosphate exchange reaction were precipitated with ammonium sulfate and applied to linear 10-30% (by volume) glycerol gradients. No more than 5 mg of protein in 1 ml of buffer was put on 32 ml of gradient solution. Centrifugation was carried out in an SW 27 rotor for 20 hours at 27,000 rpm at $0^{o}C$. Fractions of approximately 1 ml were collected.

Gel Electrophoresis: Gels usually contained 8% acrylamide and 0.45% BIS, and were run at $4^{o}C$ in 0.1 M Tris buffer at p_H 9.1, containing 0.1% 2-mercaptoethanol. In this system a good resolution of monomer and dimer is obtained.

SDS-gels, and buffer contained 0.1% SDS, and separation was done at room temperature. After destaining, gels were scanned at 590 nm. The peak area was assumed to be proportional to the amount of protein (protein units).

SDS-Treatment: At low SDS concentrations (0.01%) inactive forms of GSS dissociate stepwise and intermediates can be observed. Usually samples were incubated in the presence of 2% SDS for 15 minutes at $37^{o}C$ or 10 min at 50°C. Raising the temperature to 100°C, the time up to one hour, the SDS concentration to 2%, or including 0.1 M dithiothreitol (12) did not affect the subunit pattern. At high ionic strength (0.6 M KCl) SDS protein complexes precipitate.

Activity Tests: The ATP-pyrophosphate exchange reaction is done according to KLEINKAUF et al. (11, 13) GS-synthesis is measured by the Millipore filter assay of GEVERS et al. (10).

RESULTS AND DISCUSSION: ACTIVE AND INACTIVE FORMS OF GRAMICIDIN S-SYNTHETASE

Generally used purification methods include Sephadex G 200 filtration, DEAE-cellulose chromatography, density gradient centrifugation, and isoelectric focusing (4). Enzyme preparations, which appear homogenous on polyacrylamide gel electrophoresis (PAGE), still activate the GS-amino acids but have lost polymerizing activity. These forms of the enzyme can be dissociated by SDS into different patterns of subunits. The number and size of subunits depend to some extent on proteolysis in vitro during purification, since they are influenced by the presence of proteinase inhibitors such as phenylmethylsulfonyl fluoride (PMSF), diisopropylfluorophosphate (DFP), and EDTA. Depending on the degree of breakdown, the material still migrates as a single band in PAGE with a mobility comparable to that of bovine serum albumin (BSA)-tetramers, or it may be present only in fragments.

We have analysed profiles of glycerol gradients for subunit composition and catalytic capacity, since partially degraded material displays slightly different sedimentation behavior.

Fig.2 shows a typical profile with an almost symmetrical protein peak, with GS-synthesizing activity located in the right shoulder.

The polyacrylamide gels below show that some of the material dimerizes during electrophoresis, and the amount of dimer formed is approximately proportional to GS-formation (Fig.3). There appears to be a direct correlation between dimerization during PAGE, peptide synthesis, and resistance against SDS (Fig.4).

The asymmetric profile indicates an association equilibrium, probably a monomer-dimer system, since no higher aggregates have been observed so far.

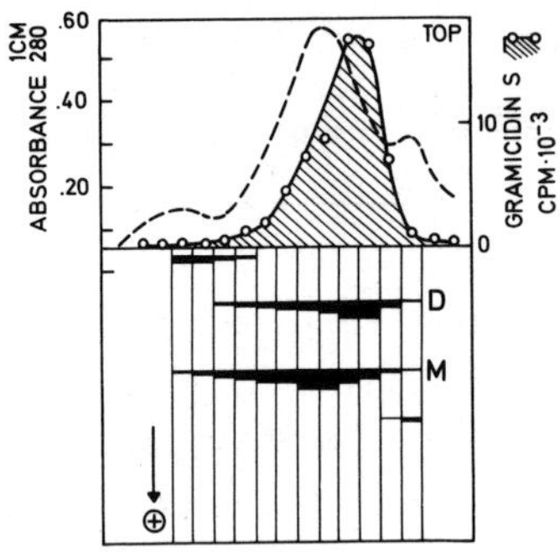

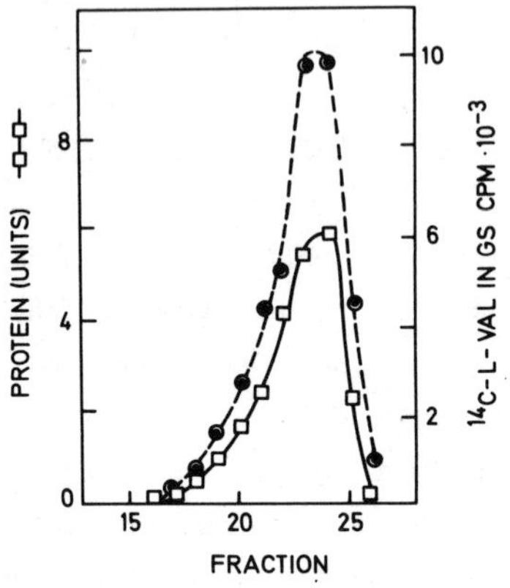

Fig.2. A 10-30% glycerol gradient profile obtained as described in Methods. Shaded area shows active GSS as determined by measuring GSS-synthesis with complementary racemase. Lower part of the figure displays PAGE-patterns of the corresponding gradient fractions in Tris-buffer at pH 9.1 and 4°C. M and D represent monomer and dimer resp.

Fig.3. Amount of dimer (in protein units derived from stained gels), and GS-synthesizing activity in the glycerol gradient profile of Fig.2. The profile of active GSS is asymmetrical, probably owing to protein interactions. The activity is partly suppressed on the left shoulder, indicating the negative effect of inactive forms.

Proteolytic degradation of active enzymes is believed to proceed stepwise and irreversibly. The removal of pantetheine may be an initiating event required for degradation *in vivo*, which accounts for loss of polymerizing activity in modified, dissociable GSS. From the observed subunits in our preparations we postulate the following degradation scheme of GSS by proteases present in crude extracts (Fig.5). The first step produces two peptides of equal size; no intermediates in the range of 280,000 to 140,000 daltons could be detected.

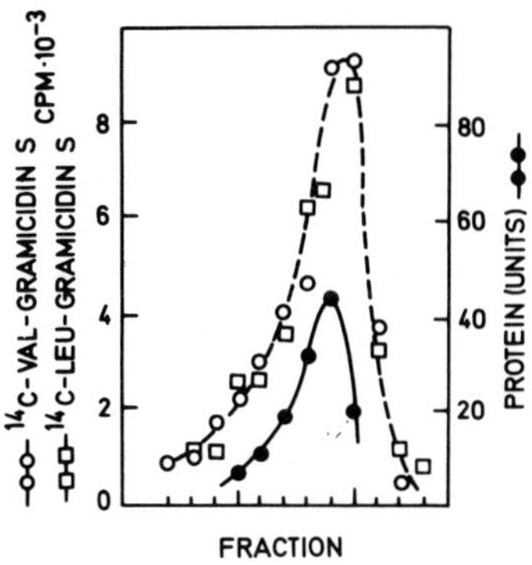

Fig.4. A glycerol gradient profile of GSS-preparations. Correlation of SDS-stable material (protein units from PAGE in Tris-SDS), and GS-formation with complementary racemase.

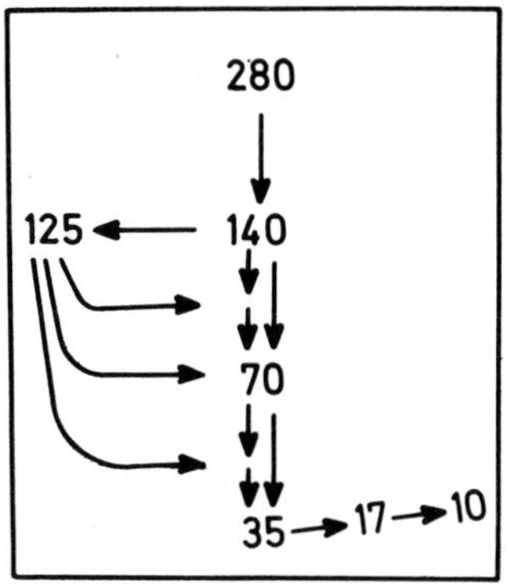

Fig.5. Enzymatic degradation of GSS I in crude extracts. Molecular weights (x 10^{-3}) of SDS-peptides are displayed; arrows indicate degradation steps.

The enzyme of 280,000 daltons remains associated in the absence of SDS even when several steps of proteolytic degradation occur. SDS treatment dissociates this enzyme into subunits not smaller than 35,000 daltons. When the 280,000 dalton complex has already been partially dissociated, then subunits smaller than 35,000 daltons can be detected. Density gradient centrifugation of these species shows at least 6 intermediates during breakdown to 70,000 dalton size.

All subunits can be detected in the presence of protease inhibitors, such as 5 mM EDTA, 1 mM DFP, and 3.45 mM PMSF. Thus, some may have formed _in vivo_.

The presence of modified forms may explain the difference in catalytic properties reported by various authors (Tab.I).

Initial velocities of the GS amino acid-dependent ATP-$^{32}PP_i$ exchange reaction.

	a	b	c	d	e	f	g
L-Phe	.04	.20	--	--	.17	--	--
L-Pro	1.00	1.00	1.00	1.00	1.00	1.00	1.00
L-Val	4.05	4.85	3.23	2.80	2.22	1.15	1.8
L-Orn	4.19	5.05	1.25	1.18	1.15	1.4	1.1
L-Leu	2.45	5.17	1.63	1.54	2.25	2.3	.9

Velocities are expressed relative to that of the L-proline-dependent reaction.

a-OTANI et al. 1970 (14), b-OTANI et al. 1969 (15), c, d-KURAHASHI et al. 1969 (16), e-GEVERS et al. 1968 (10), f-LEUNG & BAXTER 1972 (4), g-active form of GSS in this investigation.

If we consider initial velocities of ATP-$^{32}PP_i$ exchange reactions compared to the L-proline-dependent rate, large deviations can be observed. Thus, large increases in the L-valine- and the L-leucine-dependent exchanges have been found in modified forms containing 70,000 or 35,000 dalton subunits (Fig.6).

In addition to changes in reaction velocities, changes in amino acid specificity are observed. These data represent better measures of conformational changes than a comparison of the velocities of the several exchanges.

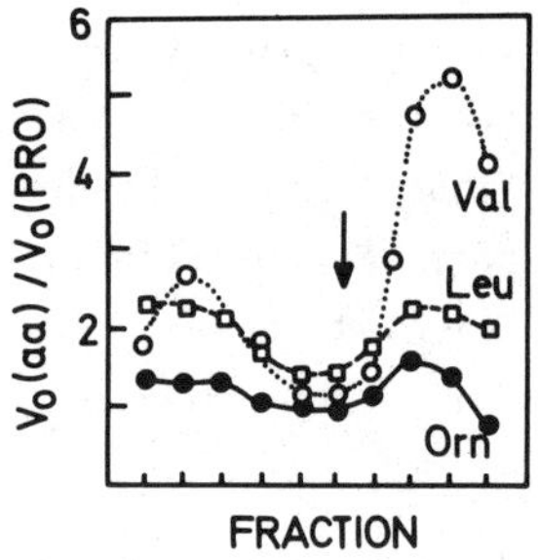

Fig.6. Comparison of initial velocities of individual ATP-$^{32}PP_i$ exchange reactions relative to the L-proline-dependent reaction. Glycerol gradient profile is taken from Fig.2. The arrow indicates maximal concentrations of active GSS. The left peak contains mostly the 35,000 dalton SDS subunits, and the right peak contains mostly the 70,000 dalton SDS-subunits.

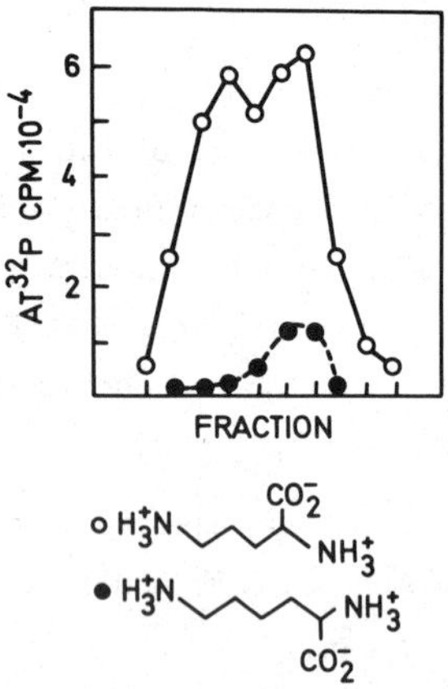

Fig.7. Glycerol gradient profile as in Fig.6. Initial velocities of the L-ornithine and L-lysine-dependent ATP-$^{32}PP_i$ exchange reactions. Only the active GSS gives measurable exchange with the analogue.

Thus we have shown that L-lysine is incorporated into GS at 65% of the rate of the natural substrate, L-ornithine. A measurable ATP-$^{32}PP_i$ exchange reaction is observed only with the active form of GSS (Fig.7).

SUMMARY

The active form of gramicidin S-synthetase is not dissociable with SDS and can be distinguished from inactive, protease-modified forms. These forms are active in the amino acid activation reaction but have lost the polymerising activity. They have slightly different sedimentation velocities depending on their subunit composition. Even in the presence of proteinase inhibitors that protected B. subtilis DNA-dependent RNA polymerase in a comparable growth period (17) against modification in vitro, GSS-preparations contain at least 50% SDS-dissociable forms. These forms have different substrate specificities from those shown by the active enzyme complex.

Acknowledgments: We thank Kristiane Schubert and Ali El-Samaraie for technical assistance, and the Schering Company for growth of cells. Parts of this work were supported by Deutsche Forschungsgemeinschaft grant Kl 148/9.

REFERENCES

(1) Lipmann, F., Gevers, W., Kleinkauf, H., Roskoski, R., Jr.: Polypeptide synthesis on protein templates: The enzymatic synthesis of gramicidin S and tyrocidines. Adv.Enzymology 35, 1-34 (1971).
(2) Bauer, K., Roskoski, R., Jr., Kleinkauf, H., Lipmann, F.: Synthesis of a linear gramicidin by a combination of biosynthetic and organic methods. Biochemistry 11, 3266-3271 (1972).
(3) Kurylo-Borowska, Z., Sedkowska, J.: Fractionation and characterization of enzymes responsible for biosynthesis of edeine A and B. Biochim. Biophys.Acta 351, 42-56 (1974).
(4) Leung, D.C., Baxter, R.M.: Substrate-derived reversible and irreversible inhibitors of the multienzyme I of gramicidin S biosynthesis. Biochim.Biophys.Acta 279, 34-47 (1972).
(5) Roskoski, R., Jr., Ryan, G., Kleinkauf, H., Gevers, W., Lipmann, F.: Polypeptide biosynthesis from thioesters of amino acids. Arch.Biochem. Biophys.143, 485-492 (1971).
(6) Stoll, E., Frøyshov, Ø., Holm, H., Zimmer, T.L., Laland, S.G.: On the mechanism of gramicidin S formation from intermediate peptides. FEBS Lett.11, 348-352 (1970).

(7) Loftfield, R.B.: The aminoacylation of transfer ribonucleic acid protein synthesis (ed.McConkey, E.H.), Vol.1, 1-88 (1971).
(8) Prescott, D.J., Vagelos, P.R.: Acyl carrier protein. Adv.Enzymology 36, 269-311 (1972).
(9) Lee, S.G., Roskoski, R., Jr., Bauer, K., Lipmann, F.: Purification of the polyenzyme responsible for tyrocidine synthesis and their dissociation into subunits. Biochemistry 12, 398-405 (1973).
(10) Gevers, W., Kleinkauf, H., Lipmann, F.: The activation of amino acids for biosynthesis of gramicidin S. Proc.Nat.Acad.Sci.USA 60, 269-276(1968).
(11) Kleinkauf, H., Gevers, W., Lipmann, F.: Interrelation between activation and polymerisation in gramicidin S biosynthesis. Proc.Nat.Acad.Sci. USA 62, 226-233 (1969).
(12) Lee, S.G., Lipmann, F.: Isolation of a peptidylpantetheine protein from tyrocidine-synthesizing polyenzymes. Proc.Nat.Acad.Sci.USA 71, 607-611 (1974).
(13) Koischwitz, H.: Isolierung des an der Gramicidin S-Biosynthese beteiligten Pantetheinträgerproteins. Diplomarbeit TU Berlin 1972.
(14) Otani, S., Jr., Yamanoi, T., Saito, Y.: Fractionation of the enzyme system responsible for gramicidin S biosynthesis. Biochim.Biophys.Acta 208 496-508 (1970).
(15) Otani, S., Yamanoi, T., Saito, Y.: Biosynthesis of gramicidin S; Ornithine activating enzyme. J. Biochem.(Tokyo) 66, 445-453 (1969).
(16) Kurahashi, K., Yamada, M., Mori, K., Fujikawa, K., Kambe, M., Imae, Y., Sato, E., Takahashi, H., Sakamoto, Y.: Biosynthesis of cyclic oligopeptide. Cold Spring Harbor Symp.Quant.Biol. 34, 815-826, (1969).
(17) Brevet, J.: Direct assay for sigma factor activity and demonstration of the loss of this activity during sporulation in Bacillus subtilis. Molec.gen.Genet.128, 223-231 (1974).

Studies on the Biosynthesis and Structure of Renal Glomerular Basement Membrane

I. Krisko and F. Gyorkey

Veterans Administration Hospital and Baylor College of Medicine, Houston, Texas 77031, USA

With respect, admiration and deep joy, we dedicate this article to Professor Fritz Lipmann to honor him on his 75th birthday. One of us had the opportunity of working in his laboratory at The Rockefeller University; I thank Dr. Lipmann for guidance. We both thank him for the inspiration which he has given us. This contribution is intended as a symbol of appreciation and to wish him well on this very special occasion.

This paper deals with the biosynthesis and structural studies of a special extracellular macromolecular material - the glomerular basement membrane (GBM). The renal glomerulus itself can be regarded as an organ. It is well demarcated from surrounding tissues by Bowman's capsule. The ultrafiltration of plasma takes place in the glomeruli, which contain three different cell types. The filtration barrier is the GBM, which is a glycoprotein (1-5). We now present evidence that isolated glomeruli synthesize glycoproteins *in vitro*, that part of the macromolecular product is GBM, and shall further characterize the system.

METHODS

Glomeruli were obtained from either young male albino rats (100-200 g) or from calves (1 yr old or younger) as described (4-6). The isolation of GBM also has been described (1, 4). The isolated GBM appeared pure when examined under the electron microscope, and its amino acid and amino sugar compositions corresponded with published data (1-3).

The assay method for protein and glycoprotein synthesis was as reported (4). Hexuronic acid was determined using the borate-carbazole method (7). To assure reproducibility and precision, we found it important to use only freshly prepared solutions of carbazole. The procedure for identification of monosaccharides by thin layer chromatography (TLC) in glomeruli has been reported (4). Glycosaminoglycans (GAG) were identified using the bidimensional electrophoresis method of Sato and Gyorkey (8).

Preparation of samples for determination of GAG content: A lyophilized aliquot of the desired fraction was digested with papain, centrifuged (2000 x g x 10 min) and the supernate dialyzed exhaustively against distilled H_2O. The dialyzed material was again lyophilized and aliquots taken for determination of hexuronic acid and analysis on bidimensional electrophoresis (using the Beckman Microsome Electrophoresis System). All data reported here have been obtained using rat glomeruli except Table 4 and Fig. 4, where bovine material was employed.

RESULTS

Using several radiolabeled amino acids, we have previously established (4-6) that isolated glomeruli are

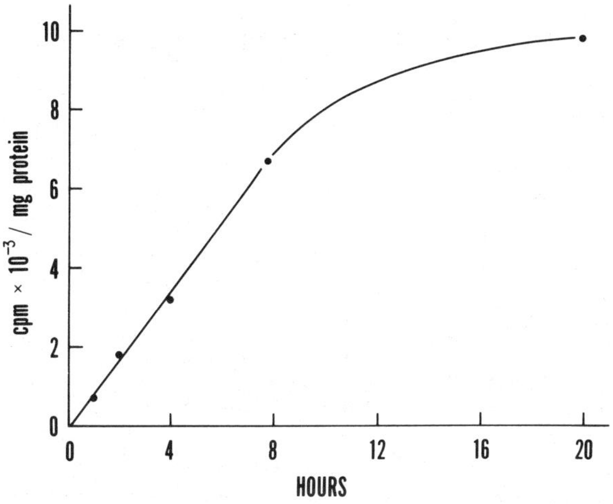

Fig. 1. Time course of ^{3}H-Pro incorporation. Each reaction mixture contained: glomeruli, ca 0.5 mg [based on protein determination by the method of Lowry (9)]; ^{3}H-Pro, 1 µCi; unlabeled Pro was omitted from the medium. The incubation was in an atmosphere of 5% CO_2-95% O_2 at 37° and was stopped by the addition of 5% trichloracetic acid (TCA) and the precipitated material washed four times in 5% cold TCA. The final pellet was hydrolyzed in 1.0 N NaOH and aliquots taken for determination of protein (9) and radioactivity (by liquid scintillation spectrometry). The ordinate represents specific activity as cpm/mg glomerular protein. (For further details, see ref. 4.)

active in protein synthesis for at least 6 hr. Since GBM has a high proline and hydroxyproline content (1-3) and since the latter can be regarded as a marker for GBM or its precursor(s), we studied proline incorporation *in vitro*. Incorporation was linear for approximately 8 hr (Fig. 1). Although proline and hydroxyproline were not separated in this experiment, we have evidence for *in vitro* hydroxylation of proline (5). In other experiments, GBM was isolated after *in vitro* incorporation of radiolabeled precursors and the GBM fraction was found to be radio-

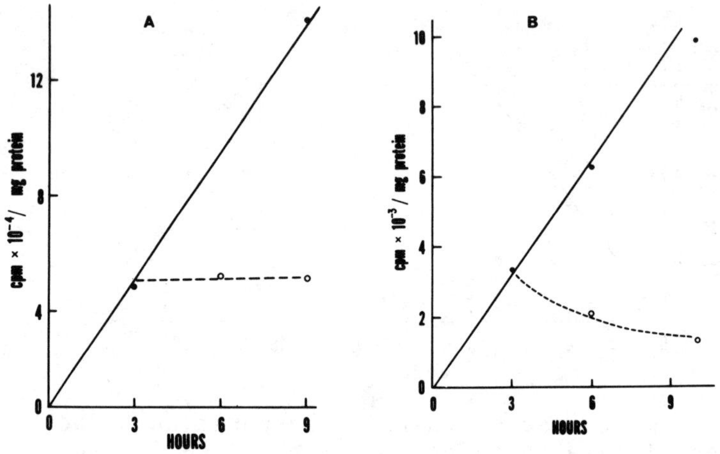

Fig. 2. Pulse labeling of glomerular proteins. A: Incubation was as described in Fig. 1, but using instead of ^{3}H-Pro five ^{3}H labeled amino acids (Pro, Lys, Glu, Asp, Leu, 2.0 µCi each per flask). The corresponding unlabeled amino acids were omitted. After 3 hr of incubation at 37°C, glomeruli in the appropriate flasks were washed 3 times (at 0-4°C). The glomeruli were finally resuspended in complete medium without labeled amino acids and incubated further at 37°C as shown on the abscissa. B: The experiment was as in A, except ^{3}H-Gal, 2 µCi per flask was used and unlabeled Glc was omitted.

active whether the label was originally in amino acids or galactose, signifying in vitro GBM synthesis (5). The above studies suggest that the glomerular in vitro system possesses an active protein and glycoprotein synthesizing machinery. To further characterize this in vitro system, we have attempted to determine the in vitro turnover rates of glomerular proteins and glycoproteins. After a 3 hr pulse using five different

Table 1. Incorporation into Glomerular Protein

	cpm per mg protein		
Precursor	No Inhibitor	Puromycin $1x10^{-4}$ M	% Inhibition
Galactose	3,160	440	86
Glucosamine	2,426	446	81
Glucose	8,140	2,508	69

The conditions of incubation were those of Fig. 1 except: Pro was not omitted; unlabeled Glc was omitted and 2 µCi of either ^{3}H-Gal, ^{3}H-GlcN, or ^{14}C-Glc was added. Puromycin was added at zero time. Incubation was for 4 hr at 37°C. (From Krisko, I. and Walker, W.G. (4), by permission.)

^{3}H labeled amino acids, no "chasing" of the label was detected in 6 hr of further incubation in presence of unlabeled medium (Fig. 2). Interestingly, under identical experimental conditions but using ^{3}H-galactose, approximately 50% of the label disappeared from macromolecules.

Puromycin inhibited the incorporation of labeled galactose, glucosamine and glucose into TCA precipitable macromolecules (Table 1). This experiment is therefore an indication that isolated glomeruli synthesize glycoproteins. Cytochalasin B significantly inhibited the incorporation into macromolecules of galactose, glucosamine and glucose; colchicine caused lesser and more variable inhibition. Neither inhibited the incorporation of amino acids, i.e., general protein synthesis (Tables 2 and 3).

Table 2. Effect of Cytochalasin B on Incorporation

Presursor	Cyto. B µg/ml	cpm/mg protein x 10^3	% Inhibition
^{3}H-Gal*	0	29.4	-
	5	4.9	83
^{14}C-GlcN*	0	11.0	-
	10	0.9	92
^{14}C-Glc*	0	42.8	-
	10	0.5	99
^{3}H Amino$^+$ acids	0	144.5	-
	10	138.0	4

* Glc was omitted from the incubation mixture.
\+ The same five ^{3}H labeled amino acids were added as in Fig. 2 and the corresponding unlabeled amino acids omitted.

The newly synthesized glycoprotein was isolated after incorporation of ^{3}H-galactose and ^{14}C-glucose and analyzed by TLC (Fig. 3). ^{3}H-galactose added to the medium was found almost entirely as galactose, while approximately 30% of ^{14}C-glucose was converted into ^{14}C-galactose.

Table 3. Effect of Colchicine on Incorporation

Precursor	Colchicine	cpm/mg protein x 10^3	% Inhibition
^{3}H-Gal*	0	29.5	-
	1×10^{-4} M	18.0	39
^{14}C-GlcN*	0	11.0	-
	1×10^{-4} M	5.2	53
^{14}C-Glc*	0	42.5	-
	1×10^{-4} M	35.4	17
^{3}H Amino acids	0	144.0	-
	1×10^{-4} M	137.0	5

* Glc was omitted from the incubation mixture.
\+ The same five ^{3}H labeled amino acids were added as in Fig. 2 and the corresponding unlabeled amino acids omitted.

Additionally, we have investigated the presence of glycosaminoglycans (GAG) in glomeruli and in the GBM itself and have evidence that at least one species of GAG, heparan sulfate, is associated with GBM. The

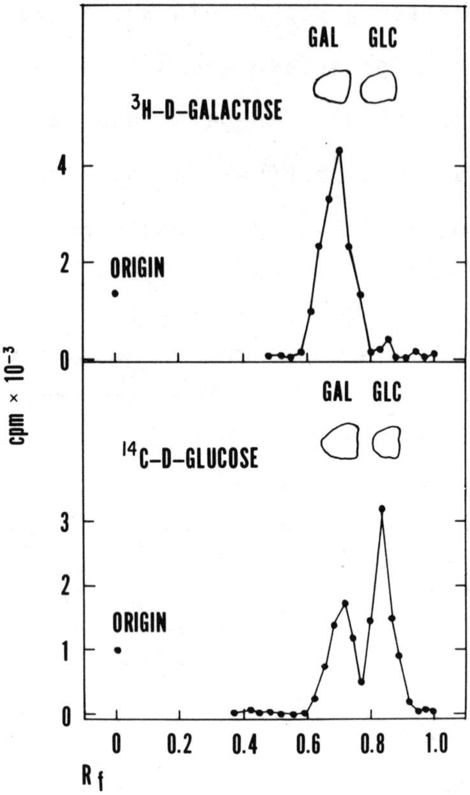

Fig. 3. Identification of monosaccharides. Thin layer chromatograms of acid hydrolysates of glomeruli after ^{3}H-Gal or ^{14}C-Glc incorporation. For further details, see ref. 4. From Krisko, I., and Walker, W.G. (4), reproduced with permission of Proceedings of the Society for Experimental Biology and Medicine.

hexuronic acid content of glomerular fractions is given in Table 4. The same fractions were subjected to bidimensional electrophoresis and compared to seven authentic GAG standards (8). Three different species of GAG's were present in the glomerular sonicate (Fig. 4). Only one of the three, heparan sulfate ("b" in Fig. 4), was found in the GBM fraction.

Table 4. Hexuronic Acid Content of Glomerular Fractions

Fraction	Hexuronic Acid, μg/mg dry wt.
Sonicate	28.4
15,000 x g pellet	9.4
"S-100"	28.1
GBM	9.3

"S-100" refers to the supernate after centrifugation at 100,000 x g for 2 hr.

DISCUSSION

Although GBM is related to ordinary vertebrate collagens, it has several unique features. Some of the salient properties of GBM are (1-5): a) It is a glycoprotein(s) consisting of ca 90% protein and 10% carbohydrate. b) It is insoluble in both aqueous and organic solvents. c) It has high glycine content (every fifth amino acid residue on the average), but not as high as collagen. d) It contains hydroxyproline and hydroxylysine, but the latter is significantly higher than in collagens. e) It has high 3-hydroxyproline content (approximately 10 times higher than interstitial collagens).

From our results we conclude that glycoproteins were synthesized by isolated glomeruli because: a) labeled

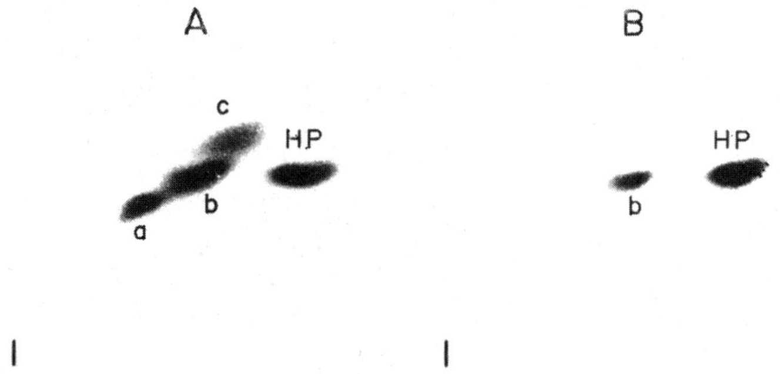

Fig. 4. Bidimensional electrophoresis was performed on cellulose acetate membranes. Approximately 0.5 µg (based on hexuronic acid content) of sample was applied at the left corner of the membrane (marked by perpendicular line). The first dimension was along the line abscissa. A = glomerular homogenate; B = GBM. HP designates the position of authentic heparin standard used as marker; b is heparan sulfate.

carbohydrates were recovered from macromolecules after *in vitro* incubation; and b) inhibitors of protein synthesis inhibited not only amino acid incorporation but also the incorporation of radiolabeled carbohydrate precursors into macromolecules, probably by inhibiting the synthesis of polypeptide intermediates which serve as carbohydrate attachment sites. The degree of inhibition varied according to precursor, which was related most likely to the variable pool size of those peptide intermediates serving as substrates for glycosylation. Depending on the pool size of these intermediates, a variable degree of inhibition occurs (10).

Cytochalasin B, a fungal metabolite, is known to inhibit several cellular processes (e.g. phagocytosis, sugar transport into cells, cytokinesis (11, 12). Ehrlich and Bornstein (11) found that it decreased collagen synthesis without any effect on general protein synthesis. In our system cytochalasin B inhibited the incorporation of galactose, glucose and glucosamine into glomerular macromolecules, while it had no effect on the incorporation of five amino acids. Colchicine had similar effects, but to a lesser and more variable extent. Ehrlich and Bornstein postulated that the inhibition of procollagen and collagen synthesis by these compounds was due to interference with the function of either microtubules or microfilaments. This in turn causes feedback inhibition due to the interruption of a step in cellular secretory processes and the resultant accumulation of biosynthetic intermediates. Assuming that galactose, glucose and glucosamine incorporation signify synthesis of secretory glycoproteins, e.g. GBM, our findings can be interpreted by invoking an hypothesis like that of Ehrlich and Bornstein (11).

Under the conditions of the experiment, the presence of hexuronic acid was presumptive evidence for GAG (exhaustive dialysis prior to the borate-carbazole reaction to remove glycosides, if any; passing of the supernate through a Dowex 50X2 column to remove any nondialyzable peptides). The presumed GAG was next subjected to bidimensional electrophoresis and compared to seven authentic GAG standards. Both the unknowns and the authentic standards were treated with chondroitinase AC, ABC and 0.04 N HCl (the latter is known to cleave the N-sulfate of GAG and thereby to decrease its mobility) and the electrophoresis was repeated. Three different GAG's were present in the glomerular sonicate. Only one of the three, heparan sulfate, was found in the GBM fraction. The nature of the association of GAG with GBM is not known at present.

SUMMARY

We have presented evidence that isolated glomeruli are active in protein and glycoprotein synthesis. When specific inhibitors of protein synthesis are used (cycloheximide, puromycin, chloramphenicol), the glomeruli behave as a typical eukaryotic system. Part of the newly synthesized macromolecules appear to be the glomerular basement membrane. Finally, we have demonstrated that the GBM contains glycosaminoglycans, most likely heparan sulfate.

Acknowledgements

We thank Dr. Clif Sato for assistance in the GAG determinations and Miss Katherine Lewis for excellent secretarial help. This work was supported by Veterans Administration Research.

References

1. Spiro, R.G.: Studies on the renal glomerular basement membrane. Preparation and chemical composition. J. Biol. Chem. 242, 1915-1922 (1967)
2. Kefalides, N.A.: Isolation and characterization of collagen from glomerular basement membrane. Biochemistry, 7, 3103-3112 (1968).
3. Kefalides, N.A.: Structure and Biosynthesis of Basement Membranes. Int. Rev. Conn. Tissue Res., 6, 63-104 (1973).
4. Krisko, I., Walker, W.G.: Protein and glyco-protein synthesis by isolated kidney glomeruli. Proc. Soc. Exper. Biol. Med., 146, 942-947 (1974).
5. Krisko, I., Walker, W.G.: Glycoprotein and glomerular basement membrane synthesis in vitro. Manuscript in preparation.
6. Krisko, I., Walker, W.G.: Protein synthesis in kidney glomeruli. Clin. Res. 20, 599 (1972).
7. Bitter, T., Muir, H.M.: A modified uronic acid carbazole reaction. Anal. Biochem. 4, 330-334, (1962).

8. Sato, C.S., Gyorkey, F.: Bidimensional electrophoresis of glycosaminoglycans on cellulose acetate membrane. Anal. Biochem., in press.

9. Lowry, O.H., Rosebrough, N.J., Farr, A.L., Randall, R.J.: Protein measurement with the Folin phenol reagent. J. Biol. Chem. 193, 265-275 (1951).

10. Spiro, R.G.: Glycoproteins. Ann. Rev. Biochem. 39, 599-638 (1970).

11. Ehrlich, H.P., Bornstein, P.: Microtubules in transcellular movement of procollagen. Nature New Biol. 238, 257-260 (1972).

12. Wessells, N.K., Spooner, B.S., Ash, J.F., Bradley, M.O., Luduena, M.A., Taylor, E.L., Wrenn, J.T., Yamada, K.M.: Microfilaments in cellular and developmental processes. Science 171, 135-143 (1971).

Ribosomes and the Target Theory

Željko Kućan
"Rugjer Bošković" Institute, Zagreb, Yugoslavia

INTRODUCTION

Irradiation of biologically active structures, ranging in size from small molecules and enzymes to viruses, results usually in exponential decline of biological activity with dose, indicating that a single "hit" of irradiation is needed to inactivate such structures. Since the energy of ionizing irradiation is absorbed in units of 60 eV (1), and at 37% survival the mean number of hits per target is 1 (the corresponding dose is designated D_{37}), one can easily calculate size of the radiation-sensitive target. The latter can be expressed in terms of volume, mass, or "target molecular weight", MW_T (for review see ref. 2). Using this approach, "target molecular weights" of numerous enzymes have been calculated, and compared with the molecular weights determined by physico-chemical methods, MW. Though the molecular weights ranged over several orders of magnitude, the agreement was usually rather good (2). This is not too surprising, since the energy of 60 eV per molecule, corresponding to 1380 $kcal \cdot mol^{-1}$, is sufficient to cause several covalent changes in the neighbourhood of the primary event. Similarly, when viruses are irradiated, MW_T usually corresponds to the MW of the infective component, RNA or single-stranded DNA (3). Viruses containing double-stranded DNA show much higher apparent radioresistance, obviously because the efficient cellular repair systems act on them during the assays of their survival.

Ribosomes were not known as objects of radiation studies at the time when I first got acquainted with them (4) during my stay at Professor Lipmann's laboratory. The fact that they are similar in size to small viruses, but with proteins rather than the nucleic acid as their functional components, imposed the question of their size as targets for ionizing irradiation. The first experiments were done with the hope to determine which fraction of the ribosome takes part in the amino acid polymerization reaction (5). Recently, additional experiments were done to clarify this question, and some molecular consequences of the irradiation of ribosomes were examined (6, 7). In this paper a brief summary of the previous results will be given, together with a systematic comparison of the inactivation probability, defined as MW_T/MW, for several macromolecular and supramolecular structures.

METHODS

Preparation of ribosomes and enzyme fraction from E. coli MRE 600, and assays of ribosome activity in the polyuridylic acid-directed polymerization of phenylalanine, have been described (7). The assay mixture contained an excess of all other components, including 50S subunits when 30S subunits were assayed, so the polymerization of phenylalanine was proportional to the concentration of the component tested.

70S ribosomes were irradiated in 10 mM Tris HCl, pH 7.4, 10 mM magnesium acetate, containing 0.1 M oxydized glutathione if indicated, at the concentration of 105 A_{260} units/ml. 30S subunits were irradiated at the concentration of 10 A_{260} units/ml in 10 mM Tris HCl, pH 7.4, 0.1 mM magnesium acetate. In most experiments bacteriophage f2 was added as an internal control to the suspension of ribosomes, at the ratio of 1 plaque-forming unit per $2 \cdot 10^4$ ribosomes; this mixing had no effect neither to the activity of ribosomes nor to the formation of plaques on E. coli K12 CR 63 F^+ (6). For some experiments, 0.1 ml-portions of the mixture were lyophilized before irradiation (5); about 80% of the ribosome activity and 2% of the phage survived this treatment. All samples were irradiated from a 3000 Ci ^{60}Co source at the exposure rate of 3760 R/min.

Thermal inactivation of ribosomes was performed at the concentration of 33 A_{260} units/ml in 10 mM Tris HCl, pH 7.4, 10 mM magnesium acetate, by 5 minutes of incubation at various temperatures, followed by quick cooling to +4 ^{o}C and the assay of phenylalanine-polymerizing activity (7). The midpoint of inactivation, T_m, is defined here as the temperature needed to inactivate 50% of the original activity under above conditions.

Preparation of RNA from ^{14}C-adenine labelled ribosomes, heat denaturation of RNA to reveal the chain breaks, sedimentation analysis of RNA, as well as determination of the number of breaks in RNA from the position of the maximum in the sedimentation curve, have been described in detail (7).

RESULTS

Several criteria have to be fulfilled in order to make the calculation of MW_T possible. First of all, a reliable assay of the biological activity has to be available. Under our assay conditions, polymerization of phenylalanine was proportional to the concentration of 70S ribosomes (5) or 30S subunits (7), unirradiated as well as irradiated. Secondly, an exponential decline of activity with dose should be obtained ("single hit" kinetics). As illustrated in Fig. 1, this is the case for the inactivation of both 70S ribosomes and the 30S subunits. For comparison, the inactivation of reproductive capacity of the bacteriophage f2, irradiated in the mixture with ribosomes (6), is shown. D_{37}--values, obtained from such plots, are used to calculate MW_T.

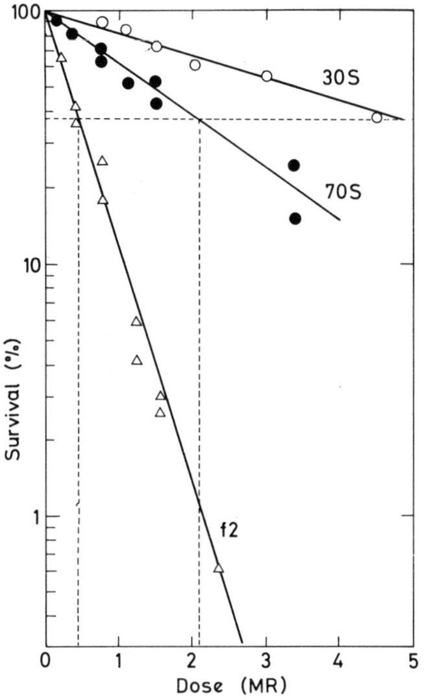

Fig 1. Inactivation of 70S ribosome, 30S ribosomal subunit, and bacteriophage f2 by gamma irradiation. A suspension of 30S subunits, or mixed suspension of 70S ribosomes and bacteriophage f2, was irradiated at -80°C. Ribosomes were assayed for their ability to support the polymerization of phenylalanine, while the phages were assayed for their plaque-forming ability on the sensitive strain of E. coli. The determination of D_{37} for 70S ribosome and the phage is shown by dashed line.

Table 1. Inactivation of ribosomes and bacteriophage f2 under various irradiation conditions

Irradiated sample	Temp. of irrad.	E. coli MRE 600 ribosomes			Bacteriophage f2		
		$D_{37} \cdot 10^{-6}$ (R)	$MW_T \cdot 10^{-6}$	MW_T/MW	$D_{37} \cdot 10^{-6}$ (R)	$MW_T \cdot 10^{-6}$	MW_T/MW
Lyophilized 70S + f2	+20°C	0.37	1.8	0.7	0.17	3.9	3.6
Lyophilized 70S + f2 in 0.1 M GSSG	+20°C	2.2	0.31	0.12	0.26	2.6	2.3
Lyophilized 70S	-80°C	2.2	0.31	0.12	-	-	-
Suspension of 70S + f2	-80°C	2.2	0.31	0.12	0.45	1.5	1.3
Suspension of f2 in 1% Bacto-Tryptone	-80°C	-	-	-	0.73	0.92	0.8
Suspension of f2 in 10% Bacto-Tryptone	-80°C	-	-	-	1.0	0.67	0.6
Suspension of 70S + f2 in 0.1 M GSSG	-80°C	2.2	0.31	0.12	1.0	0.67	0.6
Lyophilized 70S in 0.1 M GSSG	-80°C	2.2	0.31	0.12	-	-	-
Suspension of 30S	-80°C	4.8	0.14	0.14	-	-	-

D_{37} for each sample was determined from exponential inactivation curve (cf. Fig. 1), MW_T was calculated (3) from D_{37} (in roentgens) using the relationship $MW_T = 6.7 \times 10^{11}/D_{37}$. Inactivation probability, MW_T/MW, was based on the total particle weight for 70S and 30S ribosomes, and physical MW of bacteriophage RNA.

Thirdly, the inactivation has to result only from the direct effect of ionizing irradiation, i.e. effect of free radicals formed in the surrounding should be completely avoided. It was originally thought (5) that lyophilization of the suspension of ribosomes was sufficient to eliminate the indirect effect, but it was soon established that the full protection could be obtained only by the addition of glutathione as a protector, or by lowering the temperature of irradiation (6). The results of numerous experiments represented originally by exponential inactivation curves similar to that in Fig. 1, are summarized in Table 1. Under various protective conditions applied to 70S ribosomes, the inactivation probability MW_T/MW was 0.12. This low number was achieved by several protective treatments or their combinations, indicating that further protection was not possible. At the temperature providing full protection for 70S ribosomes, 30S subunits gave a comparably low inactivation probability of 0.14. On the other hand, bacteriophage f2 was not completely protected by 0.1 m GSSG at +20 °C, or by low temperature. Unexpectedly, the addition of protective substances at low temperature lowered the inactivation probability for this phage to 0.6.

Low inactivation probability for ribosomes means that most hits into their component parts do not lead to inactivation. To see whether some of these non-lethal hits influence the stability of the ribosome, midpoints of thermal inactivation were determined for ribosomes surviving certain doses of gamma irradiation. The results shown in Table 2 indicate that non-lethal hits render the ribosome significantly more susceptible to thermal inactivation.

As for the molecular consequences of ionizing irradiation in such a complex structure as ribosome, the changes in MW of ribosomal RNA are the easiest to analyze. By zone sedimentation analyzis of ^{14}C-adenine labelled RNA from irradiated 30S subunits, or from

Table 2. Midpoints of thermal inactivation of ribosomes surviving gamma irradiation

Gamma-ray dose 10^{-6} (R)	0	0.4	1.5	3.3
Midpoint of inactivation, T_m (°C)	56.5	55.6	54.5	52.2

For details see Methods.

whole irradiated E. coli cells, in both cases heat-denatured before analysis, it was possible to determine D_{37} for chain breaks in ribosomal RNA (7). The values of $D_{37} = (2.7 \pm 0.4) \cdot 10^6$ R and $D_{37} = (5.1 \pm 0.3) \cdot 10^6$ R were estimated for 23S and 16S RNA, leading to MW_T of $2.6 \cdot 10^5$ and $1.4 \cdot 10^5$, respectively. Both these values correspond to chain scission probability, MW_T/MW of about 0.25.

The summary of irradiation studies of ribosomes, together with some pertinent data on another type of nucleoproteins, i. e. viruses, is shown in Table 3.

Table 3. Summary of irradiation data on some nucleoproteins

Nucleoprotein irradiated	Irradiation conditions	Effect measured	MW_T/MW	Ref.
70S ribosome	full protection	a	0.12	
30S ribosome	-80°C	a	0.14	
Bacteriophage ØX174	-60°C, 1% B. T.	b	0.9	(8)
(ØX174 DNA)	-60°C, 1% B. T.	b	0.9	(8)
(ØX174 R. F. DNA)	-60°C, 1% B. T.	b	0.21	(3)
Bacteriophages T_2, T_4, λ	-60°C, 1% B. T.	b	0.05	(3)
TMV	-60°C, 1% B. T.	b	0.9	(8)
Bacteriophage R17	-60°C, 1% B. T.	b	0.8	(8)
Bacteriophage f2	-80°C, 1% B. T.	b	0.8	
Bacteriophage f2	-80°C, 10% B. T.	b	0.6	
(ØX174 R. F. DNA)	-60°C, 1% B. T.	c	0.75	(3)
(ØX174 DNA)	-60°C, 1% B. T.	c	0.26	(9)
50S ribosome in the cell	-80°C	c	0.25	
30S ribosome, purified	-80°C	c	0.25	

Nucleoproteins were irradiated in the intact form, except where indicated by parentheses (DNA was irradiated in these cases, R. F. stands for replicative form). B. T. stands for Bacto Tryptone. The following radiation effects were measured, using suitable assays: (a) ability to support the synthesis of polyphenylalanine; (b) viability of the virus, assayed on a suitable host; (c) chain breaks in the nucleic acid. MW_T/MW can be defined as the inactivation probability when assays (a) and (b) are used, and as the chain scission probability when effect (c) is measured. In (b) and (c) MW of the nucleic acid component, rather than MW of the whole nucleoprotein, was used to calculate MW_T/MW. References for own work are omitted.

DISCUSSION

The original goal of the investigations described in this paper was to learn more about the role of ribosome in the process of protein synthesis, using simple concepts of the target theory to estimate the size of the actual functional part of the ribosome. In the course of this work, however, it turned out that the interpretation of inactivation kinetics of complex structures, such as ribosome, is not an easy task, so that very little could be contributed towards answering the original question. On the other hand, systematic irradiation studies of ribosomes, as well as parallel studies on bacteriophage f2, have broadened, to some extent, our understanding of the irradiation action itself.

Does the low inactivation probability obtained for E. coli ribosome reveal any secret of its function? The answer would be much simpler in the case of inactivation probability of 1, which would mean that the integrity of the whole particle is essential for its function. However, since only one out of eight hits into the 70S ribosome inactivates its function (and the same is true for the 30S subunit, cf. Table 1), one has to face the next question: what does the sensitive target actually represent, or which parts are actually critical for the function of ribosome in the polymerization reaction? The target theory is, of course, helpless in such a case. It does, however, confirm the expectation of a biochemist that, once correctly assembled from undamaged components into a functional particle, the ribosome will have its active sites as well as some other sites, which were essential only in the process of assembly. It is a little surprising, nevertheless, that as much as 88% of the total mass of the ribosome is not directly involved in the polymerization reaction. It should be emphasized that the irradiation studies of enzymes have shown that hits anywhere in the protein molecule impair its activity (2). Hence, the inactivation probability of 0.12 for the ribosome obviously means that the integrity of the ribosomal components (most probably ribosomal proteins), comprizing 12% of its total mass, is critical for its function. It remains to be seen wheather this number will increase if a more complex assay is used; at least the chain initiation step is not assayed in the polyuridylic acid-directed system.

From the point of view of the target theory and molecular radiation biology, the following comments should be made:
(i) Difficulties in the interpretation of inactivation data increase with the complexity of irradiated structure, and inactivation probabilities much lower than 1 are often found (cf. Table 3). Two entirely dif-

ferent reasons could be responsible for this effect. The first is the existence of efficient cellular repair mechanisms acting on viruses containing double-stranded·DNA. Since these processes intervene between the original damage and the expression of final activity, the correlation found has no physical meaning. The second possible reason, revealed by the studies on ribosomes described here, is the existence of structural parts in multicomponent supramolecular assemblies, which are essential in the process of formation, but not in the function of these assemblies. Ribosomal RNA is an obvious example of such a component (in most of its parts, at least), though the results discussed above strongly suggest that many ribosomal proteins fall into this category.

(ii) Conditions needed to eliminate the indirect effect of irradiation vary for various targets, and, as shown in Table 3 for bacteriophage f2, unusually high concentrations of protective substances can change the inactivation probability to the value as low as 0.6. The calculation is based on MW of f2, which is, for sure, the infective component of the virus. This low value remains, therefore, rather puzzling.

(iii) The chain scission probability of the nucleic acid seems to depend on its strandedness, rather than its DNA- or RNA-nature. The probability of 0.25 determined for both 23S and 16S ribosomal RNA is identical to that determined for ØX174 DNA, but different from the value of 0.75, determined for replicative form of the same virus. The physical and chemical basis for this difference is not clear, and obviously more data on other nucleic acids and nucleoproteins, including measurements on double-stranded RNA, are needed before any general conclusions can be made.

SUMMARY

The concepts of target theory were used to estimate the size of the part of ribosome acting in the amino acid polymerization reaction. The components essential for this reaction did not exceede 12% of total mass of the ribosome, since the irradiation-induced damage in the rest of the particle did not impair its function. Comparison of irradiation data on several nucleoproteins reveals the basic difference in the reasons for low inactivation probabilities between the ribosome and some of the viruses.

ACKNOWLEDGEMENT

The experiments described in this paper were done in collaboration with Drs. Ira Pečevsky-Kučan, D. Petranović, and J. N. Herak.

REFERENCES

1. Rauth, A. M., Simpson, J. A.: The energy loss of electrons in solids. Radiat. Res. 22, 643-661 (1964).
2. Dertinger, H., Jung, H.: Molekulare Strahlenbiologie. Springer-Verlag, Berlin-Heidelberg, 1969.
3. Ginoza, W.: The effects of ionizing radiation on nucleic acids of bacteriophages and bacterial cells. Ann. Rev. Microbiol. 21, 325-386 (1967).
4. Kućan, Ž., Lipmann, F.: Differences in chloramphenicol sensitivity of cell-free amino acid polymerization systems. J. Biol. Chem. 239, 516-520 (1964).
5. Kućan, Ž.: Inactivation of isolated Escherichia coli ribosomes by gamma irradiation. Radiat. Res. 27, 229-236 (1966).
6. Petranović, D., Pečevsky-Kućan, I., Kućan, Ž.: A comparison of the direct effect of γ-rays on Escherichia coli ribosomes and bacteriophage f2. Radiat. Res. 46, 621-630 (1971).
7. Kućan, Ž., Herak, J. N., Pečevsky-Kućan, I: Functional inactivation and appearance of breaks in RNA chains caused by gamma irradiation of Escherichia coli ribosomes. Biophys. J. 11, 237-251 (1971).
8. Ginoza, W.: Radiosensitive molecular weight of single-stranded virus nucleic acids. Nature 199, 453-456 (1963).
9. Lytle, C. D., Ginoza, W.: Frequency of single-strand breaks per lethal γ-ray hit in ØX174 DNA. Int. J. Radiat. Biol. 14, 553-560 (1968).

Interrelation Between Tyrocidine Synthesis and Sporulation in Bacillus brevis

Sung G. Lee
The Rockefeller University, New York, New York 10021, USA

INTRODUCTION

The notion has long been around (1-4) that antibiotic and spore formation are interconnected. However, no fully satisfactory explanation for this has been found so far. Paulus (5) recently reported a strong inhibition of DNA-dependent RNA synthesis by tyrothrycin (tyrocidine + linear gramicidin), and quite potently by tyrocidine alone. It has long been known, however, that the enzymes that produce gramicidin or tyrocidine are produced at the end of the logarithmic growth period when the organism starts to head towards sporulation. Using the tyrocidine-forming *Bacillus brevis*, a study will be reported here that correlates the different phases of spore formation and the mode and localization of antibiotic synthesis.

Tyrocidine-Producing Enzymes in the Supernatant and Pellet Fractions.

Tyrocidine, a cyclic decapeptide antibiotic produced in *B. brevis* (ATCC 8185) during transition from the exponential to the stationary phase of growth (6,7), is synthesized by the action of three complementary enzymes (8-10). Its structure is shown in Fig. 1, which also indicates the direction of synthesis and cyclization.

Because the peak period of the tyrocidine-producing activity is only quite brief, one of the earlier problems encountered in this laboratory was determining the right moment to harvest the organism to obtain a maximum yield of these enzymes. During these attempts, it was noted that when the organism is harvested at the

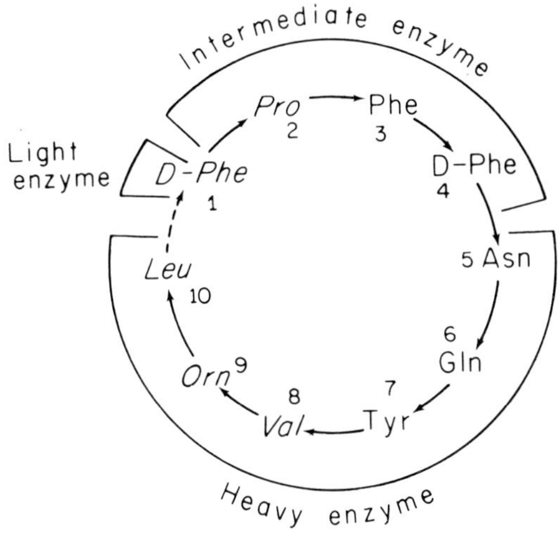

Fig. 1. Amino acid sequence in tyrocidine. Brackets embrace the amino acids activated by the three complementary enzymes. Solid arrows indicate the direction of addition; dotted arrow the direction of cyclization.

peak of in vivo tyrocidine production the enzymes are tightly bound to the 20,000 x g pellet fraction and cannot be purified. In order to obtain these enzymes from the 20,000 x g supernatant fraction, the organism has to be harvested earlier. Fig. 2 shows the growth curve and distribution of tyrocidine-producing activity in the 20,000 x g supernatant and pellet fractions. It can be seen that the tyrocidine-producing activity switches gradually from the supernatant to the pellet fraction, and that virtually all of the activity becomes associated with the pellet fraction at the peak of the combined total activity; this peak corresponds to the time of the peak of tyrocidine production in vivo.

Morphology of B. brevis Cells.

The morphological changes during the switch of tyrocidine-producing enzymes from the supernatant to the pellet fraction were examined with an electron microscope, and the results are

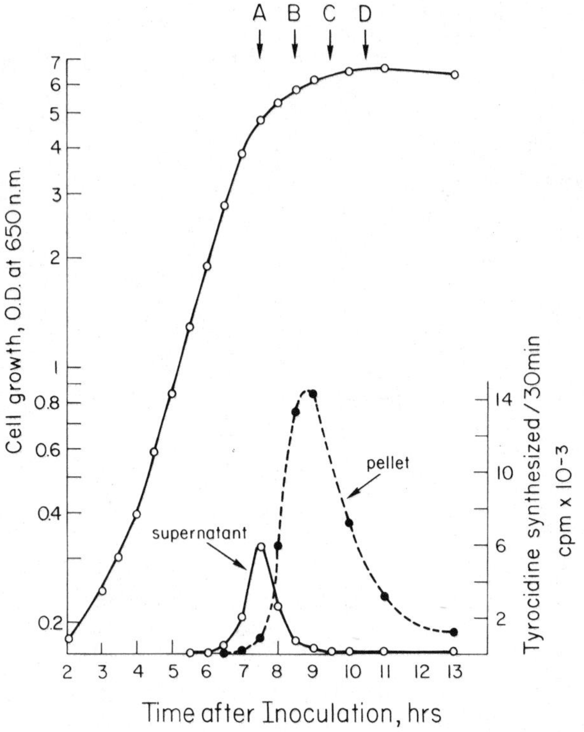

Fig. 2. The relationship between the growth of B. brevis (ATCC 8185) and the synthesis of tyrocidine by the 20,000 x g supernatant and pellet fractions. Cell lysates prepared from the cells obtained at various times were centrifuged for 20 min at 20,000 x g. Aliquots of the supernatant fraction and the suspension of the pellet fraction amounting to 5% of protein content were assayed for tyrocidine synthesis as described by Roskoski *et al*. (8).

illustrated in Fig. 3; A, B, and D correspond to the time of harvest of cells, which is indicated by arrows A, B, and D in Fig. 2. The cells show spore-septa (A) just when tyrocidine-synthesizing activity appears and is present entirely in the supernatant fraction of cell lysate. At the peak of in vivo tyrocidine-synthesizing activity, at which time the activity is

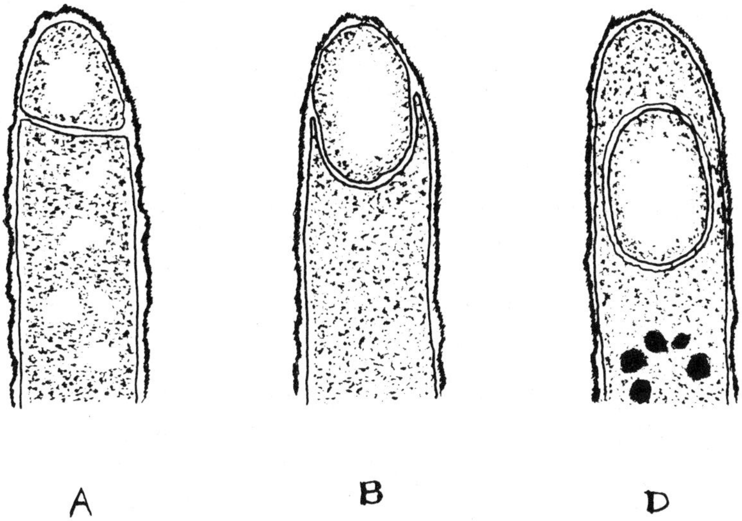

Fig. 3. Drawing of electronmicrographs of B. brevis cells at various stages of sporulation. A, B, and D correspond to cells taken at the times indicated by arrows A, B, and D in Fig. 2.

found mainly in the pellet fraction, both ends of the spore-septum membranes are pushed toward the pole (B); and when the forespore is finally released from its attachment to the cell membrane (D), virtually all of the tyrocidine-synthesizing activity is associated with the pellet fraction.

Distribution of Tyrocidine, Tyrocidine-Producing Enzymes, and Proline Within the Cell.

To examine whether tyrocidine and its constituent amino acids accumulate in any particular cellular component, e.g. forespores, the cells were briefly incubated with ^{14}C-proline, after which they were disrupted by sonification and the cellular components separated by sucrose gradient centrifugation. Fig. 4 shows the O.D. profiles of such gradients and the distribution of radioactivity. Soluble protein and ribosomes stayed at the top, the membrane fraction barely penetrated the gradient, while fore-

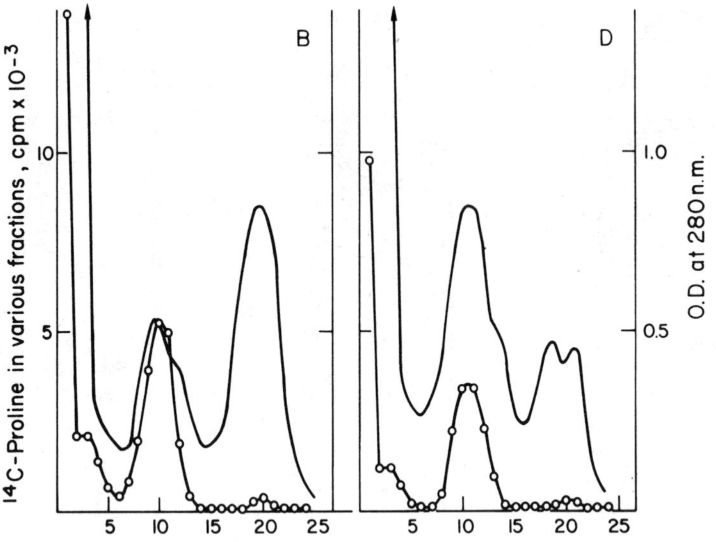

Fig. 4. Sucrose gradient centrifugation of sonicated cells. Cell samples were incubated with 5 μCi of ^{14}C-proline for 10 min at 37°; the cells were pelleted through 10% sucrose, and the pellets were suspended in 10% sucrose containing 20 mM triethanolamine buffer, pH 7.6, 5 mM $MgCl_2$, 100 mM KCl, and 1 mM dithiothreitol. The cells were then disrupted with a Branson sonifier at step 4 using a step horn. Using a SW 27 rotor, the cellular components were separated by centrifugation for 20 min at 6,000 rpm on a 25-50% sucrose gradient in the above buffer salts. Aliquots of 0.1 ml were counted for radioactivity. B and D correspond to cells taken at the times indicated by arrows B and D in Fig. 2. The peaks at fractions 10 and 20 in B and D represent forespores and broken cells.

spores and broken cells sedimented in fractions 10 and 20, and whole cells settled at the bottom of the tube. As may be seen, the forespore fraction contained over 60% of the proline radioactivity but only 16% of total protein, indicating that proline accumulated primarily in the forespores.

TABLE 1. Distribution of tyrocidine in various cell components.

Cell components	Tyrocidine (cpm)
Soluble protein[a]	1,320
Membranes[b] (presumably forespore membranes)	34,800
Forespores[c]	56,400
Broken cells[d]	5,600

[a,b] Fractions 1-4 of the gradient shown in Fig. 4 were centrifuged for 30 min to precipitate the membranes.

[c,d] Fractions 8-11 and 18-22, respectively, of Fig. 4.

Cells harvested at arrow D in Fig. 2 were incubated with 5 μCi of ^{14}C-proline for 10 min at 37°, after which the cells were collected by centrifugation and then disrupted by sonication; various components were separated as described in the legend to Fig. 4.

TABLE 2. Synthesis of tyrocidine and D-Phe-Pro-diketopiperazine by the soluble protein and forespore fractions.

Time after inoculation[a] (hours)	Soluble protein		Forespore	
	Tyrocidine	D-Phe-Pro	Tyrocidine	D-Phe-Pro
7.5	1,620	680	-	-
8.5	440	2,850	3,240	1,470
9.5	230	2,130	2,300	1,280
10.5	130	1,720	1,460	1,020

[a] Cells were harvested at various times after inoculation as indicated by arrows in Fig. 2. Soluble protein fractions (3.2 to 4.1 mg protein) and forespore fractions (0.9 to 1.1 mg protein) were assayed for tyrocidine synthesis as described by Roskoski et al. (8).

To localize tyrocidine, the membrane fraction was separated from the soluble protein fraction by a 30-minute centrifugation at 27,000 rpm; tyrocidine was extracted from these and other fractions obtained from sucrose gradient centrifugation. As may be seen in Table 1, tyrocidine was found mainly in the membrane and forespore fractions. Electron microscopy of the membrane fraction shows that it is rich in what appears to be empty forespore membranes.

Finally, tyrocidine synthesis was carried out with the supernatant and forespore fractions. As shown in Table 2, much more tyrocidine was produced by the forespore fraction than by the supernatant fraction, especially when these fractions were made from older cells. On the other hand, a similar quantity of D-Phe-Pro-diketopiperazine was produced by both the supernatant and forespore fractions. The synthesis of tyrocidine requires the light enzyme and two undissociated polyenzymes while Phe-Pro-diketopiperazine is still produced by fully dissociated polyenzymes. This result indicates that although the tyrocidine-synthesizing system had become inactive in the supernatant fraction, it remained active in the forespores.

DISCUSSION

These experiments show that tyrocidine synthesis appears in the supernatant fraction at the very beginning of spore formation. With the maturation of the spore, however, it shifts into insoluble fractions and now becomes associated with the spore, particularly the spore membrane, and eventually declines. At present, it seems preferable not to discuss the possible meaning of the shift of tyrocidine production and content into the spore compartment of the cell.

SUMMARY

Tyrocidine-synthesizing activity appears in _B. brevis_ (ATCC 8185) at the end of the exponential phase of growth, reaches a peak within an hour, then declines with the progress of sporulation.

At the onset of activity, the cells show spore-septa and all of the activity is found in the soluble protein fraction, but with the progress of sporulation the activity becomes associated with the forespores and little activity is found in the soluble protein fraction.

******

This paper is dedicated as an homage to the 75th birthday of Dr. Fritz Lipmann, a giant among biochemists and a truly great man, whose work on coenzyme A made the understanding of the intermediary metabolism complete, whose perception of high energy bonds in living systems gave a whole new vitality to biochemistry, whose elucidation of the mechanism of peptide chain elongation enlarged our knowledge of protein synthesis so immensely, and whose elegant solution of the mechanisms of the peptide antibiotic synthesis and the magic spot formation has shown us that the beauty of nature is all around us.

Acknowledgment

This study was supported by a grant to Dr. Lipmann from the United States Public Health Service.

REFERENCES

1. Bernlohr, R. W., Novelli, G. D., Biochim. Biophys. Acta 41, 541-543 (1960).
2. Bernlohr, R. W., Novelli, G. D., Arch. Biochem. Biophys. 103, 94-104 (1963).
3. Schaeffer, P., Bacteriol. Rev. 33, 48-71 (1969).
4. Sadoff, H. L., in "Spores V", Halvorson, H.O., Hanson, R., Campbell, L. L., ed., 1972, pp. 157-166.
5. Sarka, N., Paulus, H., Nature New Biol. 239, 228-230 (1972).
6. Mach, B., Reich, E., Tatum, E. L., Proc. Nat. Acad. Sci. USA 50, 175-181 (1963).

7. Fujikawa, K., Suzuki, T., Kurahashi, K., Biochim. Biophys. Acta 161, 232-246 (1968).

8. Roskoski, R., Jr., Gevers, W., Kleinkauf, H., Lipmann, F., Biochemistry 9, 4839-4845 (1970).

9. Lipmann, F., Accts. Chem. Res. 6, 361-367 (1973).

10. Lee, S. G., Roskoski, R., Jr., Bauer, K., Lipmann, F., Biochemistry 12, 398-405 (1973).

Intergeneric Complementation of RNA Polymerase Subunits

Ute I. Lill, Eva M. Behrendt and Guido R. Hartmann
Chemisches Laboratorium der Ludwig-Maximilians-Universität München, Institut für Biochemie, GFR

The protein synthesizing apparatus of all living cells seems to have a rather similar composition of proteins and nucleic acids. Among procaryotes the different components are particularly closely related. Early work of Lipmann's group has shown that the supernatant factors from various bacteria are functionally interchangeable in spite of some antigenic differences (1-4). Similarly functional correspondence has been found between the 30 S ribosomal proteins of grampositive and gramnegative bacteria (5). Much less is known about the species specificity of the proteins involved in transcription. DNA directed RNA polymerase core enzymes (EC 2.7.7.6) have been isolated from many different bacteria. Most of them are composed of two copies of a small subunit and of two large but not identical subunits. They are designated as α, β and β' respectively according to their relative electrophoretic mobility in polyacrylamide gels containing 0,1% sodium dodecylsulfate. Nothing is known about their functional correspondence. In this communication we wish to present evidence that the subunits of RNA polymerase from <u>Escherichia coli</u> and <u>Micrococcus luteus</u> are functionally interchangeable in spite of their differences in size and in mobility during cellulose acetate electrophoresis in presence of 6 M urea.

METHODS

RNA polymerase was prepared from rifampicin sensitive and from rifampicin resistant E. coli strains according to Burgess (6) and separated into subunits by ion exchange chromatography in presence of urea. The solutions containing the pure isolated subunits $^{e}\alpha$, $^{e}\beta$ and $^{e}\beta'$ respectively (the indices ^{e} and ^{m} designate the bacterial source of the subunits) were dialysed free of urea and stored in presence of 60% glycerol at -20°. Similarly RNA polymerase of M. luteus was prepared from rifampicin sensitive and rifampicin resistant strains and subsequently separated into the subunits $^{m}\alpha$, $^{m}\beta$ and $^{m}\beta'$ respectively (E. M. Behrendt and G. R. Hartmann, manuscript in preparation). The molecular weight of the M. luteus subunits was determined by polyacrylamide gel electrophoresis in presence of 0,1% sodium dodecylsulfate using the E. coli subunits for comparison (Table 1). During cellulose acetate electrophoresis in presence of 6 M urea subunit β from M. luteus

Table 1. Molecular weight of RNA polymerase subunits from M. luteus and E. coli

Species	α	β	β'
M. luteus	46 000	135 000	138 000
E. coli (7)	41 000	145 000	150 000

migrates much more rapidly to the anode than subunit β from E. coli indicating a very different overall charge. For reconstitution of enzymatically active RNA polymerase from isolated subunits in the absence of urea multiples of 0,05 nmole each of subunit β and β' and 0.1 nmole subunit α were incubated for 5 hours at 30° in presence of 10 mM Tris-HCl pH 7.9, 10 mM

$MgCl_2$, 2.5 mM $MnCl_2$, 0,1 mM ethylenediaminetetraacetate, 0,5 M KCl, 10 mM dithiothreitol and 50% glycerol.

RESULTS

Incubation of different mixtures of the isolated subunits from M. luteus and E. coli RNA polymerase led

Table 2. Enzymatic activity of incubation mixtures of RNA polymerase subunits from E. coli and M. luteus

Incubation mixture containing	Incorporation of [^{14}C]UMP into acid insoluble material in 15 min at 37° (counts x min^{-1})
$^{m}\alpha + {}^{e}\beta + {}^{e}\beta'$	22 344
$^{e}\alpha + {}^{m}\beta + {}^{e}\beta'$	270
$^{e}\alpha + {}^{e}\beta + {}^{m}\beta'$	815
$^{e}\alpha + {}^{m}\beta + {}^{m}\beta'$	783
$^{m}\alpha + {}^{e}\beta + {}^{m}\beta'$	6 119
$^{m}\alpha + {}^{m}\beta + {}^{e}\beta'$	148

to the formation of active enzyme when assayed in the standard activity assay (8) with poly[d(A-T)] as template (Table 2). Controls to each experiment were performed in which only the two subunits of one species or the single subunit of the other species were incubated. No activity higher than 186 counts x min^{-1} was observed. These results clearly proved the absence of contaminating traces of the subunit which had not been added. Their presence could have led to the formation of active enzyme composed of autologous subunits simulating the presence of an enzyme hybrid. Consequently the results of Table 2 strongly suggest the formation of the active intergeneric RNA polymerase hybrids $^{m}\alpha_2{}^{e}\beta{}^{e}\beta'$, $^{e}\alpha_2{}^{e}\beta{}^{m}\beta'$, $^{e}\alpha_2{}^{m}\beta{}^{m}\beta'$ and $^{m}\alpha_2{}^{e}\beta{}^{m}\beta'$. Subunit α can be exchanged in the enzyme of both species whereas subunit β can be replaced only in M. luteus enzyme and subunit β' only in the E. coli enzyme. Additional evidence for the formation of active enzyme hybrids is provided by experiments with subunits from rifampicin resistant and sensitive RNA polymerase. When subunit α and β' from rifampicin sensitive M. luteus RNA polymerase and subunit β from rifampicin resistant E. coli RNA polymerase were incubated active enzyme was formed which proved to be resistant to the action of the antibiotic. In contrast incubation of subunit α and β' from resistant M. luteus polymerase with subunit β from rifampicin sensitive E. coli polymerase led to the formation of rifampicin sensitive enzymatic activity. These results clearly show the formation of enzyme hybrids and the functional correspondence of subunit β from both bacteria.

In spite of their activity in the standard assay the RNA polymerase composed of heterologous subunits do not exhibit all properties of the enzyme which was reconstituted from autologous E. coli subunits. This becomes evident from the study of the de novo RNA synthesis in the absence of template (9, 10). None of the enzyme hybrids is able to catalyze the de novo synthesis of poly(I-C)·poly(I-C) or poly(A)·poly(U) in the absence of template. Only the activity to synthesize poly(A) in a DNA dependent reaction (11) survives the exchange of subunits.

DISCUSSION

The exchangeability of subunits in RNA polymerase from such different procaryotes as M. luteus (DNA base ratio (A+T)/(G+C) = 0.38) and E. coli (DNA base ratio (A+T)/(G+C) = 0.98) suggests a very close correspondence in function and three-dimensional structure of the proteins constituting the transcriptional apparatus in bacteria. The similarity is so close that even the property of rifampicin resistance of RNA synthesis, which resides in both E. coli and M. luteus strains used in the second largest subunit, can be transferred from one bacterial enzyme to the other merely by exchange of this subunit. Since corresponding subunits differ in molecular weight up to 12%

Table 3. Replacement of subunits in bacterial RNA polymerase core enzymes

subunit replaced	in E. coli enzyme	in M. luteus enzyme
α	+	+
β	-	+
β'	+	-

one has to assume that at least those parts of the subunit surface which are involved directly in the subunit-subunit interaction must be very similar in shape and in the distribution of polar and hydrophobic amino acids. Nevertheless some differences remain even in these parts because the subunit β and also the subunit β' cannot be exchanged in both directions (Table 3).

We wish to thank Miss M. Meyer and Mr. H.D. Sickinger for excellent technical assistance. These investigations have been supported by the Deutsche Forschungsgemeinschaft and by the Fonds der Chemischen Industrie.

REFERENCES

(1) Lucas-Lenard, J., Lipmann, F.: Separation of three microbial amino acid polymerization factors. Proc. Natl. Acad. Sci. USA 55, 1562 - 1566 (1966).

(2) Takeda, M., Lipmann, F.: Comparison of amino acid polymerization in B. subtilis and E. coli cell free systems; hybridization of their ribosomes. Proc. Natl. Acad. Sci. USA 56, 1875 - 1882 (1966).

(3) Gordon, J., Schweiger, M., Krisko, I., Williams, C.A.: Specificity and evolutionary divergence of the antigenic structure of the polypeptide chain elongation factors. J. Bacteriol. 100, 1-4 (1969).

(4) Sy, J., Chua, N.-H., Ogawa, Y., Lipmann, F.: Ribosome specificity for the formation of guanosine polyphosphates. Biochem. Biophys. Res. Commun. 56, 611 - 616 (1974).

(5) Higo, K., Held, W., Kahan, L., Nomura, M.: Functional correspondence between 30 S ribosomal proteins of Escherichia coli and Bacillus stearothermophilus. Proc. Natl. Acad. Sci. USA 70, 944-948 (1973).

(6) Burgess, R. R.: A new method for the large scale purification of Escherichia coli DNA dependent RNA polymerase. J. Biol. Chem. 244, 6160 - 6167 (1969).

(7) Berg, D., Chamberlin, M.: Physical studies on RNA polymerase from Escherichia coli B. Biochemistry 9, 5055 - 5064 (1970).

(8) Lill, U. I., Hartmann, G. R.: On the binding of rifampicin to the DNA directed RNA polymerase from Escherichia coli. Eur. J. Biochem. 38, 336 - 345 (1973).

(9) Krakow, J.S., Karstadt, M.: Azotobacter vinelandii RNA polymerase, IV. Unprimed synthesis of rIC co-

polymer. Proc. Natl. Acad. Sci. USA 58, 2094 - 2101 (1967).

(10) Smith, D.A., Ratliff, R.L., Williams, D.L., Martinez, A.M.: Synthesis of polyadenylate:polyuridylate catalyzed by RNA polymerase in the absence of template. J. Biol. Chem. 242, 590 - 595 (1967).

(11) Chamberlin, M., Berg, P.: Mechanism of RNA polymerase action:characterization of the DNA dependent synthesis of polyadenylic acid. J. Mol. Biol. 8, 708 - 726 (1964).

Studies on Liver Elongation Factor 1

Chen K. Liu, Andrzej B. Legocki[1] and Herbert Weissbach
The Roche Institute of Molecular Biology, Nutley, N.J. 07110 USA

INTRODUCTION

Multiple species of elongation factor 1 (EF1)[2] have been demonstrated in various mammalian tissues (1-4) and wheat (5,6), although their nature and function are not clear. We have recently presented evidence that the heavy form of EF1 ($EF1_H$) from calf brain, with molecular weights ranging from 2.5×10^5 to over 1×10^6, is an aggregate of a light form of the enzyme ($EF1_L$), the latter having a molecular weight of about 50,000 (3). Although $EF1_H$ and $EF1_L$ from calf brain are both active in aminoacyl-tRNA binding to ribosomes, $EF1_L$ reacts with GTP and aminoacyl-tRNA more efficiently than $EF1_H$ (3,7). Of interest was the finding of cholesterol and its esters in the partially purified $EF1_H$ preparations from calf brain (3). These studies on the lipid nature of EF1 were extended, and it was found that $EF1_H$ from calf brain and calf liver contained significant amounts of phospholipid, and that $EF1_H$ from these sources as well as rabbit reticulocyte EF1 could be converted to $EF1_L$ by treatment with phospholipase preparations (8).

The present study describes the properties of $EF1_H$ and $EF1_L$ from calf liver. In addition, the EF1 patterns in mouse liver have been examined under a variety of conditions.

METHODS

The materials used were obtained from commercial sources as described in previous publications (3,9).

Purification of $EF1_H$ and $EF1_L$ from Calf Liver. The procedures used were a modification of those employed by Moon *et al*. (3) for calf brain $EF1_H$ and $EF1_L$. A typical purification started with 800 gms of calf liver and involved $(NH_4)_2SO_4$ fractionation, calcium phosphate elution, Sepharose 6B and hydroxylapatite chromatography and $(NH_4)_2SO_4$ back extraction as described in detail previously (3).

A final purification of both $EF1_L$ and $EF1_H$ was achieved by sucrose gradient centrifugation. Eight tenths ml of the dialyzed $(NH_4)_2SO_4$ extracts (2500 units of enzyme) was layered on top of 30 ml of a 5-20% sucrose gradient in a Spinco SW 27 centrifuge tube. The tubes were centrifuged for 18 hrs at 25,000 rpm. After centrifugation, the bottom of the tubes were punctured and 50% glycerol was pumped into the tubes to remove the fractions. Thirty-one fractions (1 ml each) were collected. Fractions 5 and 6 (from the top) contained $EF1_L$ activity, and these were combined, dialyzed against buffer A and stored in liquid nitrogen.

It should be noted that in the preparation of $EF1_H$ from calf liver there were some significant changes compared to the brain procedure (3). The initial $(NH_4)_2SO_4$ fractionation was different than that used for brain since $EF1_H$ from liver precipitated in the 0-30% saturated $(NH_4)_2SO_4$ fraction. This precipitate, after suspending in a buffer containing 0.05 M Tris-Cl, pH 7.4 and 0.001 M dithiothreitol was centrifuged at 13,000 rpm for 1 hr, and the supernatant dialyzed. The calcium phosphate step could then be omitted, although the Sepharose and hydroxylapatite columns and $(NH_4)_2SO_4$ back extraction were as described previously for calf brain $EF1_H$ (3). In the final sucrose gradient, $EF1_H$ activity appeared in fractions 10-20.

Amino Acid Analysis. Samples of the proteins were dried under nitrogen gas, and the residue dissolved in 1 ml of 5.7 N HCl. The tube was sealed under vacuum and heated at 110° for 22 hrs. Analysis was performed using a Joel Model 6 AH analyzer.

Studies with Mouse Liver EF1. Three to four grams of mouse liver (from 2-3 month old animals) were homogenized in an equal volume of a buffer containing 0.05 M Tris, pH 7.4, 0.026 M KCl

0.005 M $MgCl_2$ and 0.25 M sucrose. The suspension was centrifuged at 12,000 x g for 10 min, and the supernatant removed and re-centrifuged at 156,000 x g for 60 min. The latter supernatant (high-speed supernatant) was diluted three times with a buffer containing 0.05 M Tris, pH 7.4, and 1 mM dithiothreitol before an aliquot (generally 0.1 ml containing 60 units of EF1) was layered on a 5-20% sucrose gradient. The gradients were performed as described previously (3).

In some cases the mice were fasted for various times before sacrificing or treated with hormones. The hormone-treated animals were injected subcutaneously with either insulin or hydrocortisone (two doses at 75 µg) at 24 and 18 hrs before sacrificing.

Assays. The assay for EF1 activity was based on the binding of Phe-tRNA to ribosomes as described previously (9). The reaction of EF1 with GTP to form an EF1·GTP complex and the formation of a Phe-tRNA·EF1·GTP complex has also been described earlier (7,10). Phospholipid and cholesterol were assayed by published procedures (11,12).

RESULTS

Purification and Properties of Calf Liver EF1. In order to obtain more information about the various forms of EF1 and the possible regulatory role of these species, it was decided to investigate EF1 from liver. For purification purposes, the enzyme was prepared from calf liver, although as described below, mice were used for the *in vivo* experiments.

$EF1_H$ and $EF1_L$ from calf liver were purified by a procedure quite similar to that used for the calf brain enzyme (3), and the results of the purification are summarized in Table 1. One important distinction between the two procedures is that the majority of $EF1_H$ from liver was separated from $EF1_L$ in the initial $(NH_4)_2SO_4$ fractionation since $EF1_H$, but not $EF1_L$, precipi-

TABLE 1

Purification of $EF1_H$ and $EF1_L$

	Total Activity (units x 10^{-3})	Specific Activity (units/mg protein)
Post-mitochondrial fraction	1100	14
$(NH_4)_2SO_4$ fraction		
0-30% ($EF1_H$)	132	21
30-70% ($EF1_L$)	530	25
Calcium Phosphate Gel		
30% $(NH_4)_2SO_4$ extract ($EF1_L$)	400	84
Sepharose 6B		
$EF1_L$	310	120
$EF1_H$	65	70
Hydroxylapatite followed by $(NH_4)_2SO_4$ back extraction		
50-45% ($EF1_L$)	16	1100
30-25% ($EF1_H$)	7.8	860
Sucrose Density Gradient		
$EF1_L$	11	1250
$EF1_H$	2.1	1400

For details see reference 3.

tates with 30% ammonium sulfate. In the calf brain purification (3), however, $EF1_L$ and $EF1_H$ were not separated at this stage but at the later calcium phosphate step (3). Specific activities of 1200-1400 have been routinely obtained for the purified preparations of $EF1_H$ and $EF1_L$. Sucrose gradient analysis (Fig. 1A) of $EF1_H$ shows that it is heterogeneous and contains several active species with molecular weights ranging from 2.5 x 10^5 to 1 x 10^6. The purified $EF1_L$ preparation appears to be homogeneous in sucrose gradient analysis (Fig. 1B) with respect to EF1 activity (molecular weight about 50,000). From the purification results and other experiments described below, it was estimated that both $EF1_H$ and $EF1_L$ were at least 90% pure.

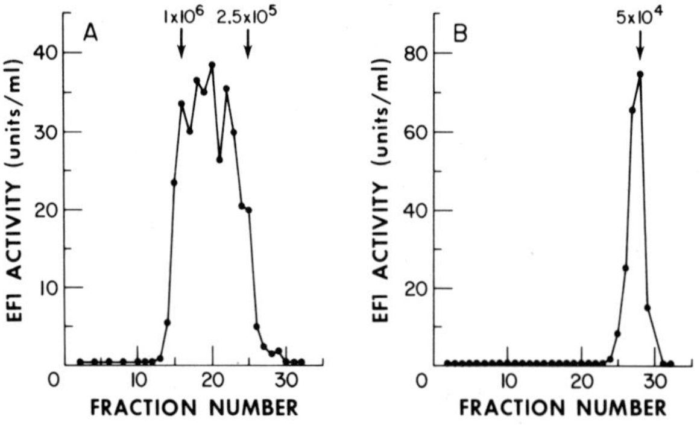

Fig. 1: Sucrose Gradient Profiles of Different EF1 Preparations. One tenth ml of the EF1 preparations (Table 1) (50-100 units of Activity) was placed on top of 4.2 ml of 5-20% sucrose gradient in a Spinco SW-56 centrifuge tube. The tubes were centrifuged for 2 hrs at 50,000 rpm. After centrifugation, the bottoms of the tubes were punctured, and 32 fractions (8 drops each) were collected. The EF1 activity was determined on each fraction.

Polyacrylamide gel electrophoresis of the $EF1_H$ and $EF1_L$ forms are seen in Fig. 2. In gel A the sucrose gradient fractions of $EF1_H$ were electrophoresed using a 5% gel at pH 8.9. A major protein band was seen at the top of the gel. Duplicate gels were sliced and eluted, and essentially all of the EF1 activity was present in the main protein band that did not enter the gel. When this $EF1_H$ preparation was electrophoresed on a 10% gel in the presence of sodium dodecyl sulfate (gel B), one major band was observed with a molecular weight of approximately 50,000 calculated from the migration of protein standards (gel C). When $EF1_L$ was electrophoresed in the presence of SDS, as seen in gel D, one major protein species was again seen, and its migration was similar to the protein species obtained by sodium dodecyl sulfate gel electrophoresis of $EF1_H$ seen in gel B. Gel E shows that $EF1_L$ is at least 90% pure using a 5% gel at pH 8.9. These data are consistent with the view that $EF1_H$ from calf liver is a complex aggregate containing one major protein species ($EF1_L$) of molecular weight about 50,000. These results are essentially the same as those obtained previously with EF1 from other studies (3,4,6).

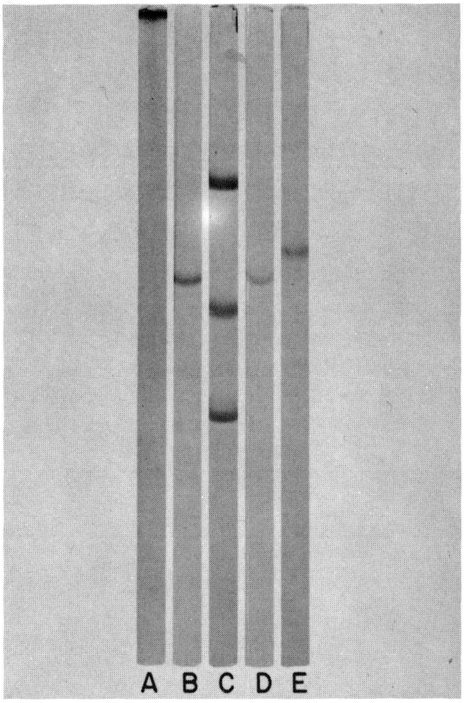

Fig. 2: Disc Gel Electrophoresis of $EF1_H$ and $EF1_L$. Disc gel electrophoresis of A) $EF1_H$ using a 5.0% polyacrylamide gel at pH 8.9. B) 10% SDS polyacrylamide gel of $EF1_H$. C) 10% SDS polyacrylamide gel of standard proteins top ⟶ bottom: bovine serum albumin, 68,000: ovalbumin, 43,000; chymotrypsinogen, 24,000. D) 10% SDS polyacrylamide gel of $EF1_L$. E) 5.0% polyacrylamide gel of $EF1_L$ at pH 8.9.

Amino Acid Composition of $EF1_H$ and $EF1_L$. To further prove that $EF1_H$ is an aggregate of $EF1_L$, the amino acid compositions of $EF1_H$ and $EF1_L$ were analyzed. Fractions from the final sucrose density gradient (Table 1) were examined by sodium dodecylsulfate gel electrophoresis. Fractions which were at least 90% pure were combined and dialyzed against water. For each amino acid analysis determination, 300 μg of either $EF1_H$ or $EF1_L$ protein were used, and the results are shown in Table 2. There was close correlation between the amino acid analysis of the two protein preparations, which strongly suggests that both $EF1_H$ and

$EF1_L$ consist of a similar polypeptide chain. It is interesting to note that neither $EF1_H$ nor $EF1_L$ contains significant amounts of cysteine.[3]

Interaction of Calf Liver EF1 with Guanosine Nucleotides and AA-tRNA. Although $EF1_H$ and $EF1_L$ are both active in AA-tRNA binding to ribosomes, it was shown previously that $EF1_L$ from calf brain (3,7) and wheat germ (5) reacts much more efficiently with GTP and AA-tRNA as compared to $EF1_H$. The experiments

Table 2

Amino Acid Compositions of $EF1_H$ and $EF1_L$

Amino Acids	$EF1_L$[a]	$EF1_H$[a]
Lys	5.94	6.97
His	2.05	2.09
Arg	4.78	4.85
Asp	9.95	9.82
Thr	5.87	5.26
Ser	6.26	5.79
Glu	10.28	11.25
Gly	9.37	7.78
Ala	10.49	9.24
1/2 Cys	trace	trace
Val	7.49	8.11
Met	1.57	1.66
Ileu	5.69	6.34
Leu	10.42	10.34
Tyr	2.98	2.89
Phe	3.45	4.98
Pro	3.41	3.53

[a]Values are given as mole percentage of total amino acids. The Analysis was performed in a Joel Model 6AH analyzer.

seen in Table 3 show that calf liver EF1 also behaves in a similar manner. The ratio of nucleotide binding per unit of EF1 activity was eight times lower with $EF1_H$ than $EF1_L$. Table 3 also shows the results of incubating the $EF1_H$ and $EF1_L$ preparations with GTP and AA-tRNA. With $EF1_H$, about 20% of EF1 participated in ternary complex formation, whereas 80% of the $EF1_L$ preparations could react with GTP and AA-tRNA to form an AA-tRNA·EF1·GTP complex.

Lipid Components in EF1. The experiments described above on the nature and reactivity of calf liver EF1 show that this enzyme is similar in its characteristics to the calf brain enzyme. From other recent results with EF1 from Krebs ascites cells (4) and wheat germ (6), it seems probably that the high molecular weight species of EF1 in all tissues are aggregates of $EF1_L$. Some fundamental questions, however, are still unanswered. These concern the exact chemical nature of the EF1 aggregates, the individual functions of $EF1_H$ and $EF1_L$ and the factors that regulate the interconversion of $EF1_H$ and $EF1_L$.

In recent communications it was shown that calf liver and brain $EF1_H$ contain both cholesterol and phospholipids (3,8). The calf liver preparations (highly purified) used in the present studies generally contained between 20-30 pmoles of total cholesterol/unit of EF1 activity and 250-450 pmoles total phospholipid/unit of EF1 activity. $EF1_L$ from liver, however, contained essentially no cholesterol and about 1/10 of the amount of phospholipid found in $EF1_H$. Chromatographic analysis (3) of the organic-extractable phosphorous material in $EF1_H$ from calf liver showed mainly the presence of phosphatidylethanolamine. In contrast, phosphatidylcholine was identified previously as the major phospholipid in $EF1_H$ from calf brain (3). Of interest is the comparison of calf liver and calf brain $EF1_H$ with regard to the ratio of total phospholipid/total cholesterol. In the former case the value was about 15-20, whereas the latter had a value of 1-2.

Table 3

Interaction of $EF1_H$ and $EF1_L$ Preparations with GTP and Aminoacyl-tRNA

Preparation	Specific Activity	pmoles GTP Bound/ Unit EF1	% EF1 Present in Ternary Complex
$EF1_L$	1200	0.63	75
$EF1_L$	1150	0.57	82
$EF1_H$	850	0.07	10
$EF1_H$	1210	0.08	33

The formation of an EF1·GTP complex and the ternary complex, the latter determined by the ability of Phe-tRNA and GTP to cause EF1 to pass through a nitrocellulose filter, have been described previously (9,10).

Studies with EF1 from Mouse Liver. A key question concerns the role of the aggregate forms of EF1 that have been found in most tissues that have been examined. A clue to their function might be obtained if it were possible to show that either the amount or distribution of the aggregates changed under certain conditions. Our initial studies have employed mice, since they are more suitable for routine laboratory studies. It was previously shown that purified preparations of calf liver $EF1_H$ could be converted to $EF1_L$ by treatment with phospholipase AB or C (8). As seen in Fig. 3, EF1 in an unfractionated mouse liver supernatant shows a typical heterogeneous pattern. However, there is almost complete conversion to $EF1_L$ after incubation with phospholipase C (600 μg). It should be noted that about 30 fold higher concentrations of phospholipase were needed to obtain the conversion of $EF1_H$ to $EF1_L$ in the crude mouse liver preparation as compared to the purified enzyme (8).

Since the crude mouse liver EF1 was heterogeneous and sensitive to phospholipase, it appeared to be typical of EF1 in other tissues and, therefore, was utilized for in vivo studies in attempts to alter the EF1 pattern. In these preliminary experi-

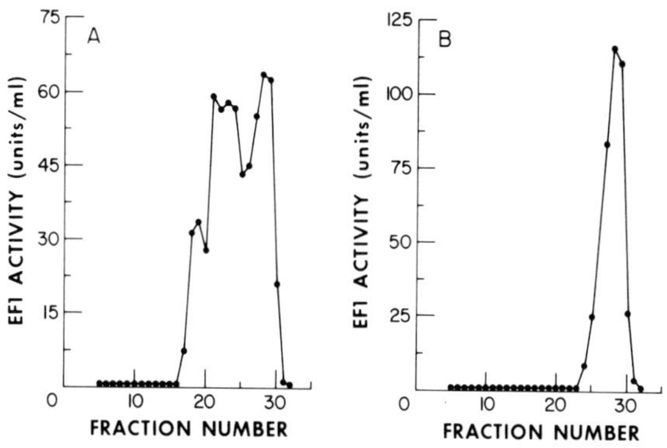

Fig. 3: Sucrose Gradient Profiles of Mouse Liver EF1 Activity. The preparation of a high-speed supernatant fraction from mouse liver is described in the Methods section, and the sucrose gradient centrifugation is described in the legend to Fig. 1. A) Profile of mouse liver EF1 (80 units) activity after incubation at 37° for 40 min. B) Eighty units of mouse liver EF1 activity after incubation with 600 µg of phospholipase C (Sigma, type 1) for 40 min at 37°.

Table 4

Activity of EF1 from Fasted Mice

Condition	Weight of liver gm	Activity unit/ml	Protein mg/ml	Specific Activity unit/mg
Fed controls	0.84	1060	41.0	25.8
24 hrs fast	0.72	672	47.8	14.2
48 hrs fast	0.60	520	43.0	12.0
72 hrs fast	0.55	512	48.6	10.6

The preparations of the liver supernatant are described in the text. The data represent the average of three mice.

ments, mice were either fasted (for periods up to 72 hrs) or treated with cortisone or insulin (see Methods) before sacrificing. Sucrose gradient profiles of liver supernatants showed no significant difference in the EF1 patterns which, in general, appeared similar to those in Fig. 3A. In addition, the EF1 patterns from male and female mice were similar. The only notable observation (Table 4) was that the livers from fasted animals, which were smaller in size, also had a lower EF1 specific activity than the control tissue. However, as noted above, this change in specific activity was not due to a change in the distribution of the EF1 species.

DISCUSSION

The properties of calf liver EF1 are similar to those of EF1 preparations from other sources (1-6). The enzyme activity is heterogeneous with the bulk of the activity in the 500-800,000 molecular weight range. These aggregates contain essentially one polypeptide chain of molecular weight of about 50,000. The purified heavy and light forms of the enzyme have similar amino acid compositions, a finding that has also been observed with the various forms of wheat germ EF1.[3]

The presence of both cholesterol and phospholipid in the purified $EF1_H$ preparations from calf liver support the view that lipids are involved in maintaining the integrity of the aggregates. The phospholipids appear to play an essential role in this respect since phospholipase treatment readily converts $EF1_H$ to $EF1_L$ as measured by sucrose gradient analysis. Recent studies (8) suggest that EF1 from all eukaryote sources are similar in that the aggregates are very likely held together by phospholipids.

The role of cholesterol in these aggregates is more puzzling. Although Hradec et al. (13) had reported several years ago that a specific cholesterol ester, cholesterol-14-methylhexadecanoate was an essential component of elongation factor 1 from two eukaryote sources, as yet, we have not confirmed this finding. Although the aggregates of EF1 always appear to contain phospholipid, they only contain significant amounts of cholesterol when the aggregates have molecular weights over 500,000. In general, based on the few tissues that have been examined, there seems to be some correlation between the amount of cholesterol present and the size of the EF1 aggregate. In the case of calf brain, where we have found the highest molecular weight aggregates, the cholesterol content has been the greatest, with values ranging between 40-100 pmoles of total cholesterol per unit of enzyme. In contrast, calf liver EF1 which has aggregates of smaller size, has cholesterol values ranging between 15 and 30 pmoles/unit of EF1, and a highly purified preparation of reticulocyte EF1[2] which has a molecular weight ranging between 180-500,000 contained only trace levels of cholesterol (less than 6 pmoles/unit). Yet all of these preparations contain significant amounts of phospholipid and can be converted to the light form of the enzyme after phospholipase treatment when examined by sucrose gradient analysis. Our current view is that phospholipids are essential for maintaining the integrity of the EF1 aggregates, whereas cholesterol is only present when the size of the aggregate exceeds a certain level. The situation with EF1 may be somewhat similar to what has been reported for the aminoacyl-tRNA synthetases in animal cells. These appear to be present as high molecular weight aggregates which have been shown to contain substantial quantities of lipid (14,15).

In the case of EF1, previous studies have indicated that the 50,000 molecular weight species is very likely the species that carries the AA-tRNA to the ribosome, since the isolation of an AA-tRNA·EF1·GTP complex has been shown to contain $EF1_L$ and not $EF1_H$ (5,9). EF1 is obviously involved in the synthesis of all of the proteins in the eukaryote cell and, therefore, one can postulate that the reversible interconversions of the heavy and light forms of the enzyme could be a mechanism by which the overall rate of elongation is regulated. Our initial attempts in this study to show changes in the EF1 pattern in mice under several physiological conditions have not been successful. In one case, namely fasting, although there was about a 50% drop in the specific activity of EF1 from the liver of fasted animals, the EF1 pattern did not significantly change over that observed in control animals. Recently Lanzani et al. (16) reported that cyclic GMP changed the profile of EF1 in a wheat extract. In these studies the cyclic nucleotide appeared to cause a small conversion of the heavy species of the enzyme to a ligher form. If as predicted the lighter species is more active than the heavy species, this could explain an earlier report that cyclic GMP stimulates protein synthesis in a eukaryote system (17). However, before definite conclusions are drawn, one should await until further data are obtained on this point.

SUMMARY

Elongation factor 1 (EF1) purified from calf liver has been studied. The active protein occurs in multiple forms of different molecular weight similar to the corresponding enzymes from other eukaryote sources. Only a single polypeptide was observed when both high molecular species of EF1 ($EF1_H$) and low molecular species of EF1 ($EF1_L$) were analyzed by electrophoresis on acrylamide gels containing sodium dodecylsulfate. In addition, the amino acid composition of $EF1_H$ and $EF1_L$ are similar, and both are devoid of cysteine. These data and earlier results support the view that the occurrence of multiple forms of EF1 is due to the association of $EF1_L$ monomers to form $EF1_H$ aggregates. Like EF1 from calf brain, heavy and light forms of EF1 from calf liver are both active in aminoacyl-tRNA binding to ribosomes, although $EF1_L$ reacts with GTP and aminoacyl-tRNA more efficiently than $EF1_H$. Calf liver $EF1_H$ contains appreciable amounts of phospholipid, mostly phosphatidylethanolamine and some cholesterol. Preliminary experiments have not shown any significant change in the relative amounts of $EF1_H$ and $EF1_L$ in mouse liver under a variety of conditions.

FOOTNOTES

1. Present address: Institute of Biochemistry, Agriculture University, 60-637 Poznan, Poland.

2. Abbreviations: EF1, elongation factor 1; $EF1_H$, heavy species; $EF1_L$, light species; AA-tRNA, aminoacyl-tRNA.

3. A similar observation on the lack of cysteine in EF1 from wheat germ has been made by Dr. Andrzej Legocki. In addition, a purified rabbit reticulocyte EF1 preparation kindly supplied to us by Drs. W. Merrick and F. Anderson also showed a similar amino acid composition as calf liver EF1 with only trace levels of cysteine present.

ACKNOWLEDGMENTS

The authors would like to express their thanks to Dr. Chun-Yen Lai for performing the amino acid analyses, Betty Redfield for her technical help, and many helpful discussions with Dr. Nathan Brot. In addition, we are particularly thankful to the Biopolymer Laboratory at Hoffmann-La Roche for help in the large-scale preparation of calf liver EF1.

REFERENCES

1. Schneir, M. and Moldave, K.: The Isolation and Biological Activity of Multiple Forms of Aminoacyl Transferase I of Rat Liver. Biochim. Biophys. Acta 166, 58-67 (1968).
2. Collins, J.F., Moon, H.-M. and Maxwell, E.S.: Multiple Forms and Some Properties of Aminoacyltransferase I (Elongation Factor 1) from Rat Liver. Biochemistry 11, 4187-4194 (1972).
3. Moon, H.-M., Redfield, B., Millard, S., Vane, F. and Weissbach, H.: Multiple Forms of Elongation Factor 1 from Calf Brain. Proc. Nat. Acad. Sci. U.S.A. 70, 3282-3286 (1973).
4. Drews, J., Bednarik, K. and Grasmok, H.: Elongation Factor 1 from Krebbs II Mouse Acites Cells: Purification, Structure and Enzymatic Properties. Europ. J. Biochem. 41, 217-227 (1974).
5. Tarrago, A., Allende, J.E., Redfield, B. and Weissbach, H.: The Effect of Guanosine Nucleotides on the Multiple Forms of Protein Synthesis Elongation Factor 1 from Wheat Embryos. Arch. Biochem. Biophys. 159, 353-361 (1973).
6. Golinska, B. and Legocki, A.B.: Purification and Some Properties of Elongation Factor 1 from Wheat Germ. Biochim. Biophys. Acta 324, 156-170 (1973).
7. Legocki, A.B., Redfield, B. and Weissbach, H.: Interaction of the Heavy and Light Forms of Elongation Factor 1 with Guanosine Nucleotides and Aminoacyl-tRNA. Arch. Biochem. Biophys. 161, 709-712 (1974).

8. Legocki, A.G., Redfield, B., Liu, C.K. and Weissbach, H.: Role of Phospholipids in the Multiple Forms of Mammalian Elongation Factor 1. Proc. Nat. Acad. Sci. U.S.A. (in press).
9. Weissbach, H., Redfield, B. and Moon, H.-M.: Further Studies on the Interactions of Elongation Factor 1 from Animal Tissues. Arch. Biochem. Biophys. 156, 267-275 (1973).
10. Moon, H.-M. and Weissbach, H.: Interaction of Brain Transferase 1 with Guanosine Nucleotides and Aminoacyl-tRNA. Biochem. Biophys. Res. Commun. 46, 254-262 (1972).
11. Chalvardjian, A. and Rudnicki, E.: Determination of Lipid Phosphorus in the Nanomolar Range. Anal. Biochem. 36, 225-226 (1970).
12. Bondjers, G. and Bjorkerud, S.: Fluorometric Determination of Cholesterol and Cholesteryl Ester in Tissue on the Nanogram Level. Anal. Biochem. 42, 363-371 (1971).
13. Hradec, J., Dusek, Z., Bermek, E. and Mattaei, H.: The Role of Cholesteryl 14-Methylhexadecanoate in Peptide Elongation Reactions. Biochem. J. 123, 959-966 (1971).
14. Hradec, J. and Dusek, Z.: Effect of Lipids, in Particular Cholesteryl 14-Methylhexadecanoate, on the Corporation of Labelled Amino Acids into Transfer Ribonucleic Acids *in vitro*. Biochem. J. 110, 1-8 (1968).
15. Bandyopadhyay, A.K. and Deutscher, M.P.: Lipids Associated with the Aminoacyl-transfer RNA Synthetase Complex. J. Mol. Biol. 74, 257-261 (1973).
16. Lanzani, G.A., Giannattasio, M., Manzocchi, L.A., Bollini, R., Soffientini, A.N. and Macchia, V.: The Influence of Cyclic GMP on Polypeptide Synthesis in a Cell-free System Derived from Wheat Embryos. Biochem. Biophys. Res. Commun. 58, 172-177 (1974).
17. Varrone, S., DiLauro, R. and Macchio, V.: Stimulation of Polypeptide Synthesis by Cyclic 3'-5'-Guanosine Monophosphate. Arch. Biochem. Biophys. 157, 334-338 (1973).

Some Thoughts on the Regulation of Arginine Biosynthesis and its Relation to Biochemistry

Werner K. Maas
Department of Microbiology, New York University School of Medicine, New York, N.Y. 10016, USA

It is now 21 years since I had the privilege of spending a year in Dr. Lipmann's laboratory in Boston, where I worked on the peptidic bond formation in the synthesis of pantothenic acid. It was shortly after returning from Boston to Dr. B.D. Davis' laboratory in New York that I became involved in the regulation of arginine biosynthesis in *Escherichia coli*. It happened in the following way: Prior to my stay in Boston I had studied pantothenate synthetase in a temperature-sensitive pantothenate auxotroph and had found that in this mutant a heat labile enzyme was formed (1). This was the first case in which it was shown that as a result of a mutation the structure of an enzyme was altered. After my return from Boston I wanted to look for mutants in which the rate of enzyme synthesis, rather than the structure of an enzyme, was altered. I used the same approach as before, to look for temperature-conditional mutants, able to form an enzyme at one temperature, but not at another. To avoid the complication of isolating mutants with heat-labile enzymes, I looked for the reverse type, "cold sensitive", mutants, able to make an enzyme at 37° but not at 25°. I started with completely blocked auxotrophs, unable to carry out a known enzymatic reaction, and selected revertants at 37° which could grow without the required growth factor. Among these I looked for strains that still had the growth factor requirement at 25°. One of my starting strains was an arginine auxotroph, blocked between ornithine and citrulline and unable to make ornithine transcarbamylase (OTC). This enzyme can be measured easily in extracts of the wild type strain. From this completely blocked auxotroph the desired revertants were obtained. OTC was produced when they were grown at 37°, but not after growth at 25°. At 25° the strains required arginine and were therefore grown with arginine; at 37° they did not require arginine and were therefore grown without arginine. The crucial and completely unexpected finding was made in control cultures, grown at 37° with arginine: here *no* OTC was produced. It was then found that in the wild-type strain the same relationship existed between OTC formation and presence or absence of arginine in the growth medium. This was my introduction to the phenomenon of enzyme repression and I have been occupied with it ever since.

Before continuing this story and to orient the reader I have indicated in Fig. 1 the steps of arginine biosynthesis in *E. coli* and in Fig. 2 the location of the genes of the arginine pathway on the linkage map of *E. coli.*

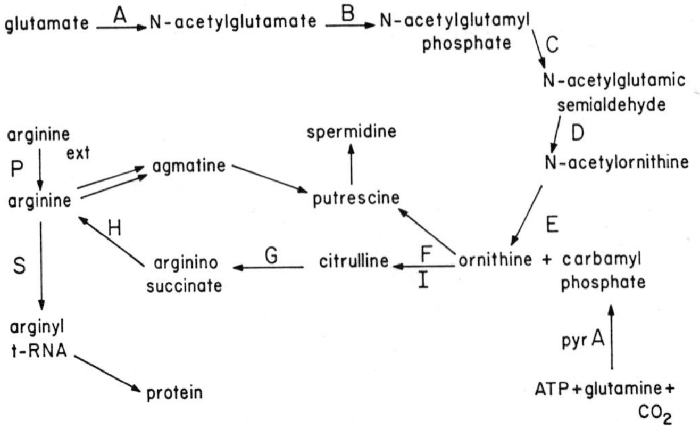

Fig. 1. Biosynthesis of arginine and related reactions. The letters denote the genes controlling the corresponding reactions. About 25% of the pathway flows into putrescine and spermidine, the remainder into protein.

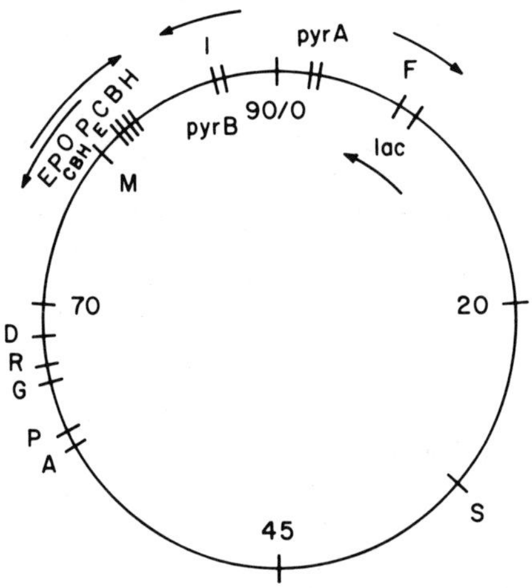

Fig. 2. Linkage map of E. coli showing genes of the arginine pathway. The arrows indicate direction of transcription. The genes of the arginine cluster are transcribed divergently, argE counter-clockwise, argCBH clockwise. A common operator region is between argE and argCBH, with the promoter for argE being on the clockwise side of the operator and the promoter for argCBH being on the counterclockwise side. The gene argM affects the formation of N-acetylornithine trans-aminase.

My initial studies in the 1950s were concerned mainly with physiological aspects of enzyme repression. In early experiments an unusual picture was found for the kinetics of OTC synthesis following the removal of arginine from a growing culture. At that time kinetics of enzyme formation had been studied mainly with inducible systems, such as β-galactosidase, and here enzyme synthesis was found to occur at a contant rate after addition of an inducer. In the case of OTC, there was an initial burst of enzyme synthesis, followed by a sharp decrease. Consideration of this unusual kinetic picture led Luigi Gorini and myself (with encouragement and advice from Aaron Novick and Leo Szilard) to carry out experiments with arginine-limited cultures in a chemostat. These showed that the rate of OTC synthesis is modulated by the intracellular concentration of arginine and that the cell has a much greater potential for synthesizing OTC than was realized from experiments with cells growing under ordinary conditions (2). Later, a similar feedback control of enzyme synthesis was found for the other enzymes of the arginine pathway. It was also shown that the activity of the first enzyme of the pathway, which converts glutamate to N-acetylglutamate, was subject to feedback inhibition by arginine (3). It should be mentioned that during this period, the same type of feedback control was demonstrated for other biosynthetic pathways, such as the formation of histidine and tryptophan. The arginine pathway differed, however, from the latter two in that the structural genes for the biosynthetic enzymes are not next to each other, but, with the exception of one cluster of 4 genes, are scattered over the linkage map (Fig. 2). This type of scattered gene arrangement is found in eukaryotic organisms and may represent a higher level of evolution than the single cluster arrangement.

I would like to pause here and point out that the finding in the 1950s of such "teleological" regulatory devices came as a surprise. I think that it was generally assumed at that time that some form of regulation existed, but that for the most part it was taken care of by the properties of the biosynthetic enzymes themselves, for example by their affinity for substrates. It was not found necessary to invoke for each pathway specialized regulatory proteins (repressors, etc.) or sites on enzymes specialized for regulation (allosteric sites). Perhaps the reason for this was that the main emphasis was on working out the exact chemical steps of intermediary metabolism and that most experiments were carried out with extracts, usually purified, in which regulatory devices could not be recognized. Relatively little attention was paid to single reaction steps as they occurred in whole cells. Besides, there was clearly something irrational, in a chemical sense, about such regulatory devices. How could one picture the inhibition by arginine of an enzyme, whose substrates and products were chemically unrelated to arginine? How could one imagine the inhibition of the production of a whole series of enzymes by arginine?

To continue, the discoveries made in the 1950s culminated in 1961 in the formulation of the Operon Model by Monod and Jacob (4). This model incorporated the available information on the regulation of enzyme synthesis into a general picture. Its heuristic value is attested by the many instances studied since 1961, in which its postulates were found to be correct. In the case of arginine, a regulatory gene, argR, which controls repressibility of the arginine synthesizing enzymes, had already been found prior to 1961. Subsequently it was shown that the product of this gene was a substance which could diffuse through the cytoplasm and, in conjugation with arginine or a derivative or arginine, could inhibit enzyme synthesis (5). The argR product was called the aporepressor, arginine (or its active derivative) were called the corepressor, in conformation with nomenclature on enzymes. It seems now that the corepressor is actually arginine itself (6). A further prediction of the Operon Model, that repression occurs at the level of transcription, has been verified recently (7). Messenger RNA (m-RNA) synthesis has been studied for the 4 genes argECBH and it has been shown that in the presence of arginine, the formation of this m-RNA is inhibited. We are now studying enzyme repression in cell-free extracts, using the system developed by Zubay to measure the synthesis of acetylornithinase, controlled by the gene argE, and argininosuccinase, controlled by the gene argH (8). Extracts of $argR^+$ strains inhibit the production of these enzymes and we are now in the process of purifying the aporepressor. With the cell-free system we expect to gain more insight into the biochemical mechanism of repression. In conjunction with these studies we are also trying to characterize the sites on the DNA of the argECBH cluster at which repression occurs.

From what has been said it can be seen that studies on regulation have opened new areas for biochemistry. By now biochemists have fully accepted the originally "strange" phenomena involved in regulation and much work is going on on the structure of regulatory sites on enzymes and on the interactions between regulatory molecules and DNA. This type of experimentation is different from that done prior to the discovery of repression, induction and feedback inhibition. There is more emphasis on noncovalent bond formation, especially bonds formed in macromolecular complexes between proteins and nucleic acids, and on changes in the configuration of macromolecules. In a more theoretical and general sense, the recognition of special biochemical regulatory devices has established the legitimacy of teleological reasoning in biochemistry. It has become clear that during evolution, mechanisms previously considered unlikely, can become established in living cells, as long as they give the organism a selective advantage over its competitors. The cell is no longer considered as a chemical machine, but as a cybernetic chemical machine. These ideas have been expressed and elaborated in the recent books by Monod (9) and Jacob (10) and can be summarized by saying that biochemistry has matured to recognize not only the "Logic of Chemistry" but also the "Logic of Life."

Acknowledgments

The investigations of the author have been supported by PHS grant GM-06048 for 16 years. The author is a PHS Career Awardee (K6-GM-15, 129).

Literature References

1. Maas, W.K. and Davis, B.D.: Production of an altered pantothenate-synthesizing enzyme by a temperature-sensitive mutant of Escherichia coli. Proc. Natl. Acad. Sci. U.S. 38, 785-797 (1952).

2. Gorini, L. and Maas, W.K.: The potential for the formation of a biosynthetic enzyme in Escherichia coli. Biochim. Biophys. Acta 25, 208-209 (1957).

3. Vyas, S. and Maas, W.K.: Feedback inhibition of acetylglutamate synthetase by arginine in Escherichia coli. Arch. Biochem. Biophys. 100, 542-546 (1963).

4. Jacob, F. and Monod, J.: On the regulation of gene activity. Cold Spring Harbor Symp. Quant. Biol. 26, 193-211 (1961).

5. Maas, W.K. and Clark, A.J.: Studies on the mechanism of repression of arginine biosynthesis in Escherichia coli. II. Dominance of repressibility in diploids. J. Mol. Biol. 8, 365-370 (1964).

6. Celis, T.F.R. and Maas, W.K.: Studies on the mechanism of repression of arginine biosynthesis in Escherichia coli. IV. Further studies on the role of arginine-tRNA in repression of the enzymes of arginine biosynthesis. J. Mol. Biol. 62, 179-188 (1971).

7. Kryzek, R. and Rogers, P.: Arginine control of transcription of argECBH messenger ribonucleic acid in Escherichia coli. J. Bacteriol. 110, 945-954 (1972).

8. Urn, E., Yang, H., Zubay, G., Kelker, N. and Maas, W.K.: In vitro repression of N-α-acetyl-L-ornithinine synthesis in Escherichia coli. Molec. Gen. Genetics 121, 1-7 (1973).

9. Monod, J.: Chance and Necessity. Random House, N.Y. (1971).

10. Jacob, F.: The Logic of Life. Random House, N.Y. (1973).

Mechanism of the Cell Cycle in DNA Synthesis in Sea Urchin Embryos

Y. Mano, N. Suzuki, K. Murakami and K. Kano
Department of Physiological Chemistry and Nutrition, Faculty of Medicine, University of Tokyo, Tokyo, Japan

DNA synthesis is well known to occur periodically during the cell cycle. In a well synchronized system from early sea urchin embryos, DNA shows typical periodic synthesis. Early sea urchin embryos are known to develop normally in sea water without added nutrients, but they can utilize endogenous deoxyribonucleosides well for DNA synthesis. A number of studies have shown that DNA synthesis in cleaving sea urchin embryos is carried out physiologically by the so-called salvage pathway (1). Thus, deoxyribonucleosides are good precursors in synthesis of DNA in sea urchin embryos. The present paper describes the mechanism of the cell cycle in DNA synthesis in sea urchin embryos.

In sea urchin embryos, DNA is first synthesized at the stage of pronuclear fusion and next in the period from telophase to the following interphase. It has been found in a number of organisms that initiation of DNA synthesis depends on protein synthesis (2 - 9). This requirement for protein synthesis has been suggested to involve formation of a factor which is needed for initiation of DNA synthesis. It was reported, however, that puromycin and cycloheximide had little effect on DNA synthesis at the stage of pronuclear fusion, although they strongly inhibited subsequent synthesis of DNA (8, 10, 11). This implies that the mechanism of initiation of DNA synthesis at the stage of pronuclear fusion is unique and differs from that in subsequent synthetic periods.

Cycloheximide at a concentration of 2 m$\underline{M}$ caused delay of cell division with prolongation of the streak stage and delay in initiation of DNA synthesis. As measured by thymidine incorporation into the DNA fraction, the decrease in the rate, and prolongation of the period of DNA synthesis caused by the inhibitor varied with its concentration and time of administration. Initiation of DNA synthesis was delayed when the inhibitor was added to suspensionof embryos between the time when DNA synthesis terminated and a definite time before the predicted time of initiation of the next period of DNA synthesis, except at the stage of pronuclear fusion. However, when the inhibitor was added after initiation of DNA synthesis, the latter proceeded normally. Addition of 10 m*M* cycloheximide immediately after fertilization or of 2 m*M* cycloheximide 60 min before fertilization also delayed DNA synthesis at the stage of pronuclear fusion, indicating that synthesis at this stage also required prior protein synthesis. These facts suggest that DNA synthesis at the stage of pronuclear fusion is not exceptional in respect to its protein requirement and that a definite amount of a particular protein must be synthesized and accumulated before initiation of DNA synthesis in each synthetic cycle (12).

It is known that synthesis of a particular protein is necessary for *in vivo* DNA synthesis, but it is still uncertain which reaction step in the cell cycle requires protein synthesis. Thus, the regulatory step causing cyclic variation in *in vivo* DNA synthesis was investigated chromatographically. Thymidine added to the medium was taken up rapidly into the acid-soluble fraction of embryos and phosphorylated, resulting in accumulation of more than 80 % as thymidine triphosphate. The phosphorylations of thymidine did not fluctuate in parallel with thymidine incorporation into the DNA fraction. The amount of thymidine triphosphate derived from exogenously added thymidine was some 10-fold more than the amount polymerized to DNA. The phosphorylations of deoxyadenosine, deoxycytidine and deoxyguanosine were somewhat different from that of thymidine, but, in general, the amount of these triphosphate derivatives originating from exogenously added deoxyribonucleosides were also much larger than the amounts of these nucleosides incorporated into the DNA fraction. The activities of thymidine kinase, thymidylate kinase and DNA polymerase, measured *in vitro*, all fluctuated regularly with time in the cell cycle, but their phases shifted sequentially according to the synthetic pathway (13). However, analysis of the acid-soluble fraction of embryos indicated that the percentage incorporation of thymidine triphosphate into the DNA fraction, measured *in vivo*, varied in parallel with thymidine incorporation into the DNA fraction during the cell cycle, while thymidine kinase and thymidylate kinase activities did not. Deoxyribonuclease activity measured with homogenates also did not fluctuate during the cell cycle. Thus, the regulatory step causing cyclic variation in *in vivo* DNA synthesis seems to be the DNA polymerizing step which involves DNA polymerase. The discrepancy between the cyclic variations of thymidine kinase and thymidylate kinase activities measured *in vivo* and *in vitro*, may largely be due to differences in assay conditions, and the accumulation of thymidine triphosphate, in particular, may cause considerable inhibition of these enzyme activities *in vivo* (14).

In parallel with the above experiments the reaction step requiring protein synthesis in DNA synthesis was investigated. Thymidine incorporation into the DNA fraction was inhibited by 0.1 mM emetine. The inhibition was mainly due to inhibition of incorporation of thymidine triphosphate into the DNA fraction, not into the acid-soluble fraction. Chromatographic analysis of the acid-soluble fraction showed that emetine caused marked accumulation of thymidine triphosphate. Analysis of the activities of DNA polymerase, thymidine kinase and thymidylate kinase *in vivo* indicated that only DNA polymerase activity was inhibited by emetine. With purified preparation DNA polymerase activity itself was scarcely inhibited by emetine. From these results the reaction step requiring protein synthesis seems to be that involving DNA polymerase in the salvage pathway of DNA synthesis. These experiments indicate that protein synthesis is required for the DNA polymerizing step,which is thought to be the regulatory step causing cyclic DNA synthesis. However, the particular protein which initiates DNA synthesis has not yet been isolated from the system. The factors described above which stimulated DNA polymerase activity were isolated from the nuclei of sea urchin embryos by fractionation with DNA cellulose. These factors could not be detected in the cytoplasm. By this fractionation procedure two kinds of DNA polymerase, eluted with 0.25 M and 0.60 M NaCl, and two kinds of stimulatory factor, eluted with 0.20 M and 0.30 M NaCl, were separated. The two polymerases showed higher activity with native DNA as a template than with denatured DNA. The stimulatory factors were named SF I and SF II, repectively. They consistently caused 3 - 4-fold stimulation and both factors were effective in stimulating both polymerases. The factors seemed to be proteins because they were heat labile and susceptible to trypsin. No deoxyribonuclease activity was detected in preparations of these factors. The factors did not stimulate purified E. coli DNA polymerase I. The template specificity and SH-sensitivity of the polymerase are vary similar to type III DNA polymerase of E. coli, which is thought to catalyze replication. The fact that the factors did not stimulate E. coli polymerase I support the idea that they may act on DNA replication (15). One of the stimulatory factors, SF I, was partially purified and characterized. Ribonucleoside triphosphates were essential for its stimulatory activity and its activity was linearly proportional to the amount of ribonucleoside triphosphate added. Of the nucleotides tested, ATP was the most effective. DNA polymerase and SF I formed an active complex consuming ATP to form AMP and pyrophosphate. When either SF I or ATP was omitted, no such complex was formed. The complex showed neither ATPase, GTPase activity nor RNA polymerase activity. Both AMP and pyrophosphate seemed to bind to the complex, detected by Sephadex G-200 filtration. The activity of this complex increased on addition of polyamines such as spermidine and spermine, and small amount of histones. From these results, SF I is suggested to act on activation of DNA polymerase (16). Similar philosophy has recently been presented by others (17, 18).

Summary

Mechanism of the cell cycle in DNA synthesis was studied with sea urchin embryos. Periodical synthesis of DNA was dependent on concomitant protein synthesis. The regulatory step in *in vivo* DNA synthesis from thymidine was suggested to be the DNA polymerizing step. As measured *in vivo*, only DNA polymerase required protein synthesis, but other phosphorylating enzymes did not. Protein factors stimulating DNA polymerase were isolated and characterized. For the stimulation, ribonucleoside triphosphate was essential. One of the stimulatory factors, SF I, ATP and DNA polymerase formed an active complex consuming ATP to form AMP and pyrophosphate, which attached to the complex. Thus, it is suggested that protein synthesis is important in periodical synthesis of DNA in the cell as a driving force for cell division.

References

1. Brachet, J., "The Biochemistry of Development", p. 90 - 101, Pergamon Press, Paris (1960).
2. Billen, D. J., J. Bacteriol. 80, 86 - 95 (1960).
3. Billen, D., Biochim. Biophys. Acta 55, 960 - 968 (1962).
4. Powell, W. F., Biochim. Biophys. Acta 55, 979 - 986 (1962).
5. Mueller, G. C., Kajiwara, K., Stubblefield, E. and Peukert, R. R., Cancer Res. 22, 1084 - 1090 (1962).
6. Littlefield, J. W. and Jacobs, P. S., Biochim. Biophys. Acta 108, 652 - 658 (1965).
7. Terasima, T. and Yasukawa, M., Exp. Cell Res. 44, 669 - 672 (1966).
8. Black, R. E., Baptist, E. and Piland, J., Exp. Cell Res. 48, 431 - 439 (1967).

9. Weiss, B. G., J. Cell Physiol. 73, 85 - 90 (1969).
10. Wilt, F. H., Sakai, H. and Mazia, D., J. Mol. Biol. 27, 1 - 7 (1967).
11. Young, C. W., Hendler, F. J. and Karnofsky, D. A., Exp. Cell Res. 58, 15 - 26 (1969).
12. Suzuki, N. and Mano, Y., Develop. Growth Differ. 15, 113 - 125 (1973).
13. Nagano, H. and Mano, Y., Biochim. Biophys. Acta 157, 546 - 557 (1968).
14. Suzuki, N. and Mano, Y., J. Biochem. 75, 1349 - 1361 (1974).
15. Murakami, K. and Mano, Y., Biochem. Biophys. Res. Commun. 55, 1125 - 1133 (1973).
16. Murakami, K. and Mano, Y., in preparation.
17. Wickner, W. and Kornberg, A., Proc. Natl. Acad. Sci. U. S. 70, 3679 - 3683 (1973).
18. Hurwitz, J. and Wickner, S., Proc. Natl. Acad. Sci. U. S. 71, 6 - 10 (1974).

Ribosome Metabolism in Starving Relaxed E. coli Cells

G. Marchis-Mouren, J. Marvaldi and A. Cozzone
Institut de Chimie Biol., Université d'Aix-Marseille,
Place Victor Hugo, 13331 Marseille Cédex 3, France.

Valyl tRNA deprivation causes an intense shift in the polysome size distribution in stringent cells but only a moderate shift in relaxed cells (1). Also the ribosomes in relaxed cells can reassemble in polysomes still functionally competent (1).

Our intention was first to examine the ribosomal protein of polysomes from starving relaxed cells by comparison with polysome from exponentially growing cells. Then we have been interested in particles assembly and ribosomal protein biosynthesis in starving cells.

We have tried to answer 3 questions :

1°) Does the ribosomal protein composition differ in polysomes from starving cells to exponentially growing cells ?

2°) What is the composition of the ribonucleoprotein particles which assemble during starvation ?

3°) Since we know that ribosomal proteins are preferentially synthesized, although at a very much reduced rate, in starving rel$^-$ cells. Is these any differential synthesis in ribosomal proteins during starvation ?

MATERIALS AND METHODS

a) Bacterial strains and growth conditions.

Strains 10B6 rel$^+$ and 10B6 rel$^-$ (2) are arginine auxotrophs and have a heat sensitive valyl-tRNA synthetase. The permissive temperature for these strains is 30°C. The doubling time is 80 min. Cells were grown at 30°C in a medium at pH 7.6 which contained the following components per liter : 12 g Tris ; 2 g KCl ; 2 g NH_4Cl ; 0.5 g $MgCl_2$ $6H_2O$; 0.02 g Na_2SO_4 ; 14.7 mg

$CaCl_2$ $2H_2O$; 178 mg Na_2HPO_4 $2H_2O$; 2 g glucose. This medium was supplemented with 50 microgram of arginine/ml (rel^- cells) or casamino acid (rel^+). Cells were grown exponentially till mid-log phase then starved by shifting the temperature up to 42°. Immediate arrest of culture growth was obtained by adding crushed ice to the medium.

The labeling conditions are as indicated in the diagram. Starving cells were exposed to ^3H uridine (25 microg/ml ; 0.5 microCi/ml) or to ^3H leucine (0.5 microg/ml ; 6 microCi/ml) and exponentially growing cells to ^3H leucine (30 microg/ml ; 3 microCi/ml). Radioactive uridine and leucine were purchased from the CEA (Saclay-France). In chase experiment leucine was added at the concentration of 2 mg/ml.

b) Preparation of polysomes.

The method of preparation was similar to that described by Klagsbrun & Rich (3). In a typical experiment, 500 ml of culture were poured over an equal vol. of crushed ice chilled at -15°C in order to reduce polysome run-off (4). All subsequent operations were carried out at 0 to 4°C. Cells were harvested by centrifugation for 8 min at 10,000 g in a Sorvall GSA rotor. The supernatant fraction was discarded and the pellet resuspended in 12 ml of sucrose-buffer solution containing 0.5 M-RNase-free sucrose, 0.016 M-Tris.HCl buffer (pH 8.1), and 0.05 M-KCl. Then 1.5 ml of freshly dissolved lysozyme solution (10 mg/ml in sucrose-buffer solution) and 0.3 ml of 10 % EDTA (pH 8.0) were added to produce protoplasts. The suspension was stirred for 4 min and 0.3 ml of 1 M-$MgCl_2$ was added to stop lysozyme action. Protoplasts were centrifuged for 6 min at 12,000 g in a Sorvall SS34 rotor and resuspended in 0.7 ml of freshly prepared lysing medium (0.05M-NH_4Cl, 0.01 M-$MgCl_2$, 0.01 M-Tris, pH 7.8) containing 0.5 % Brij-58, 0.5 % sodium deoxycholate and 5 ug RNase free DNase/ml. After 4 min, the lysate was clarified by centrifugation for 12 min at 15,000 g in an SS34 Sorvall rotor. The supernatant fraction was carefully removed and analyzed by sucrose density-gradient centrifugation. When lysates could not be analyzed immediately, they were quickly frozen and stored at -20°C.

1.3 ml of clarified lysate was layered onto 15 to 40 % sucrose gradient in 0.05 M-NH_4Cl, 0.01 M-$MgCl_2$ and 0.01 M-Tris (pH 7.8) buffer solution. Gradients were centrifuged in a Spinco SW27 rotor for 4 h at 26,000 revs/min.

c) Ribosomal protein purification and characterization

Fractions from the gradient were pooled and concentrated. Ribosomal proteins were extracted by 66 % acetic acid (5) and analyzed by bidimensional polyacrylamide gel electrophoresis (6). The carrier proteins used for chromatography and electrophoresis were extracted from ribosomes prepared according to Kurland(7). The individual protein spots were cut out from the gel slabs, and dissolved in 0.15 ml H_2O_2 at 60° for 24 h.

RESULTS

1°) The composition of ribosomal proteins in polysomes from exponentially growing and starving cells

As shown below, rel^- cells were labeled for 180 min, then half of the culture was collected and the "exponential polysomes" were prepared ; the other half of the culture was chased for leucine and starved for additional 30 min at 42° and the "starving polysomes" prepared.

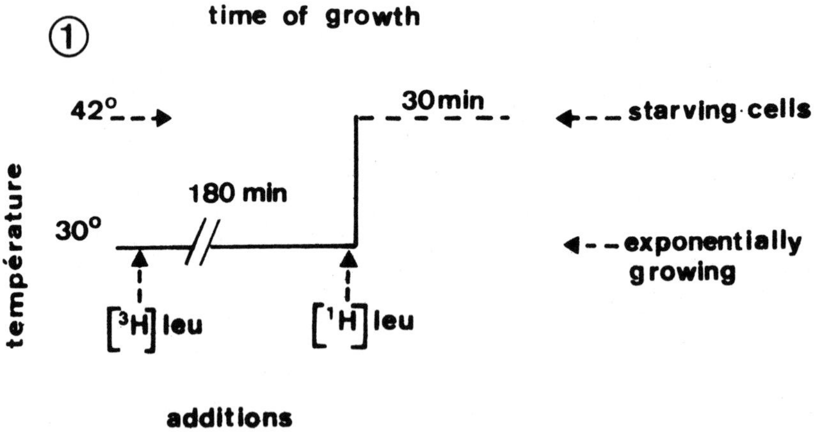

The polysomes profiles are given in Fig. 1a and 1b ; as expected "starving polysomes" are smaller than "exponential polysomes", but the amount of polysomes

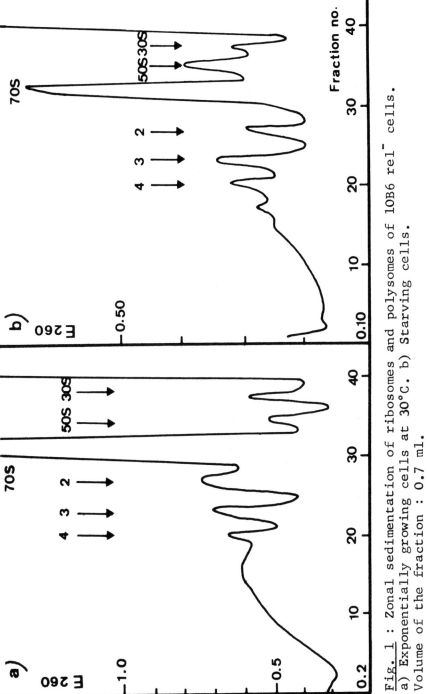

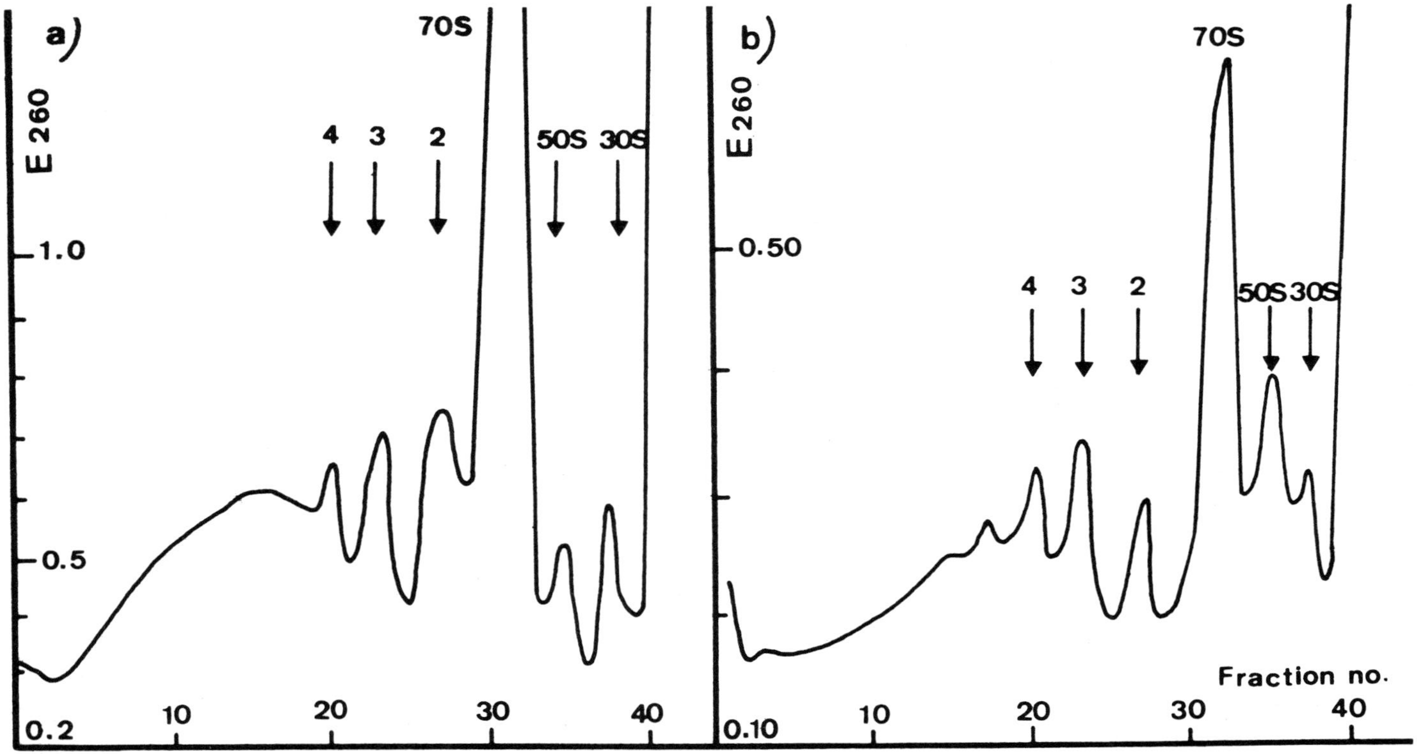

Fig. 1 : Zonal sedimentation of ribosomes and polysomes of 10B6 rel^- cells. a) Exponentially growing cells at 30°C. b) Starving cells. Volume of the fraction : 0.7 ml.

TABLE I

Polysomes in starving rel^- cells						Ribosomes in starving rel^+ cells					
S1	1.07	L1	1.04	L22	0.97	S1	1.15	L1	1.05	L22	1.02
S2	1.11	L2	0.95	L23	1.0	S2	1.10	L2	1.05	L23	0.90
S3	0.92	L3	1.08	L24	1.14	S3	1.10	L3	1.02	L24	0.96
S4	0.96	L4	1.14	L25	1.10	S4	1.10	L4	0.90	L25	0.83
S5	1.03	L5	1.09	L27	1.02	S5	1.00	L5	0.90	L27	0.90
S6	2.06	L6	1.0	L28	1.06	S6	0.90	L6	1.0	L28	0.93
S7	0.85	L7	1.0	L29	1.02	S7	0.97	L7	0.90	L29	1.02
S8	1.09	L8	0.83	L30	1.10	S8	1.03	L8	1.02	L30	1.00
S9	1.04	L9	1.38	L31	-	S9	1.10	L9	1.10	L31	0.93
S10	1.08	L10	1.38	L32	1.05	S10	1.07	L10	1.05	L32	0.96
S11	0.74	L12	1.14	L33	1.02	S11	0.94	L12	0.90	L33	1.10
S12	0.53	L13	1.0	L34	-	S12	1.0	L13	1.10	L34	-
S13	1.12	L14	-			S13	0.97	L14	1.10		
S14	0.97	L15	0.93			S14	0.94	L15	1.20		
S15	0.85	L16	1.04			S15	0.90	L16	1.10		
S16	1.00	L17	1.00			S16	0.94	L17	1.30		
S17	0.74	L18	0.94			S17	0.94	L18	1.00		
S18	1.07	L19	0.98			S18	0.97	L19	1.05		
S19	0.88	L20	1.00			S19	0.97	L20	1.02		
S20	0.90	L21	1.15			S20	0.97	L21	1.05		
S21	0.90					S21	0.94				

to total ribosomes is 55-60 % in both cases.

The ribosomal proteins from the polysome fractions were extracted, analyzed by 2-D gel electrophoresis and counted. The data given in Table I (left) are expressed as the ratio protein in "exponential polysomes" to protein in "starving polysomes". As shown all proteins are equally present except S6 which is highly reduced in starving polysomes and also but to a less extent L9 and L10. At the opposite S11, S12 and S17 seems to be present in higher amount. However the difference for these 3 last proteins may not be caused by starvation (8).
A similar experiment was carried out in rel^+ cells except that total ribosomes instead of polysomes were prepared (7). The data are given in Table I (right). Only L15 and L17 are present in lower amount in ribosomes from starving rel^+ cells.

In summary the protein composition of polysomes from starving rel^- and ribosomes from starving rel^+ cells are rather similar except for S6, the decrease of which was only observed in rel^- cells.

2°) The incorporation of ribosomal proteins in polysomes, 70S ribosomes and small particles in starving rel^- cells.

The relaxed particles were characterized by pulse labeling the cells (5 min), during starvation, with ^3H uridine. Crude ribosomes prepared by centrifugation at 150,000 g for 3 h and ribosomes prepared according to Kurland (7) were analyzed on sucrose gradient (Fig. 2). As seen relaxed particles (25-28S) are still present in purified ribosomes. The experimental conditions are as indicated below :

②

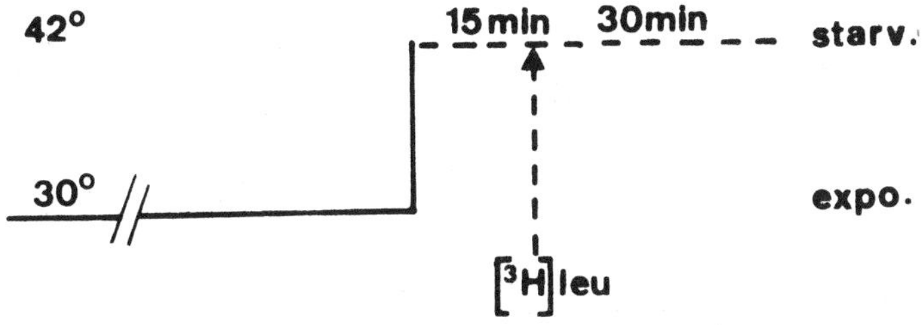

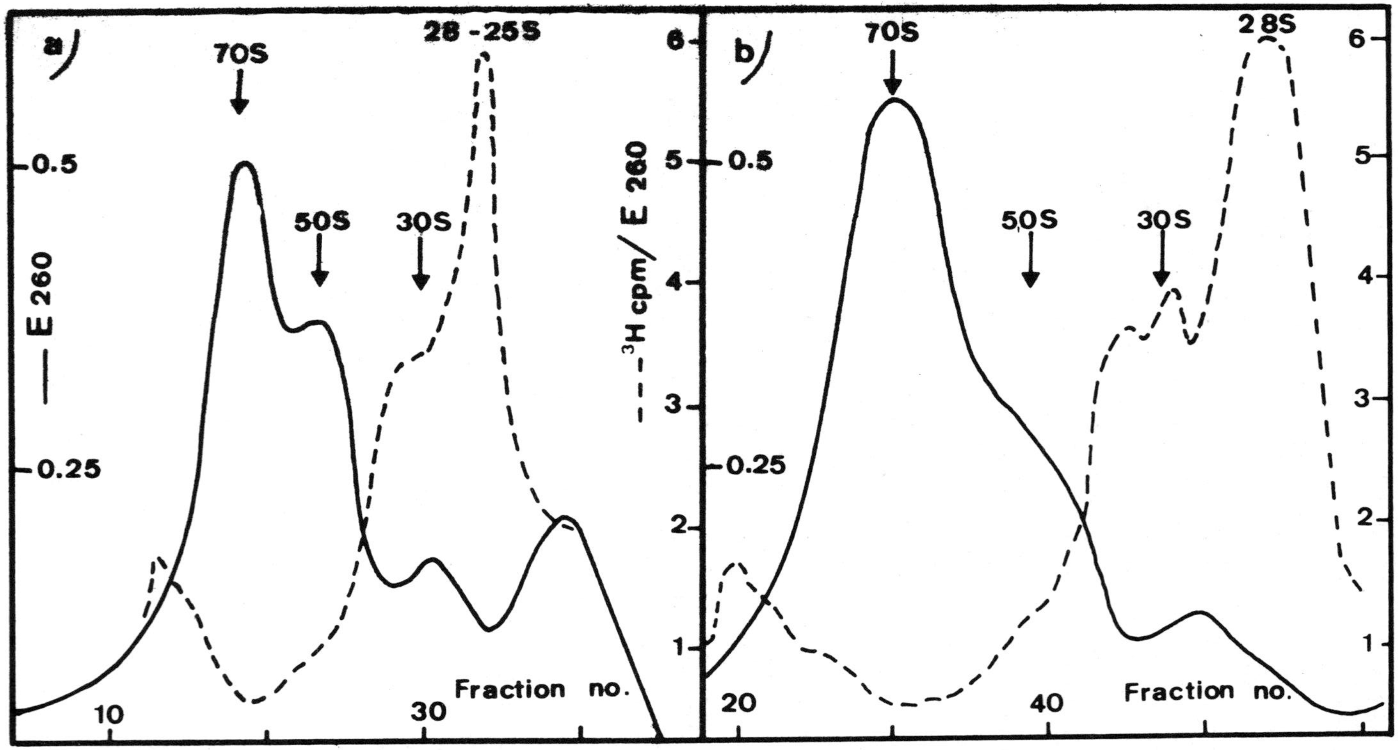

Fig. 2 : Zonal sedimentation of ribosomes, ribosomal subunits and relaxed particles from cells labeled with ^{3}H uridine during starvation. a) Crude ribosomes. b) Ribosomes purified according to Kurland (7). Zonal analysis was carried out in a 10% to 30% sucrose gradient at 26,000 revs/min for 10h30. Volume of the fraction : 0.7 ml.

After 30 min labeling the culture was stopped and 70S ribosomes and the 25-28S particles (Fig. 3) and the polysomes (Fig. 4a) were prepared by appropriate zonal sedimentation. The ribosomal proteins purified and counted. For comparison the same experiment was carried

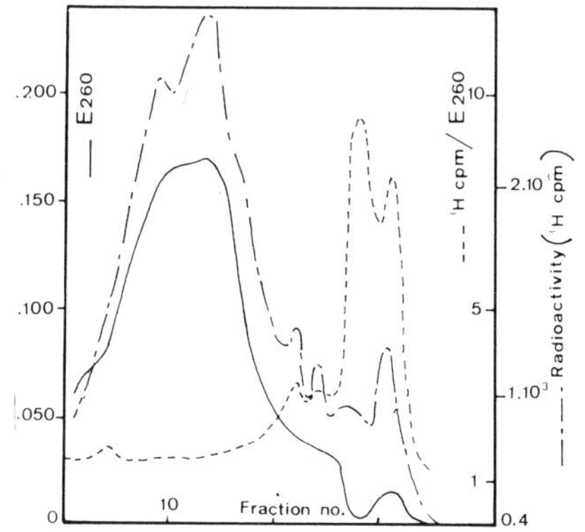

Fig. 3 : Zonal sedimentation of ribosomes, ribosomal subunits and relaxed particles from cells labeled with ^{3}H Leu during starvation. The ribosomes were prepared according to Kurland(7).
The conditions were as in Fig. 2.

out on exponentially growing cells. The data are expressed as the ratio starving to exponential (Table II). As shown a differential labeling is obtained in starving cells, which suggest that some proteins assemble (or exchange) much more rapidly than others. In the particles the highly labeled 30S proteins (S15, S20, S6) at the exception of S4, are added early in ribosome assembly (9, 10), while the poorly labeled proteins are late proteins. Such correlation is less apparent in the case of the "50S" particles. In polysomes and 70S ribosomes some early (S20, S15, S6) and late (S14, S13) proteins are highly labeled. As a control the same experiment was carried out on starving rel^{+} cells. No detectable radioactivity was found in ribosomes (Fig. 4b).

In conclusion the polysomes, the ribosomes and the particles are labeled in starving rel^{-} cells and not in

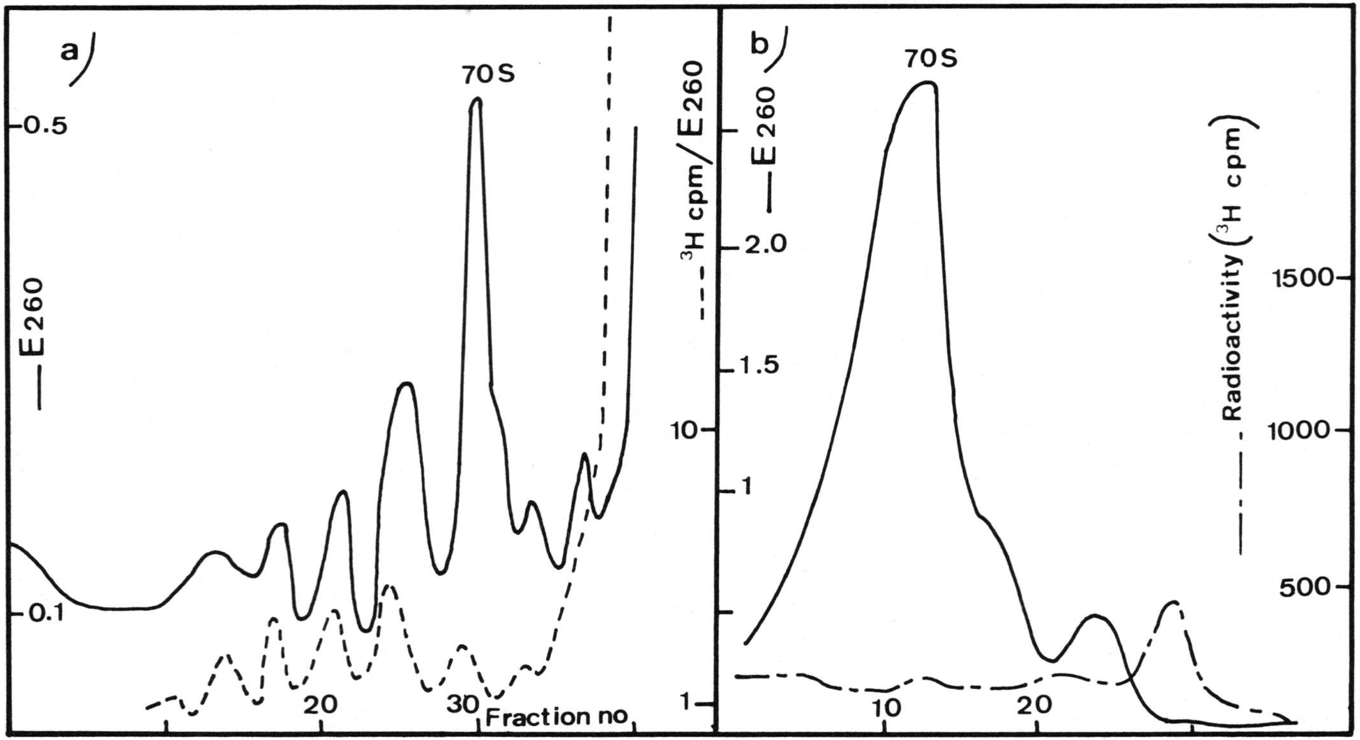

Fig. 4 : a) Zonal sedimentation of polysomes from 10B6 rel^- cells labeled with ^3H Leu during starvation. Conditions as in Fig. 1. b) Zonal sedimentation of ribosomes from 10B6 rel^+ cells labeled with ^3H Leu during starvation. Conditions as in Fig. 2.

TABLE II

POLYSOMES				70 S				PARTICLES			
S20	1	L33	.53	S20	1	L29	.77	S15	1	L34	.71
S15	.41	L32	.44	S15	.50	L34	.70	S20	.90	L29	.50
S14	.27	L29	.40	S6		L8		S6	.40	L32	.25
S13	.24	L25	.25	S14	.44	L7	.65	S18	.30	L12	.20
S18	.23	L10	.20	S10	.40	L33	.60	S16	.25	L17	.080
S6	.22	L27		S13	.29	L10	.50	S17		L25	
S19	.16	L8		S18	.24	L25	.30	S8		L20	.075
S2		L17	.19	S17	.13	L12		S13	.07	L14	
S9	.12	L30		S19	.12	L30	.25	S14		L5	
S16		L7	.11	S16	.10	L9	.20	S10		L30	
S17		L13	.09	S8		L32	.15	S19		L13	.060
S21	.11	L21		S2		L5		S9	.03	L22	.040
S10		L22	.08	S9	.08	L27	.10	S11	.02	L27	.030
S11	.10	L9		S11	.07	L17		S4		L21	
S8		L14	.07	S3	.05	L14		S12	.01	L23	.020
S3	.09	L12		S21	.04	L13	.08	S2		L33	
S5	.04	L5		S1	.03	L1		S3		L6	.015
S12	.03	L20	.06	S5		L20	.07	S5	.01	L1	
S4	.02	L24		S4	.02	L4		S7		L15	.010
S1		L28		S12		L21	.06	S1		L8	
S7	.01	L15	.05	S7	.015	L22	.05	S21		L9	
		L6	.04			L16				Others below	.010
		L16				L19	.04				
		L18				L6					
		L23				Others below	.04				
		Others below	.04								

TABLE III

Ribosomal protein biosynthesis in starving rel^- cells

S6	1	L7	0.80	L15	0.15
S20	0.85	L25	0.77	L1	0.15
S18	0.80	L33	0.74	L18	0.15
S15	0.77	L10	0.60	L19	0.15
S10	0.65	L29	0.52	L16	0.15
S14	0.60	L8	0.48	L20	0.10
S13	0.50	L30	0.47	L2	0.10
S11	0.42	L12	0.46	L3	0.10
S19	0.38	L32	0.42	L4	0.05
S9	0.38	L17	0.38		
S8	0.30	L9	0.30		
S16	0.30	L21	0.30		
S17	0.30	L5	0.30		
S12	0.25	L27	0.30		
S7	0.20	L13	0.27		
S5-S4	0.20	L14	0.27		
S2	0.20	L23	0.23		
S3	0.10	L6	0.23		
S21	0.10	L28	0.23		
S1	0.10	L24	0.23		

rel^+ cells. The proteins in these fractions are differentially labeled. This labeling may be compared with the order of ribosome assembly in exponentially growing cells.

3°) <u>Ribosomal protein synthesis in starving rel^- cells</u>

The experimental conditions are as below. The cells were labeled during starvation and chased long enough during recovery to make sure that all the proteins synthesized during starvation have joined mature ribosomes by the time.

③

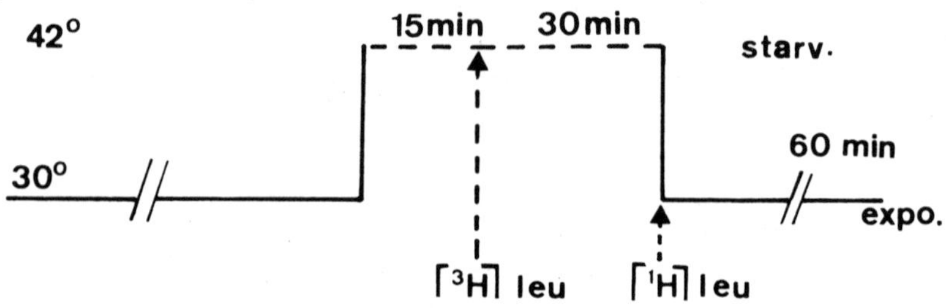

The ribosomes were purified (7) and the proteins extracted and counted. The results are given in Table III. The data are expressed again as the ratio starving to exponential. The proteins are differentially labeled and have been classified from highly labeled protein (S6, S20, S18) to poorly labeled protein (S4, S3, S21, S1).

These results indicate that the rate of synthesis of ribosomal proteins varies widely during starvation. No clear interpretation can be given at the moment. One may speculate than the order obtained would reflect the sequence of transcription for these proteins in response to starvation.

We thank Miss J. Secchi and Mr G. Issa for skilful technical assistance. This investigation was partially supported by the Centre National de la Recherche Scientifique (LA n° 202). Purchase of radioisotopes was partially supported by the "Commissariat à l'Energie Atomique".

REFERENCES

1. Cozzone, A. and Donini, P. (1973) J. Mol. Biol. 76, 149-162.

2. Kaplan, S. and Anderson, D. (1968) J. Bact. 95, 999.

3. Klagsbrun, M. and Rich, A. (1970) J. Mol. Biol. 48, 421.

4. Davis, R.D. (1971) Nature 231, 153.

5. Waller, J.P. and Harris, J.I. (1961) Proc. Nat. Acad. Sci. US 47, 18.

6. Kaltschmidt, E. and Wittmann, H.G. (1969) Anal. Biochem. 36, 401-412.

7. Kurland, C.C. (1966) J. Mol. Biol. 18, 90-108.

8. Kaltschmidt, E. and Wittmann, H.G. (1972) Biochimie 54, 167-175.

9. Marvaldi, J., Pichon, J. and Marchis-Mouren, G. (1972) Biochim. Biophys. Acta 269, 173-177.

10. Pichon, J., Marvaldi, J. and Marchis-Mouren, G. Biochem. Biophys. Res. Commun. 47, 531-538.

Subcellular Localization in the Ehrlich Ascites Cell of the Enzyme which Oxidizes Dihydroorotate to Orotate*

Takao Matsuura[+] and Mary Ellen Jones
Department of Biochemistry, School of Medicine, University of Southern California, Los Angeles, CA 90033

In rat liver the majority of the enzyme activity which can convert dihydroorotate (DHO) to orotic acid (OA) has always been observed in the mitochondria (1, 2, 3). However, smaller amounts of activity appeared to be localized in the crude nuclear pellet and in the microsomal fraction (1, 2). Recently, Kennedy (3) has demonstrated that the liver enzyme is probably localized exclusively in the mitochondria. Since liver can synthesize UMP *de novo* (4), one must assume that this mitochondrial enzyme serves to oxidize DHO to OA for the biosynthesis of UMP. The liver enzyme is membrane bound (2) and the electrons removed from DHO seem to enter the mitochondrial electron-transport system at cytochrome b or coenzyme Q_{10} (5, 6). Therefore, like the bacterial and fungal biosynthetic enzymes (7, 8), the rat liver enzyme that oxidizes DHO does not utilize NAD^+ (2) as an electron acceptor; it will be abbreviated as DHO-dehase in this paper.

Although the liver enzyme appears to exist solely in the mitochondria (3, 5, 6), this same activity in the leucocyte (9) and the Ehrlich ascites cell (10) sediments almost exclusively with the crude nuclear pellet for homogenates prepared in 0.25 M sucrose. Since mammals, like some bacteria (7), might have both a degradative enzyme as well as a biosynthetic enzyme, it seemed important to try to establish if there might be two enzymes serving different functions or whether the leucocyte and ascites cell enzyme merely appeared in the crude nuclear fraction because the mitochondria and a portion of the cytoplasm of these tissues adhere to the nuclei when they are isolated in a simple sucrose medium. From the studies reported with Ehrlich ascites cell homogenates below, it is clear that a single DHO-dehase localized in the mitochondria contains this enzyme. The regulation of de novo UMP biosynthesis is discussed in relation to this observation. It can be predicted that the rate of transfer of DHO formed in the cytosol into the mitochondria must occur efficiently to account for the fact that DHO is not recognized as a metabolite which accumulates in these cells (11).

METHODS

Reagents

New England Nuclear, Boston, MA, supplied the (6-^{14}C)-D, L-dihydroorotate, 3.03 mCi/mmole; Sigma Chemical, St. Louis, MO, provided cytochrome C, (horse heart, Type III), diNa EDTA, ADP and NADH (grade III). Menadione was purchased from ICN-K and K Laboratories, Irvine, CA; triton X-100 from Beckman, Fullerton, CA, and DNA, calf thymus (grade A) from Calbiochem, La Jolla, CA. The protein standard used for the Lowry protein assay (12) was a reconstituted serum obtained from Tavenol Laboratories, Costa Mesa, CA.

Ehrlich Ascites Cell: Growth and Homogenization

Cells were grown in Swiss-Webster male mice and then washed and swollen in water before homogenization as described previously (10). The water-swollen cells were homogenized either in 0.25 M sucrose or in 0.25 M sucrose containing 3 mM $CaCl_2$ (13) to obtain pure nuclei as observed in a phase contrast microscope, or in 0.25 M sucrose containing 1 mM EDTA, pH 7.5 (14) to obtain pure mitochondria. If the volume of the freshly collected cells was 1 ml, the volume of the water-swollen cells was about 4 ml, to which 1 ml of a 1.25 M sucrose solution containing either 15 mM $CaCl_2$ or 5 mM EDTA was added for homogenization with a Potter-type teflon and glass homogenizer.

Method for Obtaining Pure Nuclei

After homogenization in sucrose containing $CaCl_2$, the crude nuclei, which had adhering cytoplasm, were separated from the whole homogenate by centrifugation at 700g for 10 min. The 700g pellet was resuspended in 2.4 M sucrose containing 3 mM $CaCl_2$. The suspension was centrifuged at 80,000g for 60 min. The dense nuclei, which were essentially free of cytoplasm, formed a pellet, while nuclei with adhering cytoplasm (the majority) floated to the top of this dense fluid. Occasionally the 700g supernatant was centrifuged for 15 min at 9000g to collect a mitochondrial pellet and a mitochondrial supernatant. All pellets were suspended in the 0.25 M sucrose with 3 mM $CaCl_2$ medium.

Method for Mitochondria

The method used was that described by Wu and Sauer (14). This method uses 0.25 M sucrose containing 1 mM EDTA and the sedimentation of the mitochondria between 700g and 9000 or 13,000g.

Assays Methods

Dihydroorotate Dehase. For assay of homogenates and their subfactions the method of Shoaf and Jones (10) was used, except that the buffer used here was 100 mM Tris-HCl, pH 7.4. After incubation at 37° for 60 min the reaction was halted by the addition of 5 µl of 70% $HClO_4$ (added to the 0.1 ml reaction mixture) followed by the addition of 10 µl of 4 N KOH and 5 µl of 1 M $KHCO_3$ to partially neutralize the reaction mixture and to remove percholorate ion. DHO and OA were separated by thin-layer chromatography (10). The area scraped for OA (or OMP and UMP formed from OA) was that including R_f values of 0.05 to 0.45, while DHO (or CAA formed from it) was located in the area with R_f values of 0.5 to 1.0. The exact area containing radioactivity was frequently checked by scanning the plates with a Packard radiochromatogram scanner. The above assay relies on the natural electron acceptor(s) of the homogenate. For experiments with mitochondria, external electron acceptors were usually added and the buffer was 100 mM potassium phosphate, pH 7.3. DHO was always 0.5 mM (10). Other additions, when indicated, were: 0.15% triton X-100; 0.38 mM menadione; 6.7×10^{-5}M cytochrome c; or 1 mM KCN. Either the change in absorbance of cytochrome c at 550 nmeters was followed (with L-^{12}C-DHO as substrate) or the ^{14}C-OA formed from D, L-^{14}C-DHO was assessed after separation of these compounds by thin-layer chromatography as described above (10). Glutamic dehydrogenase was assayed as described by Raijman (15); cytochrome oxidase was assayed as described by Smith (16). Protein was determined by a modification of the Lowry procedure (12). DNA was measured by the Burton method (17).

RESULTS

Shoaf and Jones (10) found that when 0.25 M sucrose homogenates of Ehrlich ascites cells were separated into subcellular fractions 86% of the DHO dehase activity sedimented with the crude nuclear pellet. Upon microscopic examination most of these nuclei had a fringe of cytoplasm adhering to them, and, when the nuclei were vitally stained with Janus green, it was apparent that the adhering cytoplasm contained a large number of mitochondria. Results similar to those of Shoaf and Jones can be seen in column 2 of Table 1 of this paper, where 94% of the DHO dehase activity and 60% of the protein are precipitated with the crude nuclear pellet. Essentially none of these nuclei are free of cytoplasm so that the majority of them float when the nuclear pellet is resuspended in 2.4 M sucrose and centrifuged at 80,000g for 60 min. When 3 mM $CaCl_2$ is added to the sucrose (column 3 of Table 1), the amount of DHO-dehase in the 700g pellet is more variable but is generally high, i.e., about 70%. However, when these nuclei are rehomogenized in 2.4 M sucrose and centrifuged at 80,000g for 60 min, the precipitated nuclei have no adhering cytoplasm, the DNA/protein ratio is typical of that reported for pure nuclei from ascites hepatoma AH 130 (13) and the specific activity of the DHO-dehase ranges between 0 and 0.39 nmoles/min/mg protein. The latter value is very low in comparison to the same value for the whole homogenate, a result that suggests it is due to a few mitochondria remaining with the pelleted nuclei. In addition, the 700g supernatant now usually contains 30% of the DHO-dehase activity and the specific activity of this fraction is generally greater than 2 nmoles of OA formed/min/mg protein.

Table 1

Localization of DHO-Dehase Activity in Subcellular Fraction of Ehrlich Ascites Cell Homogenates Prepared in Various Sucrose Media.

Homogenate Fractions	0.25 M Sucrose Homogenate Plus				
	No Additions	3 mM $CaCl_2$		1 mM EDTA	
Whole homogenate (WH):					
protein, mg/ml	17.1	6.6	3.0	11.5	14.1
DHO-dehase activity:					
nmoles OA/min/ml	12.8	7.1	5.6[b]	9.0	11.6
nmoles/min/mg protein	0.75	1.1	1.9[b]	0.8	0.8
DNA: mg/ml	-	0.9	0.65	-	
mg/mg protein	-	0.14,	0.22	-	
700g pellet (crude nuclei)					
protein, % of WH value	60	54	68	37	26
DHO-dehase: % of WH value	94	79	21	33	19
nmoles/min/mg protein	1.2	1.5	0.5[b]	0.7	0.6
DNA: % of WH value	-	93	96	-	
mg/mg protein	-	0.2	0.3	-	
a) 2.4 M sucrose pellet[a]					
protein, % of WH value	1.5	10	5	-	
DHO-dehase: % of WH value	2.0	3.6	0	-	
nmoles/min/mg protein	0.96	0.4	0[b]	-	
DNA: % of WH value	-	26	9	-	
mg/mg protein	-	0.35,	0.40	-	
b) 2.4 M sucrose supernatant					
protein, % of WH value	54	27	53	-	
DHO-dehase: % of WH value	94	47	17	-	
nmoles/min/mg protein	1.3	1.8	0.6[b]	-	
DNA: % of WH value	-	56	80	-	
mg/mg protein	-	0.28,	0.34	-	
700g supernatant					
protein, % of WH value	42	44	19	-	
DHO-dehase: % of WH value	14	39	37	-	
nmoles/min/mg protein	0.25	0.95,	3.7[b]	-	
DNA: mg/ml	-	0	0	-	
9,000 (or 13,000) g precipitate					
protein, % of WH value	-	-	4	6	13[c]
DHO-dehase: % of WH value	-	-	4	38	70
nmoles/min/mg protein	-	-	2.4[b]	4.6	4.5
DNA: mg/ml	-	-	0	-	

Table 1 (continued)

9,000 (or 13,000) g supernatant					
protein, % of WH value	-	-	9.8	52	53[c]
DHO-dehase: % of WH value	-	-	0[b]	21	1
nmoles/min/mg protein	-	-		0.33,	0.2

[a] All nuclei were inspected microscopically with and without Janus green staining.

[b] The assay contained cytochrome c and menadione.

[c] This sample as centrifuged at 13,000 X g.

If the 700g supernatant is centrifuged again at 9000g all of the activity of the 700g supernatant is sedimented, as would be expected if the DHO-dehase is a mitochondrial enzyme.

Tables 1 and 2 show that when 1 mM EDTA is added to the sucrose as described by Wu and Sauer (14), and if a 700 to 13,000g fraction is collected, then 40 to 70% of the DHO-dehase precipitates in the subcellular fraction which contains the majority of two mitochondrial enzymes (see Table 2), namely, glutamate dehydrogenase (GDH) and cytochrome-oxidase. The latter enzyme and DHO-dehase fractionated nearly in parallel, while some GDH was lost from the 13,000g pellet and carried into the 13,000g supernatant. The latter result suggests, but is not definitive proof, that DHO-dehase is not localized in the mitochondrial matrix with GDH; rather it may be membrane bound as is cytochrome oxidase. Although a considerable amount of all three enzymes is precipitated with the crude nuclear pellet, i.e., the 700g precipitate, even when the sucrose contained 1 mM EDTA, the lack of enrichment of all three enzymes in this crude nuclear pellet suggests that these enzymes are not derived from the nuclei but rather are due to a cytoplasm, containing mitochondria, that still adheres to the nuclei. The amount of DHO-dehase activity recovered in the subcellular fractions was frequently greater than the activity in the whole homogenate. This may occur because some of the substrate, DHO, can be converted to carbamyl aspartate in the whole homogenate, the crude nuclear pellet, or the soluble cell sap (10) by the action of

Table 2

Subcellular Localization of DHO-Dehase, Glutamic Dehydrogenase (GDH), and Cytochrome Oxidase of an Ehrlich Ascites Cell Homogenate Prepared in 0.25 M Sucrose Containing 1 mM EDTA[a]

Marker Enzymes	Whole Homogenate	700g Precipitate	13,000g Precipitate	13,000g Supernatant
A. GDH				
(1) protein, mg/ml	10.8	3.9	1.35	5.05
(2) DHO-dehase, nmoles/min/ml	13.3	4.6	8.6	0.8
(3) DHO-dehase, S.A.[b]	1.2	1.2	6.6	0.16
(4) GDH, S.A.[c]	11.8	11.6	38	5.3
(5) DHO-dehase/GDH[d]	[1]	0.95	1.67	0.38
B. Cytochrome oxidase				
(1) protein, mg/ml	8.95	4.4	0.7	3.35
(2) DHO-dehase nmoles/min/ml	11.1	8.0	5.3	0.35
(3) DHO-dehase, S.A.[b]	1.2	1.8	7.6	0.1
(4) cytochrome oxidase, S.A.[e]	62.5	81.5	335	6.5
(5) DHO-dehase/cytochrome oxidase[d]	[1]	1.1	1.2	0.8

[a] All values are the average of two experiments.

[b] Assayed in the presence of cytochrome c and menadione; S.A. stands for specific activity expressed for all three enzymes in nmoles product formed or substrate utilized/min/mg protein.

[c] GDH control value was 130 nmoles NAD^+ formed/min/ml. (See [b] for definition of S.A.)

[d] A (3) divided by A (4) or B (3) divided by B (4) is defined as [1] for the whole homogenate.

[e] Cytochrome oxidase control value was 560 nmoles of cytochrome c reduced/min'ml. (See [b] for definition of S.A.)

dihydroorotase. This conversion would lower the true substrate concentration and would not be observed if one raised the DHO concentration. The lack of stimulation of the DHO-dehase by NAD^+ (5, 6, 9, 10) as well as the stimulation of this activity by menadione (Table 3) or other quinones (2), plus the fact that dichlorophenol indolephenol (1, 2, 5, and unpublished results by Matsuura and Jones) as well as

Table 3

Effect of Menadione, Cytochrome C, Cyanide and Triton X-100 on the DHO-Dehase Activity of Mitochondria Prepared in 0.25 M Sucrose with 1 mM EDTA.

Experiment and Added Reagent		DHO Dehase
1. Control (4.6 nmoles/min/mg protein)	=	100 %
+ KCN		48 %
+ cytochrome c		167 %
+ menadione		131 %
+ cytochrome c and KCN		162 %
+ cytochrome c and menadione		186 %
+ cytochrome c, menadione and KCN		184 %
2. Control (3.8 nmoles/min/mg protein)	=	100 %
+ triton		2.6%
+ triton and cytochrome c		2.3%
+ triton and menadione		97 %
+ triton, menadione and cytochrome c		105 %
+ triton, menadione, cytochrome c and KCN		105 %

Enzyme activity was measured by the conversion of D, L-^{14}C-DHO to ^{14}C-OA under conditions described in the methods section in the presence of phosphate buffer.

external cytochrome c (Table 3) can serve as electron acceptors strongly suggests that DHO-dehase is linked with the enzymes of electron transport of the inner membrane of the mitochondria. We now frequently measure this enzyme activity either by the formation of ^{14}C-orotate from ^{14}C-DHO or spectrophotometrically by the reduction of cytochrome c in the presence of triton X-100 and menadione. The reaction, under the conditions of Table 3, is linear up to 90 minutes and with as much as 40 μg of mitochondrial protein (for a 60 minute incubation period); however, with 80 μg of mitochondrial protein and 0.5 mM DHO the rate is 70% of the expected linear value. Since the mitochondrial pellet is enriched in this enzyme activity over the whole homogenate 5- to 10-fold, one should, with sufficient DHO, be able to use 200-400 μg of homogenate protein. The assay is thus markedly improved over the former assay (10) for this tissue where no external electron acceptor was added and no more than 70 μg of homogenate protein could be used if one wished to obtain a linear rate.

We unfortunately have not assayed the conversion of D, L-^{14}C-DHO to ^{14}C-OA and the rate of cytochrome c reduction in a single sample. We have measured these two rates in separate samples and the values with an unfractionated homogenate were 0.8 nmoles OA formed/min/mg protein vs. 2.08 nmoles cytochrome c reduced/min/mg protein. We believe the cytochrome c value is higher than the expected value of 1.6 because ^{12}C-L-DHO was the substrate for cytochrome c reduction rather than the ^{14}C-D, L-DHO. The suppliers of the radioactive D, L-DHO did not state what proportion is in the D and L form.

DISCUSSION

Previous studies on the Ehrlich ascites cell had shown that over 90% of the enzyme which oxidizes dihydroorotate to orotate sedimented in the crude nuclear fraction (10). The results given here indicate strongly that this enzyme is not located in pure nuclei but is markedly enriched when one isolates pure mitochondria. In addition, the reason the enzyme appears nearly exclusively in the 700g (crude nuclear) pellet in 0.25 M sucrose homogenates is that nearly all of the mitochondria of the Ehrlich ascites cells are trapped in what seems to be a "sticky" cytoplasm. This cytoplasm is partially freed when one adds 1 mM EDTA to the sucrose. We have hoped to see whether the same situation pertains in leucocytes, where this enzyme also sediments in the crude nuclear fraction (9). Both the leucocyte and the ascites cells have a large nucleus and the mitochondria may exist in both tissues in this special cytoplasm.

This "sticky" cytoplasm also contains 30% of the five other enzymes required for de novo UMP biosynthesis, while 70% of these five enzymes are always present in the "soluble" cytoplasm of the cell (10). The six enzymes required for UMP synthesis are illustrated in Fig. 1, which also shows the points in the pathway where regulation occurs. The first three enzymes, carbamyl phosphate synthetase II (CPSase II), aspartate transcarbamylase (ATCase), and dihydroorotase (DHOase) form an enzyme complex (complex A) whose product is DHO. The rate

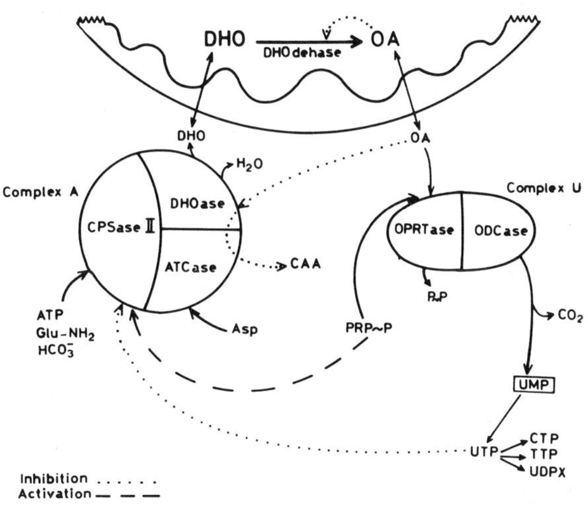

Figure 1

Illustration of Substrates, Products and Product Regluation of the Cytoplasmic Enzyme Complexes (A and U) Containing Five Enzymes Essential for de novo UMP Biosynthesis and the Mitochondrial Location of Dihydroorotase Dehydrogenase (DHO Dehase).

at which DHO is produced depends not only on the concentration of the substrates, ATP, glutamine, HCO_3^- but on the levels of UTP, an allosteric inhibitor (18), and phosphoribosyl pyrophosphate (PRP~P), an allosteric activator (19) of the CPSase II activity of complex A. In addition, orotate (OA) inhibits DHOase of complex A so that when OA levels are high carbamyl aspartate (CAA) is the sole product of complex A in intact cells (11). When extracts containing complex A are incubated *in vitro* (10) a mixture of CAA and DHO is formed, while *in vitro* incubation of intact cells with glucose, glutamine and bicarbonate (in the absence of added OA) neither CAA, DHO or OA accumulate. In the intact cell, therefore, DHO is apparently removed very efficiently into the mitochondria when it diffuses from complex A so that no equilibrium is established between CAA and DHO by dihydroorotase.

The DHO-dehase of the mitochondria has not been observed to catalyze a reversible reaction (10). This enzyme is inhibited by its product, orotate (3). The last two enzymes of the pathway are present as an extramitochondrial complex, called complex U in Fig. 1. The enzymes of complex U, orotidylate phosphoribosyl transferase (OPRTase) and orotidylate decarboxylase (ODCase), utilize OA which diffuses from the mitochondria and PRP~P to form UMP and CO_2. The first enzyme of this complex may well be the rate-limiting enzyme of the pathway (10), and OA is observed to accumulate in a number of conditions in the urine and blood (see reference 20).

SUMMARY

The enzyme that oxidizes dihydroorotate to orotate is apparently only located in the mitochondria of the mouse Ehrlich ascites cells, as had been observed for rat liver. The significance of this result is discussed.

ACKNOWLEDGMENT

It is indeed a pleasure to submit this paper to honor Dr. Fritz Lipmann, the former teacher, constant advisor and gentle critic of one of us (M. E. J.). Dr. Lipmann is also scientific great-grandfather of Dr. Takao Matsuura.

FOOTNOTES

*Supported by a grant from the National Science Foundation (GB31537).

+Permanent address: Department of Biochemistry, Kyoto Prefectural University of Medicine, Nishijin, Kyoto 602, Japan.

REFERENCES

1. Wu, R., Wilson, D.W.: Studies on the biosynthesis of orotic acid. J. Biol. Chem. 223, 195-205 (1956).
2. Miller, R.W., Kerr, C.T., Curry, J.R.: Mammalian dihydro-orotate-ubiquinone reductase complex. Canad. J. Biochem. 46, 1099-1106 (1968).
3. Kennedy, J.: Distribution, subcellular localization, and product inhibition of dihydroorotate oxidation in the rat. Arch. Biochem. Biophys. 157, 369-373 (1973).
4. Reichard, P.: The enzymatic synthesis of pyrimidines. Adv. in Enzymology 21, 263-294 (1959).
5. Miller, R.W., Curry, J.R.: Mammalian dihydroorotate-ubiquinone reductase complex. II. Correlation with cytochrome oxidase, mode of linkage with cytochrome chain, and general properties. Canad. J. Biochem. 47, 725-734 (1969).
6. Kuligowska, E., Erecinska, M.: The role of ubiquinone in dihydroorotate oxidation. Bull. L'Acad. Pol. Sci., Ser. Sci. Biol. XV, 187-190 (1967).
7. O'Donovan, G.A., Neuhard, J.: Pyrimidine metabolism in microorganisms. Bacteriol. Reviews 34, 278-373 (1970).
8. Karibian, D.: Dihydro-orotate dehydrogenase of Escherichia coli K12: effects of triton X-100 and phospholipids. Biochim. Biophys. Acta 302, 205-215 (1973).
9. Smith, Jr., L.H., Baker, F.A.: Pyrimidine metabolism in man. I. The biosynthesis of orotic acid. J. Clin. Investig. 38, 798-809 (1959).
10. Shoaf, W.T., Jones, M.E.: Uridylic acid synthesis in Ehrlich ascites carcinoma. Properties, subcellular distribution, and nature of enzyme complexes of the six biosynthetic enzymes. Biochemistry 12, 4039-4051 (1973).
11. Hager, S.E., Jones, M.E.: Initial steps in pyrimidine synthesis in Ehrlich ascites carcinoma in vitro. I. Factors affecting the incorporation of ^{14}C-bicarbonate into carbon 2 of the uracil ring of the acid-soluble nucleotides of intact cells. J. Biol. Chem.

240, 4556-4563 (1965).

12. Oyama, V.I., Eagle, H.: Measurement of cell growth in tissue culture with a phenol reagent (Folin-Ciocalteau). Proc. Soc. Exp. Biol. Med. 91, 305-307 (1956).

13. Fukuda, T., Akino, T., Amano, M., Izawa, M.: RNA synthesis in ascites hepatoma AH-130 cells of rats. Cancer Res. 30, 1-10 (1970).

14. Wu. R., Sauer, L.A.: Preparation and assay of phosphorylating mitochondria from ascites tumor cells. Methods of Enzymology X (Estabrook, R., Pullman, M.E., eds.) pp 105-110, Academic Press, New York (1967).

15. Raijman, L.: Citrulline synthesis in rat tissues and liver content of carbamyl phosphate and ornithine. Biochem. J. 138, 225-232 (1974).

16. Smith, L.: Cytochromes a, a_1, a_2 and a_3. Methods of Enzymology II (Colowick, S.P., Kaplan, N.O., eds.) pp. 732-740, Academic Press, New York (1955).

17. Burton, K.: Determination of DNA concentration with diphenylamine. Methods of Enzymology XII (Grossman, L., Moldave, K., eds.) pp 163-166, Academic Press, New York (1968).

18. Tatibana, M., Ito, K.: Control of pyrimidine biosynthesis in mammalian tissues. Partial purification and characterization of glutamine-utilizing carbamyl phosphate synthetase of mouse spleen and its tissue distribution. J. Biol. Chem. 244, 5403-5413 (1969).

19. Tatibana, M., Shigesada, K.: Activation by 5-phosphoribosyl-1-pyrophosphate of glutamine dependent carbamyl phosphate synthetase from mouse spleen. Biochem. Biophys. Res. Commun. 46, 491-497 (1972).

20. Jones, M.E.: Regulation of uridylic acid biosynthesis in eucaryotic cells. Curr. Topics Cellular Reg. 6 (Horecker, B.L., Stadtman, E.R., eds.) pp 227-265, Academic Press, New York (1972).

Comparison of Polypeptide Chain Initiation Factors from Artemia salina and Rabbit Reticulocytes

C. Nombela, N. A. Nombela and S. Ochoa
Department of Biochemistry
New York University School of Medicine, New York, N. Y. 10016 USA

The first systematic study of eukaryotic initiation factors was carried out by Anderson and collaborators (1, 2). They isolated from 0.5 M KCl ribosomal washes of rabbit reticulocytes three protein factors designated as M1, M2 and M3. M1 and M2 stimulated poly (U) translation at low Mg^{2+} concentrations; M3 was required for translation of natural messengers (e.g., globin mRNA) and, in this regard, appeared to be the counterpart of the prokaryotic initiation factor IF-3 (3). Some of the properties of partially purified preparations of M1 were reminiscent of those of the prokaryotic initiation factor IF-2. Ribosomal binding of the eukaryotic initiator Met-$tRNA_f$ (4, 5), from reticulocytes or other sources, required both GTP and M2 besides M1 (6). These same preparations of M1 promoted the AUG-dependent binding of fMet-$tRNA_f$ (6), or the poly (U)-dependent binding of acPhe-tRNA (7) to reticulocyte ribosomes but in this case, in contrast to IF-2, there was no GTP requirement. There was no requirement for M2 either. Moreover, the requirements for fMet- or Met-puromycin synthesis appeared to be more complex than in prokaryotic systems (8, 9).

In this laboratory (10-12) we isolated from the cytosol of Artemia salina embryos a protein factor that catalyzed the AUG-dependent binding of E. coli fMet-$tRNA_f$ and Met-$tRNA_f$, or the poly (U)-dependent binding of acPhe-tRNA and Phe-tRNA, to 40S ribosomal subunits of eukaryotic origin. The bound aminoacyl-tRNA's reacted with puromycin to form the corresponding aminoacyl-puromycins upon addition of 60S subunits, with no requirement for GTP. This factor has been referred to as EIF-1 (eukaryotic initiation factor 1). The isolation of a similar factor from mammalian sources had in fact been reported from other laboratories (13-15). Because of a generous gift of M1 and M2 from Dr. W. F. Anderson we were able to study the effect of partially purified preparations of these factors on the AUG-directed binding of fMet-$tRNA_f$ and Met-$tRNA_f$ to A. salina 40S subunits and aminoacyl-puromycin synthesis[1]. In contrast to A. salina EIF-1 the M1 preparations were inactive with Met-$tRNA_f$ whether with or without GTP and/or M2.

Since it was possible that some contaminant of M1 interfered with the Met-$tRNA_f$, but not the fMet-$tRNA_f$ reaction, the above studies were repeated with homogeneous preparations of both A. salina EIF-1 (16) and rabbit reticulocyte IF-M1 (17). In this paper we show that, like A. salina EIF-1, pure reticulocyte IF-M1 catalyzes the binding of both fMet-$tRNA_f$ and Met-tRNA$_f$ to A. salina 40S ribosomal subunits and, following the addition of 60S subunits, the synthesis of the corresponding aminoacyl puromycins occurs with no requirement for GTP. Moreover, reticulocyte IF-M1 can be inactivated by antibody against A. salina EIF-1. Both proteins thus appear to be closely related in function and structure.

Preparations and Methods. Essentially homogeneous A. salina EIF-1 was prepared as previously described (16) except for the substitution of chromatography on QAE-Sephadex for the hydroxylapatite step. A. salina EIF-1

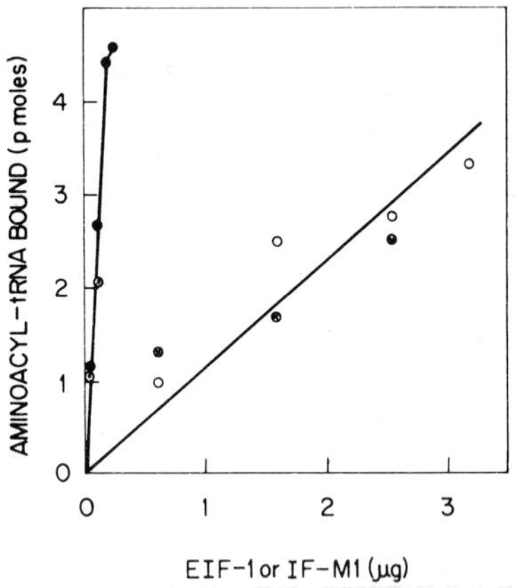

FIG. 1. Binding of f[^{14}C]Met-tRNA$_f$ and [^{14}C]Met-tRNA$_f$ to A. salina 40S ribosomal subunits with EIF-1 from A. salina supernatant or IF-M1 from rabbit reticulocyte ribosomal wash. Samples contained, in a final volume of 60 μl, Tris-HCl buffer, pH 7.4, 80 mM; KCl, 150 mM; $Mg(OAc)_2$, 4.5 mM; dithiothreitol (DTT), 2 mM; 40S subunits, 0.34 A_{260} unit; AUG, 0.05 A_{260} unit; either f[^{14}C]Met-tRNA or [^{14}C]Met-tRNA (545 cpm/pmole; methionine charging ratio of tRNA, 54.5 pmoles/A_{260} unit), each 24 pmoles; and varying amounts of EIF-1 or IF-M1. After incubation for 20 min at 0° the ribosome-bound radioactivity was determined by the Millipore filtration procedure (21). A blank of 0.59 pmole with f ^{14}C Met-tRNA and 0.45 pmole with [^{14}C]Met-tRNA, in the absence of added factor, was subtracted throughout. (● ●), A. salina EIF-1, f[^{14}C]Met-tRNA; (⊙ ⊙), A. salina EIF-1, [^{14}C]Met-tRNA; (o o), reticulocyte IF-M1, f[^{14}C]Met-tRNA; (⊗ ⊗) reticulocyte IF-M1, [^{14}C]Met-tRNA.

antibody was obtained by injection of 25 μg of the protein into the foot pads of each of two rabbits at three week intervals until a high titer was reached (about 9 weeks) as judged by Ouchterlony plate assay (18). Immunoglobulin was prepared from the immune serum by standard procedures (19). We are indebted to Dr. I. Schenkein, of this Medical Center, for help with the antibody preparation and for a gift of normal rabbit immunoglobulin. Virtually homogeneous rabbit reticulocyte IF-M1 was the generous gift of Drs. W. F. Anderson and W. C. Merrick of the National Institutes of Health. A. salina 40S and 60S ribosomal subunits were prepared as in previous work (10). ApUpG (AUG) was from the Miles Laboratories, GTP from Sigma, puromycin from Cal Biochem, and E. coli B tRNA, as well as [^{14}C] methionine, from Schwarz/Mann. Charging of the tRNA with [^{14}C] methionine and formylation of [^{14}C] Met-tRNA were done as described (10). Protein was determined by the Lowry procedure (20) with crystalline serum albumin as standard. Other methods are described in the figure and table legends.

Results. As shown in Fig. 1 both A. salina EIF-1 and reticulocyte IF-M1 catalyzed the AUG-dependent binding of either fMet-$tRNA_f$ or Met-$tRNA_f$ to A. salina 40S subunits although their specific activities were very different. Table 1 shows, in addition, that the binding reaction catalyzed

Table 1. Formation of initiation complex with A. salina ribosomes. Comparison of EIF-1 from A. salina supernatant and IF-M1 from rabbit reticulocyte ribosomal wash

Factor	Aminoacyl-tRNA	GTP (mM)	Aminoacyl-tRNA bound (pmol)	Aminoacyl-puromycin synthesis (pmol)	(% of aa-tRNA bound)
A. salina	f [^{14}C] Met	0	5.70	5.20	91
"	"	0.3	5.42	4.45	82
Reticulocyte	"	0	4.00	4.20	100
"	"	0.3	4.00	3.95	99
A. salina	[^{14}C] Met	0	2.82	1.65	59
"	"	0.3	2.01	1.09	54
Reticulocyte	"	0	2.58	1.88	73
"	"	0.3	2.50	1.36	54

Duplicate samples containing either A. salina EIF-1 (0.17μg) or reticulocyte IF-M1 (2.25μg) and the remaining components of the standard assay for AUG-directed aminoacyl-tRNA binding to A. salina 40S ribosomal subunits (see legend to Fig. 1) were incubated for 30 min at 0°. This was followed by the addition of A. salina 60S subunits (0.75 A_{260} unit) to all samples, puromycin (75μg) to each duplicate sample, and an equal volume of water to the other. After a further 15 min incubation at 0°, ribosomal binding of aminoacyl-tRNA and aminoacyl-puromycin synthesis were determined as described (10).

by either factor and the synthesis of fMet- or Met-puromycin proceeded to a similar extent in the presence or absence of added GTP. The efficiency of conversion to aminoacyl-puromycin was in both cases greater with fMet-tRNA; this had already been noted for the A. salina factor (12).

Fig. 2 shows the inactivation of A. salina EIF-1 by its specific antibody and the cross-reactivity of this antibody with rabbit reticulocyte IF-M1[2]. The amount of antibody needed to inactivate the latter was 4-5 times greater than that required to inactivate the former.

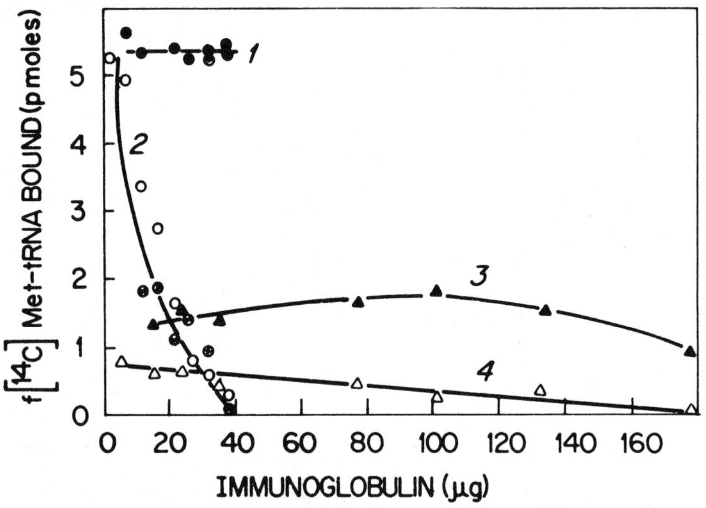

FIG. 2. Reactivity of anti-EIF-1 (A. salina) with EIF-1 from A. salina supernatant and cross-reactivity with IF-M1 from rabbit reticulocyte ribosomal wash. A. salina EIF-1 (0.25 µg) or reticulocyte IF-M1 (1.7 µg) were each incubated with increasing amounts of either control or anti-EIF-1 immunoglobulin, for 30 min at 24° followed by 30 min at 0°, in a reaction mixture containing 100 mM Tris-HCl buffer, pH 7.4, 188 mM KCl, 5.6 mM $Mg(OAc)_2$, and 2.5 mM DTT. The samples were then supplemented with A. salina 40S ribosomal subunits, AUG, and f[^{14}C]Met-tRNA, to give for all components the same final concentrations as in the standard aminoacyl-tRNA binding assay (see legend to Fig. 1). The samples were incubated for 20 min at 0° and assayed for ribosome-bound radioactivity in the usual way. Curve 1, A. salina EIF-1, control immunoglobulin (the symbols ● and ⊙ correspond to separate experiments); curve 2, A. salina EIF-1, anti-EIF-1 immunoglobulin (the symbols o and ⊗ correspond to separate experiments). Curves 3 (▲) and 4 (△), as 1 and 2, respectively, but with reticulocyte IF-M1.

Discussion. The above results show that the supernatant factor EIF-1 from A. salina embryos and the ribosomal wash factor IF-M1 from rabbit reticulocytes are closely related in function and structure. This is so regardless of the difference in specific activities between the two preparations. The

A. salina preparation was roughly 10-15 times more active on a weight basis than the one from reticulocytes. This difference could be due to one or more causes, e.g., basic difference in the catalytic activity of the two proteins, lesser efficiency of the reticulocyte factor with heterologous A. salina ribosomes and/or E. coli Met-tRNA, or lesser stability of the reticulocyte factor. It may be noted that whereas A. salina EIF-1 appears to consist (10) of two equal subunits (each about 74,000 daltons), reticulocyte IF-M1 consists (22) of a single polypeptide chain (about 65000 daltons).

Our results suggest that a given initiation factor may be partly ribosome-bound and partly free in the cytosol depending on its affinity for the ribosomes under the conditions employed for its isolation. EIF-1 has been found in both the cytosol and the 0.5 M KCl ribosomal wash (in a ratio about 15-20:1) of A. salina embryos and wheat germ[3] (23). Similar factors have been isolated from rat liver supernatant (15, 24), rabbit liver ribosomal wash (25), rabbit reticulocyte (26) and mouse L cell (11) supernatant, ascites cell supernatant (27) and ribosomal wash (28). The rat liver supernatant factor has been highly purified (29) and proved to be functionally identical to A. salina EIF-1 and reticulocyte IF-M1.

Our results emphasize that it is often necessary to achieve considerable purification of a protein before its function can be properly understood. It seems clear that the partially purified reticulocyte IF-M1 used in earlier work (6-8) contained contaminants that interfered with its function. The interference may have been caused by a hydrolase in reticulocyte ribosomal washes that releases methionine from 40S-bound Met-$tRNA_f$ but does not attack bound N-acetyl-Met-$tRNA_f$ (30) and therefore is probably inactive on 40S-bound fMet-$tRNA_f$.

Summary. Homogeneous preparations of the initiation factors EIF-1 from the cytosol of Artemia salina embryos and IF-M1 from rabbit reticulocyte ribosomal washes catalyze the AUG-dependent binding of E. coli f[^{14}C]Met-$tRNA_f$ or [^{14}C]Met-$tRNA_f$ to A. salina 40S ribosomal subunits with no requirement for GTP. The corresponding puromycin derivatives f[^{14}C]Met- and [^{14}C]Met-puromycin are formed in good yields upon the subsequent addition of 60S subunits. Antibody against A. salina EIF-1 shows significant cross-reactivity with rabbit reticulocyte IF-M1. Thus, the two proteins are closely related functionally and structurally.

We are indebted to Dr. W. F. Anderson and Dr. W. C. Merrick, National Institutes of Health, Bethesda, Maryland, for a gift of homogeneous rabbit reticulocyte IF-M1. We also thank Dr. W. G. Robinson for reading the manuscript and Mr. Horace Lozina for excellent technical assistance. The work was aided by grants from the National Institutes of Health (AM 01845), U. S. Public Health Service, and the American Cancer Society (NP-58B).

Footnotes

[1]R. P. McCroskey and S. Ochoa (1972), unpublished experiments.

[2]When these experiments were done our supply of IF-M1 was running low and the factor had lost some activity through repeated thawing and freezing.

[3]B. V. Treadwell, L. Goldstein and W. G. Robinson, personal communication.

REFERENCES

1. Shafritz, D.A., Prichard, P.M., Gilbert, J.M., Anderson, W.F.: Separation of two factors, M1 and M2, required for poly U dependent polypeptide synthesis by rabbit reticulocyte ribosomes at low magnesium concentration. Biochem.Biophys.Res.Commun. 38, 721-727 (1970).
2. Prichard, P.M., Gilbert, J.M., Shafritz, D.A., Anderson, W.F.: Factors for the initiation of hemoglobin synthesis by rabbit reticulocyte ribosomes. Nature 226, 511-514 (1970).
3. Ochoa, S., Mazumder, R.: Polypeptide chain initiation. In The Enzymes (ed. P.D.Boyer) vol. X, pp. 1-51 (Academic Press. New York and London, 1974).
4. Smith, A.E., Marcker, K.A.: Cytoplasmic methionine transfer RNAs from eukaryotes. Nature 226, 607-610 (1970).
5. Brown, J.C., Smith, A.E.: Initiator codons in eukaryotes. Nature 226 610-612 (1970).
6. Shafritz, D.A., Anderson, W.F.: Factor dependent binding of methionyl-tRNAs to reticulocyte ribosomes. Nature 227, 918-920 (1970).
7. Shafritz, D.A., Anderson, W.F.: Isolation and partial characterization of reticulocyte factors M_1 and M_2. J. Biol. Chem. 245, 5553-5559 (1970).
8. Shafritz, D.A., Laycock, D.G., Anderson, W.F.: Puromycin-peptide bond formation with reticulocyte initiation factors M_1 and M_2. Proc.Nat.Acad.Sci.USA 68, 496-499 (1971).
9. Shafritz, D.A., Laycock, D.G., Crystal, R.G., Anderson, W.F.: Requirement for GTP in the initiation process on reticulocyte ribosomes and ribosomal subunits. Proc.Nat.Acad.Sci.USA 68, 2246-2251 (1971).
10. Zasloff, M., Ochoa, S.: A supernatant factor involved in initiation complex formation with eukaryotic ribosomes. Proc.Nat.Acad.Sci.USA 68, 3059-3063 (1971).
11. Zasloff, M., Ochoa, S.: Polypeptide chain initiation in eukaryotes: Functional identity of supernatant factor from various sources. Proc.Nat.Acad.Sci.USA 69, 1796-1799 (1972).

12. McCroskey, R. P., Zasloff, M., Ochoa, S.: Polypeptide chain initiation and stepwise elongation with Artemia ribosomes and factors. Proc.Nat.Acad.Sci.USA 69, 2451-2455 (1972).

13. Leader, D.P., Wool, I.G., Castles, J.J.: A factor for the binding of aminoacyl transfer RNA to mammalian 40S ribosomal subunits. Proc.Nat.Acad.Sci.USA 67, 523-528 (1970).

14. Gasior, E., Rao, P., Moldave, K.: The interaction of aminoacyl-tRNA and N-acylaminoacyl-tRNA with ribosomes and ribosomal subunits. Biochim.Biophys.Acta 254, 331-340 (1971).

15. Gasior, E., Moldave, K.: Evidence for a soluble protein factor specific for the interaction between aminoacylated transfer RNA's and the 40 S subunit of mammalian ribosomes. J. Mol. Biol. 66, 391-402 (1972).

16. Zasloff, M., Ochoa, S.: Polypeptide chain initiation in eukaryotes IV. Purification and properties of supernatant initiation factor from Artemia salina embryos. J. Mol. Biol. 73, 65-76 (1973).

17. Merrick, W.C., Safer, B., Adams, S., Kemper, W.: Purification and properties of rabbit reticulocyte initiation and elongation factors. Federation Proc. 33, 1262 (1974).

18. Ouchterlony, Ö.: Antigen-antibody reactions in gels. IV. Types of reactions in coordinated systems of diffusion. Acta Pathol. Microbiol. Scand. 32, 231-240 (1953).

19. Fudenberg, H.H., Deutsch, H.F., Fahey, J.L.: Purification of antibody. In Methods in Immunology and Immunochemistry (ed. C. A. Williams and M.W.Chase) vol. I, pp. 307-332 (Academic Press, New York and London, 1967).

20. Lowry, O.H., Rosebrough, N.J., Farr, A.L., Randall, R.J.: Protein measurement with the Folin phenol reagent. J. Biol. Chem., 193, 265-275 (1951).

21. Nirenberg, M., Leder, P.: RNA codewords and protein synthesis. The effect of trinucleotides upon the binding of sRNA to ribosomes. Science 145, 1399-1407 (1964).

22. Anderson, W.F.: Purification and properties of rabbit reticulocyte initiation and elongation factors. EMBO Workshop on Initiation of Protein Synthesis in Prokaryotic and Eukaryotic Systems, Noordwijkerhout, The Netherlands, April 17-19, 1974.

23. Treadwell, B. V., Goldstein, L., Robinson, W.G.: Supernatant and ribosomal factors required for initiator tRNA binding and poly(Phe) synthesis at low [Mg^{++}] in a eukaryotic system. Federation Proc. 33, 1286 (1974).

24. Leader, D.P., Wool, I.G.: Partial purification and characterization of an initiation factor from rat liver which promotes the binding of

Phenylalanyl-tRNA to 40S ribosomal subunits. Biochim.Biophys.Acta 262, 360-370 (1972)

25. Picciano, D.J., Prichard, P.M., Merrick, W.C., Shafritz, D.A., Graf, H., Crystal, R.G., Anderson, W.F.: Isolation of protein synthesis initiation factors from rabbit liver. J. Biol. Chem. 248, 204-214 (1973).

26. Cimadevilla, J.M., Morrisey, J., Hardesty, B.: A functional interaction between methionyl-transfer RNA hydrolase and a transfer RNA binding factor. J. Mol. Biol. 83, 437-446 (1974).

27. Leader, D. P., Klein-Bremhaar, H., Wool, I.G.: Distribution of initiation factors in cell fractions from mammalian tissues. Biochem. Biophys.Res.Commun. 46, 215-224 (1972).

28. Eich, F., Drews, J.: Isolation and characterization of a peptide chain initiation factor from Krebs II ascites tumor cells. Biochim.Biophys.Acta 340, 334-338 (1974).

29. Moldave, K.: Properties of a soluble factor from rat liver that stimulates the binding of acetylphenylalanyl-tRNA and methionyl-tRNA to 40S ribosomal subunits. EMBO Workshop on Initiation of Protein Synthesis in Prokaryotic and Eukaryotic Systems, Noordwijkerhout, The Netherlands, April 17-19, 1974.

30. Morrisey, J., Hardesty, B.: Met-tRNA hydrolase from reticulocytes specific for Met-tRNA$_f^{Met}$ on 40S ribosomal subunits. Arch. Biochem. Biophys. 152, 385-397 (1972).

Thiamine-Binding Protein of Escherichia coli

Y. Nose, A. Iwashima and A. Matsuura-Nishino
Department of Biochemistry, Kyoto Prefectural University of Medicine, Kamikyoku, Kyoto, Japan

A thiamine-binding protein which is thought to be participated in the thiamine transport through cell membranes of *E. coli* has been isolated from the osmotic shock fluid (1, 2). It has been purified to a homogeneous extent by affinity chromatography using agarose coupled with thiamine diphosphate (3).
On the other side, studies on interactions of tryptophan residue in the protein with thiamine and thiamine diphosphate have been reported (4-6). On this basis, the effects of chemical modification of the thiamine-binding protein on its binding activity was investigated to obtain an information concerning the binding site of thiamine.
This paper describes some properties of the purified thiamine-binding protein and evidence for probable involvement of a tryptophan residue in the binding site of the protein.

EXPERIMENTAL

Bacteria —— A mutant of *E. coli* K12 auxotrophic for thiamine thiazole (KG 33) was used in the experiment. The cells grown on Davis-Mingioli's minimal medium (7) containing 0.04 μM thiazole were harvested at an early stationary phase, washed once with the minimal medium.
Chemicals —— Sepharose 6 B was obtained from Pharmacia, cyanogen bromide from Wako Pure Chemical Industries. ^{14}C-thiamine hydrochloride (thiazole-2-^{14}C), specific activity 18.9 mCi/m mole was purchased from Radiochemical Centre, England. TMP[1], TDP, oxythiamine hydrochloride and pyrithiamine hydrobromide were the products of Sigma Chemical Company. Thiamine sulfuric acid ester and dimethialium were kindly supplied from Dr. S. Yurugi,

Takeda Chemical Industries Ltd., TTP and chloroethylthiamine from Dr. T. Yusa, Sankyo Central Research Laboratory. Dimethyl (2-hydroxy-5-nitrobenzyl)-sulfonium bromide was purchased from Wako Pure Chemical Industries. All other chemicals were of analytical grade available from commercial sources.

Purification of thiamine-binding protein —— Harvested cells from the medium were subjected to the cold osmotic shock by the method of Neu and Heppel (8). The thiamine-binding protein from the shock fluid was purified by ammonium sulfate precipitation, DEAE-cellulose column chromatography and TDP-sepharose application as described in the previous paper (3).

^{14}C-thiamine uptake by whole cells —— Washed cells preincubated with or without chemical modifiers at 37°C for 15 min were re-suspended in the minimal medium containing 0.4 % glucose at 0.20 of absorbancy at 560 nm which corresponds to 0.14 mg of dry weight per ml. Ten milliliters of the suspensions were used for measuring the uptake of ^{14}C-thiamine by the method of Kawasaki *et al* (9).

Thiamine binding activity —— Thiamine binding activity was determined by equilibrium dialysis procedure as follows. The dialysis bag containing 10 μg of protein, as measured by the method of Lowry *et al* (10), in a total volume of 1.0 ml was dialysed against 500 ml of 0.2 M potassium phosphate buffer (pH 7.0) containing 0.02 μM ^{14}C-thiamine at 4°C for 20 hr. The amount of ^{14}C-thiamine bound to the protein was calculated by subtracting the radioactivity of the dialysate from that in the bag.
The thiamine binding activity was expressed as the amount of thiamine bound per milligram of protein. For examination of effects of thiamine derivatives and analogues on the activity, the compounds were added to the dialyzing buffer simultaneously with ^{14}C-thiamine at the desirable concentrations.

Polyacrylamide gel electrophoresis —— Gel electrophoresis was performed in a vertical gel electrophoresis apparatus according to the method of Werber and Osborn (11). The gel consisted of

7 % acrylamide, 0.2 % bisacrylamide, 0.1 M sodium phosphate buffer (pH 7.1) and 0.15 % ammonium persulfate in the presence and absence of 8 M urea. Of the purified protein sample, 25 μg of the protein was incubated at room temperature with or without urea. The native and urea-treated samples were applied on the gel column, respectively and the electrophoresis was run for 1.5 hr at a constant current of 8 ma per tube. The gel slab was stained with 0.25 % Coomassie blue in 7 % acetic acid for 5 hr. Destaining was performed in 7 % acetic acid with occassional shaking for 2 days.

SDS-acrylamide gel electrophoresis and molecular weight determination —— Polyacrylamide gel electrophoresis in the presence of SDS was performed as follows. The protein was incubated at 37°C for 2 hr in 0.1 M sodium phosphate buffer (pH 7.1), 1 % in SDS and 1 % in β-mercaptoethanol, then run in a polyacrylamide gel containing 1 % SDS in place of 8 M urea under the same condition as described above. After staining and destaining, the mobility of the protein was plotted against the logarithm of molecular weights of the marker proteins which were run in the same way.

Chemical modification of thiamine-binding protein —— A series of experiments using the following chemical reagents known to react with some of amino acid residues in the protein was carried out. The purified protein, each approximately 10-15 μg was treated with the reagent in a final volume of 1.0 ml under the appropriate conditions.

Iodination was carried out at a concentration of 2.5×10^{-4}M of iodine in 0.1 M KI solution, pH 7.0, 4°C for 4 hr according to the method of Hayashi *et al* (12). N-Bromosuccinimide treatment was performed at 2.5×10^{-6}M concentration, 25°C, in 0.05 M acetate buffer (pH 4.5) for 1.5 hr by the method of Spande and Witkop (13). N-Acetylimidazole was reacted at 1×10^{-4}M concentration, 25°C, in 0.05 M sodium borate buffer (pH 7.5) for 2.5 hr (14). SH-inhibitors such as iodoacetate, *p*-chloromercuribenzoate and N-ethylmaleimide were also reacted under the conditions as indicated in Table II. Dimethyl (2-hydroxy-5-nitrobenzyl)-sulfonium bromide (HNB-$(CH_3)_2SBr$) freshly prepared was reacted at the three different concentrations, 22°C, pH 3.0 adjusted with 6 N HCl for 4 hr under a dark condition (15).

RESULTS AND DISCUSSION

Effect of urea on thiamine-binding protein —— As earlier reported (3), the binding of thiamine-binding protein to the TDP-sepharose was fairly strong and the protein was barely eluted with a high concentration of urea (8 M) in 0.05 M potassium phosphate buffer (pH 7.0) resulting in a reversible inactivation. We therefore tested the influence of urea on the thiamine-binding protein using polyacrylamide gel electrophoresis. Fig. 1 shows that the protein, previously incubated with 8 M urea at room temperature for 2 hr gave a slower moving component in the acrylamide gel containing urea (B). In contrast, the native sample (A) (the gel did not contain urea) migrated faster suggesting a probable conformational change of the protein with urea.

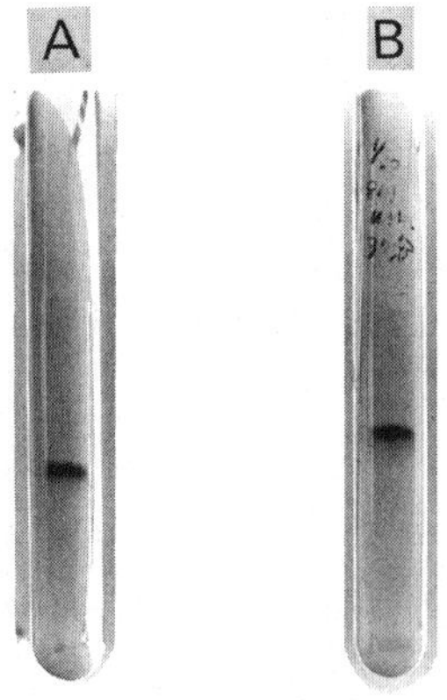

Fig. 1. Polyacrylamide gel electrophoresis in the absence and presence of urea of the thiamine-binding proetin
The direction of electrophoretic migration was from top to bottom.
A : Control sample without urea
B : Protein was pretreated with 8 M urea for 30 min at room temperature

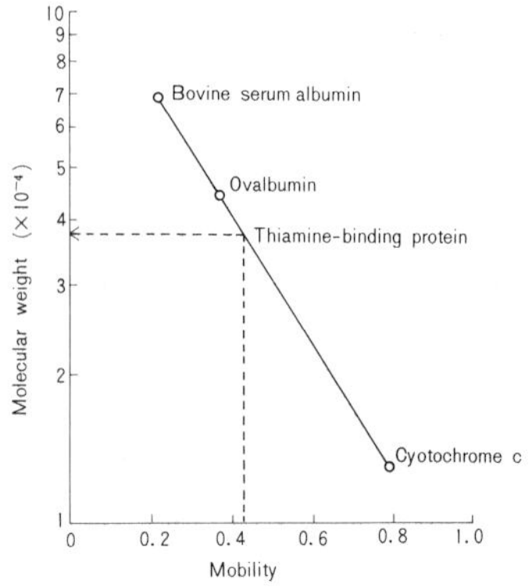

Fig. 2. Determination of the apparent molecular weight of the thiamine-binding protein by polyacrylamide gel electrophoresis in SDS

Acrylamide gel electrophoresis on SDS —— The thiamine-binding proteins exhibiting a different movement with urea treatment were subjected to electrophoresis on SDS-acrylamide gel which can be used for determination of the subunit structure of protein.
The two proteins, urea-treated and untreated samples, both showed a same movement as a single band (with a mobility of 0.65 with respect to bromophenol blue) on the acrylamide gel containing 1 % SDS. In addition, as shown in Fig. 2, the molecular weight of the protein determined by the SDS-method, using cytochrome c (mol. wt. 12,300), ovalbumin (mol. wt. 43,000) and bovine serum-albumin (mol. wt. 68,000) as marker was found to be 38,000 in agreement with the value of 39,000 in the native protein previously obtained by gel filtration (3).
pH optimum and stability —— A typical curve showing the effect of pH on thiamine binding activity is shown in Fig. 3. The pH optimum was between 8 and 9, when assayed in 0.1 M citrate buffer. A similar pH curve was obtained with 0.1 M Tris-HCl buffer although the activity was slightly lower. The binding was sharply decreased at a higher or lower pH, but the loss of activity was almost recovered by returning the pH. Although the purified thiamine-binding protein is thermolabile and completely

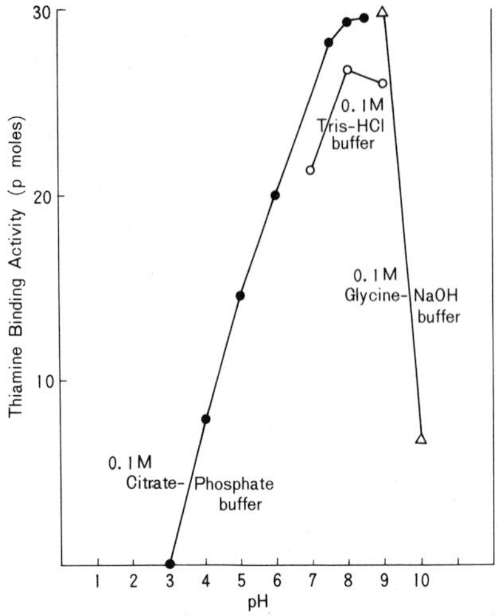

Fig. 3. Effect of pH on thiamine binding activity

Table I. Inhibition by thiamine derivatives and analogues of thiamine binding activity

The standard assay conditions were used except for the addition of thiamine analogues.

Addition	Thiamine analogues/thiamine 1	100
	% Inhibition	
Thiamine sulfuric acid ester	100	
TMP	88.6	
TPP	69.8	
TTP	50.8	
Chloroethylthiamine		66.0
Dimethialium		44.4
Hydroxyethylthiamine		44.7
Oxythiamine		25.3
Pyrithiamine		0

inactivated by heating at 100°C for 10 min, it could be kept at -20°C without any activity loss for several months and no significant loss of activity was found by freezing and thawing.

Effect of thiamine derivatives and analogues on thiamine binding activity —— It has been reported that the uptake of thiamine by *E. coli* cells was inhibited by some of thiamine analogues and the thiamine binding activity of the partially purified protein was by thiamine phosphates (9, 16). Therefore, effects of thiamine derivatives and analogues on the activity of the purified thiamine-binding protein were examined (Table I).

At a ratio of thiamine derivative to thiamine of 1 : 1, marked inhibitions were observed with thiamine sulfuric acid ester and thiamine phosphates at the almost same order of magnitude, being completely inhibited with thiamine sulfuric acid ester. However, the inhibitions by thiamine analogues were fairly low

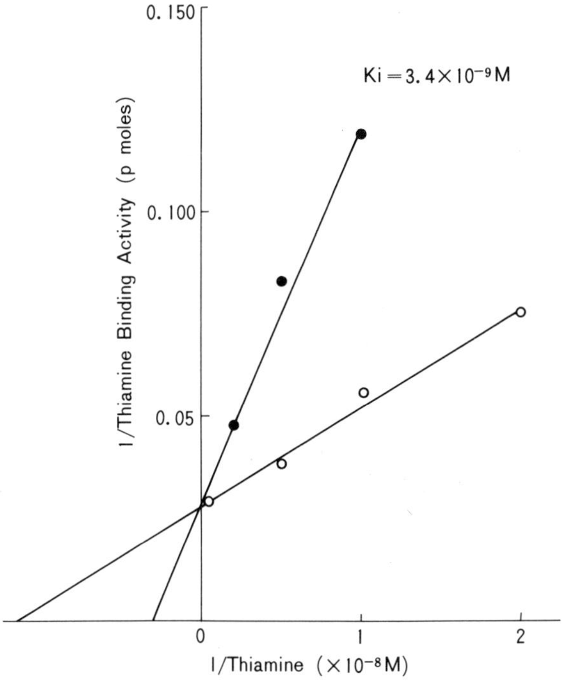

Fig. 4. Effect of thiamine monophosphate on thiamine binding activity

○ No thiamine monophosphate

● with 0.01 μM thiamine monophosphate

The thiamine-binding protein (7 μg) was used.

compared with the thiamine derivatives. The inhibition by pyrithiamine and oxythiamine known to be antimetabolites of thiamine were rather weak. These results suggest that the binding of thiamine to the protein is highly specific for its chemical structure and it is of interest that the thiamine derivative containing an acidic group appeared to bind more easily to the protein than thiamine itself. This is compatible with the reason of binding of the protein to TDP-sepharose.

As shown in Figs. 4 and 5 Lineweaver-Burk plots of the binding as a function of thiamine concentrations gave a value of apparent Km for thiamine, 9.2×10^{-9}M and the inhibitions by TMP and chloroethylthiamine were both competitive and showed Ki values of 3.4×10^{-9}M and 6.8×10^{-7}M, respectively.

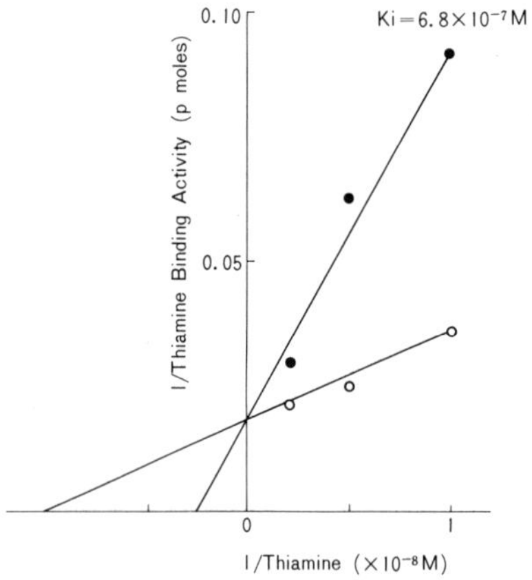

Fig. 5. Effect of chloroethylthiamine on thiamine binding activity

○ No chloroethylthiamine

● with 2 μM chloroethylthiamine

The thiamine-binding protein (7 μg) was used.

Chemical modification of thiamine-binding protein —— Effect of various chemical treatments on thiamine binding activity of the protein were examined (Table II).

Iodine treatment, under a fairly violent condition, caused a

Table II. Effects of various chemical treatments on thiamine binding activity

The thiamine-binding protein (10 μg) was incubated with each reagent under the indicated condition, then thiamine binding activity was measured by the standard assay procedure.

Reagent	Condition of treatment	Concentration of reagent in treatment (M)	Inactivation (%)
Iodine (KI_3)	pH 7.0 4°C, 4 hr	5×10^{-4}	100
N-Bromosuccinimide	pH 4.5, 25°C, 1.5 hr	2.5×10^{-6}	50
N-Acetylimidazole	pH 7.5, 25°C, 2.5 hr	1×10^{-4}	0
Iodoacetic acid	pH 5.5, 25°C, 27 hr	1×10^{-5}	2.6
N-Ethylmaleimide	pH 7.0 4°C, 17 hr	1×10^{-3}	0
p-Chloromercuribenzoate	pH 7.0 4°C, 17 hr	1×10^{-3}	0
Dimethyl (2-hydroxy-5-nitrobenzyl)-sulfonium bromide (water soluble)	pH 3.0, 22°C, 4 hr	2×10^{-5}	16.9
		2×10^{-4}	39.4
		2×10^{-3}	78.7

complete loss of the activity. N-Bromosuccinimide known to attack tryptophan- and tyrosine-peptide bonds brought 50 % inactivation under the condition indicated. N-Acetylimidazole, a selective acetylating agent of tyrosine had no effect. It was of interest that SH-inhibitors such as iodoacetic acid, N-ethylmaleimide and *p*-chloromercuribenzoate did not affect the binding activity, while the thiamine uptake by whole cells were markedly inhibited by these compounds at a same concentration as shown in Table IV.

Horton and Tucker (15) have observed that HNB-$(CH_3)_2SBr$ is specific in the reaction with tryptophan residue in acidic or neutral solutions. Cysteine was the only other amino acid which was modified with the reagent. However, SH-inhibitors had no effect on the activity and the HNB-$(CH_3)_2SBr$ gave a significant inhibitory effect on the thiamine binding activity. From these results, it seems likely that a tryptophan residue is involved in the thiamine binding site of the protein. When cold thiamine was added to the binding protein prior to addition of HNB-$(CH_3)_2SBr$, followed by measuring thiamine binding activity, inhibitory effect was partially reduced (Table III).

Table III. Reaction of HNB-$(CH_3)_2SBr$ with the thiamine-binding protein in the absence and presence of thiamine

Thiamine (5×10^{-7}M) was added to the thiamine-binding protein (10 µg) prior to the addition of the reagent.
Thiamine binding activity was measured by the standard assay procedure.

Concentration of reagent (M)	Inactivation (%) Thiamine (-)	Thiamine (+)
2×10^{-5}	17.5	——
2×10^{-4}	47.9	16.9
2×10^{-3}	76.6	66.2

Although the concentration of cold thiamine added was limited to 5×10^{-7}M to avoid an exchange reaction with ^{14}C-thiamine during equilibrium dialysis, it partially protects the protein from inactivation by the sulfonium reagent.

Effect of chemical reagents for modification on ^{14}C-thiamine uptake by *E. coli* cells —— As shown in Table IV, the thiamine uptake by the cells is markedly inhibited by the SH-inhibitors at the concentrations which had no effect on the binding activity of the protein. The fact suggests an existence of another factor beside the thiamine-binding protein in the transport system of *E. coli* KG 33. On the other hand, HNB-$(CH_3)_2$SBr inhibited the thiamine binding as well as thiamine uptake.

Table IV. Effect of several reagents for chemical treatment of the cells on ^{14}C-thiamine uptake by *E. coli* KG 33

Addition (M)		^{14}C-thiamine uptake (nmoles/g dry weight)	% Inhibition
None		149.8	——
Iodoacetic acid, 1×10^{-3}		73.4	51.0
N-Ethylmaleimide, 1×10^{-3}		11.1	92.6
N-Bromosuccinimide, 1×10^{-4}		5.7	96.2
Dimethyl (2-hydroxy-	1×10^{-3}	97.3	35.0
5-nitrobenzyl)-	2×10^{-3}	29.6	80.2
sulfonium bromide,	5×10^{-3}	11.4	92.4

SUMMARY

Investigations were made of the thiamine-binding protein purified to a homogeneous state from the osmotic shock fluid of *E. coli* and the following results were obtained.

1. The molecular weight calculated by SDS-polyacrylamide gel electrophoresis method was approximately 38,000 which was in agreement with that of the native protein obtained by gel filtration.

2. Urea-treated protein showed a slower migration in the electrophoresis by urea containing acrylamide gel than that of native protein.

3. Apparent Km of the binding for thiamine was 9.2×10^{-9}M and optimal pH was between 8 and 9.

4. Binding activity of the protein was markedly inhibited by thiamine sulfuric acid- and phosphoric acid esters, while to

a lesser extent by thiamine analogues. The inhibitions by thiamine monophosphate and chloroethylthiamine were both competitive with thiamine.

5. Binding activity was sensitive to dimethyl (2-hydroxy-5-nitrobenzyl)-sulfonium bromide and N-bromosuccinimide among various chemical modifiers tested, but not to the SH-inhibitors. Evidence for probable involvement of a tryptophan residue in the binding site of the protein was presented.

[1]. The abbreviations used are : TMP, thiamine monophosphate ; TDP, thiamine diphosphate ; TTP, thiamine triphosphate ; SDS, sodium dodecyl sulfate ; HNB-$(CH_3)_2$SBr, dimethyl (2-hydroxy-5-nitrobenzyl) sulfonium bromide.

REFERENCES

1. Nishimune, T., Hayashi, R.: Thiamine-binding protein and thiamine uptake by *Escherichia coli*. Biochim. Biophys. Acta 244, 573-583 (1971)
2. Iwashima, A., Matsuura, A., Nose, Y.: Thiamine-binding protein of *Escherichia coli*. J. Bacteriol. 108, 1419-1421 (1971)
3. Matsuura, A., Iwashima, A., Nose, Y.: Purification of thiamine-binding protein from *Escherichia coli* by affinity chromatography. Biochem. Biophys. Res. Commun. 51, 241-246 (1973)
4. Biaglow, J. E., Mieyal, J. J., Suchy, J., Sabel, H. Z.: Coenzyme interactions. III. Characteristics of the molecular complexes of thiamine with indole derivatives. J. Biol. Chem. 244, 4054-4062 (1969)
5. Kochetov, G. A., Usmanov, R. A.: Charge transfer interactions in transketolase-thiamine pyrophosphate complex. Biochem. Biophys. Res. Commun. 41, 1134-1140 (1970)
6. Heinrich, C. P., Noack, K., Wiss, O.: Chemical modification of tryptophan at the binding site of thiamine pyrophosphate

in transketolase from baker's yeast. Biochem. Biophys. Res. Commun. 49, 1427-1432 (1972)

7. Davis, B. D., Mingioli, E. S.: Mutants of *Escherichia coli* requiring methionine or vitamin B_{12}. J. Bacteriol, 60, 17-28 (1950)
8. Neu, H. C., Heppel, L. A.: The release of enzymes from *Escherichia coli* by osmotic shock and during formation of spheroplasts. J. Biol. Chem. 241, 3055-3065 (1965)
9. Kawasaki, T., Miyata, I., Esaki, K., Nose, Y.: Thiamine uptake in *Escherichia coli*. I. General properties of thiamine uptake system in *Escherichia coli*. Arch. Biochem. Biophys. 131, 223-230 (1969)
10. Lowry, O. H., Rosenbrough, N. J., Farr, A. L., Randall, R. J.: Protein measurement with the Folin phenol reagent. J. Biol. Chem. 193, 265-275 (1951)
11. Weber, K., Osborn, M.: The reliability of molecular weight determination by dodecyl sulfate-polyacrylamide gel electrophoresis. J. Biol. Chem. 244, 4406-4412 (1969)
12. Hayashi, K., Shimoda, T., Yamada, K., Kumai, A., Funatsu, M.: Iodination of lysozyme. I. Differential iodination of tyrosine residues. J. Biochem. 64, 239-245 (1968)
13. Spande, T. F., Witkop, P. in Hirs, C. H. W. (Editor) : Tryptophan involvement in the function of enzymes and protein hormones as determined by selective oxidation with N-bromosuccinimide. Methods in Enzymology, Vol. 11, Academic Press, New York, 1967, p. 506-522
14. Riordan, J. F., Vallee, B. L. in Hirs, C. H. W. (Editor) : O-Acetyltyrosine. Methods in Enzymology, ibid., p. 570-576
15. Horton, H. R., Tucker, W. P.: Dimethyl (2-hydroxy-5-nitrobenzyl) sulfonium salts. Water-soluble environmentally sensitive protein reagents. J. Biol. Chem. 245, 3397-3401 (1970)
16. Matsuura, A., Iwashima, A., Nose, Y.: Inhibition by thiamine phosphates of thiamine uptake in *Escherichia coli*. J. Vitaminol. 18, 29-33 (1972)

Photoaffinity Labelling of tRNA Binding Sites on E. coli Ribosomes

James Ofengand and Ira Schwartz
Roche Institute of Molecular Biology, Nutley, N.J. 07110, USA

A detailed analysis of the topography of tRNA binding sites in ribosomes is being pursued in our laboratory by the use of chemically derivatized tRNA as a photoaffinity probe. Chemical (1-3) and photochemical (4) affinity labelled tRNA has been utilized by other workers to identify the ribosomal components of the peptidyl transferase center by means of AA-tRNA suitably derivatized at the amino group of the aminoacyl moiety but conflicting results have been obtained, due at least in part to the use of affinity labels of different chemical specificity. For example, chemical affinity labelling identified the 50S proteins, L2 and L27 (1) or L27, L15, and L16 (2-3) as being at or near the peptidyl-transferase center while photoaffinity labelling from the same place on the tRNA gave reaction only with 23S rRNA (4).

For our experiments we sought a reagent which would not only show a general reactivity for a wide variety of functional groups and thus avoid biasing our results toward either RNA or proteins, but which could also be placed at defined sites in tRNA other than at the aminoacyl end. The first requirement dictated the use of photoaffinity labelling derivatives while the second suggested derivatization of minor nucleotides in tRNA since they occur at known loci and in appropriate cases, only once in a given tRNA.

Photoaffinity labelling has several advantages over chemical affinity labelling. Since the photoaffinity probe is not reactive until irradiated a tRNA derivatized with such a label can first be bound to the ribosome under the appropriate conditions with subsequent activation of the probe by irradiation. This allows one to control the point in time at which covalent link formation occurs. The carbene or nitrene formed upon irradiation is very reactive as well as non-specific (5), and is therefore capable of reaction with almost any component of either the ribosomal RNA or protein which might be within range of the probe. Aryl azides are generally preferred to alkyl azides or diazo compounds because of the longer lifetime of arylnitrenes in aqueous solution (5) and because the red-shifted absorption spectrum of the parent aryl azide means that photo-activation can be accomplished at wavelengths that are not deleterious to either the ribosome or tRNA.

For these initial studies, we derivatized the 4-thiouridine residue since many workers have successfully modified this residue in tRNA without loss of biological activity (lit. cited in ref. 6). The minimal length probe that would incorporate the needed bifunctional groups, p-azidophenacyl bromide, was linked to the ^{4}Srd residue at position 8 in purified *E. coli* valine tRNA, and cross-linking to the ribosome at both the A and P sites was examined.

MATERIALS AND METHODS

General. *E. coli* $tRNA_1^{Val}$, $tRNA^{fMet}$, unfractionated tRNA, mixed *E. coli* aminoacyl-tRNA synthetase, elongation factor EFTu from *E. coli*, and 1.0 M NH_4Cl-washed ribosomes from *E. coli* B were prepared according to previously described methods (7,8). The ribosome preparation contained 60-65% 70S, 35-40% free 50S and < 2% free 30S particles. Phenacyl bromide was from Aldrich Chemical Co., p-azidophenacyl bromide (Fig. 1) was synthesized by the Chemical Research Division of Hoffmann-La Roche according to an unpublished procedure of Dr. S. Hixson, University of Massachusetts, and the bromoacetic ester analog (Fig. 5) was a gift from Dr. Hixson.

Phenacyl or p-Azidophenacyl $tRNA_1^{Val}$. Twenty A_{260} units of purified $tRNA_1^{Val}$, in 1.0 ml of 0.1 M phosphate buffer, pH 7.4 was added to 9.5 ml of 4.9 mM (2000 x molar excess) phenacyl bromide or p-azidophenacyl bromide in 100% DMSO. The resultant reaction mixture (90% in DMSO) was homogeneous. The reaction was terminated by addition of a 25-fold molar excess of mercaptoethanol over phenacyl bromide. After dialysis, the derivatized valyl-tRNA was isolated by ethanol precipitation.

Valyl-tRNA. Control and derivatized valyl-tRNA were prepared by scaling-up the standard charging assay (see below) using [^{3}H]valine (30.6 Ci/mmole).

Functional Assays for Valyl-tRNA Activity. Assays of aminoacylation, valyl-tRNA-EFTu-GTP ternary complex formation, EFTu-dependent binding to ribosomes, non-enzymatic binding to the P site, and synthesis of a (Val,Phe) copolypeptide were performed as described previously (8,9) with small modifications. The extent of reaction was measured in all of the above assays, and was proportional to the amount of tRNA or valyl-tRNA added to the mixture.

RESULTS

Reaction of $tRNA_1^{Val}$ with Phenacyl or p-Azidophenacyl Bromide. Reaction of ^{4}Srd with primary halides adjacent to a carbonyl or vinyl group results in the type of addition product illustrated in Fig. 1. When ^{4}Srd was mixed with a 2-fold excess of p-azidophenacyl bromide at pH 7.4 in 60% methanol, a rapid reaction occurred to produce a new compound with R_f on silica gel TLC intermediate between ^{4}Srd and the bromide. The structure of the product was characterized by its absorption spectrum which was invariant with pH between 2 and 10 and showed no evidence of the band at 330 nm characteristic of N_3-substituted 4-thiouridine (6). Azide (2120 cm^{-1}), acetophenone ketone (1690 cm^{-1}), and ribose hydroxyl (33-3400 cm^{-1}) infrared absorption bands were also present. A more

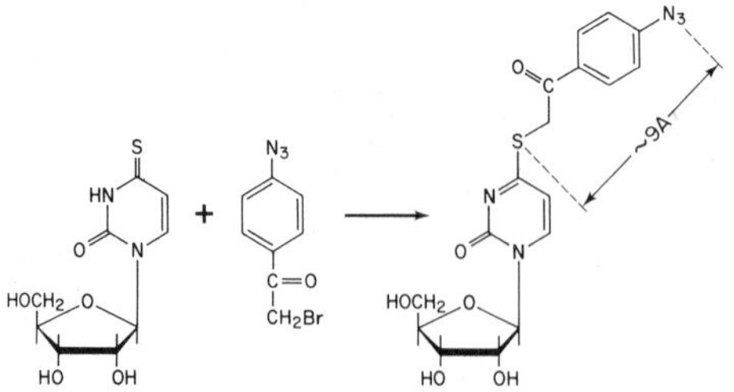

Fig. 1. Scheme for the reaction of p-azidophenacyl bromide with 4-thiouridine. In the maximally extended form the distance from the sulfur atom to the azido group is 8.5A and in the direction indicated.

complete characterization of the product will be presented elsewhere.

Reaction of ^{4}Srd at position 8 in E. coli valine tRNA with the p-azidophenacyl group could not be monitored spectrally due to interference from the tRNA, and the probe was not available in radioactive form at this time. Consequently, an indirect method was used to measure the extent of reaction. Photochemically-induced crosslinking of $^{4}Srd_{8}$ with Cyd_{13} in tRNA yields a binucleotide which, after chemical reduction, is highly fluourescent. If, however, the sulfur atom of the ^{4}Srd is blocked, cross-link formation cannot occur. A decrease in the specific fluorescence of tRNA after irradiation is, therefore, a measure of the loss of ^{4}Srd.

Reaction with ^{4}Srd in tRNA was very rapid, being over in less than 5 min at room temperature (Table 1). The Table also shows that there was no loss of ^{4}Srd if phenacyl bromide was omitted, and that $tRNA_{1}^{Val}$, $tRNA^{fMet}$, and unfractionated tRNA all behaved similarly. The extent of reaction was 83-84% by this assay in all 3 cases, and could not be further increased by increasing the reaction time, reagent concentration, or by denaturation of the tRNA before reaction. We do not have a ready explanation for the apparent lack of reactivity of 16% of the tRNA.

TABLE 1
REACTION OF PHENACYL BROMIDE WITH 4-THIOURIDINE IN tRNA

Reaction Mixture	Reaction Time[a]	Specific Fluorescence		
		$tRNA^{Val}$	$tRNA^{fMet}$	$tRNA^{Mixed}$
Complete	0	13.6	9.8	9.9
Complete	5	2.4	1.7	1.5
Minus Phenacyl Br	0	14.6	11.6	-
Minus Phenacyl Br	5	14.5	11.7	-
% Reaction		83	83	84

[a] For 0t, tRNA was added after 2 min reaction with mercaptoethanol.
Derivatization was performed as described in Materials & Methods. 30 µl samples were diluted to 300 µl with 50 mM Bicine, pH 7.5, 10 mM $Mg(OAc)_2$, and irradiated, reduced and assayed for fluorescence (7).

The above results showed that ^{4}Srd in tRNA could react with p-azidophenacyl bromide. Model studies with similar compounds (10-12) have shown that the major nucleotides do not react, and pseudouridine is also unreactive under these conditions (unpublished experiments). Nevertheless, a direct test of the specificity of the reaction with nucleotides in tRNA using radioactive p-azidophenacyl bromide would be desirable, and such experiments are in progress.

Functional Activity of Derivatized tRNAs. In order to be sure that reaction of ^{4}Srd with this probe did not alter the function of the tRNA molecule, the activity assays listed in Table 2 were examined. Amino acid acceptor activity, binding to the ribosomal P site, EFTu-dependent binding to the ribosomal A site, AA-tRNA-EFTu-GTP ternary complex formation and the ability to transfer valine into polypeptide were all 90-100% of untreated tRNA control values. It is interesting that the presence of the zwitterionic azido group on the phenacyl moiety restored fully the ability of the derivatized tRNA to ac ept amino acid.

Covalent Link Formation Between Ribosomes and S-(p-azidophenacyl)-[^{3}H]Val-tRNA Bound at the P Site. The kinetics of covalent cross-link formation between the modified Val-tRNA and ribosomes is presented in Fig. 2. Covalent link formation was dependent on the presence of both polynucleotide and ribosomes, showing that the tRNA was not activated while free in solution with subsequent re-

TABLE 2
FUNCTIONAL ACTIVITY OF PHENACYL-tRNAs

Assay	Phenacyl-tRNA	p-azidophenacyl tRNA
	Per Cent of Control	
Amino Acid Acceptance	71	100
P site Binding	89	100
Enzymatic A site binding	90	-
EFTu-GTP Binding	91	-
Polypeptide Synthesis	96	-

Assays were performed as described in Materials and Methods.

action randomly with the ribosome, polynucleotide, or itself to give a filter-adsorbable product. Moreover, phenacyl-tRNA, which lacks the photolyzable azido group, did not show any crosslinking to ribosomes on irradiation of a complete reaction mixture (data not shown). Crosslinking was essentially complete at 4 hrs, with an efficiency of covalent bond formation of 15-20% of the non-covalently bound tRNA in 20 mM Mg^{++}. The slowness of the reaction is probably due to inefficient absorption of energy since the absorption maximum for the 4-thiouridine derivative is at 303 nm, while the lamp emission maximum is at 350 nm and Pyrex glass was used to filter out light below 305 nm to avoid radiation damage to the ribosome or tRNA. The relatively low yield (20%) may

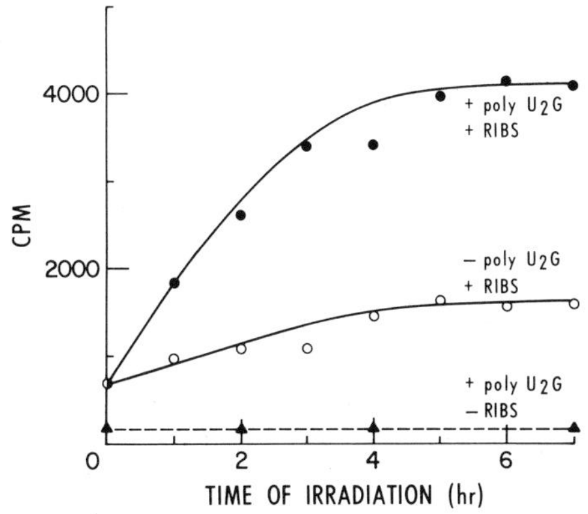

Fig. 2. Kinetics of covalent linking of S-(p-azidophenacyl)-[^3H]-valyl-tRNA to ribosomes. Ribosomal P site binding mixtures at 20 mM Mg^{++} in the absence of tetracycline were incubated for 15 min at 30° with a 5-10 fold excess of ribosomes. 70-75% of the Val-tRNA was complexed under these conditions. The addition of 4 x 10^{-4} M tetracycline did not affect the extent of binding. The mixtures were then irradiated at 0° in a Rayonet RPR-100 photochemical reactor equipped with 350 nm lamps and Pyrex filters so that less than 0.05% of the light energy was transmitted below 310 nm. At the indicated times two samples were removed. One was filtered through millipore filters and washed with 20 mM $MgCl_2$, 50 mM Tris, pH 7.4, 50 mM KCl, and the second was diluted to 0.1 mM Mg^{++}, filtered, and washed with 0.1 mM $MgCl_2$, 50 mM Tris, pH 7.4, 50 mM KCl. Radioactivity retained on the filters in 0.1 mM Mg^{++} was taken as covalently bound tRNA. The amount of polynucleotide-dependent radioactivity retained in 20 mM Mg^{++} was 30,000 cpm and did not change during the course of irradiation. ●—●, complete incubation mixture; O—O, incubation mixture minus poly U,G (2:1); ▲—▲, incubation mixture minus ribosomes.

reflect competition between buffer molecules and ribosomal components for the reactive nitrene.

The S-(p-azidophenacyl)-[^{3}H]Val-tRNA-70S ribosome covalent complex was dissociated into 30S and 50S subunits and separated on a sucrose gradient as shown in Fig. 3. When polynucleotide was present in the original reaction mixture all the radioactivity migrated with the 30S peak. There was no detectable radioactivity associated with the 50S subunit. If polynucleotide was omitted from the reaction, or the complete incubation mixture was not irradiated, no radioactivity was associated with either subunit.

The 30-subunit-S-(p-azidophenacyl)-[^{3}H]Val-tRNA covalent complex was isolated and applied to a sucrose gradient which was 0.5% in SDS in order to separate the 30S ribosomal proteins from the 16S RNA. Under these conditions, the proteins complex with the SDS and do not migrate into the gradient. The results are shown in Fig. 4. The A_{260} peak corresponds to 16S RNA and all the radioactivity sedimented with this peak.

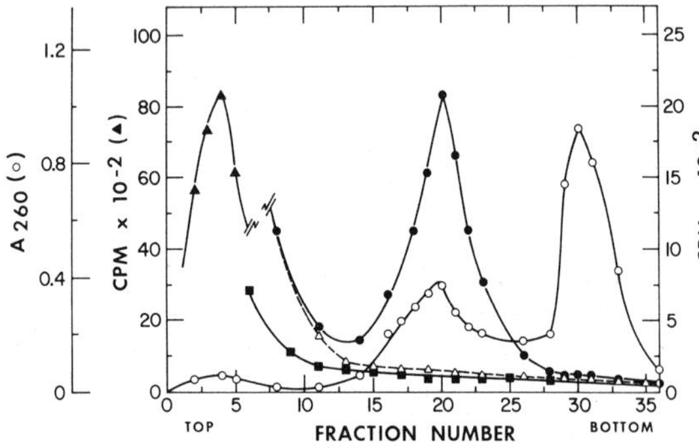

Fig. 3. Sucrose gradient separation of 30S and 50S ribosomal subunits following covalent attachment of S-(p-azidophenacyl)-[^{3}H]-valyl-tRNA. Irradiated incubation mixtures of tRNA and ribosomes were dissociated to 30S and 50S subunits by overnight dialysis against 10 mM Tris, pH 7.0, 50 mM KCl, 6 mM mercaptoethanol, and 0.3 mM $MgCl_2$, applied to a 10-30% sucrose gradient in the same buffer and centrifuged. Fractions were collected and the A_{260} and radioactivity of each fraction were determined. O——O, A_{260}; ●——●, complete incubation mixture; Δ——Δ, incubation mixture minus poly U,G (2:1); ■——■, complete incubation mixture but unirradiated.

Effect of increasing the length of the probe. In order to see if ribosomal proteins might be just beyond the range (8.5Å) of this probe, a longer probe was attached to the 4-thiouridine residue of the tRNA in the same way (Fig. 5). In its maximally extended form, this probe is 3.5 Å longer than the previous one and makes a somewhat different angle with the 4-thiouridine C_4-S_4 vector. Nevertheless, the same results were obtained when a tRNA carrying this probe was linked to the ribosome at the P site by irradiation. In this experiment N-acetyl-valyl-tRNA was used to insure the specificity of P site binding. About the same efficiency (18%) of covalent linking was observed, and all of the Val-tRNA was attached to the 16S rRNA.

Covalent link formation between ribosomes and S-(p-azidophenacyl)-[^{3}H]Val-tRNA bound at the A site. A site binding was defined as binding in 10 mM Mg^{++} dependent on added EFTu. By adding excess deacylated $tRNA^{Val}$ to block the P site (13), a good dependency was obtained (Fig. 6). Using these conditions, the kinetics of covalent link formation were examined, and found to be essentially

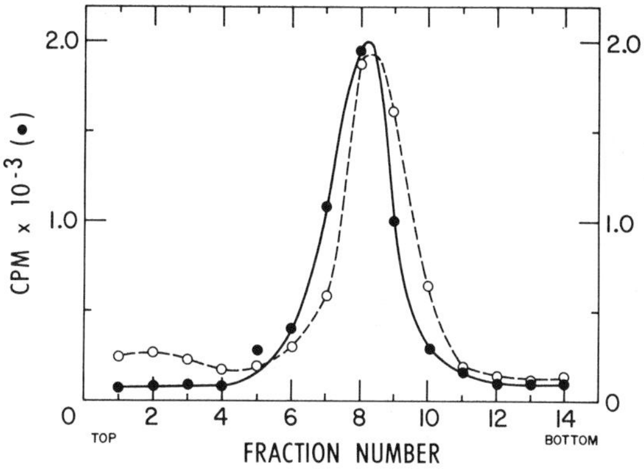

Fig. 4. SDS gradient separation of 30S proteins and 16S RNA. The 30S subunit-tRNA covalent complex was isolated from the sucrose gradient of Fig. 3 by adding $MgCl_2$ to 10 mM, and centrifuging at 55,000 rpm for 4 hrs. The pelleted 30S-tRNA complex was resuspended in 10 mM Tris, pH 7, 10 mM $MgCl_2$ to an A_{260} of 30, and layered on a 5-20% sucrose gradient containing 20 mM cacodylate, pH 5.8, 100 mM NaCl, 1 mM EDTA, and 0.5% SDS. After centrifugation, fractions were collected and the A_{260} and TCA-precipitable radioactivity determined. O—O, A_{260}; ●—●, cpm.

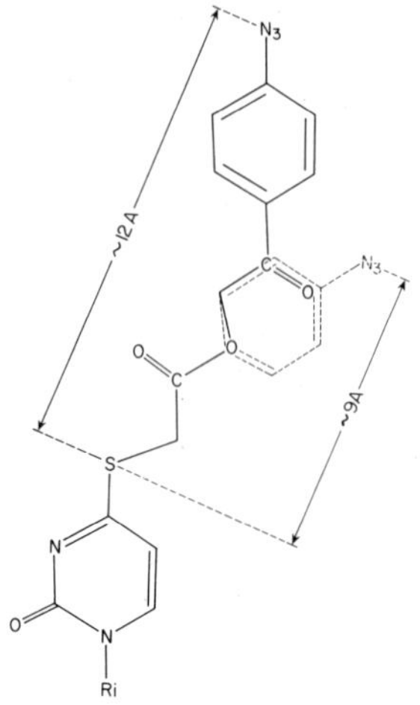

Fig. 5. Structure and orientation of the adduct of 4-thiouridine with the bromoacetic ester of p-azidophenacyl bromide. The ester is shown in solid lines and the p-azidophenacyl adduct in dotted lines.

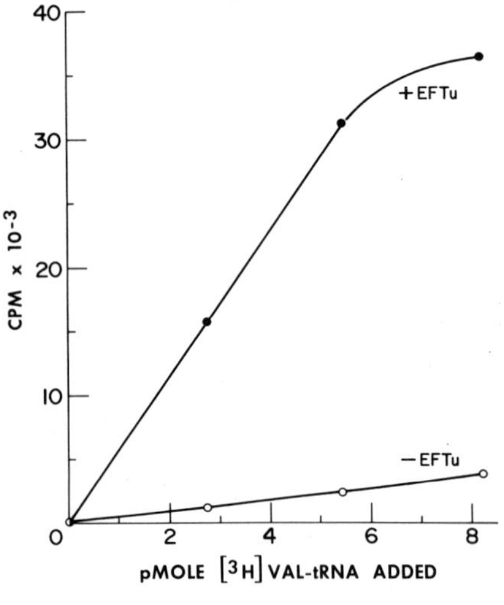

Fig. 6. Elongation factor-dependent binding of S-(p-azidophenacyl)-[^{3}H]valyl-tRNA to ribosomes. Ribosomal A site binding in 10 mM Mg^{++} was carried out as described in Materials and Methods using the indicated amounts of derivatized [^{3}H]Val-tRNA in the presence and absence of elongation factor Tu. 0.4 A_{260} units of uncharged $tRNA^{Val}$ was also added to block non-enzymatic binding at the P site. EFTu-dependent binding was unchanged in the absence (shown above) or presence of 10^{-4} M sparsomycin.

the same as for the P site reaction of Fig. 2. The efficiency of EFTu-dependent covalent linking, however, was much lower, being only 5.7% of that non-covalently bound.

When the covalent complex was dissociated into 30S and 50S subunits and separated on a sucrose gradient, the results shown in Fig. 7 were obtained. Not only was the overall efficiency of crosslinking reduced at the A site compared to the P site, but now

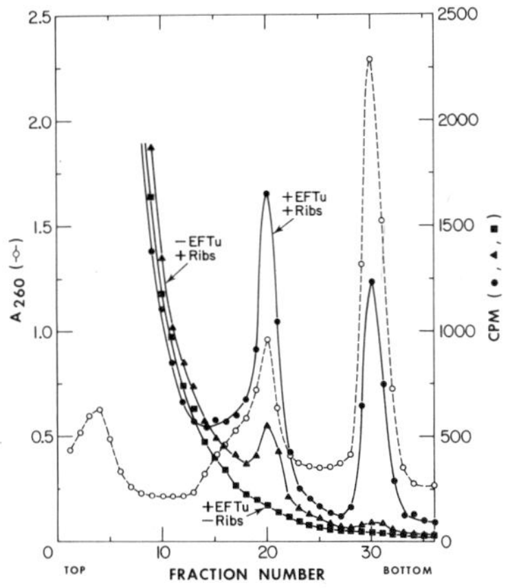

Fig. 7. Sucrose gradient separation of 30S and 50S ribosomal subunits following covalent attachment of S-(p-azidophenacyl)-[^{3}H]Val-tRNA bound at the A site. An A site binding reaction was carried out as in Fig. 6. After 30 min at 30° C, the reaction mixture was irradiated as described in Fig. 2 at 0° for 4 hrs, the optimal time. Sucrose gradient analysis was performed as described in Fig. 3. ○—○, A_{260}; ●—● , complete incubation mixture; ▲—▲ , minus EFTu; ■—■ , minus ribosomes.

the tRNA was about equally partitioned between the two subunits (54% with the 30S, and 46% with the 50S, after correction for EFTu-independent binding).

Both subunit-tRNA covalent complexes were isolated from the sucrose gradient and re-run on an SDS gradient as above in order to separate the ribosomal proteins from the RNA. The results are shown in Fig. 8. In contrast to the above results, some tRNA was now found attached to 30S ribosomal proteins. If all of the EFTu-independent binding is assumed to be due to P site binding to the

16S RNA, then 30% of the total 30S-bound tRNA was attached to protein. On the 50S, however, >95% of the tRNA bound was linked to the ribosomal protein and/or 5S rRNA. No linking to the 23S rRNA was found.

DISCUSSION

The probable orientation of the photoaffinity probe with respect to the three-dimensional structure of tRNA is illustrated in Fig.9. As shown in Fig. 5, the nitrene atoms of the probes in their

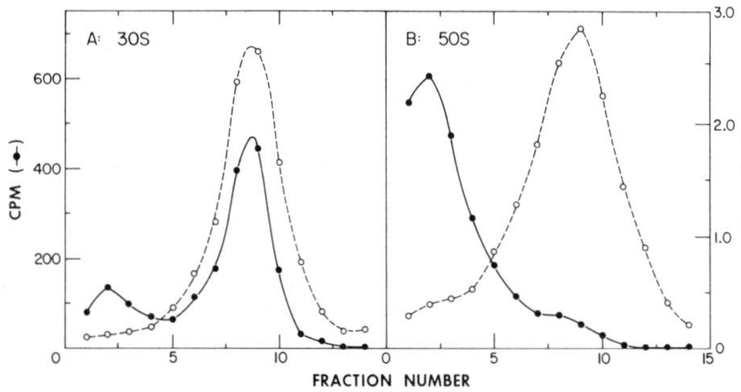

Fig. 8. SDS gradient separation of protein and RNA from the 50S and 30S ribosomal subunits. 30S and 50S subunit-tRNA covalent complexes were isolated from the sucrose gradient of Fig. 7, treated with SDS, and analyzed as described in Fig. 4. A: 30S subunit; B, 50S subunit. O—O, A_{260}; ●—●, cpm.

maximally extended states are 8.5 and 12A from the sulfur atom of the thiouridine and in a line angled approximately 120° from the C_4-S_4 vector. In the three-dimensional model (14), rotation about this vector would be restricted when the probe is attached, so that the active nitrene is probably constrained to the region illustrated in the Figure. However, before this data can be used to construct detailed models of tRNA-ribosome complexes, evidence is needed that this structure is maintained when tRNA is bound to the ribosome. Irrespective of the actual structure on the ribosome, exclusive labelling of the 16S rRNA from tRNA bound at the P site means that neither the RNA nor any proteins of the 50S subunit are within the 12A range of the probe, and also that none of the 30S proteins are within the reactive zone.

The marked difference between these results and those obtained when the tRNA was bound at the A site underscores the topological differences between these two sites and illustrates the sensitivity of the method to tRNA orientation on the ribosome. The fact that both 30S and 50S components were about equally labelled implies that at least part of the A site for tRNA is at or near the junction of the two subunits. Interpretation of the label distribution between RNA and protein would, at this stage, be premature.

The only other non-selective affinity probe attached to tRNA which has been used up to now was at the aminoacyl end and resulted in exclusive labelling of the 23S rRNA of the 50S subunit (4). Although there is too much flexibility inherent in the unpaired terminal nucleotides plus the derivatized aminoacyl group to allow any strong conclusions to be drawn about tRNA from a comparison of our results with this finding, it does seem clear that more of the tRNA structure must be in contact with the 30S subunit than has been supposed up to now.

Many other studies of tRNA binding sites on ribosomes have been carried out, using a variety of techniques such as specific chemical or immunological blocking, deletion or supplementation of spe-

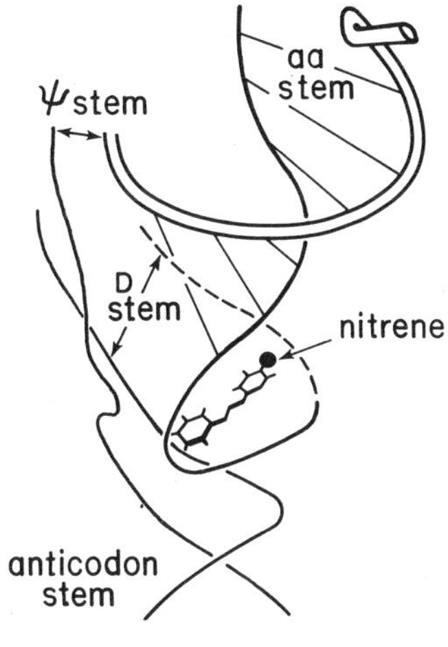

Fig. 9. Perspective drawing of the three-dimensional crystal structure of yeast phenylalanine tRNA (14) with the p-azidophenacyl probe inserted in its most probable orientation.

cific proteins, and protection by tRNA from some inactivating process. In our view, all these approaches suffer from the inability to distinguish between a direct effect on the tRNA binding site and an allosteric effect which indirectly alters the binding site. In addition, none of these approaches give any information about what region of the tRNA is involved. We believe the approach described here is direct, and within the normal geometric constraints, unambigious, and will prove to be of considerable utility. We are presently in the process of further characterizing the sites of attachment described above, and in addition, extending our studies to other binding sites, varying the length of the photoaffinity probe, placing these probes at other defined sites in tRNA, and examining other tRNAs such as $tRNA^{fMet}$

ABSTRACT

The S-(phenacyl-p-azide) of 4-thiouridine in E. coli $tRNA_1^{Val}$ was prepared for use as a photo-affinity probe of tRNA binding sites on ribosomes. The derivatized tRNA was 90-100% as active as control tRNA for aminoacylation, non-enzymatic binding to the ribosomal P site, EFTu-dependent binding to the A site, EFTu-GTP-AA-tRNA ternary-complex formation, and transfer of valine into polypeptide. Irradiation of S-(p-azidophenacyl)-[^{3}H]valyl-tRNA bound non-covalently to the ribosomal P site resulted in covalent attachment of 15-20% of the non-covalently bound tRNA to the ribosomes. Linking occurred exclusively to the 16S rRNA of the 30S ribosomal subunit. The same result was obtained when the bromoacetic ester of p-azidophenacyl bromide, a probe 3.5 A longer than the p-azidophenacyl group, was used in its place. Thus, that portion of the ribosome within 12A of the 4-thiouridine of tRNA, when it is bound in the P site, is solely 16S rRNA.

In contrast, irradiation of S-(p-azidophenacyl)-[^{3}H]valyl-tRNA bound non-covalently to the ribosomal A site under the influence of EFTu resulted in covalent attachment of 6% of the non-covalently bound tRNA. The tRNA was approximately equally distributed between both 30S and 50S subunits. Linking to the 30S subunit was 70% to the 16S rRNA and 30% to ribosomal protein but attachment to the 50S subunit was solely to the ribosomal proteins and/or 5S rRNA <5% being bound to the 23S rRNA.

REFERENCES

1. Oen, H., Pellegrini, M., Eilat, D., & Cantor, C.R. Identification of 50S proteins at the peptidyl-tRNA binding site of Escherichia coli ribosomes. Proc. Nat. Acad. Sci. USA 70,2799-2803 (1973).

2. Czernilofsky, A.P., Collatz, E.E., Stöffler, G. & Kuechler, E. Proteins at the tRNA binding sites of Escherichia coli ribosomes. Proc. Nat. Acad. Sci. USA 71, 230-234 (1974).

3. Hauptmann, R., Dzernilofsky, A.P., Voorma, H.O., Stöffler, G., & Kuechler, E. Identification of a protein at the ribosomal donor site by affinity labeling. Biochem. Biophys. Res. Commun. 56, 331-337 (1974).

4. Bispink, L. & Matthaei, H. Photoaffinity labeling of 23S r-RNA in Escherichia coli ribosomes with poly(U)-coded ethyl-2-diazomalonyl-Phe-tRNA. FEBS Lett. 37, 291-294 (1973).

5. Knowles, J.R. Photogenerated reagents for biological receptor-site labeling. Accounts Chem. Res. 5, 155-160 (1972).

6. Kumar, S.A., Krauskopf, M. & Ofengand, J. Effect of intramolecular photochemical crosslinking and of alkylation of 4-thiouridine in E. coli $tRNA^{Val}_{I}$ on the heterologous mischarging by yeast phenylalanyl-tRNA synthetase. J. Biochem. (Tokyo) 74, 341-353 (1973).

7. Ofengand, J., Delaney, P., & Bierbaum, J., in Methods in Enzymology, eds. Moldave, K. & Grossman, L. (Academic Press, New York), Vol. 29, pp. 673-684 (1974).

8. Ofengand, J., Bierbaum, J., Horowitz, J., Ou,C.-M. & Ishaq, M. Protein synthetic ability of Escherichia coli valine transfer RNA with pseudouridine,ribothymidine,and other uridine-derived residues replaced by 5-fluorouridine. J. Mol. Biol.,in press. (1974).

9. Krauskopf, M., & Ofengand, J. The function of pseudouridylic acid in transfer ribonucleic acid.Irradiation and cyanoethylation of E. coli valine tRNA fragments. FEBS Lett. 15,111-115 (1971).

10. Secrist, J.A., Barrio, J.R. & Leonard, N.J. Attachment of a fluorescent label to 4-thiouracil and 4-thiouridine. Biochem. Biophys. Res. Commun. 45, 1262-1270 (1971).

11. Sato, E. & Kanaoka, Y. Reaction of thiouracil and thiouridine with 2-hydroxyl-5-nitro-benzyl bromide. Biochim. Biophys. Acta 232, 213-216 (1971).

12. Hara, H., Horiuchi, T., Saneyoshi, M., & Nishimura, S. 4-thiouridine-specific spin-labeling of E. coli transfer RNA. Biochem. Biophys. Res. Commun. 38, 305-311 (1970).
13. Ofengand, J. & Henes, C. The function of pseudouridylic acid in transfer RNA. II. Inhibition of aminoacyl transfer RNA by T-ψ-C-G. J. Biol. Chem. 244, 6241-6253 (1969).
14. Kim, S.H., Quigley, G.F., Suddath, F.L., McPherson, A., Sneden, D., Kim, J.J., Weinzierl, J., & Rich, A. 3-dimensional structure of yeast phenylalanine transfer RNA: folding of the polynucleotide chain. Science 179, 285-288 (1973).

Identification of Central Transmitters with Electrically Stimulated Brain Slices

Fernando Orrego
Instituto Nacional de Cardiología, México 7, D. F.

It is quite surprising that the chemical nature of the transmitters that operate in the vast majority of neocortical synapses (i.e. 98 per cent of them), is unknown. Only acetylcholine (ACh), norepinephrine (NE), and 5 - hydroxy-tryptamine (5-HT) (1-3) seem to have qualified unambiguously as cortical transmitters, while dopamine (DA), that acts as a transmitter at lower levels, still has to be proven as such in the cortex. (4). However, these potent amines are released only in a small fraction of cortical synapses. The major excitatory and inhibitory transmitters being unknown.

Because of their powerful effects on neuronal function, several amino acids that occur in cortical neurons have become transmitter suspects (5). The acidic ones, glutamate and aspartate, exert a strong excitatory effect on all neurons of the central nervous system (CNS), in a similar manner to the natural excitatory transmitter(s), while small neutral amino acids like GABA, glycine, taurine and others, inhibit neuronal function. The electrophysiological effects of the natural inhibitory transmitter that is preponderant in higher levels of the CNS, being well imitated by GABA (6), while natural post-synaptic inhibition seen in the spinal cord, medulla and pons is mimicked by glycine (7). Pharmacological antagonism of exogenously applied GABA and glycine are also very similar to that of the corresponding natural transmitters (8-10). However, of all the amino acid suspects, only a relatively strong case has been made for GABA. The main arguments in its favour being: its presence in the central nervous system (11,12), where it seems to be located in neurons (13). The similarity of its action with that of the natural inhibitory transmitter in the cerebral cortex, Deiters nucleus and other structures (6,8-10), and the similar antagonism that bicuculline and picrotoxin exert on natural and GABA - mediated inhibition (8-10). GABA is released from the cerebral cortex *in vivo* following stimulation of it (14), and from neocortical chopped tissue preparations electrically stimulated *in vitro* (15), (however, on both these points, *vide infra*). The GABA synthesizing enzyme, glutamate decarboxylase I is present in the CNS (12) and, to a certain extent, concentrated in presynaptic axon terminals (16). Inactivation mechanisms, mainly through high affinity uptake, have also been described for GABA (17)

Our approach to the identification of cortical transmitters has been through the use of neocortical thin slices. This preparation, which we started to use while in the Lipmann Lab., has the advantages that incubation fluid may be rapidly modified, that the preparation may be electrically stimulated, and that, when properly obtained (18), the slices are capable of maintaining their respiration and capacity to synthesize macromolecules for several hours (19). In addition, slice neurons maintain good membrane potentials, that

may be depolarized by a variety of stimuli (20) and, even give rise to more complex electrical activity (21).
The release of neurotransmitters may be readily elicited by electrical stimulation of brain slices (2,3, 22), however, in several of such studies, amino acids that cannot be considered as transmitter suspects, such as arginine, cycloleucine, glutamine, leucine, lysine and proline are also released (22,23). This indicated that, although this preparation could be valuable for studying transmitter release, appropriate criteria for differentiating the release of transmitters and of non-transmitter amino acids had not been found. One notable exception being the work of McIlwain and Snyder (3), who showed that superfused piriform slices, when electrically stimulated, released NE and 5 - HT, but not glycine.

The purpose of the present study was to try to find, using superfused neocortical thin slices, criteria that would allow a discrimination to be made between the release of a known transmitter and of a known non-transmitter amino acid, and then apply these criteria to the study of transmitter suspects, mainly neuroactive amino acids.

METHODS

Slices were obtained from 120g Sprague - Dawley rats, with blade and blade-guide (18), and initially incubated in 1 ml. of a glucose-salt solution (24) containing 1 mM calcium ions and radioactive test substances (indicated in Figure Legends), in a Dubnoff shaker at 37°, with 5% CO_2 in O_2 as gas phase, for

30 to 60 min. After this period, in which the cells actively transport the test substances, the slices were placed in quick-transfer electrodes, and these in 30 ml. beakers, containing 4 ml. of glucose-salt solution, and appropriate connections for aspirating and replenishing the fluid, which was done every 2 min. Each aspirated fraction being collected separately in test tubes. The slice was continuously gassed with 5% CO_2 in O_2. Electrical stimuli were 50 Hz sine-wave current or pulses generated by a Grass S 4 A stimulator. All applied potentials were measured with an oscilloscope, and are expressed as maximum instantaneous voltages. After the superfusion, the slices were homogenized and the radioactivity remaining in it, as well as that present in each fraction counted in a scintillation counter (For details see ref. 25). Efflux is expressed as fractional rate constants (the fraction of radioactivity present in the slices, at any given time that is released per min)(26). This way of expressing the results, or other equivalent ones, being essential for meaningful comparison between substances.

RESULTS AND DISCUSSION

Electrically Induced Release of ^{3}H-NE and ^{14}C-AIB.

Initially, ^{3}H-L-NE was chosen as representative of known transmitters, and ^{14}C-α- aminoisobutyrate (AIB), a non-metabolizable amino acid, as representative of non-transmitter amino acids. AIB also has the advantage of being transported by the same transport system as small neutral amino acids, such as glycine, alanine and serine (27). When superfused slices, that

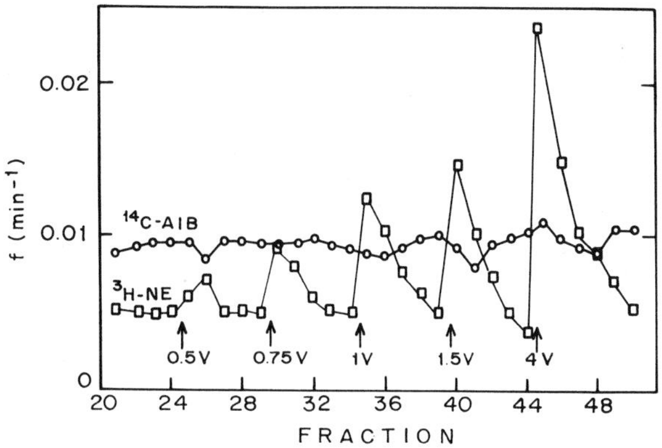

Figure 1. Electrically Induced Release of NE and AIB. Efflux is expressed as the fraction of c.p.m. present in the slice released per min. At the arrows, 50 Hz, sine wave stimulation was applied for 30 sec. In the incubation preceding superfusion, the slices were incubated with ^{3}H-L-NE (5μCi, 8 x 10^{-7} M) and ^{14}C-AIB (1 μCi, 10^{-4} M) for 60 min.

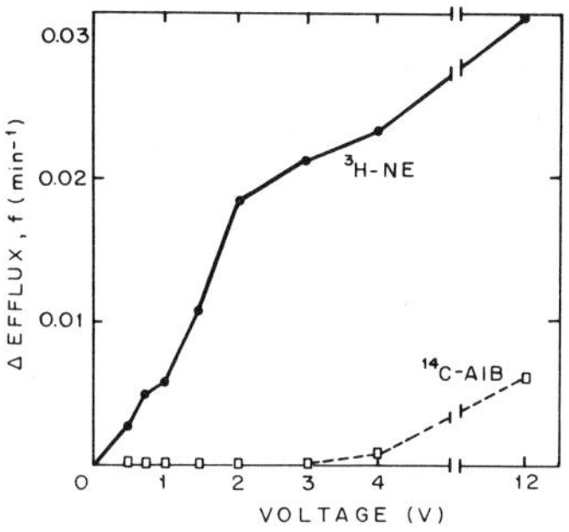

Figure 2. Voltage Profile of the Release of NE and AIB. The curves represent 8 experiments performed as in Fig. 1.

had previously accumulated ^{3}H-NE and ^{14}C-AIB, were electrically stimulated in a repetitive manner, starting with low applied potentials (Fig. 1), it was found that NE could be released with stimuli as weak as 0.5 V; and that the amount released was roughly proportional to stimulus strength, while the release of AIB became evident only with stimuli larger than 3 V (Fig. 2).

However, because of the nature of active transport, at the cell membrane level, of both NE and amino acids, these findings are not in themselves conclusive, and calcium dependency, a <u>sine qua non</u> requisite for transmitter secretion (28) had to be shown. This is so, because active transport depends on the asymmetric distribution of sodium ions on both sides of plasma membranes (29). The high external concentration of sodium, $(Na)_o$, increases the affinity of the transport system for its specific ligand on the outside of the membrane, while the low internal sodium, $(Na)_i$, by decreasing the affinity of the transport carrier on the internal side of the membrane, causes the release of the small ligand. Shuttling of the carrier from the internal to external surfaces of the membrane, and vice versa, leads to accumulation of substances against a concentration gradient. If $(Na)_i$ is increased, as is known to occur when brain slices are electrically stimulated (24), an increase in the efflux rate of all substances that are actively transported by sodium-dependent mechanisms, should be expected. However, because of the non-linear relationship between Na ions and affinity (Km) of the membrane carriers for their ligands (30), this may not be evident with small increases in $(Na)_i$. Also, important differences between carriers for different substances, with respect to their Na-affinity relationships may be predicted. That is, small increases in $(Na)_i$ may affect efflux rates of different substances to quite different degrees.

When the efflux of NE and AIB was studied in slices superfused with medium with no added calcium (Fig.3), a very different pattern was seen. No electrical-

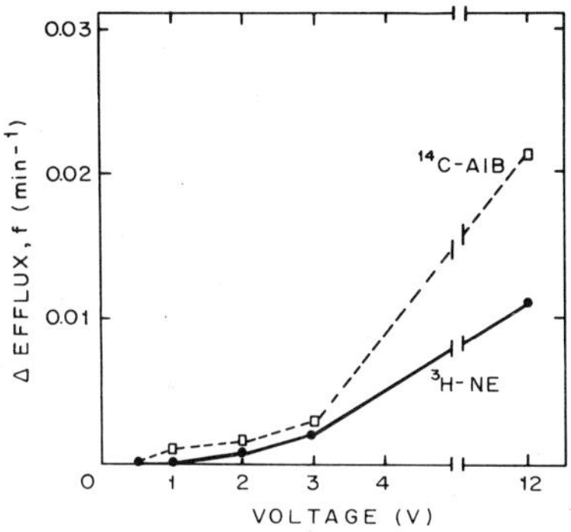

Figure 3. Voltage Profile of NE and AIB Efflux Induced in the Absence of Calcium. Calcium was omitted from superfusion but not from incubation fluid.

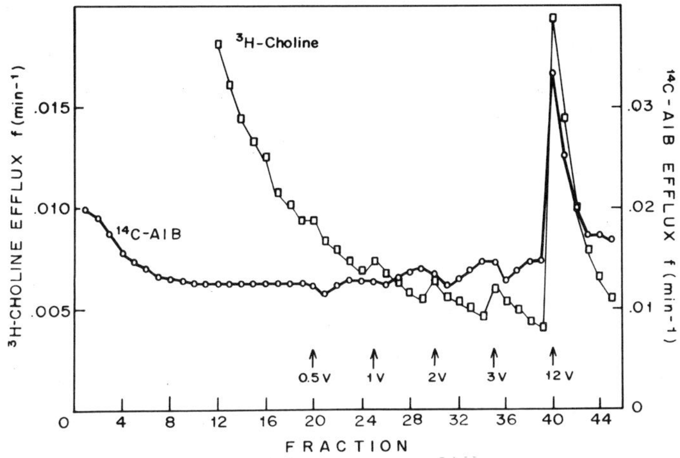

Figure 4. Electrically Induced Release of Acetylcholine. Slices were initially incubated with AIB and ^{3}H-Choline (5 μCi 2 x 10^{-6}M). The procedure is basically that of ref. 33.

ly-induced release of NE could be evidenced with 0.5 and 1 V, while that obtained with higher potentials was drastically reduced. With AIB, efflux now appeared at 1-2 V, and the one seen at higher potentials was noticeably increased. These findings are, to our knowledge, the first to show an absolute requirement for calcium in transmitter release from the electrically stimulated CNS _in vitro_. The increase in

AIB efflux, and its presence in the lower voltage region, may be interpreted as due to an increased Na influx, which is known to occur when electrical stimulation is applied in calcium-deficient media (31). Similarly, the small NE induced-efflux seen in this case, which parallels the efflux of AIB, may be interpreted as due to the release of "free" NE present in the cytoplasm, that is released by the reverse operation of the sodium-activated amine "pump", present in the plasma membrane of noradrenergic cells (32). The coexistence of these sodium-dependent and calcium-dependent processes, with the former enhanced when nominally no calcium is present (31), have been the cause of the great difficulty in demonstrating an absolute calcium requirement in these cases.

By the use of NE and AIB as model compounds, it seems that adequate criteria have been reached for classifying substances as "transmitter-like" or as "amino acid-like." The former are released in the region of lower applied potentials, and show an important dependence on calcium ions, while the latter are released with higher potentials, and their efflux is enhanced, or at least, not diminished, when no calcium is present. Because of the small diameter of noradrenergic axons in the mammalian neocortex, their excitability is probably rather low, and the release of other cortical transmitters possibly may be seen at the same or even lower potentials as those that induced NE release. Preliminary experiments on the release of ACh (Fig. 4), are consistent with such interpretation. However, because of the different relationships of sodium to the affinity of amino acid transport systems, it also seems possible that sodium-dependent

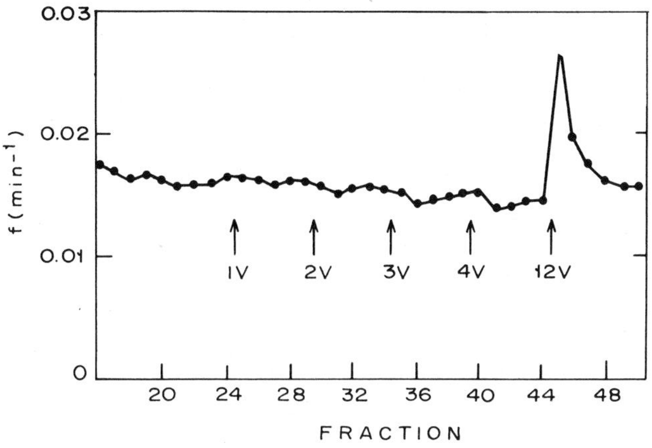

Figure 5. Release of GABA by sine-wave stimuli. Slices were incubated with ^{3}H-GABA (5 μCi, 5 x 10^{-7}M). Stimuli were 50Hz A.C. for 30 sec.

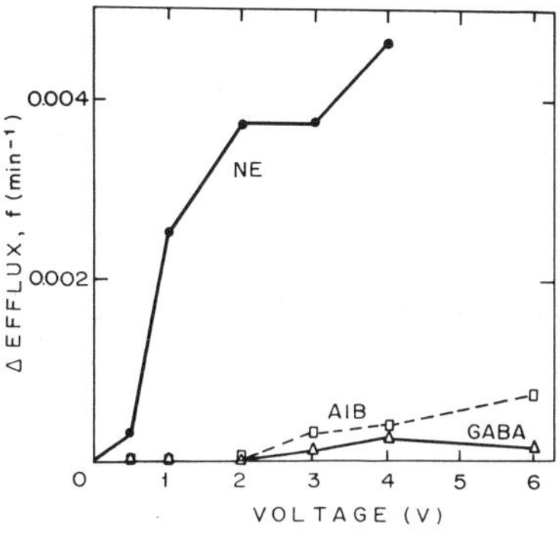

Figure 6. Voltage Profile of the Release of NE, GABA, and AIB by Rectangular Biphasic Pulses (80/sec, 3msec, 30sec). Data of 6 experiments. Note amplification of ordinate scale.

amino acid release may be, in some cases, seen at lower potentials than with AIB. Discrimination with transmitter secretion being then possible only on the basis of calcium dependency.

The Case for GABA.

When we started this series of experiments with GABA,

we were intimately convinced that it was the major, if not the only, cortical inhibitory transmitter. It was rather surprising to find that electrical stimulation was unable to release it in the transmitter region, and that increased efflux was only apparent in the same region where that of AIB is seen (Fig. 5). This was invariably the case when sine-wave stimuli were applied. However, the failure of this type of stimuli had already been reported by Srinivasan, Neal and Mitchell (15) working with chopped neocortical tissues. They also indicated that the release could only be obtained with stimuli that had a rapid potential rise and relatively fast frequency (more than 45/sec). When the slices were stimulated by us with rectangular, biphasic pulses (80/sec, 3 m sec), these were also unable to release GABA at potentials where NE is released (Fig. 6), and increased efflux only became evident at potentials where AIB was also released. Monophasic, square wave pulses (100 sec^{-1}, 5 msec), that are quite inconvenient in this type of studies, because they polarize the electrode wires and generate toxic electrolytic products (34-36) were also unable to release GABA without also releasing AIB.

It was also noticed that in other studies (15), GABA efflux was measured in the presence of amino-oxyacetic acid (AOAA), an inhibitor of GABA transaminase, while that of other amino acids, was studied in its absence. This additional difference could possibly explain some of the discrepancies between the work of Srinivasan et al. (15) and our own. When the electrically induced release of GABA and AIB was studied simultaneously in the presence of 10^{-5} M AOAA (Fig. 7),

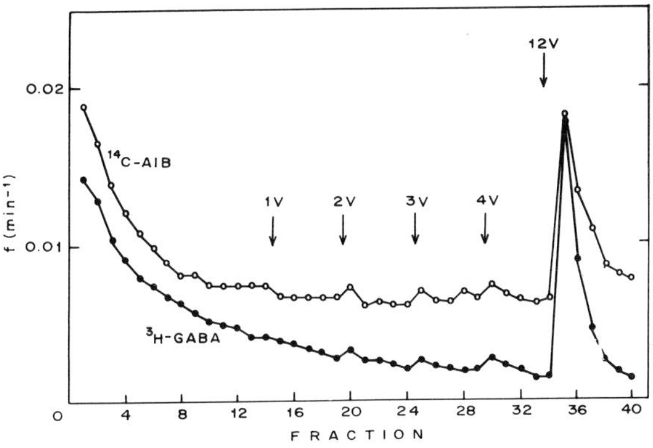

Figure 7. Release of GABA and AIB in the Presence of Amino-oxyacetic Acid. AOAA (10^{-5}M) was present during incubation and superfusion. Stimuli were 50Hz A.C. for 30 sec.

it was now found that <u>both</u> GABA and AIB were now released in the transmitter region. The induced efflux of GABA appearing at 2 Volts, and that of AIB at 2-3 Volts. With 5 x 10^{-5} M AOAA, the release was further displaced to lower voltages, GABA now being released at 1 V and AIB at 2 V, while peaks at 4 and 12 V of both substances also showed an increase. This release seen in the presence of AOAA did not diminish in media without calcium. Our interpretation of this phenomenon is that AOAA increases neuronal excitability in the slices, probably by increasing sodium influx during stimulation. This effect of AOAA also explains why the release of GABA has been readily seen when tested in the presence of AOAA, while that of non-transmitter amino acids, tested in its absence, was of much lesser magnitude (15).

These findings indicate that GABA is behaving in this system as a non-transmitter amino acid. Several other arguments such as the subcellular distribution of GABA (37,38), and the lack of effect of cytochalasin B, a blocker of neurotransmitter release, on stimulus-induced efflux of GABA from synaptosomes (39), also support a non-transmitter role for GABA. Kaczmarek and Adey (40). have been recently unable to release GABA *in vivo*, by electrical stimulation of the cortex, however, see also (14).

The present system, and the criteria we have obtained from it, that allow a rather clear discrimination to be made between transmitter and non-transmitter substances, seems to have an interesting potentiality. The clarification of the role of GABA in the CNS hopefully being a first step in the elusive problem of central transmitter identification.

SUMMARY

Criteria that allow a discrimination to be made between the release of transmitters and of non-transmitter amino acids, have been obtained with electrically stimulated superfused neocortical slices. When these criteria were applied to GABA, this amino acid behaved as a non-transmitter amino acid.

Acknowledgements.

I am grateful to O. Petit and M. Ortiz for their fine technical assistance. To Dulce María Ortiz for secretarial help. And, especially, to Leonor Ceruti, J. Jankelevich, E. Ferrera and R. Miranda who have collaborated in different aspects of this work.

REFERENCES

1. Hebb, C.: Central Nervous System at the cellular level: Identity of transmitter agents. Ann. Rev. Physiol. 32, 165-192 (1970).

2. Baldessarini, R.J. and Kopin, I.J., The effect of

drugs on the release of norepinephrine-3-H from central nervous system tissues by electrical stimulation in vitro, J. Pharmacol. Exp. Ther. 156, 31-38 (1967).

3. McIlwain, H. and Snyder, S.H., Stimulation of piriform and neocortical tissues in an in vitro flow-system: Metabolic properties and release of putative neurotransmitters, J. Neurochem. 17, 521-530 (1970).

4. Thierry, A.M., Blanc, G., Sobel, A., Stinus, L. and Glowinski, J., Dopaminergic terminals in the rat cortex, Science 182, 499-501 (1973).

5. Curtis, D.R. and Johnston, G.A.R., Amino acid transmitters, in Handbook of Neurochemistry (A. Lajtha, ed), 4, 115-134, Plenum Press, New York 1970.

6. Dreifuss, J.J., Kelly, J.S. and Krnjević, K., Cortical inhibition and γ-aminobytyric acid, Exp. Brain Res. 9, 137-154 (1969).

7. Werman, R., Davidoff, R.A. and Aprison, M.H., Inhibitory action of glycine on spinal neurons in the cat. J. Neurophysiol 31, 81-95 (1968).

8. Curtis, D.R., Duggan, A.W., Felix, D. and Johnston, G.A.R., GABA, bicuculline and central inhibition, Nature 226, 1222-1224 (1970).

9. Galindo, A., GABA-picrotoxin interaction in the mammalian central nervous system, Brain Res. 14, 763-767 (1969).

10. Curtis, D.R., Duggan, A.W. and Felix, D., GABA and inhibition of Deiters neurones, Brain Res. 23, 117-120 (1970).

11. Awapara, J., Landua, A.J., Fuerst, R. and Seale, B., Free γ-aminobutyric acid in brain, J. Biol. Chem. 187, 35-39 (1950).

12. Roberts E. and Frankel, S., γ-Aminobutyric acid in brain: its formation from glutamic acid, J. Biol. Chem. 187, 55-63 (1950).

13. Benjamin, A.M. and Quastel, J.H., Locations of Amino acids in brain slices from the rat, Biochem.

J. 128, 631-646 (1972).

14. Iversen, L.L., Mitchell, J.F. and Srinivasan, V., The release of γ-aminobutyric acid during inhibition in the cat visual cortex, J. Physiol. (Lond.) 212, 519-534 (1971).

15. Srinivasan, V., Neal, M.J. and Mitchell, J.F., The effect of electrical stimulation and high potassium concentrations on the efflux of ^{3}H-γ-aminobutyric acid from brain slices, J. Neurochem. 16, 1235-1244 (1969).

16. Fonnum, F., The distribution of glutamate decarboxylase and aspartate transaminase in subcellular fractions of rat and guinea-pig brain, Biochem J. 106, 401-412 (1968).

17. Iversen, L.L. and Johnston, G.A.R., GABA uptake in rat central nervous system: comparison of uptake in slices and homogenates and the effects of some inhibitors, J. Neurochem. 18, 1939-1950 (1971).

18. McIlwain, H. and Rodnight, R., Practical Neurochemistry, p.112-118, Little, Brown, Boston. 1962.

19. Orrego, F. and Lipmann, F., Protein synthesis in brain slices. Effect of electrical stimulation and acidic amino acids, J. Biol. Chem. 242, 665-671 (1967).

20. Hillman, H.H., Campbell, W.J. and McIlwain, H., Membrane potentials in isolated and electrically stimulated mammalian cerebral cortex, J. Neurochem. 10, 325-339 (1963).

21. Yamamoto, C. and Kawai, N., Origin of the direct cortical response as studied in vitro in thin cortical sections, Experientia 23, 821 (1967).

22. Katz, R.I., Chase, T.N. and Kopin, I.J., Effect of ions on stimulus-induced release of amino acids from mammalian brain slices J. Neurochem. 16, 961-967 (1969).

23. Hammerstad, J.P., Murray, J.E. and Cutler, R.W.P., Efflux of amino acid neurotransmitters from rat spinal cord slices. Factors influecing the electrically induced efflux of ^{14}C-glycine and

^{3}H-GABA, Brain Res 35, 357-367 (1971).

24. Keesey, J.C., Wallgren, H. and McIlwain, H., The sodium, potassium and chloride of cerebral tissues: Maintenance, change on stimulation and subsequent recovery, Biochemical J. 95, 289-300 (1965).

25. Orrego, F., Jankelevich, J., Ceruti, L. and Ferrera, E., Differential effects of electrical stimulation on the release of ^{3}H-norepinephrine and ^{14}C-α-aminoisobutyrate from brain slices, Nature, in the press.

26. Hopkin, J. and Neal, M.J., Effect of electrical stimulation and high potassium concentrations on the efflux of ^{14}C-glycine from slices of spinal cord, Brit. J. Pharmacol. 42, 215-223 (1971).

27. Levi, G., Blasberg, R. and Lajtha, A., Substrate specificity of cerebral amino acid exit in vitro, Arch. Biochem. Biophys. 114, 339-351 (1966).

28. Douglas, W.W., Stimulus-secretion coupling: the concept and clues from chromaffin and other cells, Brit. J. Pharmacol. 34, 451-474 (1968).

29. Schultz, S.G. and Curran, P.F., Coupled transport of sodium and organic solutes, Physiol. Rev. 50, 637-718 (1970).

30. Martin, D.L. and Smith, A.A., Ions and the transport of gamma-aminobutyric acid by synaptosomes, J. Neurochem. 19, 841-855 (1972).

31. Frankenhaeuser, B. and Hodgkin, A.L., The action of calcium on the electrical properties of squid axons, J. Physiol. (Lond.) 137, 218-244 (1957).

32. Hamberger, B., Reserpine resistant uptake of catecholamines in isolated tissues of the rat. Acta Physiol. Scand. Suppl. 295, 1-56 (1967).

33. Somogyi, G. and Szerb, J.C., Demonstration of acetylcholine release by measuring efflux of labelled choline from cerebral cortical slices, J., Neurochem. 19, 2667-2677 (1972).

34. Shipley, J.W. and Goodeve, C.F., The law of alternating current electrolysis and the electro-

lytic capacity of metallic electrodes, Trans. Am. Electrochem. Soc. 52, 375-402 (1927).

35. McIlwain, H., Metabolic responses in vitro to electrical stimulation of sections of mammalian brain, Biochem. J. 49, 382-393 (1951).

36. Miyamoto, H. and Kasai, M., Reexamination of electrical stimulation on sarcoplasmic reticulum fragments in vitro, J. Gen. Physiol. 62, 773-786 (1973).

37. Mangan, J.L. and Whittaker, V.P., The distribution of free amino acids in subcellular fractions of guinea-pig brain, Biochem. J. 98, 128-137 (1966).

38. De Belleroche, J.S. and Bradford, H.F., Amino acids in synaptic vesicles from mammalian cerebral cortex: a reappraisal, J. Neurochem. 21, 441-451 (1973).

39. Nicklas, W.J. and Berl, S., Effects of cytochalasin B on uptake and release of putative transmitters by synaptosomes, Nature 247, 471-473 (1974).

40. Kaczmarek, L.K. and Adey, W.R., The efflux of ^{45}Ca and ^{3}H-γ-aminobutyric acid from cat cerebral cortex, Brain Res. 63, 331-342 (1973).

The Acetylation of Cysteine-281 in Glyceraldehyde-3-Phosphate Dehydrogenase by an S→S Transfer Reaction

Jane Harting Park and Blanche P. Meriwether
Department of Physiology and Neuromuscular Disease Center
Vanderbilt University Medical School
Nashville, Tennessee 37232

INTRODUCTION

In glyceraldehyde-3-phosphate dehydrogenase, cysteine-149 is the catalytically active site for the dehydrogenase, transferase, and esterase activities of the enzyme (1,2). This cysteine residue is acetylated at pH 7.0 by the substrates, acetyl phosphate and p-nitrophenyl acetate (1,2). When the pH is raised to 8.5, the acetyl group migrates to lysine-183 in an S→N transfer reaction (3,4). In the course of the studies on the S→N transfer, the formation of a small amount of a third acetyl-enzyme complex was noticed by Park and Meriwether (5,6). In order to prepare larger quantities of the new complex, experimental conditions were altered so as to stoichiometrically acetylate another cysteine moiety - namely, cysteine-281. This thioester bond appears to be formed by an S→S transfer reaction. The data indicate that cysteine-149 and cysteine-281 can be approximated when the three dimensional structure of the enzyme is labilized by removal of the bound NAD and an increase in the temperature from 0° to 37°. The implications of the S→S transfer reaction will be discussed in terms of enzyme catalysis, cellular physiology, and the interesting differences of glyceraldehyde-3-phosphate dehydrogenases crystallized from various species.

MATERIALS AND METHODS

Crystalline glyceraldehyde-3-phosphate dehydrogenase from rabbit muscle and yeast, ^{14}C-p-nitrophenyl acetate, and ^{14}C-acetyl phosphate were prepared as described in previous papers (1,2). The analytical procedures are also detailed in prior publications (1,2,5,6) or in the text.

RESULTS

When glyceraldehyde-3-phosphate dehydrogenase is acetylated with ^{14}C-p-nitrophenyl acetate (PNPA) at pH 7.0 and 0°, a thioester bond is formed at the catalytically active site, cysteine-149. Three to four moles of acetyl groups are bound per mole of enzyme (Table I).

Table I: The Acetylation of Glyceraldehyde-3-Phosphate Dehydrogenase with ^{14}C-p-nitrophenyl acetate at 0° and 37°.

Conditions and Observations	Source of Enzyme			
	Rabbit		Yeast	
Temperature of Acetylation	0°	37°	0°	37°
Moles ^{14}C-Acetyl per Mole Enzyme	3.4	9.2	3.2	4.1
Physical State	soluble	precipitates in 2 mins.	soluble	faint turbidity

The charcoal treated, NAD free glyceraldehyde-3-phosphate dehydrogenase (0.1 µmole, 14 mg) was incubated with ^{14}C-p-nitrophenyl acetate (2.0 µmoles) at pH 7.0 for 20 minutes at 0° or 37° as indicated above. The moles of ^{14}C-acetyl group bound per mole of enzyme was determined as previously described (1). The dehydrogenase from both rabbit and yeast became slightly turbid when incubated at 37° with an amount of alcohol equivalent to that used in dissolving the p-nitrophenyl acetate.

The acetylated enzyme was digested with pepsin and the radioautograph of the ^{14}C-acetyl peptides (Fig. 1) was obtained after electrophoresis at pH 3.5. The peptides S1 to S4 are fragments

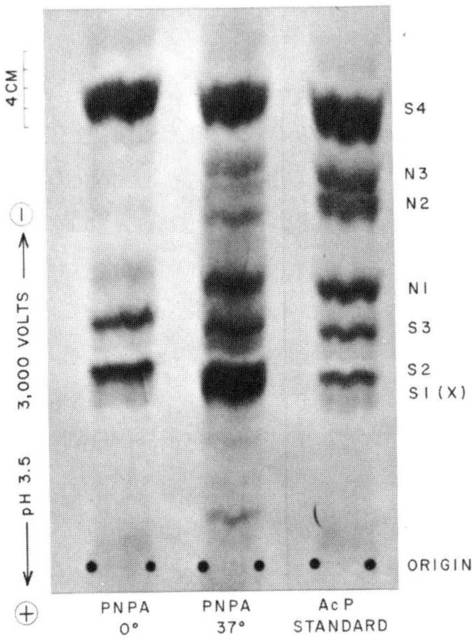

Fig. 1. The Effect of Temperature on the Acetylation of Glyceraldehyde-3-phosphate dehydrogenase with ^{14}C-p-nitrophenyl acetate.

The two left hand columns show the ^{14}C-acetyl peptides obtained from the pepsin digestion of the acetyl enzymes prepared from rabbit muscle glyceraldehyde-3-phosphate dehydrogenase and ^{14}C-p-nitrophenyl acetate at 0° or 37° as outlined in Table I. The procedures for the electrophoresis at pH 3.5 have been described previously (2). For identification of the peptides, the right hand column shows the pattern obtained from the pepsin digestion of an acetyl enzyme labeled with ^{14}C-acetyl phosphate at pH 7.0 and room temperature (3). The S-acetyl peptides, S1 to S4, are fragments of the active center octadecapeptide surrounding cysteine-149. The N-acetyl peptides, N1 to N3, are components of the tridecapeptide containing lysine-183. Acetylation at 37° produces a marked darkening of film in the area of the S1 peptide which is designated (X).

of an octadecapeptide with the sequence as previously determined in collaboration with J.I. Harris (2):

$$
\begin{array}{c}
{}^{14}COCH_3 \\
| \\
\text{Lys-Ile-Val-Ser-Asn-Ala-Ser-}\underset{\underline{149}}{\text{Cys}}\text{-Thr-Thr-Asn-}\underset{\underline{153}}{\text{Cys}}\text{-Leu-Ala-Pro-Leu-Ala-Lys.}
\end{array}
$$

When the enzyme is acetylated at 37° instead of 0°, there are three striking differences. First, the moles of acetyl groups bound increases to about 9.0 (Table I). Secondly, the enzyme precipitates within two minutes. Thirdly, a number of new peptides appear in the radioautograph (Fig. 1). There is a substantial amount of radioactivity in the area of the bands N1, N2, and N3. These three peptides are seen in the acetyl phosphate standard and are known to contain the amino acids surrounding ^{14}C-N-acetyl-lysine-183. In addition, there is increased activity in the area of the S-1 peptide, which has been designated (X). This acetylated residue was preliminarily identified as cysteine since the ^{14}C-label was removed by performic acid oxidation.

There are four cysteine residues per monomer of rabbit muscle dehydrogenase in positions 149, 153, 244 and 281. The likelihood of additional acetylation of residue 153 could be ascertained with the yeast enzyme which has only two cysteine residues per monomer, namely, cysteine 149 and 153. When the yeast enzyme was treated with p-nitrophenyl acetate at 37°, the number of bound ^{14}C-residues only increased from 3.2 to 4.1 (Table I); therefore, cysteine-153 is not an easily accessible site. Moreover, there was no significant precipitation of the yeast protein, and the faint turbidity was equivalent to that produced by the alcohol used to dissolve the p-nitrophenyl acetate.

In order to determine the amino acid sequence of the radioactive residue in the (X) band, several large scale acetylations were carried out with 150 mgs of rabbit muscle enzyme. The precipitation of the labeled enzymes and the proteolytic digestion were performed as previously described (1,2). The peptides were partially separated by passage over a G-25 Sephadex column using 2% acetic acid as the eluant. The radioactive peptides which appeared in two major peaks were further purified by electrophoresis at pH 3.5, 2.1 and 6.5. The amino acid composition of the eluted bands indicated that the major portion of the radioactivity was in peptides containing cysteine-281. The following sequence was determined from characterization of the peptide fragments:

$$\text{Asp-Gln-Val-Val-Ser-}\underset{281}{\overset{^{14}COCH_3}{\overset{|}{\text{Cys}}}}\text{-Asp}$$

There was a minor contamination by small fragments of the active center octadecapeptide. No significant amount of impurity was noted from peptides containing cysteine-244 with a known sequence:

$$\text{Asp-Leu-Thr-}\underset{244}{\text{Cys}}\text{-Arg-Leu-Glu}$$

In order to determine the moles of ^{14}C-acetyl-cysteine-281 per mole of dehydrogenase, the bands from radioautographs were eluted and counted. After suitable corrections were made for the losses occurring during this procedure, it was found that three to four moles of cysteine-281 were labeled per mole of enzyme.

Mechanism of the S-acetylation of Cysteine-281.

By analogy with the S→N migration, it appeared that cysteine-281 was probably acetylated by a transfer of the ^{14}C-acetyl group from cysteine-149 to cysteine-281 in an S→S transfer reaction. This possibility was investigated by blocking cysteine-149 and then attempting to acetylate cysteine-281 at pH 7.0 and 37°. For this experiment, the active site cysteine was first specifically carboxymethylated with iodoacetic acid

(2), and then ^{14}C-p-nitrophenyl acetate was added. Carboxymethylation of the dehydrogenase lowered the number of bound ^{14}C-acetyl groups from 8.7 to 2.0 (Table II). As shown in the radioautograph (Fig. 2), there was some faint acetylation of cysteine-281 and lysine-183. Almost no radioactivity was observed in the active site peptides. p-Nitrophenyl acetate is a very reactive substrate, and a partial acetylation of the lysine-183 residues in the carboxymethylated dehydrogenase has been previously observed at pH 8.5 and 0° or room temperature (6). Thus it was not unexpected that one might see some direct acetylation reactions at pH 7.0 and 37°. However, the very marked impairment of the S-281 and N-183 acetylations by carboxymethylation could indicate that the major pathway for labeling of these two residues is the S⇀S or S⇀N transfer reaction. Alternatively, the carboxymethylated enzyme might be sterically hindered in a manner analogous to that of the histidine in ribonuclease (7).

Table II: The Inhibition of the Acetylation of Glyceraldehyde-3-phosphate Dehydrogenase by Carboxymethylation or NAD

Temperature	Additions	Moles of ^{14}C-Acetyl Bound per Mole of Enzyme
0°	none	2.6
37°	none	8.7
37°	IAA	2.0
37°	NAD	1.7

The enzyme (0.1 µmole) was preincubated at 0° for 10 minutes with iodoacetic acid (4.0 µmoles) or NAD (1.0 µmole) where indicated in the above Table. The protein was then acetylated for 10 minutes with p-nitrophenyl acetate (2.0 µmoles) at 0° or 37°. The determination of the moles of ^{14}C-acetyl groups bound to the enzyme was carried out as described in Table I.

These results were confirmed in a second experiment using NAD to block the acetylation of cysteine-149 (1). It has been assumed that the bound coenzyme sterically hinders the approach of the large aromatic ring of p-nitrophenyl acetate (1). In the presence of coenzyme, acetylation at 37^0 was again inhibited, and just 1.7 moles of ^{14}C-acetyl groups were bound per mole of enzyme (Table II). As predicted, only a trace of radioactivity appeared in the active site peptides (Fig. 2). There was a weak and approximately equal labeling of both cysteine-281 and lysine-183. In the presence of NAD or IAA there was no precipitation of the enzyme, indicating that acetylation of more than one residue of cysteine-281 per tetramer or simultaneous labeling of three different residues is necessary for denaturation.

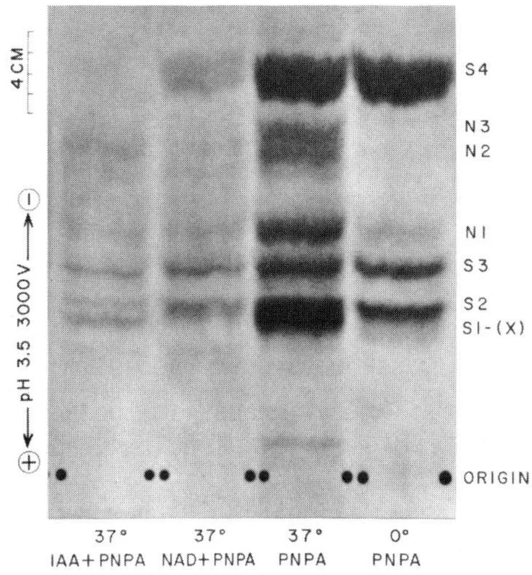

Fig. 2. Inhibition of the Acetylation of the Dehydrogenase by Carboxymethylation or NAD.

The ^{14}C-acetyl enzymes of this radioautograph are those which were analyzed in Table II. The procedures and the identification of the fragments are outlined in Fig. 1. The significance of the peptide pattern is explained in the text.

Function of Cysteine-281

It is difficult to determine whether cysteine-281 is involved in the catalytic mechanism because of the rapid denaturation after the acetylation of about two of these residues. However, it was possible to prepare a soluble S-281 acetylated enzyme with only a small amount of radioactivity in this position. For this preparation, the dehydrogenase was acetylated at pH 7.0 and 0° and then passed over a Sephadex column. The S→S and S→N transfers occurred at minimal rates during Sephadex chromatography (6). Radioautography and quantitative measurements of the ^{14}C-bands demonstrated that the enzyme contained nearly one residue of ^{14}C-acetyl-cysteine-281 per tetramer. NAD and arsenate (6) were added to the Sephadex-treated enzyme in an attempt to deacetylate this residue. However, neither the S-acetyl-cysteine-281 nor the N-acetyl-lysine-183 could be deacetylated as measured by ^{14}C-counting and confirmed by radioautography. This is in direct contrast to experiments with the readily deacetylated cysteine-149 (6).

The Sephadex treatment itself demonstrates the stability of the acetyl moiety on cysteine-281 which acquired almost one ^{14}C-acetyl group during the passage through the column. On the other hand, 60% of the ^{14}C-acetyl groups at the active site were either lost or transferred. The acetyl group on cysteine-281 is probably too stable for catalytic reactivity and behaves like a model thioester, i.e. S-acetylgluthathione. Thus one can conclude that this residue is involved in maintaining the structural conformation.

DISCUSSION

Glyceraldehyde-3-phosphate dehydrogenase from a variety of sources has been shown to have an identical sequence around the active site cysteine-149 and lysine-183 (8). The overall identity of amino acid residues in the dehydrogenases from pig, lobster, and yeast is 60% (8). Rossmann and his collaborators have proposed an even broader homology in the crystalline structure of the NAD binding area (residues about 1-160) in four dehydrogenases, glyceraldehyde-3-phosphate, lactate, malate, and alcohol dehydrogenase (9). Such similarities are indeed interesting from both a functional and an evolutionary point of view. However, just as fascinating are the differences between the triose phosphate dehydrogenases crystallized from different species. For example, the fact that S→S and S→N transfers occur more readily in the rabbit than in the yeast enzyme may be related to the tighter binding of NAD to the mammalian enzyme as required for protein stabilization. The lobster enzyme has three entirely different cysteine residues at positions 22, 130, and 250; however, the effect of acetylation at 37° has not been tested with this enzyme (8). Another illustration of differences in the significance of amino acid residues is histidine-38. This histidine moiety is required for maximal rates of the phosphorylation step as catalyzed by the rabbit muscle enzyme (10). Photooxidation of the lobster enzyme indicates this is not the case with the arthropod muscle protein since histidine-38 is replaced by glutamic acid (11,8). These variations, which occur outside the active center peptide, may well be responsible for the species variability in properties such as binding constants, reaction rates, crystallographic and immunological characteristics.

To date, it has not been definitively determined whether the S→S transfer reaction is intermolecular, intramolecular, or intramonomeric. The data strongly suggest that it is at least intramolecular, and that the participating residues are probably within a 5 Å distance. In this regard, the crystallographic data may be helpful unless the three dimensional structure is severely altered by NAD removal or a high temperature of 37°. As in the case of the S→N transfer (6), preliminary data indicate that histidine-38 is not involved in the S→S transfer. The participation of other histidine residues has not been excluded.

In addition to residue proximity, the transfer reactions may occur so readily because of the special properties of cysteine-149 which is 800 times more reactive than cysteine itself (12). The acetylation of lysine-183 and cysteine-281 are essentially irreversible transfers--the former, by virtue of the stability of the N-acetyl bond and the latter by precipitation of the protein. Dual acetylation of cysteine-149 and lysine-183 does not cause precipitation (5). Moreover, at pH 6.4, the rate of the S→N transfer was decreased more than the S→S transfer, but the enzyme was nonetheless denatured. This again indicates that the cysteine-281 was responsible for precipitation.

Both the S→S and S→N transfer reactions inactivate the enzyme and may, therefore, be considered as the catalytic equivalents of a "dead-end street" (5,6). It has been suggested that the N-acetylation of lysine-183, as well as the photooxidation of histidine-38, may inhibit the dehydrogenase activity by inducing conformational changes (6,10). Such a change may provide at least a partial explanation for these two modifications, but it is obviously the primary mechanism for inhibition by alteration of cysteine-281. With regard to the biological consequences of these inhibitions, the dehydrogenase has a natural defense against self-denaturation in the strongly bound NAD, which would prevent inactivation. Glyceraldehyde-3-phosphate dehydrogenase is unique among dehydrogenases in crystallizing with three moles of NAD bound per mole of enzyme and in having a higher binding constant for NAD than NADH. Since the NAD concentration in many cells is about fifteen times higher than NADH, the intracellular conditions appear optimal for protection of the enzyme *in vivo*.

SUMMARY

Under physiological conditions, pH 7.0 and 37°, glyceraldehyde-3-phosphate dehydrogenase can be acetylated by p-nitrophenyl acetate at cysteine-149, lysine-183 and cysteine-281. The newly observed acetylation at cysteine-281 occurs by an S→S transfer reaction of the acetyl group from cysteine-149 to cysteine-281. This substrate migration rapidly inactivates the enzyme by causing precipitation of the protein. Cysteine-281, therefore, appears to be involved in maintaining the stability of the protein structure. NAD protects against the inactivation and thereby assumes the unique role of channeling the substrate down the proper catalytic pathway, and it may also aid in preventing the formation of aberrant enzyme complexes.

ACKNOWLEDGEMENTS

This work was supported by grants from the National Science Foundation, U.S. Public Health Service, and the Muscular Dystrophy Association of America. We are grateful to Miss M. Hopfer and Mrs. P. Stelling for assistance in the preparation of this manuscript.

REFERENCES

1. Park, J.H., Meriwether, B.P., Clodfelder, P., and Cunningham, L.W.: The hydrolysis of p-nitrophenyl acetate catalyzed by 3-phosphoglyceraldehyde dehydrogenase. J. Biol. Chem., 236, 136 (1961).
2. Harris, J.I., Meriwether, B.P., and Park, J.H.: Chemical nature of the catalytic sites in glyceraldehyde-3-phosphate dehydrogenase. Nature, 198, 154 (1963).
3. Mathew, E., Agnello, C.F., and Park, J.H.: N-acetylation of 3-phosphoglyceraldehyde dehydrogenase by substrates. J. Biol. Chem., 240, 3232 (1965).

4. Park, J.H., Agnello, C.F., and Mathew, E.: S→N transfer and dual acetylation in the S-acetylation and N-acetylation of 3-phosphoglyceraldehyde dehydrogenase by substrates. J. Biol. Chem., 241, 769 (1966).
5. Park, J. H.: The catalytic significance of S-acetylation and N-acetylation of 3-phosphoglyceraldehyde dehydrogenase. in N.O. Kaplan and E. Kennedy (Editors), Current aspects of biochemical energetics, Academic Press, New York, 1966, p. 299.
6. Mathew, E., Meriwether, B.P., and Park, J.H.: The enzymatic significance of S-acetylation and N-acetylation of 3-phosphoglyceraldehyde dehydrogenase. J. Biol. Chem., 242, 5024 (1967).
7. Crestfield, A.M., Stein, W.H., and Moore, S.: The properties and conformation of the histidine residues at the active site of ribonuclease. J. Biol. Chem., 238, 2421 (1963).
8. Jones, G.M.T., and Harris, J.I.: Glyceraldehyde 3-phosphate dehydrogenase: Amino acid sequence of enzyme from baker's yeast. FEBS Letters, 22, 185 (1972).
9. Buehner, M., Ford, G.C., Moras, D., Olsen, K.W. and Rossman, M.: D-glyceraldehyde-3-phosphate dehydrogenase: Three-dimensional structure and evolutionary significance. Proc. Nat. Acad. Sci., 70, 3052 (1973).
10. Francis, S.H., Meriwether, B.P., and Park, J.H.: Effects of photooxidation of histidine-38 on the various activities of glyceraldehyde-3-phosphate dehydrogenase. Biochemistry 12, 346 (1973).
11. Park, J.H., Meriwether, B.P., and Hill, E.: unpublished results.
12. Olsen, E., and Park, J.H.: Studies on the mechanism and active site for the esterolytic activity of 3-phosphoglyceraldehyde dehydrogenase. J.Biol.Chem.,239, 2316 (1964).

Regulation of Elongation Factor G-Ribosomal GTPase Activity

A. Parmeggiani, G. Sander, R. C. Marsh, J. Voigt, K. Nagel, and G. Chinali
Gesellschaft für Molekularbiologische Forschung, Abteilung Biochemie,
3301 Stöckheim über Braunschweig, German Federal Republic

Elongation factor G (EF-G) plays an essential role in the process of polypeptide chain growth. As already suggested in 1966 by Nishizuka and Lipmann, the EF-G-ribosomal GTPase is involved in the translocation of peptidyl-tRNA from the acceptor- to the peptidyl-site of the ribosome (1-4). It is not yet known whether the catalytic center for GTP hydrolysis is primarily localized on EF-G or on the ribosome; however, the requirement of these two components for GTPase activity underlines the essential role of both EF-G and ribosomes for expression of this reaction. Unlike elongation factor T (EF-T)-ribosomal GTPase, which requires aminoacyl-tRNA and is stimulated by mRNA (5), EF-G GTPase is also active in the absence of these two components of protein biosynthesis (6). The ribosomal center for EF-G-dependent GTP hydrolysis has been localized on the 50S subunit (7-8) but the 30S subunit plays an important role in the expression of turnover GTPase activity. Indeed, proteins from both ribosomal subunits have been identified as being involved (9-17). At high monovalent cation concentrations, where optima for polyphenylalanine synthesis (18,19) or for EF-G-dependent GTPase activity (20-22) have been observed, association of ribosomal subunits is fundamental for expression of the turnover activity of the GTPase reaction (6,13, 23). There has, however, been disagreement among authors as to the amount of turnover GTPase activity which can be supported by the 50S subunit (1,6, 23-25). To reconcile these differences and to gain a better insight into the regulation of this activity we have examined the interdependence of several parameters in this reaction (21,22).

MATERIALS AND METHODS

Pure EF-G, EF-T (EF-Tu + EF-Ts) and NH_4Cl-washed ribosomes from E. coli B $T2^r$

or A19 were prepared essentially as described (21,26). Ribosomal subunits were isolated by sucrose density gradient centrifugation at 0.5 mM $MgCl_2$ using a Spinco 15 Ti zonal rotor. The 30S were 98% pure and the 50S 95-96%, as measured by sedimentation in analytical sucrose density gradients. $tRNA^{Phe}$ was purified to 50% and charged as described (27). N-acetylation of Phe-tRNA was carried out according to Haenni and Chapeville (28). To obtain maximal activity with the resulting 70S ribosomes, 30S subunits were present in the assays in a sixfold molar excess over 50S. Details of the individual experiments are described in the legends.

RESULTS AND DISCUSSION

It occurred to us that the concentration of monovalent cations, which plays an important role in protein biosynthesis in vitro (6,18,19), might also be a major determinant for the turnover of 50S-EF-G GTPase activity. Fig. 1 shows the striking effect of NH_4^+ and K^+ on EF-G GTPase activity with the 50S subunit alone. This ribosomal subunit became increasingly able to support this reaction when the concentration of NH_4^+ and K^+ was lowered below that required for optimal GTPase activity in the presence of added 30S subunits. Maximal activity was obtained with only 2 mM NH_4^+ or K^+, the amount carried over with the ribosomes into the assay. Increasing concentrations of these cations caused progressive inhibition. A different effect of NH_4^+ and K^+ was observed in the presence of 30S. Here inhibition by NH_4^+ and K^+ was relieved and these cations, particularly NH_4^+, became stimulatory. The optimum with NH_4^+ (70-100 mM) was about twice as high as with K^+ (25-40 mM). The molar ratio of EF-G to 50S subunits greatly affected both 50S-dependent GTPase activity and the stimulatory action of 30S at low monovalent cations. At a ratio of 5 to 1 no stimulation by 30S was observed, but at a ratio of 1 to 1, which is approximately that observed in _E. coli_ cells (29), the stimulation was two- to threefold. 50S-EF-G GTPase was very low at concentrations of NH_4^+ or K^+ higher than 100 mM, and most of this activity could be explained by the 5% contamination with 30S. Our results show that 50S are able to efficiently support EF-G GTPase activity at concentrations for monovalent cations far below the optimum for in vitro protein biosynthesis. At 14 mM Mg^{2+} optimal activity in poly(U)-directed polyphenylalanine synthesis

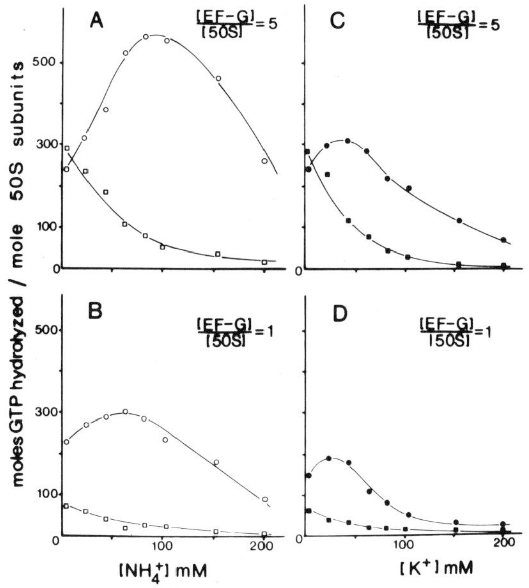

Fig. 1. Effect of NH_4^+ and K^+ on the EF-G-dependent GTPase activity of 50S or 50S plus 30S. The 75 µl reaction mixtures contained: 20 mM Tris-HCl, pH 7.8 - 14 mM $MgCl_2$ - 1 mM dithiothreitol - 2% glycerol (from ribosomes) - 25 nmoles of γ-[^{32}P]GTP (2-3 Ci/mole) - 10 pmoles of 50S - 60 pmoles of 30S when present. 50 pmoles of EF-G were added in A and C and 10 pmoles in B and D. Prior to the addition of EF-G and GTP the mixtures were incubated for 15 min at 30^0. Hydrolysis of GTP was measured after a second 10 min incubation at 30^0 as the amount of P_i liberated (37). Results were not corrected for contamination of the 50S subunits by 30S. The 30S blank varied between 2 and 16 pmoles GTP hydrolyzed/mole 30S. These values could be entirely attributed to the contaminating 50S and were subtracted. 50S with NH_4^+ (□), 50S plus 30S with NH_4^+ (○), 50S with K^+ (■), and 50S plus 30S with K^+ (●).

is obtained with [NH_4^+] or [K^+] above 100 mM (18,19). In this regard it is instructive to note that in optimal growth conditions E. coli intracellular K^+ concentration is 200-300 mM (30,31). This suggests that the ability of free 50S subunits to support EF-G GTPase activity is strongly inhibited in vivo.

Because of the effect of the EF-G to 50S ratio on GTPase activity, we studied this parameter in more detail at 2 and 80 mM NH_4^+ (Fig. 2). At the lower [NH_4^+], 30S had a stimulatory effect when this ratio was below 5. At higher ratios 30S became inhibitory. At 80 mM NH_4^+, addition of 30S stimulated the reaction over the whole range of EF-G to 50S ratios tested. With

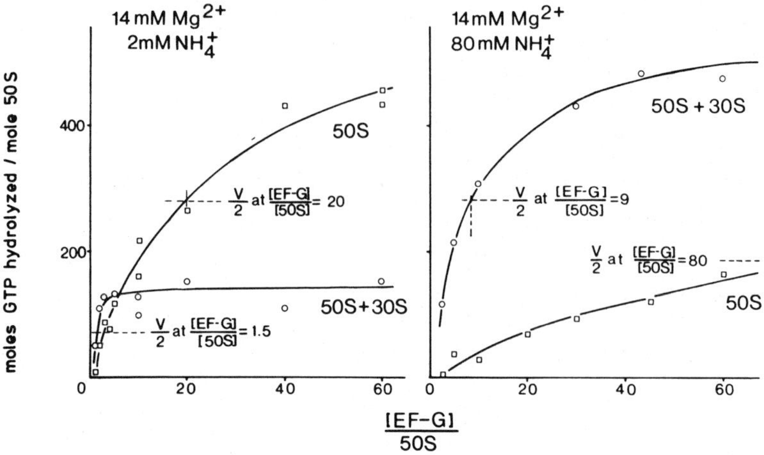

Fig. 2. Effect of increasing concentrations of EF-G on the GTPase activity of 50S with and without 30S subunits at 2 and 80 mM NH_4^+. Assay conditions were as in Fig. 1 except that 5 pmoles of 50S and 30 pmoles of 30S were used. The [EF-G]/[50S] values for half-maximum GTP-hydrolysis ($\frac{V}{2}$) were calculated from a double reciprocal plot of the data.

50S alone activity was as expected strongly inhibited. Nevertheless even at 80 mM NH_4^+ maximal velocity (V) with 50S alone reached about two thirds of the value obtained in the presence of 30S. The smaller subunit also greatly reduced the EF-G to 50S ratio needed for $\frac{V}{2}$ at both NH_4^+ concentrations. The most striking result was the ability of 50S alone to reach at 2 mM NH_4^+ a similar V as that obtained by addition of 30S at 80 mM NH_4^+. At the same time this experiment points out the central regulatory role of the 30S subunit on the kinetics of this reaction in vitro and again shows that the conditions which are presumably present in the bacterial cell favor the inhibition of the free 50S activity.

We wish here to point out the different behavior of EF-T-ribosomal GTPase with respect to monovalent cations. Although the same or a closely overlapping region of the 50S subunit seems to support both EF-T and EF-G GTPases (11,32,33), 50S alone did not promote EF-T-dependent GTPase activity at any NH_4^+ or K^+ concentration tested (Fig. 3). The presence of Phe-$tRNA^{Phe}$ was a requirement in all ionic conditions (21).

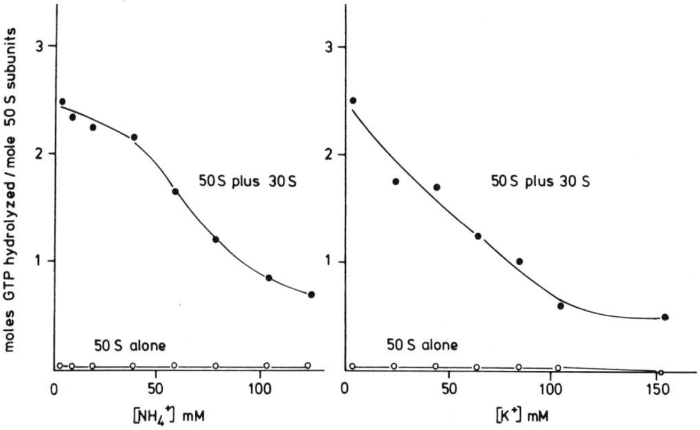

Fig. 3. Effect of NH_4^+ and K^+ on EF-T-ribosomal GTPase activity. The assay system contained in 75 µl: 20 mM Tris-HCl, pH 7.8 - 14 mM $MgCl_2$ - 0.5% glycerol (from ribosomes) - 5 pmoles of 50S subunits - 10 pmoles of 30S subunits where indicated - 50 pmoles of EF-T - 100 pmoles of Phe-$tRNA^{Phe}$ when present - 3 µg of poly(U) and 120 pmoles of γ-[^{32}P]GTP (1000 Ci/mole). The reaction mixture was incubated for 2 min at 30° and the GTP hydrolysis measured. 50S and 30S plus Phe-$tRNA^{Phe}$ (●), 50S alone plus Phe-$tRNA^{Phe}$ (○). 30S alone plus Phe-$tRNA^{Phe}$ or 50S and 30S in the absence of Phe-$tRNA^{Phe}$ supported no EF-T GTPase activity.

K_m values for GTP hydrolysis at a 1:1 EF-G to 50S ratio are listed in Table 1. The K_m with 70S ribosomes remained essentially the same at both 2 and 80 mM NH_4^+ and was about half that with the 50S subunit. This indicates that substrate affinity is hardly affected by the NH_4^+ concentration.

Table 1. K_m for GTP of the EF-G-dependent GTPase activity of 50S with and without 30S at 2 and 80 mM NH_4^+.

ribosomal particles	2 mM NH_4^+	80 mM NH_4^+
50S	1.4×10^{-4}M	1.3×10^{-4}M
50S plus 30S	0.75×10^{-4}M	0.6×10^{-4}M

The experimental conditions were the same as described in the legend of Fig. 1. In these experiments a molar ratio of EF-G to 50S of 1 was used. [GTP] was varied from 0.3 to 4.5×10^{-4}M, and K_m values were determined from double reciprocal plots.

Bogatyreva et al. (19) have shown an interdependence in the Mg^{2+} and monovalent cation concentrations in poly(U)-directed polyphenylalanine synthesis. As Fig. 4 illustrates, an increase in $[NH_4^+]$ resulted in a shift towards higher values of the Mg^{2+} optimum for EF-G-dependent GTPase with both 50S alone and 50S plus 30S. Gordon and Lipmann have observed a similar shift by using 70S (20).

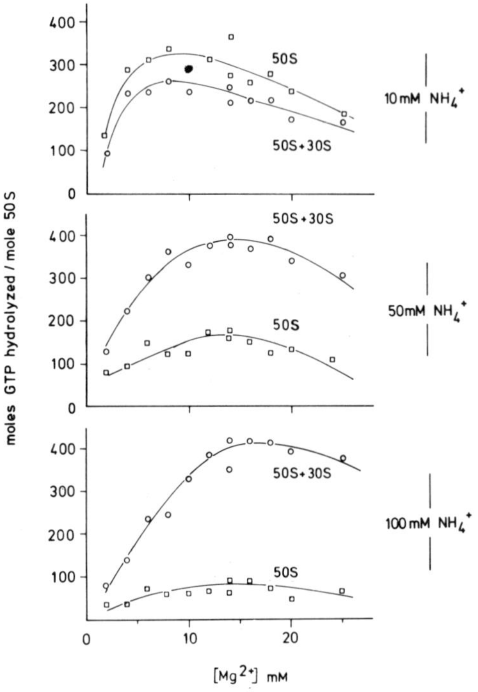

Fig. 4. Interdependence of the Mg^{2+} and NH_4^+ requirements for the EF-G GTPase activity of 50S with and without 30S. Assay conditions were as in Fig. 1. In these experiments a molar ratio of EF-G to 50S of 5 was used. 50S (□), and 50S plus 30S (○).

The effect of poly(U) and $tRNA^{Phe}$ on EF-G-ribosomal GTPase activity as a function of cation concentrations was next investigated. Conway and Lipmann (6) first observed in 1964 a strong stimulation by poly(U) plus charged or uncharged tRNA at 8 mM Mg^{2+}. Later other authors (34,35) failed to observe more than a 10% stimulation at $[Mg^{2+}]$ less than 15 mM and described strong inhibition at higher $[Mg^{2+}]$. With our system stimulation occurred with poly(U) and $tRNA^{Phe}$, Phe-$tRNA^{Phe}$ or Ac-Phe-$tRNA^{Phe}$ (Fig. 5), the effectiveness of these components following this order, and was particularly evident at Mg^{2+} concentrations between 5-10 mM. At higher Mg^{2+} concentrations stimulation progressively disappeared and was replaced by an inhibition. We would like to note that at the 80 mM NH_4^+ or K^+ used in the GTPase assay we obtained a Mg^{2+} optimum of 7 mM for poly(U)-directed polyphenylalanine synthesis, i. e. at the concentration for maximal stimulation by poly(U) plus tRNA. This underlines the physiological importance of the stimulation by tRNA in combination with mRNA.

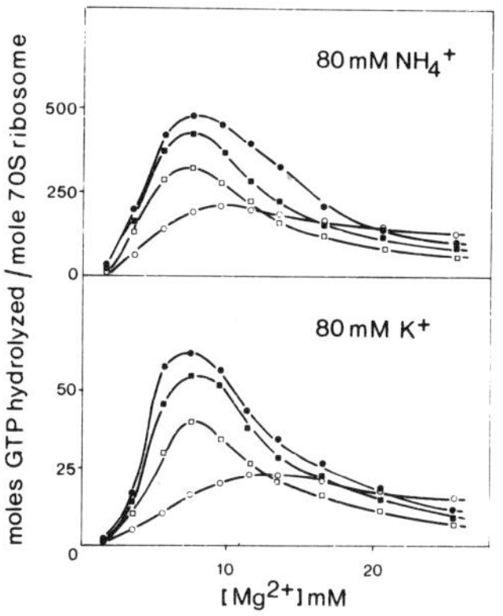

Fig. 5. Effect of poly(U) plus charged or uncharged $tRNA^{Phe}$ on EF-G-ribosomal GTPase as a function of Mg^{2+} concentration. The 75 µl reaction mixture contained: 20 mM Tris-HCl, pH 7.8 - 1 mM dithiothreitol - 80 mM NH_4Cl or 80 mM KCl - $MgCl_2$ as indicated - 0.3% glycerol - 30 (with NH_4^+) or 7.5 nmoles (with K^+) of γ-[^{32}P]GTP (2-8 Ci/mole) - 15 pmoles of 70S ribosomes - 15 pmoles of EF-G - 4 µg of poly(U) - 20 pmoles of $tRNA^{Phe}$ (●), Phe-$tRNA^{Phe}$ (■) or Ac-Phe-$tRNA^{Phe}$ (□), respectively. In the control poly(U) and $tRNA^{Phe}$ species were omitted (○). Reaction mixtures were incubated for 10 min at 30° prior to addition of GTP. Hydrolysis of GTP was measured after a second 10 min incubation at 30° as in Fig. 1. Poly(U) and $tRNA^{Phe}$ species added individually to the reaction mixtures had little effect on the EF-G GTPase activity of 70S ribosomes (not shown).

Fig. 6 illustrates the importance of NH_4^+ and K^+ concentration on the stimulation of EF-G GTPase by poly(U) plus $tRNA^{Phe}$ at 6.5 mM Mg^{2+}. With both cations there was a critical concentration below which the combination of poly(U) plus $tRNA^{Phe}$ was inhibitory. For NH_4^+ this was 5 mM and for K^+ 25 mM. The dependence of EF-G GTPase on poly(U) plus $tRNA^{Phe}$ increased with in-

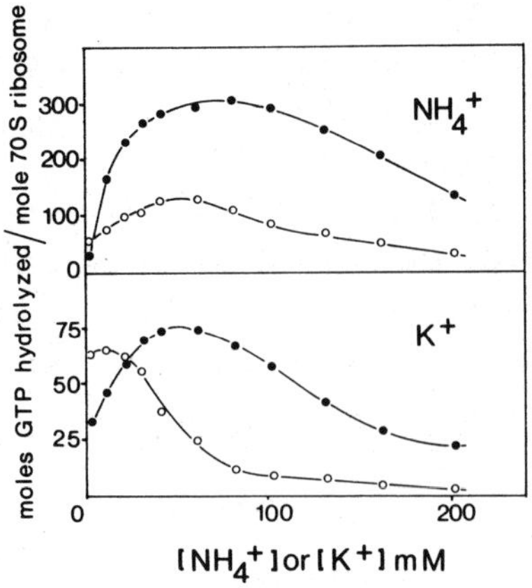

Fig. 6. Effect of poly(U) plus $tRNA^{Phe}$ on EF-G-ribosomal GTPase as a function of NH_4^+ and K^+ concentration. Assay conditions were as described in Fig. 5 except that the $MgCl_2$ was 6.5 mM, NH_4Cl and KCl concentrations were varied as indicated and 22.5 or 6 nmoles of γ-[^{32}P]GTP were used in the presence of NH_4^+ and K^+, respectively. Without (○) or with 4 µg of poly(U) plus 25 pmoles of $tRNA^{Phe}$ (●).

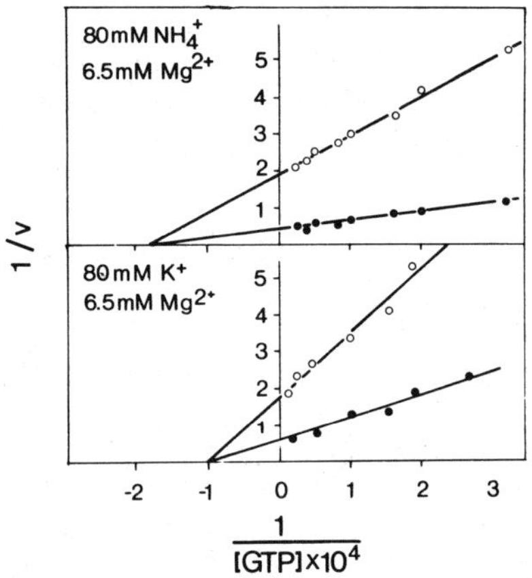

Fig. 7. Effect of poly(U) plus $tRNA^{Phe}$ on the K_m of the EF-G-ribosomal GTPase. In the upper panel, the 75 µl reaction mixtures contained: 20 mM Tris-HCl, pH 7.8 - 1 mM dithiothreitol - 80 mM NH_4Cl - 6.5 mM $MgCl_2$ - 0.3% glycerol - GTP as indicated - 5 pmoles of 70S ribosomes - 5 pmoles of EF-G - 2 µg of poly(U) plus 8 pmoles of $tRNA^{Phe}$, when present. In the lower panel the reaction mixtures had the same composition except that the concentrations of ribosomes, EF-G, poly(U) and $tRNA^{Phe}$ were three times higher and 80 mM KCl was present instead of NH_4Cl. Assay conditions were as described in Fig. 5. Without (○) or with poly(U) plus $tRNA^{Phe}$ (●).

creasing monovalent cation concentrations. Poly(U) plus $tRNA^{Phe}$ did not, however, alter the K_m of the EF-G-ribosomal GTPase which was 0.6 x 10^{-4}M with NH_4^+ and 1.0 x 10^{-4}M with K^+ (Fig. 7).

All these experiments point out the critical importance of the concentration of the monovalent and divalent cations in the coordination of the EF-G GTPase activity with the action of other components in protein biosynthesis. Their interdependence gives an idea of the complexity of the regulation of this reaction. In the bacterial cell undoubtedly many more numerous parameters control EF-G-ribosomal GTPase. One of these is apparently elongation factor T which competes with EF-G for interaction with ribosomes (33).

EF-G is able to support in vitro much higher GTPase activity than would seem required for polypeptide synthesis (6,20,36). The ultimate goal of our investigations is to try to reproduce in vitro the conditions which allow full coordination of EF-G-ribosomal GTPase with polypeptide synthesis on the ribosome.

SUMMARY: The interdependence of several parameters which regulate EF-G-ribosomal GTPase is described. Turnover activity of the 50S subunit is inhibited by increasing concentrations of monovalent cations. This effect is reversed by the presence of the 30S subunit which shifts the optimal concentration of monovalent cations towards higher values. The molar ratio of EF-G to 50S is critical for the kinetics of the reaction. Both 50S- and 50S plus 30S-dependent GTPase activities appear to be able to reach a similar V when assayed at the respective NH_4^+ optima with saturating amounts of EF-G. The 30S subunit greatly increases the affinity of the 50S subunit for EF-G in the GTPase reaction and reduces the K_m of GTP. The range of the ionic conditions present in the bacterial cell seems to be inhibitory for a free 50S-dependent GTPase activity. The study of the effect of poly(U) plus $tRNA^{Phe}$ as a function of the cation concentration underlines the physiological importance of the stimulation of EF-G-ribosomal GTP hydrolysis by these two components of protein synthesis.

This work was supported by grant Pa 106 of the Deutsche Forschungsgemeinschaft.

REFERENCES

1. Nishizuka, Y., Lipmann, F.: The interrelationship between guanosine triphosphatase and amino acid polymerization. Arch.Biochem.Biophys. 116, 344-351 (1966).
2. Haenni, A.-L., Lucas-Lenard, J.: Stepwise synthesis of a tripeptide. Proc.Nat.Acad.Sci. U.S.A. 61, 1363-1369 (1968).
3. Leder, P., Skogerson, L.E., Roufa, D.J.: Translocation of mRNA codons, II. Properties of an anti-translocase antibody. Proc.Nat.Acad.Sci. U.S.A. 62, 928-933 (1969).
4. Erbe, R.W., Nau, M.M., Leder, P.: Translocation and translocation of defined RNA messengers. J.Mol.Biol. 38, 441-460 (1969).
5. Gordon, J.: Hydrolysis of guanosine 5'-triphosphate associated with binding of aminoacyl transfer ribonucleic acid to ribosomes. J.Biol. Chem. 244, 5680-5686 (1969).
6. Conway, T.W., Lipmann, F.: Characterization of a ribosome-linked guanosine triphosphatase in Escherichia coli extracts. Proc.Nat.Acad. Sci. U.S.A. 52, 1462-1469 (1964).
7. Bodley, J.W., Zieve, F.J., Lin, L.: Studies on translocation, IV. The hydrolysis of a single round of guanosine triphosphate in the presence of fusidic acid. J.Biol.Chem. 245, 5662-5667 (1970).
8. Brot, N., Spears, C., Weissbach, H.: The interaction of transfer factor G, ribosomes, and guanosine nucleotides in the presence of fusidic acid. Arch.Biochem.Biophys. 143, 286-296 (1971).
9. Kischa, K., Möller, W., Stöffler, G.: Reconstitution of a GTPase activity by a 50S ribosomal protein from E. coli. Nature New Biol. 233, 62-63 (1971).
10. Hamel, E., Koka, M., Nakamoto, T.: Requirement of an Escherichia coli 50S ribosomal protein component for effective interaction of the ribosome with T and G factors and with guanosine triphosphate. J.Biol. Chem. 247, 805-814 (1972).
11. Sander, G., Marsh, R.C., Parmeggiani, A.: Isolation and characterization of two acidic proteins from the 50S subunit required for GTPase activities of both EF-G and EF-T. Biochem.Biophys.Res.Commun. 47, 866-873 (1972).
12. Schrier, P.I. Maassen, J.A., Möller, W.: Involvement of 50S ribosomal proteins L6 and L10 in the ribosome-dependent GTPase activity of elongation factor G. Biochem.Biophys.Res.Commun. 53, 90-98 (1973).
13. Marsh, R.C., Parmeggiani, A.: Requirement of proteins S5 and S9 from 30S subunits for the ribosome-dependent GTPase activity of elongation factor G. Proc.Nat.Acad.Sci. U.S.A. 70, 151-155 (1973).
14. Parmeggiani, A., Sander, G., Voigt, J., Marsh, R.C.: Role of the 30S ribosomal subunit in the elongation factor-dependent GTPases in Escherichia coli. First Symposium on Ribosomes and Ribonucleic Acid Metabolism, Smolenice Castle, Slowak Academy of Science, Proceedings 1, 247-258 (1973).
15. Cohlberg, J.A.: Activity of protein-deficient 30S ribosomal subunits in elongation factor G-dependent GTPase. Biochem.Biophys.Res.Commun. 57, 225-231 (1974).
16. Highland, J.H., Ochsner, E., Gordon, J., Bodley, J.W., Hasenbank, R., Stöffler, G.: Coordinate inhibition of elongation factor G function and ribosomal subunit association by antibodies to several ribosomal proteins. Proc.Nat.Acad.Sci. U.S.A. 71, 627-630 (1974).

17. Brot, N., Yamasaki, E., Redfield, B., Weissbach H.: The properties of an E. coli ribosomal protein required for the function of factor G. Arch.Biochem.Biophys. 148, 148-155 (1972).
18. Conway, T.W.: On the role of ammonium or potassium ion in amino acid polymerization. Proc.Nat.Acad.Sci. U.S.A. 51, 1216-1220 (1964).
19. Bogatyreva, S.A., Trifonov, E.N., Spirin, A.S.: Dependence of poly(U)-directed cell-free system on proportions of divalent and monovalent cations. Dokl.Akad.Nauk SSSR 195, 213-216 (1970).
20. Gordon, J., Lipmann, F.: Role of divalent ions in poly U-directed phenylalanine polymerization. J.Mol.Biol. 23, 23-33 (1967).
21. Voigt, J., Sander, G., Nagel, K., Parmeggiani, A.: Effect of NH_4^+ and K^+ on the activity of the ribosomal subunits in the EF-G- and EF-T-dependent GTP hydrolysis. Biochem.Biophys.Res.Commun. 57, 1279-1286 (1974).
22. Parmeggiani, A., Voigt, J., Nagel, K., Sander, G.: Effect of NH_4^+ and K^+ on the activity of the ribosomal subunits in the EF-G-dependent GTPase reaction. Fed.Proc. 33, 1403 (1974).
23. Voigt, J., Parmeggiani, A.: Action of methanol on the association of ribosomal subunits and its effect on the GTPase activity of elongation factor G. Biochem.Biophys.Res.Commun. 52, 811-818 (1973).
24. Modolell, J., Vazquez, D., Monro, R.E.: Ribosomes, G-factor and siomycin. Nature New Biology 230, 109-112 (1971).
25. Ballesta, J.P.G., Montejo, V., Vazquez, D.: Reconstitution of the 50S ribosome subunit. Localization of G-dependent GTPase activity. FEBS Lett. 19, 79-82 (1971).
26. Parmeggiani, A., Singer, C., Gottschalk, E.M.: Purification of the amino acid polymerization factors from Escherichia coli. Methods Enzymol. 20, 291-302 (1971).
27. Chinali, G., Parmeggiani, A.: Properties of the elongation factors from Escherichia coli. Exchange of elongation factor G during elongation of polypeptide chain. Eur.J.Biochem. 32, 463-472 (1973).
28. Haenni, A.-L., Chapeville, F.: The behaviour of acetylphenylalanyl soluble ribonucleic acid in polyphenylalanine synthesis. Biochim.Biophys.Acta 114, 135-148 (1966).
29. Gordon, J.: Regulation of the in vivo synthesis of the polypeptide chain elongation factors in Escherichia coli. Biochemistry 9, 912-917 (1970).
30. Schultz, S.G., Solomon, A.K.: Cation transport in Escherichia coli, I. Intracellular Na^+ and K^+ concentrations and net cation movement. J.Gen. Physiol. 45, 355-369 (1961).
31. Epstein, W., Schultz, S.G.: Cation transport in Escherichia coli, V. Regulation of cation content. J.Gen.Physiol. 49, 221-234 (1965).
32. Modolell, J., Cabrer, B., Parmeggiani, A., Vazquez, D.: Inhibition by siomycin and thiostrepton of both aminoacyl-tRNA and factor G binding to ribosomes. Proc.Nat.Acad.Sci. U.S.A. 68, 1796-1800 (1971).
33. Richter, D.: Inability of E.coli ribosomes to interact simultaneously with the bacterial elongation factors EF-Tu and EF-G. Biochem.Biophys. Res.Commun. 46, 1850-1856 (1972).
34. Modolell, J., Vazquez, D.: Inhibition by aminoacyl transfer ribonucleic acid of elongation factor G-dependent binding of guanosine nucleotide to ribosomes. J.Biol.Chem. 248, 488-493 (1973).
35. Ballesta, J.P.G., Vazquez, D.: Ribosomal activities dependent on elongation factors T and G. Effects of methanol. Biochemistry 12, 5063-5068 (1973).

36. Nishizuka, Y., Lipmann, F.: Comparison of guanosine triphosphate split and polypeptide synthesis with a purified E. coli system. Proc.Nat.Acad. Sci. U.S.A. 55, 212-219 (1966).
37. Wahler, B.E., Wollenberger, A.: Zur Bestimmung des Orthophosphats neben säure-molybdat-labilen Phosphorsäureverbindungen. Biochem.Z. 329, 508-520 (1958).

Deoxynucleotide-Polymerizing Enzymes of Murine Myelomas*

C. Pénit, A. Paraf[+] and F. Chapeville

Laboratoire de Biochimie du Développement et [+]Laboratoire d'Immunodifférenciation, Institut de Biologie Moléculaire du C.N.R.S. et de l'Université Paris VII, 2 Place Jussieu, 75005 Paris, FRANCE

INTRODUCTION

Mammalian cells contain several deoxynucleotide-polymerizing enzymes. Numerous studies have been published concerning these enzymes in different organisms (1-6). Information on their molecular and catalytic properties *in vitro* is accumulating, but little is known about their role *in vivo*. Table 1 is a summary of a few properties of the mammalian DNA polymerases. A common nomenclature for these enzymes has not yet been adopted; we shall use the nomenclature proposed by McCaffrey *et al.* (7) because of its simplicity.

Table 1. Summary of some properties of mammalian DNA polymerases

Polymerase	MW	Template	Subcellular location
C	150,000	DNA	Cytoplasmic
N	50,000	DNA or RNA	Cytoplasmic and nuclear
A	70,000	Homopolyribonucleotides	Cytoplasmic and nuclear

Murine myelomas exhibit several interesting properties: 1. These tumors, easily induced in the animal, are transplantable (8) and show a great stability of their differenciated state, producing

* We dedicate this paper to Dr. Fritz Lipmann with our sincere admiration.

a well-defined type of immunoglobulin. 2. Myeloma cells produce A-type intracisternal virus-like particles, as well as C-type particles (9). 3. These cells can be cultured *in vitro* (10), and, in the case of the MOPC 173 tumor, several variants have been obtained (11, 12). The variations affect cell morphology, capacity to produce virus particles and oncogenicity. 4. Myeloma cells derive from immunocytes, multipotent cells that play a very important biological role. All these properties have led us to suppose that the study of the deoxynucleotide-polymerizing activities in murine myelomas might be of interest.

METHODS

The methods used have been published elsewhere (13, 14) and will be only summarized here.

Tumors and Cell Lines

All experiments were performed with Balb/C mice. Subcutaneous tumors MOPC 173 (11), TEPC 15 (15) and MOPC 315 (16) were harvested two weeks after injection of a suspension of cells. Two variants of MOPC 173 cells obtained in culture by Paraf *et al.* (11, 12) were also used: MF_2 fibroblastic cells, that show no contact inhibition, produce high quantities of A-type and C-type particles and are transplantable, and ME_2 epitheloid cells, which are contact-inhibited, are not virus-produc ng and are unable to induce tumors in the animal. Dissociated tumor cells, as well as MF_2 cells induce ascites when injected intraperitoneally. Thymus of 5 week-old mice, spleens from athymic nude mice and L-cells served as controls.

Preparation of Cell Extracts.

1. *Soluble extracts*. Detergent-treated cell homogenates were centrifuged at 100,000 *g* for 1 hr. The pH of the supernatant was

lowered to 5 and the precipitate obtained collected and dissolved in a buffer containing 0.05 M Tris-HCl pH 7.9, 1 mM EDTA, 2.8 mM 2-mercaptoethanol and 20% glycerol (TEMG buffer).

2. *Particulate extracts*. Homogenates suspended in a buffer containing 0.01 M Tris-HCl pH 7.5, 1 mM EDTA, and 0.25 M sucrose were treated with detergent in the presence of 10 mM dithiothreitol and centrifuged at 8,000 *g* for 20 min. The resulting supernatant was centrifuged at 200,000 *g* for 90 min. The pellet obtained (high speed pellet) was resuspended in the same buffer without detergent.

Purification of C-Type Particles

The supernatant fluid from MF_2 cell cultures was used. The C-type particles were purified using differential centrifugation and equilibrium density centrifugation in 15-65% sucrose gradients.

Analytical Techniques

1. *Phosphocellulose chromatography and gel filtration.* The protein concentration of the pH 5 extract was adjusted to 2 mg/ml and 10 ml were applied to a P11 phosphocellulose column (10 x 1.2 cm) After a 40 ml wash by TEMG buffer containing 500 µg/ml bovine serum albumin (TEMG-BSA), the adsorbed proteins were eluted by an 80 ml gradient of 0-1 M KCl in TEMG-BSA, at a flow rate of 10 ml/hr. Fractions of 1 ml were collected, and assayed for DNA polymerase activities.

In some cases, the phosphocellulose column was eluted by stepwise addition of 0.25 M KCl and of 0.45 M KCl in TEMG-BSA. For a better separation of the terminal transferase from DNA polymerase C, the 0.45 M KCl eluate was filtered through a Sephadex G-150 column.

2. *Sucrose gradient centrifugation.* All gradients were made in a

buffer solution containing 0.01 M Tris-HCl pH 7.5, 1 mM EDTA and 10 mM DTT. *a)* Velocity sedimentation. Aliquots (200 μl) of the 200,000 *g* pellet suspension were layed on 5-20% sucrose gradients established on a cushion of 200 μl of 65% sucrose and centrifuged at 50,000 *g* for 45 min in an SW 50 L rotor. Fractions (250 μl) were collected from the bottom. *b)* Equilibrium density gradient centrifugation. Aliquots (200 μl) of cell extract were layered on 5 ml 15-65% sucrose preformed gradients, and centrifuged for 3 hr at 200,000 *g* in an SW 50 L rotor.

<u>Polymerase Assays:</u>

1. *Endogenous reaction.* Crude preparations (50-100 μg of protein) were tested in an incubation volume of 100 μl. For density gradient purified preparations, as little as 1 μg of protein was tested in the same volume. The incubation medium contained 5 μmoles of Tris-HCl pH 7.9, 0,2 μmoles of DTT, 1 μmole of $MgCl_2$, 2 μmoles of KCl, 20 nmoles of ATP, 5 nmoles of dATP, dGTP, dCTP and 70 pmoles of ^{3}H-dTTP (Radiochemical Center, Amersham), 6250 cpm/pmole. Incubations were at 37°C and the radioactivity in acid-precipitable material was determined. For the study of the influence of pancreatic ribonuclease A (Worthington Biochemical Co.), samples containing 50-100 μg of protein preincubated in a total volume of 50 μl either for 60 min at 0°C in water, or for 15 min at 37°C in 0.02 M Tris-HCl pH 7.9 and 0.1 M KCl. The ribonuclease was pretreated by heating at 80°C for 10 min.

2. *Exogenous primer-directed reaction.* The incubation mixture contained in a final volume of 100 μl : *a)* $d(pT)_4$-primed reaction : 5 μmoles of Tris-HCl pH 7.4, 0.2 μmoles of dithiothreitol, 0.08 μmole of $MnCl_2$, 0.3 OD_{260} unit of $d(pT)_4$ (Collaborative Research) and 250 pmoles of ^{3}H-dTTP (2,000 cpm/pmole). *b)* Activated DNA-primed reaction : 5 μmoles of Tris-HCl pH 7.9, 0.2 μmoles of dithiothreitol, 1 μmole of $MgCl_2$, 0.2 OD_{260} unit of activated calf thymus DNA, 5 mμmoles each of dATP, dGTP, dCTP and

200 pmoles of ^{3}H-dTTP (4,000 cpm/pmole). *c)* $rA_n \cdot dT_{10}$-primed reaction : 5 μmoles of Tris-HCl pH 7.9, 0.2 μmole of dithiothreitol, 0.1 μmole of $MnCl_2$, 0.02 OD_{260} unit of $rA_n \cdot dT_{10}$ (Boehringer) and 1 nmole of ^{3}H-dTTP (500 cpm/pmole). *d)* $rA_n \cdot dT_n$-primed reaction: idem with 0.025 unit of $rA_n \cdot dT_n$ (Miles) in place of $rA_n \cdot dT_{10}$. The acid-precipitable material was collected by filtration on Millipore filters and the radioactivity was determined.

RESULTS AND DISCUSSION

Polymerases A, C and N

These three enzymes are present in all the mammalian cells studied. Enzymes C and N are very similar to the DNA polymerases described in avian systems (17, 18), in which no enzyme of type A has been detected. Their activities can be distinguished by using different primers (19). They can be separated by various methods; a single column chromatography on a phosphocellulose of a soluble extract leads to a good resolution of enzymes C and N, and to a lower resolution of enzymes C and A. Table 2 and Figure 1 show the results obtained in the case of MOPC 173 tumors.

We have found no important variations of activity of polymerase N in relation with the growth rate of the cells, while the activity of polymerase C is higher in dividing cells. Polymerase A is not easily detectable in tumoral cells by a simple chromatography since it is eluted from phosphocellulose at the same KCl molarity as polymerase C.

Reverse Transcriptase (Enzyme V)

The fact that myeloma cells produce A and C-type particles, and the existence of a non-producing variant led us to investigate the possible presence of reverse transcriptase activity (20, 22) in the soluble and in the particulate extracts.

Table 2. Primer specificities of DNA polymerases from MOPC 173 tumors

Primer	Nucleotide added	MOPC 173 DNA polymerases (molarity of elution from phosphocellulose) enzyme V (0.28 M)	enzyme C (0.35 M)	enzyme N (0.55 M)	C-type particles
$rA_n \cdot dT_n$	(^{3}H-dTTP)	5.75	1.6	6.7	52
$rA_n \cdot dT_n$	(^{3}H-dATP)	0.06	0.70	0.02	0.00
$rA_n \cdot dT_{10}$	(^{3}H-dTTP)	12.20	0.47	0.00	65
$dA_n \cdot dT_{10}$	(^{3}H-dTTP)	0.02	1.42	7.20	0.00
$rC_n \cdot dG_{10}$	(^{3}H-dGTP)	4.60	0.45	0.00	50

The assays were performed under the conditions described in Methods, with 20 µl of each phosphocellulose fraction. Incubations were at 37°C for 30 min. C-type particles were tested with 50 µg of virus proteins per assay. It must be noted that the 0.55 M phosphocellulose enzyme can use $rA_n \cdot dT_{10}$ as template if the temperature is lowered to 30°C, but even under these conditions it can not use $rC_n \cdot dG_{10}$. The values are expressed in pmoles/30 min.

Using $rA_n \cdot dT_{10}$ and $rC_n \cdot dG_{10}$ as primers, we detected reverse transcriptase activity in soluble extracts from solid tumors and from ME_2 as well as from MF_2 cells; thus, the presence of this enzyme in soluble form in tumoral tissues is demonstrable even in the absence of actual virus production. The reverse transcriptase activity can not be detected in non tumoral tissues (e.g. thymus or spleen), but it is present in L cells, which are known to produce virus-like particles.

The analysis of the high speed pellets from different cells showed that only those from virus-producing cells contain a reverse transcriptase activity associated with particles whose density in sucrose gradients is similar to that of C-type viruses (1.16-1.18). However, when such particles were incubated in the absence of synthetic primer, and in the presence of all four deoxyribonucleotide triphosphates, only a very low polymerizing

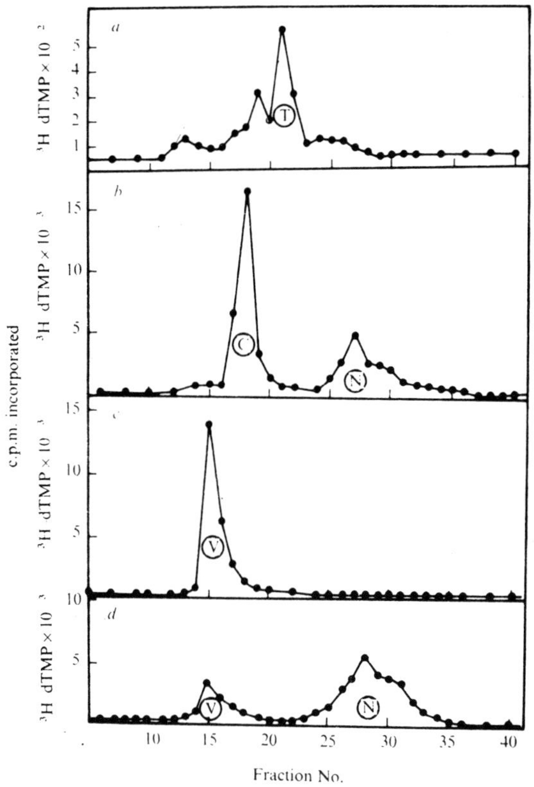

Figure 1. Phosphocellulose chromatography of DNA polymerase activities from murine myeloma MOPC 173 extracts. Each panel represents the result of the assays of the same column with different primers: a) $d(pT)_4$-primed reaction. b) Activated DNA-primed reaction. c) poly A·$d(pT)_{10}$-primed reaction. d) poly A·poly dT-primed reaction. 25 µl of each phosphocellulose fraction were incubated in conditions described in Methods.

activity was detected. This endogenous activity was not sensitive to pancreatic ribonuclease. The same results were obtained with extracellular C-type particles produced by MF_2 cells.

Endogenous Ribonuclease-Sensitive DNA Polymerase Activity

As mentioned above, the exogenous primer-directed reverse transcriptase activity present in high speed pellets of virus-producing cells was associated with particles of density 1.16-1.18.

The same high speed pellets possess an endogenous ribonuclease sensitive DNA polymerase activity, and a careful analysis by sucrose gradients showed that this activity was associated with a material whose density was 1.08 (ref.13). The same results were obtained using non virus-producing cells.

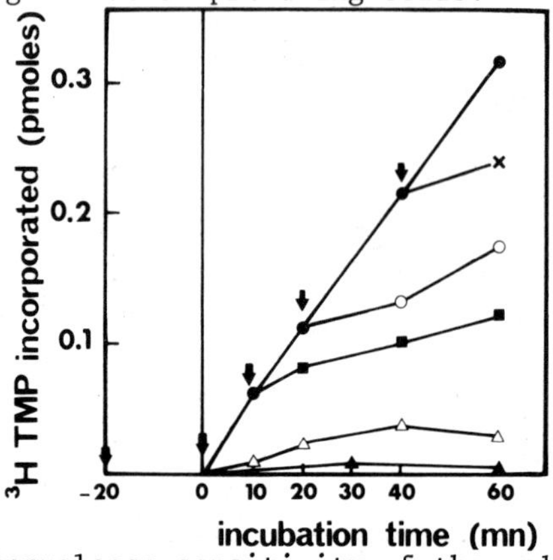

Figure 2. Ribonuclease-sensitivity of the endogenous reaction in the presence of the 200,000 *g* pellet from MF_2 cells. Each point represents the number of pmoles of ^{3}H-dTMP incorporated into acid insoluble material, in 50 µl samples. The arrows represent the time of addition of ribonuclease (2 µg for 100 µl incubation mixture). (●) no addition of RNase. Addition of RNase at time: (▲) - 20 min (preincubation), (△) 0 min, (■) 10 min, (○) 20 min, (x) 40 min.

This activity is similar to that described in chick embryo cell extracts (23). The nature of the enzyme catalyzing this reaction has not yet been well established. In our case, the ribonuclease-sensitivity is maintained after the onset of polymerization (Figure 2), suggesting that RNA acts as template; however, further experiments are necessary to confirm this and to exclude the possibility that RNA is used only as initiator a DNA-dependent DNA-polymerase.

Terminal Deoxynucleotidyltransferase (Enzyme T)

In the course of our studies on DNA polymerases in myeloma cells, we have found that these cells also contain the terminal deoxynucleotidyltransferase (14). All the myelomas or derived cells tested (solid tumors MOPC 173, TEPC 15 and MOPC 315, both types of MOPC 173-derived cultured cells, and ascites MOPC 173) show the corresponding activity. In Figure 1 are shown the profiles of the deoxynucleotide-polymerizing activities obtained by chromatography on a phosphocellulose column of a soluble extract from MOPC 173 tumors. The T activity was also found in mouse thymus, but not in extracts from spleen and L cells. This result showing the specificity of the thymus is consistent with those obtained in other animal species (24). The terminal transferase is not completely separated from polymerase C and A by a single phosphocellulose chromatography; a complete separation was achieved by gel filtration on Sephadex G-150: the elution volume of terminal transferase is consistent with a molecular weight of 30,000, which is close to that observed for the calf thymus enzyme (25).

The fact that the only animal tissue so far known to possess the terminal transferase activity is the thymus may be interpreted by assuming that this activity is specific of thymocytes, that is of T cells in the course of their differenciation. McCaffrey *et al.* (7) considered that the presence of the enzyme in lymphoblasts of a child suffering of acute lymphoblastic leukemia denoted their thymic origin. The discovery of the terminal transferase in myelomas can not be simply explained in the same manner. Indeed, by cytological and immunological criteria, myeloma cells are considered as B cells; they also produce high quantities of immunoglobulins. A possible explanation of the observed phenomenon is that the tumoral process by itself leads to the derepression of a mechanism which normally functions only in the thymus. Another hypothesis is that myeloma cells are not true B cells, but represent an intermediary stage of differenciation between B and T cells.

The physiological function of terminal deoxynucleotidyl transferase is unknown. Recently, Baltimore suggested that the enzyme might act as a mutagenic agent and could be responsible for antibody variability (26).

CONCLUSION

Murine myeloma cells possess a complex system of deoxynucleotide-polymerizing enzymes, some of which have been described by other authors (27). In addition to the three known enzymes present in all mammalian cells, we have found two other polymerases: reverse transcriptase and terminal deoxynucleotidyltransferase. The presence of reverse transcriptase was established by its ability to use poly C oligo dG as a primer (this activity is specific of the reverse transcriptase (22). However, we were unable to demonstrate the capacity of this enzyme to transcribe the endogenous RNA of virus-like particles (type A and type C); this result may be due to the state of the RNA rather to the enzyme itself.

The discovery of the presence of terminal deoxynucleotidyl transferase is an interesting result because it might help to elucidate the origin of myelomas in mice, and the physiological role of the enzyme.

SUMMARY

The deoxynucleotide-polymerizing activities of murine myelomas have been studied using two different methods: 1) Sucrose density gradients analysis of particulate extracts; 2) Phosphocellulose chromatography of soluble extracts.

The results obtained by the first method had led us to conclude that an enzyme similar to reverse transcriptase (as judged by its specificity for synthetic primers) is associated with virus-like particles (of density 1.16-1.18) produced by these tumors and by cultured myeloma cells, and that other particles (of density 1.08) contain an endogenous ribonuclease-sensitive DNA polymerase activity. No activity of this latter type could be detected in the virus-like particles.

Using the second method we have demonstrated in murine myeloma tumors as well as in cultured virus-producing or not producing cells, beside the 3 "normal" DNA polymerases, the presence of soluble reverse transcriptase and of deoxynucleotide transferase, an enzyme which was previously found only in thymus cells.

ACKNOWLEDGEMENTS

We wish to thank Miss E. Faivre for preparation of cultured myeloma cells, A. Dru and G. Jaureguiberry for providing L cells, J.C. Salomon for providing nude mice, and G. Brun, A.L. Haenni and F. Rougeon for fruitful discussions and help during this work.

This work was supported by a grant from the Institut National de la Santé et de la Recherche Médicale, contrat n° 73142818 "Enzymologie des virus oncogènes".

REFERENCES

1. Weissbach, A., Schlabach, A., Friedlender, B., Bolden, A. (1971) Nature New Biology 231, 167-170.
2. Chang, L.M.S., Bollum, F.J. (1971) J. Biol. Chem., 246 5835-5837.
3. Smith, R.G., Gallo, R.C. (1972) Proc. Nat. Acad. Sci. USA 69, 2879-2884.
4. Baril, E.F., Jenkins, M.D., Brown, O.E., Laszlo, J., Morris, H.P. (1973) Cancer Res. 33, 1187-1193.

5. Chang, L.M.S., McKay Brown, Bollum, F.J., (1973) J. Mol. Biol. 74, 1-8.

6. Fry, M., Weissbach, A. (1973) J. Biol. Chem. 248, 2678-2683.

7. McCaffrey, R., Smoler, D.F., Baltimore, D. (1973) Proc. Nat. Acad. Sci. USA 70, 521-525.

8. Cohn, M. (1967) Cold Spring Harbor Symp. Quant Biol. 32, 211-221.

9. Watson, J., Ralph, P., Sarkar, S., Cohn, M. (1970) Proc. Acad. Sci. USA 66, 344-351.

10. Namb, Y., Hanaoka, M. (1969) J. Virol. 102, 1486-1497.

11. Paraf, A., Moyne, M.A., Duplan, J.F., Scherrer, R., Stanislawski, M., Bettane, M., Lelievre, L., Rouze, P., Dubert, J.M. (1970) Proc. Nat. Acad. Sci. USA 67, 983-990.

12. Legrand, E., Moyne, M.A., Paraf, A., Duplan, J.F. (1972) Ann. Inst. Pasteur 123, 641-660.

13. Pénit, C., Paraf, A., Rougeon, F., Chapeville, F. (1974) FEBS Letters 38, 191-196.

14. Pénit, C., Paraf, A., Chapeville, F. (1974) Nature, 249, 755-757.

15. Potter, M. Liberman, R. (1970) J. Exptl. Med. 132, 737-751.

16. Eisen, H.M., Sims, E.S., Potter, M. (1968) Biochemistry 7, 4126-4134.

17. Brun, G., Rougeon, F., Lauber, M., Chapeville, F. (1974) Eur. J. Biochem. 41, 241-251.

18. Rougeon, F., Brun, G., Chapeville, F. (1974) Eur. J. Biochem 41, 253-261.

19. Robert, M.S., Smith, R.G., Gallo, R.C., Sarin, P.S., Abrell, J.W. (1972) Science 176, 798-800.

20. Leis, J.P., Hurwitz, J. (1972) J. Virol. 9, 130-142.

21. Temin, H.M., Baltimore, D. (1972) in "Advances in Virus Research" Smith, K.M., Lauffer, M.A., Bang, F.B. ed. 17, 129-186.

22. Baltimore, D., McCaffrey, R., Smoler, D.F. (1973)"Virus Research", 51-59.

23. Kang, C.Y., Temin, H.M. (1972) Proc. Nat. Acad. Sci. USA 69 1550-1554.

24. Chang, L.M.S. (1971) Biochem. Biophys. Res. Commun. 44, 124-131.

25. Chang, L.M.S., Bollum, F.J. (1971) J. Biol. Chem. 246, 909-916.

26. Baltimore, D. (1974) Nature 248, 409-411.

27. Persico, F.J., Nicholson, D.E., Gottlieb, A.A. (1973) Cancer Res. 33, 1210-1216.

Properties and Functions of Ethanol-Potassium Chloride Extractable Proteins from 80S Ribosomes and their Interchangeability with the Bacterial Proteins L7/L12

Dietmar Richter and Wim Möller

Max-Planck-Institut für Molekulare Genetik, Abteilung Wittmann, Berlin-Dahlem, Germany

Laboratory for Physiological Chemistry, State University of Leiden, Leiden, The Netherlands

INTRODUCTION

The specificity of the elongation factors and ribosomes from pro- and eukaryotic organisms has been studied with the result that within one class there seems to be no restriction of interchangeability whereas between the two classes only the prokaryote type elongation factor Tu (EF-Tu) can cross-react with 80S ribosomes (for review see ref. 1). These results suggest that EF-Tu and EF-G from bacteria have non-identical ribosomal recognition regions on 80S ribosomes, and that species specificity is most pronounced for the bacterial translocation factor EF-G.

So far no attempt has been reported to interchange ribosomal proteins from 70S and 80S ribosomes. The former contain two structural acidic proteins, L7 and L12, that are involved in the recognition of the initiation, elongation and termination factors of protein synthesis (2). The eukaryotic functional counterpart of L7/L12 has not yet been reported although 80S ribosomes do contain specific ribosomal recognition region(s) for elongation factor 1 and 2 (3-5). We will report here on experiments with 80S ribosomes from yeast and reticulocyte cells that contain two acidic 60S ribosomal proteins comparable to the bacterial L7 and L12. Cross recombination experiments revealed that hybrid ribosomes could be formed that consisted of yeast 60S core particles and *Escherichia coli* L7/L12. With yeast elongation factors the hybrid ribosomes are active in all peptide chain elongation reactions.

METHODS

The yeast *Saccharomyces fragilis* from the American Type Culture Collection was grown as described (6); 80S ribosomes (6) and elongation factors 1 and 2 from yeast (6) and reticulocyte cells (7, 8) were obtained by standard procedures. Assays for polyphenylalanine synthesis (6), for GTP hydrolysis (9) and for ribosomal binding of GTP (4) were carried out as reported. 60S and 40S ribosomal subunits were prepared by the method of Blobel and Sabatini (10).

Preparation of Ribosomal Core Particles and Extracted Protein Fraction: Basically the method of Hamel et al. (11) was used. 1.5 mg/ml of 80S ribosomes, 60S or 40S ribosomal subunits were dialyzed against 10 mM imidazole buffer pH 7.2, 1 mM $Mg(OAc)_2$, 1 mM DTT, (buffer 1). 1 ml ribosome solution adjusted with 10% acetid acid to pH 5-6, was stirred with 0.34 ml of 4 M KCl for 10 min at $4^{o}C$; then 1,300 ml of ice-cold abs. ethanol were added and the solution stirred for additional 5 min. The mixture was centrifuged at 15,000 rpm for 10 min, the supernatant withdrawn, the pellet dissolved in 1.0 ml buffer 1 and re-extracted with KCl and ethanol as indicated above. The precipitate was centrifuged at 15,000 rpm for 10 min and the supernatant fraction combined with the previous one. The ribosomal pellet was dissolved in 1 ml buffer 2 (20 mM Tris-HCl, pH 7.4, 5 mM $Mg(OAc)_2$, 1 mM DTT) and dialyzed against the same buffer. The supernatant fractions from the first and second centrifugation step were treated with 2.5-fold volume of $-20^{o}C$ acetone. The precipitate was collected by centrifugation at 15,000 rpm for 15 min, dialyzed against buffer 2, and analyzed by polyacrylamide gel electrophoresis. Two acidic proteins were detected contaminated with basic proteins. This protein fraction is referred to as extracted protein fraction (E.P. fraction).

RESULTS

Ocurrence of Two Very Acidic Proteins in 80S Ribosmes: 80S ribosomal proteins from yeast were analysed by two-dimensional polyacrylamide gel electrophoresis according to Kaltschmidt and Wittmann (12) and were shown to yield a pattern with two distinct

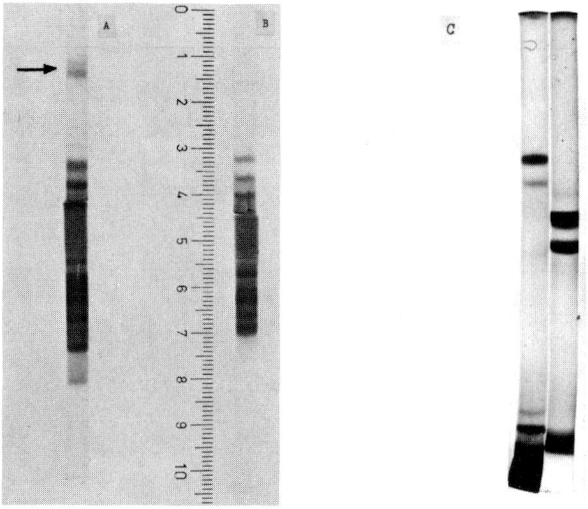

Figure 1: Polyacrylamide Gel Electrophoresis of 60S and 40S Ribosomal Subunits from Yeast.
600 ug of 60S and 400 µg of 40S ribosomal subunits were applied to polyacrylamide gels. The bottom gel consisted of 5%, the sample gel of 4% and the top gel of 10% polyacrylamide. Plastic tubes (10 x 0.6 cm) were filled with 0.6 ml bottom gel, then 0.2 ml of sample gel and finally 0.6 ml of the top gel were applied. Electrophoresis was carried out at pH 8.6 and 60 volts for 18 to 20 hrs at 4°C. For staining gels were presoaked for 1 hr in 50% trichloroacetic acid, then stained for 1 hr in coomassie blue (0.1% solution in 50% trichloroacidic acid) and destained in 7.5% acetic acid (13). 80S reticulocyte ribosmes were prepared according to standard methods and the ribosomes treated with ethanol-NH_4Cl. The extracted proteins were passed through Sephadex G-25, equilibrated with 0.05% formic acid. The proteins were lyophilized and isoelectric focussing pH 4-6 performed as reported (14).
A) 60S; B) 40S; C) Isoelectric focussing pH 4-6 of E.P. fraction from 80S reticulocyte ribosomes (heavy band on bottom contains multiple proteins). Left E.P. fraction ret., right L7/L12 MRE 600. Proteins under A and B were acetic acid extracted.

acidic proteins in the region even more acidic than the *E. coli* proteins L7 and L12. In addition when 60S or 40S subunit proteins of yeast were analysed on one-dimensional polyacrylamide gels, the 60S but not the 40S protein sample gave two fast migrating, acidic protein spots (Figure 1 A and B).
When 80S or 60S ribosomes from yeast were extracted according to a modified procedure of Hamel et al. (11) as described in the experimental section, the corresponding core particles did not show two fast migrating spots but the E.P. fractions did. Simularly the ethanol extractable fraction from 80S reticulocytes also showed two very acidic proteins both in two-dimensional gel electrophoresis and in isoelectric focussing experiments. The E.P. fraction from reticulocytes yielded two bands which are more acidic than those corresponding to L7/L12 together with a number of more basic proteins at the bottom of the gel (Figure 1 C). Although there was still contamination by basic proteins, the 60S E.P. fraction was shown to be replaceable by the bacterial proteins L7/L12 in a number of functional assays. The amount of the two acidic proteins present in the 80S ribosomes, 60S subunits and E.P. fraction has not yet been determined, although the number of copies per 60S subunits seems well below that present in the ribosomes of *E. coli*.

Functional Properties of the E.P. Fraction: Yeast 60S subunits were treated with ethanol and salt and studied in the peptide chain elongation process. 60S core particles that lack the E.P. fraction complemented with untreated 40S subunits were inactive in the protein synthetic reaction, whereas in the presence of the E.P. fraction about 60% of the polyphenylalanine synthesis was regained.
Figure 2 depicts experiments where the functions of both enzyme factors, elongation factor 1 and 2, were studied separately with 60S core particles, with and without the E.P. fraction. Not unlike the bacterial protein L7/L12 the yeast E.P. fraction was fully required to restore the functions of elongation fac-

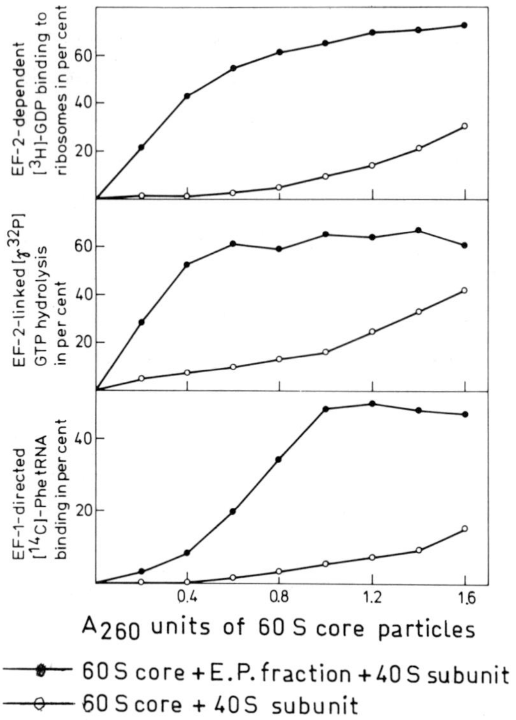

Figure 2: Dependency of Elongation Factor-Linked Functions on 60S Core Particles, E.P. Fraction and 40S Ribosomal Subunits in a Yeast System.

In all assays 100% activity corresponds to experiments where 60S (0.5 A_{260} units) and 40S (0.3 A_{260} units) were used. EF-1-directed Phe-tRNA binding to ribosomes (6) was carried out in 100 µl reaction mixture with 0.7 A_{260} units of 40S, 9 pmoles of ^{14}C-Phe-tRNA (spec. act. 700 pmoles/mg tRNA), 50 µg of poly (U), 7 µg of extracted protein, 40 µg of EF-1 and 60S core particles as indicated. 100% activity corresponded to 3.8 pmoles of ^{14}C-Phe-tRNA bound per A_{260} unit of ribosomes.

EF-2-linked (γ-^{32}P)-GTP hydrolysis was carried out in 100 µl reaction volume (9) with 0.7 A_{260} units of 40S, 10 µg of extracted protein, 20 µg of EF-2, 60S core particles as indicated and 200 pmoles of (γ-^{32}P)-GTP (spec. act. 1.5 Ci/mmole). 100% activity corresponded to 31 pmoles of hydrolysed GTP per A_{260} unit of ribosomes.

EF-2-dependent (^{3}H)-GDP binding was assayed with 1.1 A_{260} units of 40S, 10 µg of extracted protein fraction, 50 pmoles of (^{3}H)-GDP and 60S core particles as indicated. Total reaction volume was 50 µl (4). 100% activity corresponded to 3.5 pmoles of bound (^{3}H)-GDP per A_{260} units of ribosomes.

tor 1-dependent ribosomal Phe-tRNA binding and elongation factor 2-linked GTP hydrolysis as well as ribosomal GDP binding. Restoration of the activity of the 60S core particles by the E.P. fraction was only partially observed.

Formation of Hybrid Ribosomes: The data presented above support the possibility that the yeast E.P. fraction from the larger subunit contains a component which is functionally equivalent to L7/L12 from *E. coli*. Therefore an attempt was made to study their functional interchangeability from one class to another. Yeast hybrid ribosomes that consisted of yeast 60S core particles, *E. coli* L7/L12 and 40S subunits were complemented with yeast elongation factor 2 and found to be active in binding GDP to hybrid ribosomes (Figure 3).

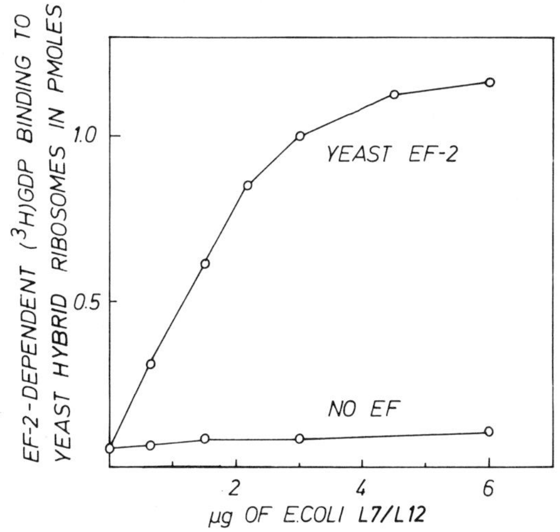

Figure 3: Yeast Elongation Factor 2-Dependent GDP Binding to Yeast Hybrid Ribosomes.

Assay conditions for (^{3}H)-GDP binding were the same as described in Figure 2 except that 1.5 A_{260} units of yeast 60S core particles, 0.9 A_{260} units of 40S subunits, and 12 μg of yeast EF-2 were used. The concentrations of *E. coli* L7/L12 were as indicated in the figure.

Although the activity of the hybrid ribosomes was somewhat less than that with the homologous ribosome preparation, elongation factor 2 apparently did not strongly discriminate between bacterial and non-bacterial acidic protein fractions. Considering that eukaryotic elongation factors are inactive on 70S ribosomes and considering the lack of discrimination between L7/L12 and its eukaryotic counterpart, the results indicate that in addition to these acidic proteins, other ribosomal regions in the core are involved in the species specific "recognition process" of the elongation factors. Similarly, elongation factor 1-directed Phe-tRNA binding to hybrid ribosomes was also observed, although this activity was 30% less than that activity found with homologous yeast ribosome preparations.

Table 1 : Functional Interchangeability of Bacterial L7/L12.

Conditions	Polyphenylalanine synthesized pmoles/A_{260} units of ribosomes
Yeast 60S Core + Yeast 40S	0.8
Yeast 60S Core + Yeast 40S + Yeast E.P. fr.	3.1
Yeast 60S Core + Yeast 40S + *E. coli* L7/L12	2.3

The assay contained 15 µg of yeast EF-1 and 10 µg of yeast EF-2. Polyphenylalanine synthesis was carried out as described (6).

The results in Table 1 indicate that hybride ribosomes from yeast synthesize polyphenylalanine; again the homologous ribosome preparation was slightly more active than the hybride ribosomes.

DISCUSSION

The main conclusion of this report is that 60S ribosomes contain a protein fraction which is functionally equivalent to bacterial L7/L12, and that in addition to the acidic protein

fraction other ribosomal regions are necessary for the species specific process of recognition of the elongation factors. Since the E.P. fraction used in our experiments still were contaminated by basic proteins, there is as yet no absolute prove whether the two acidic proteins present in the E.P. fraction are solely responsible for the functional restoration of the yeast ribosomal core particles. Cross-combination tests reveal that the *E. coli* L7/L12 can functionally replace the eukaryotic E.P. fraction, which implies that the two acidic proteins present in the E.P. fraction are comparable with the bacterial L7/L12. These results are complemented by experiments of Wool and Stöffler, who found that antibodies against two acidic proteins from yeast cross-react with bacterial L7/L12 (15). The same group also noticed two acidic proteins in rat liver ribosomes (16).
Our unpublished results have indicated the presence of two very acidic proteins in 60S ribosomes of *Artemia salina*. The two proteins gave similar amino acid compositions and molecular weights of about 13,000.

SUMMARY

60S but not 40S ribosomal subunits from yeast contain two acidic proteins that are extractable with ethanol and KCl. The remaining 60S core particles are inactive in all elongation factor-dependent reactions. By adding back the extracted proteins (E.P. fraction) the overall protein synthetic reaction can be restored. Hybrid ribosomes formed with 60S core particles and *E. coli* L7/L12 and yeast 40S subunits are active with yeast elongation factors in peptide chain elongation reactions.

Footnote: New address of D. R.: Abteilung Zellbiochemie am Institut für Physiologische Chemie, Universität Hamburg, 2 Hamburg 20, Martinistr. 52

ACKNOWLEDGEMENTS

Part of this work was carried out in the laboratories of Dr. F. Lipmann and Dr. H.G. Wittmann whose continuous support and stimulating criticism is gratefully acknowledged. We thank H. von Seydlitz and J. Kriek for excellent technical help.

REFERENCES

1. Lucas-Lenard, J., and Lipmann, F. (1971) Ann. Rev. Biochem. 40, 434.
2. Möller, W. (1974) in "The Ribosome" (Ed. M. Nomura), in press.
3. Baliga, B.S., and Munro, H.N. (1971) Nature New Biol. 233, 257.
4. Richter, D. (1973) J. Biol. Chem. 248, 2853.
5. Nombela, C., and Ochoa, S. (1973) Proc. Nat. Acad. Sci. USA 70, 3556.
6. Richter, D. (1971) Biochemistry 10, 4422.
7. Hardesty, B., and Mckeehan, W. (1971) Methods in Enzymology 20C, 330.
8. Hardesty, B., McKeehan, W., and Cilp, W. (1971) Methods in Enzymology 20C, 316.
9. Richter, D., and Klink, F. (1971) Methods in Enzymology 20C, 319.
10. Blobel, G., and Sabatini, D. (1971) Proc. Nat. Acad. Sci. USA 68, 390.
11. Hamel, E., Koka, H., and Nakamoto, T. (1972) J. Biol. Chem. 247, 805.
12. Kaltschmidt, E., and Wittmann, H.G. (1970) Anal. Biochem. 36, 401.
13. O'Farrell, P., and Gold, L. (1973) J. Biol. Chem. 248, 5499.

(10) Blobel, G. and Sabatini, D. (1971) Proc. Nat. Acad. Sci. USA 68, 390.

(11) Hamel, E., Koka, H., and Nakamoto, T. (1972) J. Biol. Chem. 247, 805.

(12) Kaltschmidt, Ė. and Wittmann, H. G. (1970) Anal. Biochem. 36, 401,

(13) O'Farrell, P., Gold, L., and Huang, W. H. (1973) J. Biol. Biochem. 248, 5499.

(14) Möller, W., Groene, A., Terhorst, C., and Amons, R. (1972) Eur. J. Biochem. 25, 5.

(15) Wool, I. (1974) in "The Ribosome" (Ed. M. Nomura), in press.

(16) Sherton, C. C. and Wool, I. G. (1972) J. Biol. Chem. 247, 4460.

Choline Acetyltransferase: Reactions of the Active Site Sulfhydryl Group

Robert Roskoski, Jr.
Department of Biochemistry
The University of Iowa, Iowa City, Iowa 52242, USA

INTRODUCTION

The theory of chemical neurotransmission states that a nerve impulse is transmitted from the presynaptic neuron to the post-synaptic cell (nerve, muscle, or gland) by means of a chemical substance. Acetylcholine is an established transmitter agent at the vertebrate neuromuscular junction, the preganglionic sympathetic and parasympathetic divisions of the autonomic nervous system, and the postganglionic division of the parasympathetic system (1). In addition to mediating voluntary movement by triggering skeletal muscle contraction, acetylcholine plays a role in the regulation of heart rate, blood pressure, digestion, reproduction, and other vegetative functions by transmitting the cholinergic nerve impulses of the autonomic system. Acetylcholine is also found in the vertebrate central nervous system where it is a probable, but not proven, chemical transmitter agent. Norepinephrine and gamma-aminobutyric acid (GABA) are additional examples of established neurotransmitter substances (1).

In addition to its importance in neurobiochemistry, the elucidation of the mechanism of acetylcholine biosynthesis helped lay the foundations for modern bioenergetics. The first demonstration that the energy of ATP is used for biosynthesis (outside of glycolysis) was made by Nachmansohn and Machado in 1943 (2). They showed that cell-free extracts prepared from brain and from electric organ of eel catalyze the ATP-dependent biosynthesis of acetylcholine. Nowadays, the concept of the ATP-dependent biosynthesis of many biological compounds is firmly engrained in biochemistry in large part due to the guidance provided by Dr. Lipmann's 1941 review (3). The mechanism of the ATP-dependence remained obscure until the elucidation of the structure and properties of coenzyme A by Dr. Lipmann and his coworkers (4-6). Lynen then identified the thioester bond in acetyl coenzyme A (6), the first demonstration of an energy-rich linkage not containing phosphate. The mechanism of acetate activation was finally mapped out by Berg (7). He showed that acetate thiokinase (EC 6.2.1.1) catalyzes the formation first of acetyladenylate and then acetyl coenzyme A as outlined:

$$\text{ATP} + \text{acetate} \rightleftharpoons \text{acetyladenylate} + \text{PPi}$$

$$\text{acetyladenylate} + \text{coenzyme A} \rightleftharpoons \text{acetyl coenzyme A} + \text{AMP}$$

Acetyl coenzyme A is at the crossroads of many catabolic and anabolic reactions (6). An important source of extramitochondrial acetyl coenzyme A is derived from the citrate-cleavage reaction, discovered by Srere and Lipmann (8). ATP citrate lyase (EC 4.1.3.8) catalyzes the following reaction:

$$\text{ATP} + \text{citrate} + \text{coenzyme A} \rightleftharpoons \text{acetyl coenzyme A} + \text{oxaloacetate} + \text{ADP} + \text{Pi}$$

Choline acetyltransferase or acetylcholine synthetase (EC 2.3.1.6) catalyzes the bioformation of acetylcholine with the stoichiometry given in the following chemical equation:

$$\text{acetyl coenzyme A} + \text{choline} \rightleftharpoons \text{acetylcholine} + \text{coenzyme A}$$

According to the theory of chemical neurotransmission, the nerve action potential promotes the release of acetylcholine into the synaptic cleft. The liberated neurotransmitter interacts with the postsynaptic cell, triggers the appropriate response, and is then inactivated. Acetylcholine esterase (EC 3.1.1.7), a membrane-bound enzyme, catalyzes the hydrolytic inactivation.

The present studies on the mechanism of the choline acetyltransferase reaction were carried out in the Iowa City laboratory. Thiol reagents had been shown to inhibit the enzyme prepared from squid head ganglia (9), primate placenta (10), torpedo (11), and mammalian brain (12,13). The studies in the present paper support the hypothesis that the active site of choline acetyltransferase contains a sulfhydryl group. This group mediates the transfer of the acetyl from coenzyme A to choline. First, the enzymic -SH reacts with acetyl coenzyme A to form an acetylthioenzyme intermediate and coenzyme A. Choline then reacts with the acetyl-enzyme to form acetylcholine and the regenerated enzyme.

EXPERIMENTAL PROCEDURE

The bovine brain enzyme was prepared as previously described (14). The radiochemical enzyme assays, coenzyme A bioassay, isolation of the acetyl-enzyme by Sephadex gel filtration and sources of materials have been previously documented (14-16). Unless otherwise noted, the incubation of thiol reagents and enzyme was carried out at pH 7.4 for 15 min at 37°.

The following Km values (apparent) were determined by Lineweaver-Burk plots:acetyl coenzyme A, 15 μM; choline, 0.75 mM, coenzyme A, 20-200 μM; and acetylcholine 1 to 5 mM. In the latter cases, the Km depends upon the concentration of the other substrate.

RESULTS

Isolation of the Acetyl-Enzyme Complex by Sephadex Gel Filtration

When [^{14}C] acetyl coenzyme A (or [^{14}C] acetylcholine) is incubated with the partially purified acetylcholine synthetase and the reaction mixture is Sephadexed, radiolabel is associated with the protein in the eluant (Figure 1).

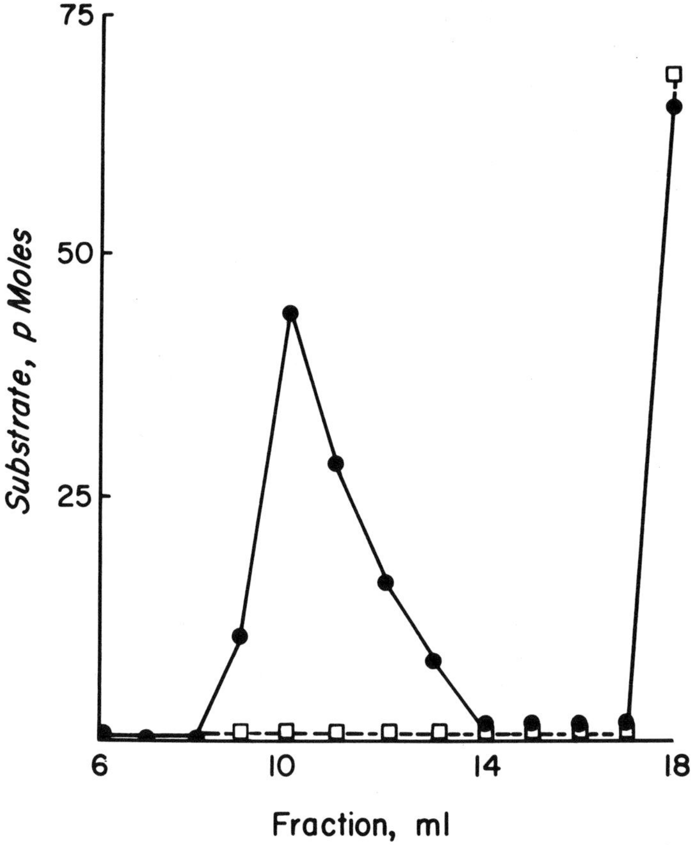

Figure 1. Isolation of the [^{14}C] acetyl-enzyme by Sephadex gel filtration. The enzyme extract was incubated with [^{14}C] acetyl coenzyme A for 15 min at 37°, chilled, and Sephadexed. The [^{14}C] was measured by liquid scintillation spectrometry and the coenzyme A, by bioassay as previously documented (14). (●) ^{14}C label; (□), coenzyme A.

The initial experiments showed that coenzyme A is associated with the postulated acetyl-enzyme intermediate using acetyl coenzyme A as donor substrate (14). When N-ethylmaleimide is added to the reaction mixture, followed by gel filtration, then coenzyme A is not associated with the acetyl-enzyme (15). This excludes the isolation of a noncovalent acetyl coenzyme A· enzyme complex. The N-ethylmaleimide may react with coenzyme A with subsequent dissociation of the adduct from the enzyme.

Using doubly-labeled acetylcholine, it was possible to show that the acetyl, but not the choline is associated with the eluant protein (14). It is therefore possible to prepare the alleged acetyl-enzyme intermediate from either acetyl donor free from coenzyme A or choline.

Bioreactions of the [^{14}C] Acetyl-Enzyme

If the alleged acetyl-enzyme is a bona fide intermediate in the enzymic reaction, it ought to react with the acceptor substrates to form the reaction products. Using either acetyl coenzyme A or acetylcholine as donor, when the acetyl-enzyme is isolated by Sephadex gel filtration and incubated with either coenzyme A or choline, the corresponding product forms (Table 1). The transfer from the acetyl-enzyme to acceptor substrate argues

Table 1: Chemical Competence of the [^{14}C]Acetyl-Enzyme. From (14).

	Product (pmoles)	
Donor for [^{14}C]Acetyl-Enzyme Formation	Acetyl coenzyme A	Acetyl-choline
[^{14}C]Acetyl coenzyme A	36	42
[^{14}C]Acetylcholine	44	43

against a concerted reaction between acetyl coenzyme A and choline and supports the notion of a two-step chemical mechanism. Further studies are required to show that this is the mechanism under conditions of turnover. These experiments, however, demonstrate that the isolated intermediate is chemically competent.

Properties of the Acetyl-Enzyme Link

A series of experiments to determine the nature of the chemical bond between the acetyl group and the transferase were initiated. The labeled acetate is not discharged by hot 10% trichloro-acetic acid (90°, 20 min), nor by 6 M guanidinium-Cl. These results indicate that the acetyl group is covalently linked to the enzyme protein. Treatment of the acetyl-enzyme with neutral, salt-free 3 M hydroxylamine (50°, 15 min) liberates the acetate as the corresponding acetylhydroxamate. Furthermore, dilute alkali (pH 10) discharges the acetyl group. These results argue against an N-acetylimidazole derivative, which is acid and alkali labile, and against O-acetylserine or threonine derivatives, which are cleaved by hot trichloroacetic acid. These experiments are consistent with the notion that the acetyl group is bound to the enzyme as a thio ester. To substantiate this hypothesis, the acetyl-enzyme was subjected to performic acid oxidation. Thio esters are cleaved by this procedure, but oxygen esters are not (17). It was found that all the protein-bound acetate is liberated by performic acid oxidation. In control experiments, 100% of the acetate was liberated from acetyl coenzyme A and none was liberated from acetylcholine. Although not quantitatively recovered because of volatility, the liberated product comigrates with acetate on silica gel thin layer chromatograms (14). Table 2 summarizes these results.

Table 2: Identification of the Acetyl-Enzyme Link as Thio Ester From (14).

Treatment	Product
1. Hot trichloroacetic acid	Protein-bound acetate
2. Alkali, pH 10	Acetate
3. 3 M Hydroxylamine, pH 5.7	Acetylhydroxamate
4. Performic acid oxidation	Acetate

Thiol Reagent Inhibition of Choline Acetyltransferase

The isolation and identification of a presumptive thio ester intermediate prompted an investigation of the effects of thiol reagents on the enzyme. These experiments showed that N-ethylmaleimide, iodoacetate, iodoacetamide, p-chloromercuribenzoate and Ellman's reagent inhibit transferase activity. Inhibition by the latter two reagents was reversed by incubating the enzyme with 1 mM dithiothreitol (37°, 10 min).

Table 3: Effect of Substrates on N-ethylmaleimide Inactivation of Choline Acetyltransferase. From (15).

Addition	Concentration	Enzyme activity % control
None		4
Acetyl coenzyme A	50 μM, 15 μM	105
Acetyl coenzyme A	5 μM	47
Acetylcholine	5 mM	67
Acetylthiocholine	1 mM, 50 μM	103
Acetylthiocholine	25 μM	60
Choline	5 mM, 10 mM	5
Coenzyme A	50 μM	23

The N-ethylmaleimide reaction was chosen for further study. It inactivates the enzyme with a second-order rate constant of 32 $M^{-1}sec^{-1}$. Preincubation of the enzyme with 15 μM (Km) acetyl coenzyme A fully protects against N-ethylmaleimide inhibition (Table 3). Acetylcholine partially protects and acetylthiocholine fully protects against thiol reagent inactivation. On the other hand, choline fails to protect.

Resistance of the Acetyl-Enzyme to N-Ethylmaleimide Inactivation

When the [^{14}C]acetyl-thioenzyme is isolated by Sephadex gel filtration free from coenzyme A, and then is incubated with choline in the presence of N-ethylmaleimide, acetyl transfer is not inhibited (Table 4). Subsequent to the transfer, however, enzyme activity is markedly decreased. In the absence of acceptor substrate, and therefore transfer, the acetyl-enzyme is not inhibited by the thiol reagent. Similarly, the preparation of the postulated acetyl-enzyme from the acetylcholine donor converts the enzyme to a form resistant to N-ethylmaleimide inactivation (not shown). In this latter case acetyl transfer is also associated with concomitant susceptibility to N-ethylmaleimide inhibition.

Table 4: Formation of [^{14}C]Acetylcholine from [^{14}C]Acetyl-enzyme in the Presence of N-Ethylmaleimide. From (15).

Incubation with [^{14}C]Acetyl-Enzyme	Product	Enzyme activity after incubation
	pmoles	nmoles/5 min
Choline, 2 mM	47	40.3
Choline, 2 mM; N-ethylmaleimide; 5×10^{-5} M	42	4.1
Choline, 0 mM; N-ethylmaleimide 5×10^{-5} M	0.5	36.7

The effect of N-ethylmaleimide was tested when the enzyme is catalyzing the turnover of substrates. At saturating concentrations of acetyl coenzyme A and choline (100 μM and 7.5 mM, respectively), N-ethylmaleimide does not inactivate the transferase (Table 5). This indicates that the enzyme is piled up in a form not susceptible to thiol reagent inhibition. At lower substrate concentrations (25 μM acetyl coenzyme A and 0.75 mM choline), the transferase is still resistent to thiol reagent inhibition. Raising the choline concentration to 7.5 mM now converts the enzyme to a form susceptible to N-ethylmaleimide inactivation (Table 5).

Table 5: Substrate Protection Against N-Ethylmaleimide Inactivation During Turnover. From (15).

Addition	Transferase activity
	% control
100 μM acetyl coenzyme A + 7.5 mM choline	103
25 μM acetyl coenzyme A + 0.75 mM choline	98
25 μM acetyl coenzyme A + 7.5 mM choline	44
25 μM acetyl coenzyme A	100
7.5 mM choline	4

DISCUSSION

The choline acetyltransferase reaction proceeds through a two-step chemical mechanism. First, an enzymic -SH reacts with acetyl coenzyme A to form an acetylthioenzyme and coenzyme A. The acetyl group, activated as thio ester, then undergoes a nucleophilic attack by choline to form acetylcholine and the regenerated enzyme. The mechanism can be outlined as follows:

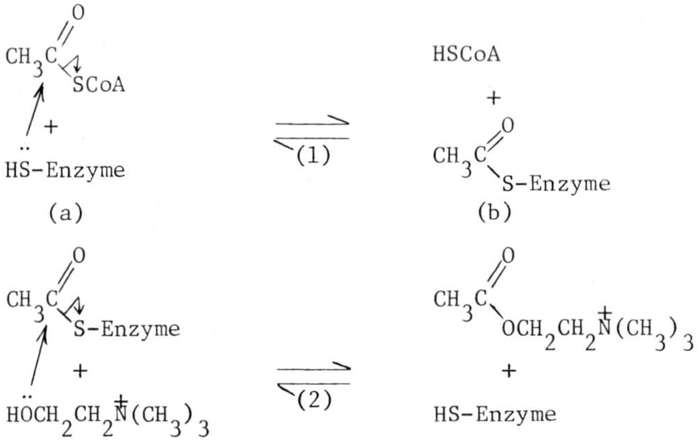

The evidence supporting this mechanism includes: (1) isolation of the acetyl-enzyme intermediate, (2) demonstration of the chemical competence of the postulated intermediate, (3) demonstration that the acetyl-enzyme link is a thioester and (4) protection against thiol reagent inhibition by low concentrations of acetyl donor substrate, and deprotection by choline during turnover.

The free enzyme (a) is susceptible to thiol reagent inactivation. The acetyl-thioenzyme (b) is resistant to inactivation (Table 4). Incubation of (b) with choline deprotects the enzyme by converting it to (a). Incubation of the free enzyme with acetyl coenzyme A converts it to a form which is resistant to N-ethylmaleimide inhibition. At high concentrations of acetyl coenzyme A and choline (during turnover), the enzyme is resistant to thiol reagent inhibition. This suggests that the transferase is piled up in a form where the thiol group is shielded. Increasing the concentration of choline does not deprotect the enzyme and argues that deacylation is slow. When the concentration of both substrates is lowered to the Km (apparent) value, the enzyme is still resistant to thiol reagent inhibition. Increasing the choline concentration now converts the transferase to a form susceptible to inhibition. Under conditions of low acetyl coenzyme A and high choline concentrations, deacylation and acylation proceed at similar rates.

One question often asked is how can a two-step enzyme mechanism be proposed when the steady state kinetic data are not of the parallel line type (18,19). Although the adherence of an enzyme to parallel line kinetics point to a two-stage or Ping-Pong mechanism (20), such a mechanism is not excluded when the kinetics are not of this type. For example, *E. coli* succinate thiokinase does not exhibit Ping-Pong kinetics. Rather, the kinetic studies are consistent with a sequential mechanism in which all three substrates combine with the enzyme before the release of product (21), even though the reaction intermediate involves a kinetically competent phosphoenzyme intermediate (22). The article by Lenard Spector may be consulted for more exhaustive documentation of the contention that a two-step enzyme reaction need not exhibit parallel line kinetics (23).

Finally, it must be mentioned that the postulated acetyl-thio-enzyme intermediate might be an adventitious product. Further experiments are required to demonstrate that the rate constants for its formation and further reaction are adequate to account for the observed rate of the enzymic reaction.

SUMMARY

The active site of choline acetyltransferase contains a sulfhydryl group. This group mediates the transfer of the acetyl from coenzyme A to choline. The chemically competent acetylthioenzyme is isolable by Sephadex gel filtration. The free enzyme is susceptible, but the acetyl-enzyme is resistent to thiol reagent inactivation.

ACKNOWLEDGEMENT

This research is supported by Grant NS 11310 by the U.S. Public Health Service.

REFERENCES

1. Iverson, L. L.: Neurotransmitters, Neurohormones, and Other Small Molecules in Neurons. In Schmitt, F. O., Ed. The Neurosciences (2nd Edition) (Rockefeller University Press, New York, 1970) pp. 768-781.
2. Nachmansohn, D. and Machado, A. L.: The Formation of Acetylcholine. A New Enzyme: "Choline Acetylase". J. Neurophysiol. 6, 397-404 (1943).
3. Lipmann, F.: Metabolic Generation and Utilization of Phosphate Bond Energy. Adv. Enzymology 1, 99-162 (1941).
4. Lipmann, F. and Kaplan, N. O.: A Common Factor in the Enzymatic Acetylation of Sulfanilamide and of Choline. J. Biol. Chem. 162, 743-744 (1946).
5. Lipmann, F., Kaplan, N. O., Novelli, G. D., Tuttle, L. C. and Guirard, B. M.: Coenzyme for Acetylation, A Pantothenic Acid Derivative. J. Biol. Chem. 167, 869-870 (1947).
6. Jaenicke, L. and Lynen, F.: Coenzyme A. In Boyer, P. D., Lardy, H. and Myrbäck, K., Eds. The Enzymes (2nd edition) (Academic Press, New York, 1960) Vol. 3, pp 3-103.
7. Berg, P.: Acyl Adenylates: An Enzymatic Mechanism of Acetate Activation. J. Biol. Chem. 222, 991-1013 (1956).
8. Srere, P. A. and Lipmann, F.: An Enzymatic Reaction between Citrate, Adenosine Triphosphate and Coenzyme A. J. Amer. Chem. Soc. 75, 4874 (1953).
9. Reisberg, R. B.: Sulfhydryl Groups of Choline Acetylase.

Biochim. Biophys. Acta 14, 442-443 (1954).

10. Schuberth, J.: Choline Acetyltransferase. Purification and Effect of Salts on the Mechanism of the Enzyme-Catalyzed Reaction, Biochim. Biophys. Acta 122, 470-481 (1966).

11. Morris, D.: The Effect of Sulfhydryl and Other Disulfide Reducing Agents on Choline Acetyltransferase Activity Estimated with Synthetic Acetyl-CoA, J. Neurochem. 14, 19-27 (1967).

12. Potter, L. T., Glover, V. A. S. and Saelens, J. K.: Choline Acetyltransferase from Rat Brain, J. Biol. Chem. 243, 3864-3870 (1968).

13. Choa, L. P. and Wolfgram, F.: Purification and Some Properties of Choline Acetyltransferase (EC 2.3.1.6) from Bovine Brain, J. Neurochem. 20, 1075-1081 (1973).

14. Roskoski, R. Jr.: Choline Acetyltransferase. Evidence for an Acetyl-Enzyme Reaction Intermediate, Biochemistry 12, 3709-3714 (1973).

15. Roskoski, R. Jr.: Choline Acetyltransferase: Inhibition by Thiol Reagents, J. Biol. Chem. 249, 2156-2159 (1974).

16. Roskoski, R. Jr.: Choline Acetyltransferase. Reversible Inhibition by Bromoacetyl Coenzyme A and Bromoacetylcholine, Biochemistry 13, 2295-2298 (1974).

17. Harris, J. I., Meriwether, B. P. and Park, J. H.: Chemical Nature of the Catalytic Sites in Glyceraldehyde 3-Phosphate Dehydrogenase, Nature (London) 198, 154-157 (1963).

18. Glover, V. A. S. and Potter, L. T.: Purification and Properties of Choline Acetyltransferase from Ox Brain Striate Nuclei, J. Neurochem. 18, 571-580 (1971).

19. White, H. L. and Wu, J. C.: Kinetics of Choline Acetyltransferases (EC 2.3.1.6) from Human and Other Mammalian Central and Peripheral Nervous Tissues, J. Neurochem. 20, 297-307 (1973).

20. Arion, W. J. and Nordlie, R. C.: Liver Microsomal Glucose 6-Phosphatase, Inorganic Pyrophosphatase, and Pyrophosphate-Glucose Phosphotransferase, J. Biol. Chem. 239, 2752-2757 (1964).

21. Moffet, F. J. and Bridger, W. A.: The Kinetics of Succinyl Coenzyme A Synthetase from Escherichia coli, J. Biol. Chem. 245, 2758-2762 (1970).

22. Bridger, W. A., Millen, W. A. and Boyer, P. D.: Substrate Synergism and Phosphoenzyme Formation in Catalysis of Succinyl Coenzyme A Synthetase, Biochemistry 7, 3608-3616 (1968).

23. Spector, L. B.: Covalent Enzyme-Substrate Intermediates in Transferase Reactions, Bioorganic Chemistry 2, 311-321 (1973).

Control of Gene Expression by E. coli Virus T7

M. Schweiger, P. Herrlich, H.J. Rahmsdorf, S.H. Pai,
H. Ponta, M. Hirsch-Kauffmann
Max-Planck-Institut für Molekulare Genetik,
Berlin-Dahlem, Germany

In Verehrung unserem Lehrer Professor Fritz Lipmann gewidmet.

One of the basic problems of modern biology is the mechanism of control of genetic expression. A human cell for example contains an amount of DNA that would suffice for the synthesis of several million proteins [1]. Uncontrolled expression of this information would lead to chaos. Life requires the ordered use of genetic information, ordered both in time and amount of specific gene product. Considering the wealth of available information, the negative control of genetic expression seems to predominate. Most genes are inactive at any given moment of a cell's life cycle. During development of the differentiated organism from the omnipotent germ cell, a whole cascade of gene activation and inactivation is required.

It is sufficient in this context to illustrate the significance of gene regulation by a few examples: (i) hormones trigger specific target cells which then synthesize characteristic proteins[2], (ii) antigens lead to the activation of specific genes in the appropriate lymphocytes which then produce specific antibodies[3], (iii) messages from the cell surface of mammalian cells to the genome then produce contact inhibition[4], and (iv), as already mentioned, cell differentiation is accompanied by gene activation and inactivation[5].

Although gene regulation is of great fundamental importance, its biochemical analysis has been neglected. Ignorance of gene regulatory mechanisms is apparent particularly with complex eukaryotic systems but also with prokaryotic systems. The lac operon of E.coli has captured the attention of molecular biologists. Although a great deal is known about its biochemical regulation[6], several basic questions have not been answered. For example, catabolite and transient repression of the lac operon is not understood even though experiments on the transcriptional effectors have been carried out by the method of cell-free enzyme synthesis[7]. The mechanism for regulating the ratios of the three individual proteins of the lac operon is also unknown. Initial experiments on the regulation of other operons of E.coli look promising[8] but the biochemical analysis is still in a preliminary state.

The successful investigation into a new field requires powerful methods and a suitable subject. Gene expression in vitro is the method of choice. The efficiency of transcription and translation in the in vitro system is amazingly high[9]. The possibility of omitting components and separating transcription from translation, for example, make the method ideal for the study of gene control.

In combination with the methodology, a suitable system is also very important. One such system is E.coli infected with the virus T7. Two independent chromosomes exhibit such refined control mechanisms that the originally omnipotent genetic equipment of the cell is directed toward the production of only a few structural virus proteins. This process of differentiation is directed by virus-coded proteins synthesized early after infection. Several positive and negative control mechanisms are involved. We will describe recent studies on the biochemistry of gene control in this T7 system.

Early transcription

The molecular weight of virus T7 DNA is 2.5×10^7 [10]. The information is sufficient for coding for 25 to 30 proteins. Nineteen genes have been mapped[10]. The virus compensates for the loss of other gene functions which have, therefore, been considered to be "non-essential". The genes transcribed by the host RNA polymerase are clustered in one region of the T7 chromosome[11;12]. T7 DNA carries but one promoter for *E.coli* RNA polymerase. The promoter and a cluster of control genes is located at one end (left end) of the genome. Transcription and translation of these genes initiates the differentiation process.

Transcription by *E.coli* RNA polymerase results in one large polygenic RNA (molecular weight 2.2×10^6 [12]). Termination of this transcription occurs at a signal at 20.2% of genome length[10]. The signal is recognized by the enzyme without the help of other factors[12;13]. Termination is not complete, however, and 20-30% of the enzyme molecules (varying from one strain to the other) transcribe a larger product of 3.4×10^6 daltons[9]. A minute amount of polymerase molecules seem to transcribe to the right end of the genome. We do not know whether there are additional specific termination sites.

The polygenic early RNA has originally been isolated

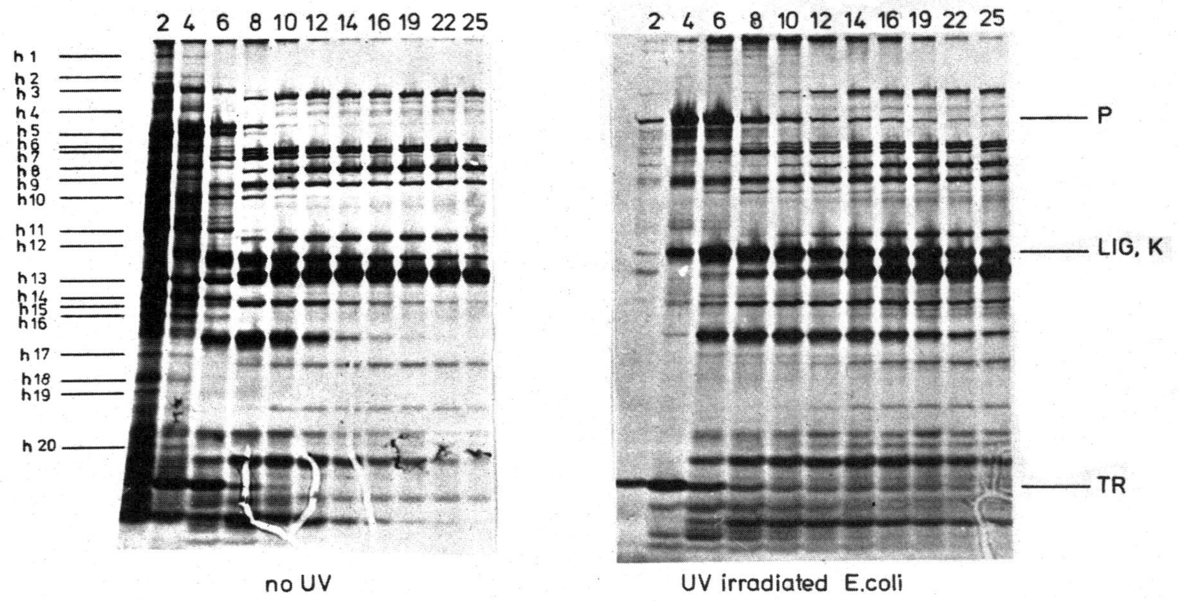

Figure 1: Time course of protein synthesis after $T7^+$ infection. 0.2 ml aliquots of a culture of E.coli B_{s-1} were pulse labeled (2 min followed by 2 min chase) with ^{14}C amino acids at various times after infection and the proteins were separated by SDS gel electrophoresis (10-20% exponential gradient acrylamide). The beginning of each pulse is indicated. The gels were dried and autoradiographed. Techniques as in ref.21 and 23.

from *in vitro* reactions with purified RNA polymerase [12] and has not been detected *in vivo*[14]. It is, however, present in an RNase III negative host after T7 infection[15]. RNase III cleaves the RNA into monogenic species[15] and the cleavage is required for translation[16].

The viral control functions

A careful analysis of the kinetics of appearance of virus proteins (figure 1; schematic diagram in figure 2) and viral RNAs leads to the distinction of various control phenomena: The control proteins (symbols: P = polymerase, TR = translational repressor, LIG = DNA ligase, K = protein kinase) appear in a defined order ranging from 2 (K and TR) to 6 (LIG) minutes.

Host protein synthesis (symbols: h1 - h20 = representative examples) ceases at 2-5 minutes.
The synthesis of the control (early) proteins is dis-

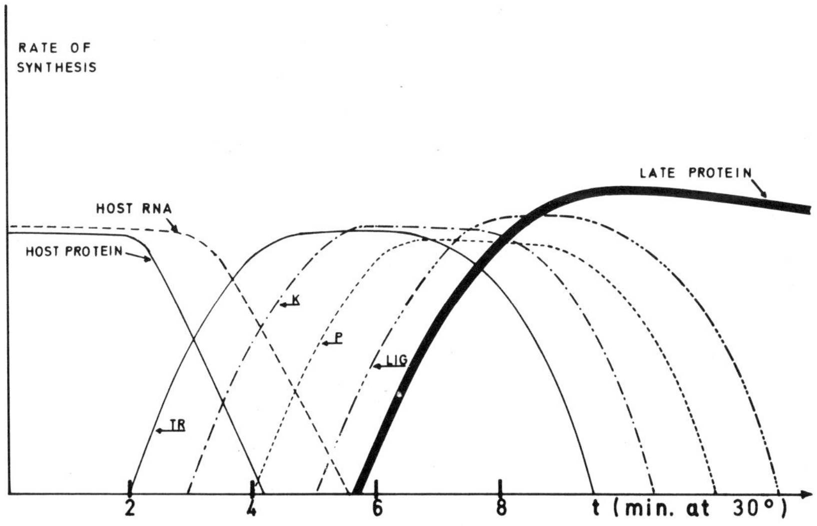

Figure 2: Schematic evaluation of figure 1.

continued at 10-12 minutes.
The production of most virus proteins (non-early) starts in a burst at 6 minutes.
Among the non-early proteins, two classes are recognized. The rate of synthesis of those proteins whose genes are adjacent to the early region (early-late) decreases at 15 minutes and the late proteins (virus structural proteins) continue to be synthesized until lysis. At about 20 minutes, the end point of differentiation is reached: only the structural proteins are synthesized. All other genes have been inactivated by a series of negative control mechanisms.

1) Defined order of early protein synthesis

The schedule of the appearance of early proteins is a consequence of the early transcription which starts from one promoter. The linear array of genes specifies the time course of their transcription and translation [9]. RNase III cleavage of the early transcription product probably occurs in the same sequential order as does transcription itself, and does not appear to be limiting.

2) New virus-coded RNA polymerase

Associated with virus autonomy is the synthesis of a virus-coded RNA polymerase[17]. Historically this was the first detected control function associated with T7 [18;19]. This polymerase differs in many properties from the host enzyme. Among these properties are the molecular weight (one peptide chain of 100 000 daltons), the resistance against rifampicin and the preference for the template T7 DNA[17]. Virus RNA polymerase magnifies the transcription of the late genes[9]. A promoter in the early-late portion of the genome is also recognized. But we do not know whether any transcription of early genes by virus RNA polymerase occurs in vivo[9].

3) Early transcriptional control protein

If one calculated the nucleotide and energy requirement for the synthesis of 100 virus particles per cell, it is clear that nothing can be wasted. As expected from figure 1, the synthesis of cellular macromolecules unnecessary for phage T7 production is discontinued. Thus, host DNA[20], RNA[21] and protein synthesis[22] are blocked. Moreover, the synthesis of virus-coded products is controlled to optimize the yields.

Control of transcription resisted the experimental approach for some time because three mechanisms are required to obtain the desired results. The mechanisms have subsequently been separated by genetic and physical methods and their biochemistry been elucidated[23].

T7 induces a protein kinase[24]. The kinase is one of the earliest gene products. As a consequence of kinase induction, a large number of E.coli proteins become phosphorylated between 2 and 4 minutes (figure 3). Mutants which lack virus RNA polymerase and, therefore, induce little or none of most T7 proteins, still produce the same number of phosphorylated proteins between 2 and 4 minutes. The phosphorylated proteins must, therefore, be E.coli proteins. One additional phosphorylated peptide chain, however, appears later, at 4-6 minutes (arrow in figure 3). The appearance of this peptide coincides with the inactivation of protein kinase and with a change in migration of the kinase peptide chain corresponding to a larger molecular weight comigrating with this phosphate band (figure 3). Treatment with alkaline phosphatase reverses the size increase with concomitant loss of the phosphate. Homo-

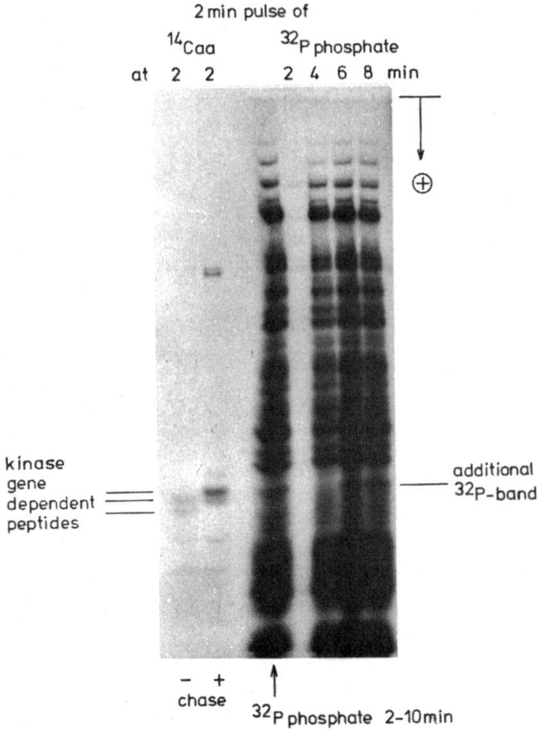

Figure 3: Time course of T7 induced protein phosphorylation. 0.2 ml aliquots of a culture of E.coli B_{s-1} in TG medium were either pulse labeled with ^{14}C amino acids or with ^{32}P-phosphate at various times after infection with T7 am342 LG3. The cells were solubilized by SDS-mercaptoethanol-buffer and the proteins separated by SDS gel electrophoresis (10% acrylamide) after treatment with RNase. See also ref. 24.

geneous preparations of protein kinase in vitro lose activity when incubated with ATP. These data suggest that protein kinase activity may be regulated by autophosphorylation.

The analysis of kinase defective virus (by deletion or UV irradiation, of T7 which also lacks the virus-specific RNA polymerase) indicates that the kinase gene may be involved in gene expression[23]. In cells infected with virus carrying a defect in the kinase gene region or with partially UV irradiated virus, host RNA species, whose synthesis normally discontinues at 2-4 minutes, are synthesized at unchanged rate for longer periods [23;25]. Moreover, early T7 mRNA synthesis also occurs for longer times. The biochemical mechanism is possibly the phosphorylation of the ß' subunit of E.coli RNA polymerase which has recently been detected[26].

The biological significance of this early transcriptional control is the block of host and excess early RNA synthesis.

4) T7 transcriptional inhibitor

In kinase deletion mutants, host and early T7 transcription are ultimately discontinued with a delay (at 10-12 min)[23]. The time lag and UV mapping indicate that the gene responsible for this control is promoter distal to the kinase (map position in the neighborhood of the ligase gene). As in the case of kinase, the protein responsible for this second shutoff mechanism (the transcriptional inhibitor) was purified to homogeneity and its action was analyzed by in vitro RNA and protein synthesis[27]. The transcriptional inhibitor specifically blocks the action of E.coli RNA polymerase by binding to the enzyme and inhibiting its initiation. It fails to inhibit the T7 RNA polymerase. Thus, the transcriptional inhibitor fulfills the requirements deduced from the regulatory pattern in vivo: host and early T7 transcription ceases even in the absence of the early transcriptional control while T7 late transcription is not affected.

Subsequently, a third mechanism becomes operative. T7 coded DNases break down host DNA[28] (also at 10-12 min) thereby mobilizing the nucleotide precursors required for T7 replication and at the same time preventing further host transcription.

5) Translational repressor

The earliest event after phage infection is a specific translational block[25]. This action can be distinguished in vivo from the transcriptional mechanisms: in the absence of the early transcriptional control, host RNA synthesis continues to late times but ß-galactosidase synthesis is still inhibited at 2-4 minutes. By cell-free enzyme synthesis the translational repressor has been characterized and partially purified from the ribosomal wash. It specifically inhibits the initiation of host messenger[25].

6) mRNA degradation

T7 not only induces control mechanisms that strongly favor T7 specific message synthesis, but saves its own messenger from degradation. The halflife of T7 messenger RNA - determined by SDS gel electrophoresis, hybridization or just by acid precipitation - is prolonged (20-30 min at 30°)[29]. In the same infected cells, trp operon messenger has the same short halflife as in the uninfected host[30]. Thus, the RNA degrading apparatus must be able to distinguish between two kinds of mRNA. Beside this degrading mechanism, a second mechanism has been detected in T7: it inactivates messenger RNA without apparent change in RNA length. The halflife of functional T7 mRNA is tested by isolation of the RNA and translation in a cell-free system[22;23]. The functional halflife is shorter than the physical halflife: functional ligase mRNA decays with $t\,1/2$ of 4.5-6 min (30°). Again the halflife of functional T7 RNA is longer than the halflife of host messenger RNA.

7) Other gene control mechanisms

In the late part of the T7 cycle we see additional genetic control phenomena: T7 DNA replication does not enhance the production of late proteins. The mechanism is not known. Several data also suggest the existence of translational control in the late phase. At times when no more ligase enzyme synthesis occurs, ligase mRNA can still be isolated and translated in vitro. Both, gene 1 and ligase RNA are pulse-labeled and detected in SDS-gels at times when no T7 RNA polymerase or ligase are synthesized. Also the mechanism of shut-off of synthesis of those proteins that originate from the early-late region, has not been elucidated.

8) The pattern of gene control in T7 (figure 4)

The information for the control of gene expression becomes available by the transcription of the first 20% of T7 DNA and subsequent cleavage of the transcript by RNase III (figure 4, (a)). The control proteins then convert - in the order of their appearance - gene expression from the host to T7 only. The translational repressor blocks translation at the host messengers that exist at this moment (figure 4, (b)); the T7 protein kinase phosphorylates E.coli RNA polymerase which then possibly no longer initiates at promoters on E.coli DNA (figure 4 (c)), also further initiation at the unique promoter on T7 DNA is blocked, thereby, controlling excessive early transcription. The transcription of DNA is completely under T7 control with the synthesis of T7 RNA polymerase which starts at 3-4 specific promoters. DNA ligase is involved in the replication of T7 DNA. The initiation point for replication is located also in the early region at 17%. The transcriptional inhibitor binds to the host polymerase molecules (figure 4 (d)). Thus, all transcription by E.coli RNA polymerase is definitely blocked. The action of nucleases makes all host nucleotides available for T7 DNA replication (figure 4 (e)). The cell be-

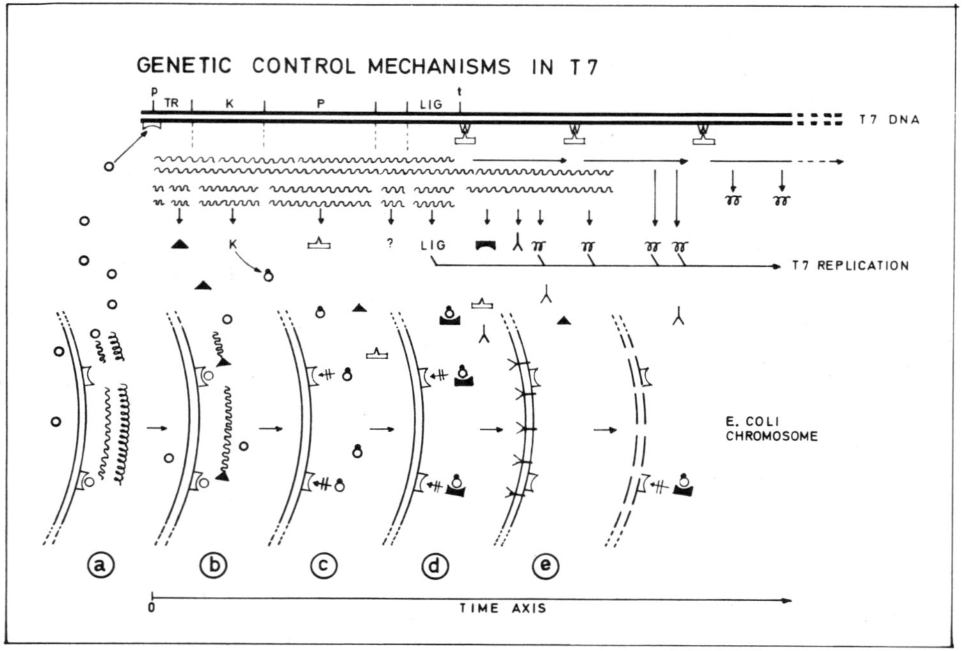

Figure 4: Genetic control mechanisms in T7.
═ DNA; ∼ RNA; protein; promoter for E.coli RNA polymerase; promoter for T7 virus RNA polymerase; O = E.coli RNA polymerase; virus RNA polymerase; = phosphorylated E.coli RNA polymerase; K = kinase; T7 transcriptional inhibitor; ▲ T7 translational repressor; ⅄ = virus-induced DNase; P = polymerase gene; LIG = DNA ligase; p = promoter; t = termination signal; TR = T7 translational repressor gene.

comes fully differentiated and produces only the structural proteins until T7 maturation is complete and lysis occurs.

The T7 system has yielded a more complete analysis and understanding of the biochemistry of gene control. Part of what was learned from the T7 system has already been applied successfully to other systems. It is reasonable to hope that the T7 system will help in the understanding of gene control in biology.

M.S. und P.H. erinnern sich in Dankbarkeit an die Jahre 1968-1970 in Prof. Lipmann's Labor, an die "Lipmann Familie" dieser Jahrgänge, deren freundschaftlicher Zusammenhalt noch heute besteht, und an die stimulierenden Diskussionen im Rahmen der Rockefeller Universität.
Wir danken herzlich Dr. Robert Roskoski für die kritische Durchsicht unseres Manuskripts.

References

(1) Sober, H.A. (editor): Handbook of Biochemistry, The Chemical Rubber Co., Cleveland 1970, pp. H104-H116.

(2) Schimke, R.T., Palacios, R., Palmiter, R.D., Rhoads, R.E.: Hormonal Regulation of Ovalbumin Synthesis in Chick Oviduct. In F.T. Kenney et al. (editors) "Gene expression and its regulation". pp. 123-135. Plenum New York 1973.

(3) Feldman, M., Globerson, A.: Reception of immunogenic signals by lymphocytes. Current Topics in Developmental Biology 8, 1-40 (1974).

(4) Stoker, M.: Contact and short-range interactions affecting growth of animal cells in culture. Current Topics in Developmental Biology 2, 107 - 128 (1967).

(5) Rutter, W.J., Pictet, R.L., Morris, P.W.: Toward molecular mechanisms of developmental processes. Ann. Rev. Biochem. 42, 601-646 (1973).

(6) Beckwith, J.R., Zipser, D. (editors): "The lactose operon". Cold Spring Harbor Laboratory 1970.

(7) Zubay, G., Schwartz, D., Beckwith, J.: Mechanism of activation of catabolite-sensitive genes: A positive control system. Proc. Natl. Acad. Sci. U.S. 66, 104-110 (1970).

(8) Zubay, G.: In vitro synthesis of protein in microbial systems. Ann. Rev. Genetics 7, 267-287 (1973).

(9) Schweiger, M., Herrlich, P.: DNA-directed enzyme synthesis in vitro. Current Topics in Microbiology and Immunology 65, 59-132 (1974).

(10) Studier, F.W.: Bacteriophage T7. Science 176, 367-376 (1972).

(11) Davis, R.W., Hyman, R.W.: Physical locations of the in vitro RNA initiation site and termination site of T7 M DNA. Cold Spring Harbor Symp. Quant. Biol. 35, 269-281 (1970).

(12) Millette, R.L., Trotter, C.D., Herrlich, P., Schweiger, M.: In vitro synthesis, termination, and release of active messenger RNA. Cold Spring Harbor Symp. Quant. Biol. 35, 135-142 (1970).

(13) Schweiger, M., Herrlich, P., Millette, R.L.: Gene expression in vitro from deoxyribonucleic acid of bacteriophage T7. J. Biol. Chem. 246, 6707-6712 (1972).

(14) Siegel, R.B., Summers, W.C.: The process of infection with coliphage T7. III. Control of phage-specific RNA synthesis in vivo by an early phage gene. J. Mol. Biol. 49, 115-123 (1970).

(15) Dunn, J.J., Studier, F.W.: T7 early RNAs and Escherichia coli ribosomal RNAs are cut from large precursor RNAs in vivo by ribonuclease III. Proc. Natl. Acad. Sci. U.S. 70, 3296-3300 (1973).

(16) Hercules, K., Schweiger, M., Sauerbier, W.: Cleavage by RNase III converts T3 and T7 early precursor RNA into translatable message. Proc. Natl. Acad. Sci. U.S. 71, 840-844 (1974).

(17) Chamberlin, M., McGrath, J., Waskell, L.: New RNA polymerase from Escherichia coli infected with bacteriophage T7. Nature 228, 227-231 (1970).

(18) Hausmann, R., Gomez, B.: Amber mutants of bacteriophages T3 and T7 defective in phage-directed deoxyribonucleic acid synthesis. J. Virol. 1, 779-792 (1967).

(19) Studier, F.W.: The genetics and physiology of bacteriophage T7. Virology 39, 562-574 (1969).

(20) Messer, W., Ponta, U., Ponta, H., Rahmsdorf, H.J., Pai, S.H., Hirsch-Kauffmann, M., Herrlich, P., Schweiger, M., unpublished.

(21) Simon, M.N., Studier, F.W.: Physical map of the early region of bacteriophage T7 DNA. J. Mol. Biol. 79, 249-265 (1973).

(22) Schweiger, M., Herrlich, P., Scherzinger, E., Rahmsdorf, H.J.: Negative Control of protein synthesis after infection with bacteriophage T7. Proc. Natl. Acad. Sci. U.S. 69, 2203-2207 (1972).

(23) Ponta, H., Rahmsdorf, H.J., Hirsch-Kauffmann, M., Pai, S.H., Zillig, W., Herrlich, P., Schweiger,M.: Control of gene expression in T7. Role of T7 protein kinase in the regulation of protein synthesis. Molec. Gen. Genetics, submitted.

(24) Rahmsdorf, H.J., Pai, S.H., Ponta, H., Herrlich, P., Roskoski, Jr., R., Schweiger, M., Studier, F.W.: Protein kinase induction in Escherichia coli by bacteriophage T7. Proc. Natl. Acad. Sci. U.S. 71, 586-589 (1974).

(25) Herrlich, P., Rahmsdorf, H.J., Pai, S.H., Schweiger, M.: Translational control induced by bacteriophage T7. Proc. Natl. Acad. Sci. U.S. 71, 1088-1092 (1974).

(26) Zillig, W., Rahmsdorf, H.J., Ponta, H., Pai, S.H., Hirsch-Kauffmann, M., Herrlich, P., Schweiger, M., in preparation.

(27) Ponta, H., Rahmsdorf, H.J., Pai, S.H., Herrlich, P., Schweiger, M.: Control of gene expression in T7. Transcriptional Inhibitor: isolation of a new control protein and mechanism of action. Molec. Gen. Genetics, submitted.

(28) Sadowski, P.D., Kerr, C.: Degradation of Escherichia coli B deoxyribonucleic acid after infection with deoxyribonucleic acid-defective amber mutants of bacteriophage T7. J. Virol. 6, 149-155 (1970).

(29) Summers, W.C.: The process of infection with coliphage T7. IV. Stability of RNA in bacteriophage-infected cells. J. Mol. Biol. 51, 671-678 (1970).

(30) Marrs, B.L., Yanofsky, C.: Host and bacteriophage specific messenger Degradation in T7-infected Escherichia coli. Nature New Biology 234, 168-170 (1971).

Covalent Enzyme-Substrate Intermediates in Carboxyl Activation[1]

Leonard B. Spector
The Rockefeller University, New York, New York 10021

INTRODUCTION

A case can be made for the proposition that the enzymatic activation of a carboxylic acid, requiring the consumption of a molecule of ATP, proceeds generally via a stage in which the enzyme is phosphorylated or adenylylated.[2] It happens that phosphorylated enzymes do indeed mediate the action of a number of well-known enzymes. While the number of such enzymes is still relatively small, the range of structural diversity of their carboxylic acid substrates is broad. This fact encourages the view that ATP-dependent carboxyl activation does, in general, require the mediation of a phospho-enzyme.

But for the enzymes acting on more than two substrates in a reaction the phospho-enzyme intermediate is insufficient to carry the reaction to completion. Indications are that in such cases the phospho-enzyme is succeeded by yet another covalent enzyme-substrate intermediate - an acyl-enzyme - which bears an equally prominent share in the chemical mechanism of the reaction. And as a connecting link between these two covalent intermediates is interposed an acyl phosphate, which is non-covalently bound to the enzyme. Together, these three intermediates constitute a three-stage process for the activation of carboxylic acids in multi-substrate reactions.

Two-Substrate Enzymes.

Since its discovery in 1944 by Lipmann (1), acetate kinase has stood as the prototypic carboxylate kinase, much as acetyl phosphate stands as the prototypic acyl phosphate. The two-substrate reaction which the enzyme catalyzes,

$$CH_3COO + ATP \rightleftharpoons CH_3COOP + ADP$$

is somewhat more complex than is expressed in the above equation. What the equation fails to express is the covalent participation of the enzyme in the process. Reacting with either ATP or acetyl phosphate, the enzyme is phosphorylated at an early stage of the reaction (2-4). The phospho-enzyme can actually be isolated, and has been shown to partake in the two half-reactions:

$$E + ATP \rightleftharpoons E{\sim}P + ADP$$

$$E{\sim}P + CH_3COO \rightleftharpoons CH_3COOP$$

In similar vein, the enzyme 3-phosphoglycerate kinase catalyzes the reaction,

$$P\text{-}OCH_2CHOHCOO + ATP \rightleftharpoons P\text{-}OCH_2CHOHCOOP + ADP$$

which we know now is more accurately represented as occurring in two stages:

$$E + ATP \rightleftharpoons E{\sim}P + ADP$$

$$E{\sim}P + P\text{-}OCH_2CHOHCOO \rightleftharpoons E + P\text{-}CH_2CHOHCOOP$$

with the phosphorylated enzyme intervening as a stage in the total reaction (5). For both acetate kinase and 3-phosphoglycerate kinase the phosphoryl group in the respective phospho-enzymes appears to be fixed to a carboxyl of the enzyme as an acyl phosphate:

$$E\text{-}COO + ATP \rightleftharpoons E\text{-}COOP + ADP$$

Three-Substrate Enzymes.

Three-substrate enzymes, unlike the two-substrate ones, do not catalyze the production of acyl phosphates free in solution. Rather, the acyl phosphate is formed in the condition of an enzyme-bound acyl phosphate as a stage in the ultimate formation of product. A case in point is succinic thiokinase which catalyzes the reaction,

$$OOCCH_2CH_2COO + ATP + CoA \rightleftharpoons OOCCH_2CH_2COSCoA + P_i + ADP$$

in which a phosphorylated enzyme joins in the reaction as an intermediate (6,7). The enzyme is, in fact, actually isolated from tissue as the phospho-enzyme (6). Succinate is activated by the reaction with the phospho-enzyme to form enzyme-bound succinyl phosphate (8,9), and the latter, with coenzyme A, generates succinyl-CoA (9,10):

$$\mathrm{E + ATP \rightleftharpoons E{\sim}P + ADP}$$
$$\mathrm{E{\sim}P + OOCCH_2CH_2COO \rightleftharpoons E\cdots\cdots OOCCH_2CH_2COOP}$$
$$\mathrm{E\cdots\cdots OOCCH_2CH_2COOP + CoA \rightleftharpoons E + OOCCH_2CH_2COSCoA + P_i}$$

A more complex three-substrate enzyme is ATP citrate lyase, which catalyzes the splitting of citrate into oxaloacetate and acetyl-CoA, following the preliminary activation of a citrate carboxyl:

$$\mathrm{OOCCH_2\text{-}\underset{OH}{\overset{COO}{C}}\text{-}CH_2COO + ATP + CoA \rightleftharpoons OOCCH_2COCOO + CH_3COSCoA + ADP + P_i}$$

Here again, an isolable phospho-enzyme takes part in the reaction (11,12). And, as with succinate activation, the phospho-enzyme acts upon citrate to form enzyme-bound citryl phosphate (13,14). The latter moves the reaction forward by acting at a catalytic group of the enzyme to form the second covalent intermediate, citryl-enzyme, which can easily be isolated (11,14). It is the citryl-enzyme which undergoes the lyase reaction to yield oxaloacetate and acetyl-enzyme (15). The acetyl-enzyme, which is also isolable by gel filtration (15), thereupon reacts with coenzyme A to generate acetyl-CoA. The sequence of partial reactions catalyzed by ATP citrate lyase is shown below:

$$\mathrm{E + ATP \rightleftharpoons E{\sim}P + ADP}$$
$$\mathrm{E{\sim}P + citrate \rightleftharpoons E\cdots\cdots citryl\text{-}P}$$
$$\mathrm{E\cdots\cdots citryl\text{-}P \rightleftharpoons E{\sim}citryl + P_i}$$
$$\mathrm{E{\sim}citryl \rightleftharpoons E{\sim}acetyl + oxaloacetate}$$
$$\mathrm{E{\sim}acetyl + CoA \rightleftharpoons E + acetyl\text{-}CoA}$$

ATP citrate lyase may well be unequalled among enzymes in the complexity of the total reaction which it catalyzes. The five partial reactions delineated above require the participation of three covalent enzyme-substrate intermediates, all of which are isolable; and citryl phosphate, which can be synthesized chemically (13), enters into the reaction as an enzyme-bound intermediate. It is clear that the third covalent intermediate - the acetyl-enzyme - stems directly from citryl-enzyme through lyase action, and that the acyl bond to the enzyme is the same one in both intermediates. Also clear is that the energy-rich bond in the phospho-enzyme is conserved in the citryl-enzyme which succeeds it, and that the gulf between these two covalent intermediates is bridged by the enzyme-bound citryl phosphate.

Recurring briefly to the succinic thiokinase reaction, we recall that a phospho-enzyme and enzyme-bound succinyl phosphate are both participants therein. If the carboxyl activation catalyzed by ATP citrate lyase is a pattern for such activations (wherein a phospho-enzyme is succeeded by an acyl-enzyme via an enzyme-bound acyl phosphate), then succinic thiokinase ought, accordingly, to have a succinyl-enzyme stage in its reaction pathway. A succinyl-enzyme, however, has not so far been isolated. Yet indirect evidence for its existence is at hand in the coenzyme A—succinyl-CoA exchange which the enzyme can catalyze in the absence of the other reaction components (16,17). On the whole it seems fair to conclude that succinic thiokinase and ATP citrate lyase, enzymes for which so many intermediates have been confirmed by isolation and chemical synthesis, follow the same general pattern in the catalysis of carboxylic acid activation.

A multi-substrate enzyme for which the final acceptor of the activated carboxylate is something other than coenzyme A is γ-glutamylcysteine synthetase. Here the acceptor is the amino acid cysteine:

$$\text{L-OOCCHNH}_2\text{CH}_2\text{CH}_2\text{COO} + \text{ATP} + \text{L-HSCH}_2\text{CHNH}_2\text{COO} \rightleftharpoons \text{OOCCHNH}_2\text{CH}_2\text{CH}_2\text{CONH}\underset{\substack{|\\ \text{CH}_2\text{SH}}}{\text{C}}\text{HCOO} + \text{ADP} + \text{P}_i$$

Like the carboxyls of succinate and citrate, the γ-carboxyl of glutamate undergoes a three-stage activation process before its ultimate linkage to acceptor, in this case the amino group of cysteine. The purified enzyme catalyzes a powerful ADP-ATP exchange in the absence of all reaction components except Mg, which signals the participation of a phospho-enzyme in the reaction

(18,19). Also, a variety of chemical evidence points to an enzyme-bound γ-glutamyl phosphate as an intermediate (20). The enzyme appears, moreover, to follow a ping pong mechanism with glutamyl-enzyme as an intermediate (21); and, supportive of this, the enzyme catalyzes a cysteine—γ-glutamylcysteine exchange in the absence of ATP and glutamate at a rate equal to that in the presence of these two substrates (21). It follows then that the γ-glutamylcysteine synthetase reaction, like the succinic thiokinase and ATP citrate lyase reactions, is mediated by a pair of covalent enzyme-substrate intermediates and a connecting, non-covalent, enzyme-bound acyl phosphate.

The enzymes so far considered use ATP in such fashion as to produce ADP and P_i from it. These comprise the first half of the enzymes listed in Table 1. The other enzymes split ATP into AMP and pyrophosphate. They, too, show strong signs of catalyzing carboxyl activation by a three-stage process, with an adenylylated-enzyme and an enzyme-bound acyl adenylate as the first two of the three anticipated intermediates. It will be clear from the Table that in no case have all three intermediates been demonstrated for any one enzyme catalyzing an AMP-PP split of ATP. Listed are only those enzymes for which at least two of the three anticipated intermediates have some experimental basis. It would be tiresome to recite here the details of this basis. The lacunae in the Table suffice to indicate what is still wanting to a more complete proof of the three-stage activation process for these enzymes. In each case, an enzyme-bound acyl adenylate has an acknowledged role in the action of the enzyme. Lacking is either the E~AMP or the E~acyl intermediate. But given the <u>chemical</u> <u>sameness</u> of the reactions, that is to say,

$$\text{R-COO} + \text{ATP} + \text{acceptor} \rightleftharpoons \text{R-CO-acceptor} + \text{AMP} + \text{PP}$$

the finding of an E~AMP (or E~acyl) for several of these enzymes warrants expectation of the same for the rest.

TABLE 1. Enzymes for which E~P, E~acyl, and enzyme-bound acyl phosphate intermediates have been identified.

	E~P(AMP)	Acyl~P(AMP)	E~acyl	Product of ATP split	Refs.
Acetate kinase[a]	+	+	———	ADP	1-4
3-P-glycerate kinase[a]	+	+	———	"	5
Succinic thio-kinase	+	+	+	"	6-10, 16,17
ATP citrate lyase	+	+	+	"	11-15
γ-Glutamylcysteine synthetase	+	+	+	"	18-21
Glutathione synthetase	+	+		"	22-24
Pyruvate carboxylase[b]	+		+	"	25,26
Asparagine synthetase	+	+		AMP	27,28
Tryptophanyl-tRNA synthetase	+	+		"	29-31
Acetyl-CoA synthetase[c]		+	+	"	32
Carnosine synthetase	+	+		"	33,34
Gramicidin S synthetase		+	+	"	35,36
Tyrocidine synthetase		+	+	"	36-38

A plus mark signifies that evidence exists for the participation, in the indicated reaction, of the intermediate specified at the head of the column.

[a] For this two-substrate enzyme the acyl phosphate is a final product of the reaction and not an enzyme-bound intermediate. The enzyme therefore has no E~acyl intermediate.

[b] The "carboxyl" group which is activated by this enzyme is the bicarbonate ion. The corresponding enzyme-bound acyl phosphate ought to be carboxyl phosphate, which has yet to be isolated or synthesized. The E~acyl for this enzyme is the carboxy-biotin-E.

[c] Acetyl-CoA synthetase catalyzes a transfer of acetyl between CoA and dephospho-CoA in the absence of other reaction components, which is regarded as indirect evidence for the existence of acetyl-enzyme (H. Anke and L. B. Spector, unpublished experiments)

DISCUSSION

Acetate kinase and 3-phosphoglycerate kinase are transferases which catalyze the reversible transfer of a phosphoryl group between ATP and the carboxyl of the second substrate. A transferase seems generally to catalyze its reaction by attaching itself covalently to the atom, or group of atoms, undergoing transfer from donor to acceptor. The arguments upholding this principle of transferase action have been set out elsewhere (39). Acetate kinase and 3-phosphoglycerate kinase are two-substrate enzymes which conform closely to this principle, since the existence of their respective phospho-enzymes is easily demonstrated (1-5). The other enzymes of Table 1 are three-substrate enzymes and are classified by the Enzyme Commission not as transferases, but as synthetases or, in the case of ATP citrate lyase, as a lyase. Yet each enzyme acts first to phosphorylate (or adenylylate) the carboxyl group of its substrate to yield an acyl phosphate - which holds fast to the enzyme. Except for this latter feature, the action of the three-substrate enzymes is no different in its <u>chemical form</u> from that of the two-substrate enzymes. Thus the three-substrate enzymes, like the two-substrate enzymes, are genuine transferases. As such they ought to adhere to the transferase principle (39). That they tend indeed to do so is borne out by the many plus marks in the first column of Table 1. A phospho-enzyme intervenes thus in the activation of carboxylic acids whose structures range in complexity from bicarbonate and acetate, through succinate and citrate, to the diverse amino acids, β-alanine and tryptophan. The inference is that carboxylic acids of all structures may require the same mode of activation.

Having synthesized the enzyme-bound acyl phosphate, the multisubstrate enzyme of Table 1 has now to transfer the activated acyl group to its ultimate acceptor. Once more, the enzyme must fulfill the function of a transferase. Expectation is that it will do so by transferring the acyl group from its linkage to phosphate into covalent linkage with a catalytic group on the enzyme (39), in which condition the acyl is primed for reaction with acceptor. This view of the second transferase reaction finds support in the many plus marks shown in the third column of Table 1.

From the foregoing it follows that each multi-substrate enzyme of Table 1 - and probably the many others of which each is representative - is in reality two enzymes packed into one; in effect, a double transferase. The double transferase character of these enzymes is doubtless dictated by the instability of the acyl phosphate which is formed as the product of the first transferase reaction. Whereas acetyl phosphate and other simple acyl phosphates have a rather wide range of pH stability (40,41), the more complex succinyl phosphate (40), citryl phosphate (14), γ-glutamyl phosphate (42), acetyl adenylate (43), tryptophanyl adenylate (30), and the like, have each a narrower range of pH stability, and a marked tendency toward instability under pH conditions not far removed from neutrality. Maintaining such compounds bound to the enzyme seems to preserve them from destruction. But this, we see, forces upon the enzyme the necessity of catalyzing two distinct transferase reactions - with three distinguishable intermediates.

SUMMARY

Enzymatic activation of carboxylic acids with ATP seems generally to be mediated by a phospho- (or adenylyl-) enzyme. Multi-substrate enzymes catalyze activation through a sequence of two simple transferase reactions, and appear to require, in addition to the phospho-enzyme, an acyl-enzyme intermediate. The route from the phospho-enzyme to the acyl-enzyme is over a non-covalent, enzyme-bound acyl phosphate, which completes the triad of intermediates necessary to the operation of the enzyme.

Footnotes.

[1] This essay was written to honor Fritz Lipmann as he enters the seventy sixth year of a singularly creative life. His past is surely but prelude.

[2] It will be obvious that an adenylylated enzyme is a phosphorylated enzyme which is additionally substituted on a phosphoryl oxygen by an adenosyl group.

Acknowledgment.

I thank the National Science Foundation and the American Cancer Society for support.

References.

1. Lipmann, F., J. Biol. Chem. 155, 55-70 (1944).
2. Anthony, R. S., Spector, L. B., J. Biol. Chem. 245, 6739-6741 (1970).
3. Anthony, R. S., Spector, L. B., J. Biol. Chem. 246, 6129-6135 (1971).
4. Anthony, R. S., Spector, L. B., J. Biol. Chem. 247, 2120-2125 (1972).
5. Walsh, C. T., Jr., Spector, L. B., J. Biol. Chem. 246, 1255-1261 (1971).
6. Kreil, G., Boyer, P. D., Biochem. Biophys. Res. Commun. 16, 551-555 (1964).
7. Bridger, W. A., Millen, W. A., Boyer, P. D., Biochemistry 7, 3608-3616 (1968).
8. Nishimura, J. S., Meister, A., Biochemistry 4, 1457-1462 (1965).
9. Hildebrand, J. G., Spector, L. B., J. Biol. Chem. 244, 2606-2613 (1969).
10. Nishimura, J. S., Grinnell, F., Adv. Enzymol. 36, 183-202 (1972).
11. Inoue, H., Suzuki, F., Tanioka, H., Takeda, Y., Biochem. Biophys. Res. Commun. 26, 602-608 (1967).

12. Inoue, H., Suzuki, F., Tanioka, H., Takeda, Y., J. Biochem. (Tokyo) 63, 89-100 (1968).
13. Walsh, C. T., Jr., Spector, L. B., J. Biol. Chem. 243, 446-448 (1968).
14. Walsh, C. T., Jr., Spector, L. B., J. Biol. Chem. 244, 4366-4374 (1969).
15. Suzuki, F., in Y. Ogura, Y. Tonomura, T. Nakamura, eds., "Molecular Mechanisms of Enzyme Action", University Park Press, Baltimore, 1972, pp. 265-279.
16. Upper, C. D., Doctoral Dissertation, University of Illinois, (1964).
17. Cha, S., Cha, C. M., Parks, R. E., Jr., J. Biol. Chem. 242, 2582-2592 (1967).
18. Strumeyer, D. H., Bloch, K., J. Biol. Chem. 235, PC27 (1960).
19. Webster, G. C., Varner, J. E., Arch. Biochem. Biophys. 52, 22-32 (1954).
20. Orlowski, M., Meister, A., J. Biol. Chem. 246, 7095-7105 (1971).
21. Davis, J. S., Balinsky, J. B., Harington, J. S., Shepherd, J. B., Biochem. J. 133, 667-678 (1973).
22. Snoke, J. E., Bloch, K., J. Biol. Chem. 213, 825-835 (1955).
23. Wendel, A., Flohé, L., Z. Physiol. Chem. 353, 523-530 (1972).
24. Nishimura, J. S., Dodd, E. A., Meister, A., J. Biol. Chem. 239, 2553-2558 (1964).
25. Scrutton, M. C., Keech, D. B., Utter, M. F., J. Biol. Chem. 240, 574-581 (1965).
26. Scrutton, M. C., Utter, M. F., J. Biol. Chem. 240, 3714-3723 (1965).
27. Chou, T. C., Handschumacher, R. E., Fed. Proc. 29, 407 (1970).
28. Horowitz, B., Meister, A., J. Biol. Chem. 247, 6708-6719 (1972).
29. Kisselev, L. L., Kochkina, L. L., Doklady Akad. Nauk 214, 215-217 (1974).

30. Kingdon, H. S., Webster, L. T., Jr., Davie, E. W., Proc. Nat. Acad. Sci. USA 44, 757-765 (1958).

31. Karasek, M. A., Castelfranco, P., Krishnaswamy, P. R., Meister, A., J. Amer. Chem. Soc. 80, 2335-2336 (1958).

32. Berg, P., J. Biol. Chem. 222, 991-1014 (1956).

33. Stenesh, J. J., Winnick, T., Biochem. J. 77, 575-581 (1960).

34. Kalyankar, G. D., Meister, A., J. Biol. Chem. 234, 3210-3218 (1959).

35. Gevers, W., Kleinkauf, H., Lipmann, F., Proc. Nat. Acad. Sci. USA 63, 1335-1342 (1969).

36. Lipmann, F., Science 173, 875-884 (1971).

37. Roskoski, R., Jr., Kleinkauf, H., Gevers, W., Lipmann, F., Biochemistry 9, 4846-4851 (1970).

38. Kleinkauf, H., Roskoski, R., Jr., Lipmann, F., Proc. Nat. Acad. Sci. USA 68, 2069-2072 (1971).

39. Spector, L. B., Bioorganic Chem. 2, 311-321 (1973).

40. Walsh, C. T., Jr., Hildebrand, J. G., Spector, L. B., J. Biol. Chem. 245, 5699-5708 (1970).

41. Koshland, D. E., Jr., J. Amer. Chem. Soc. 74, 2286-2292 (1952).

42. Levintow, L., Meister, A., Fed. Proc. 15, 299 (1956).

43. Berg, P., J. Biol. Chem. 222, 1015-1023 (1956).

The Role of Pantothenate in Citrate Lyase

Paul A. Srere and Manoranjan Singh
Pre-Clinical Science Unit, Veterans Administration Hospital and Department of Biochemistry, The University of Texas Health Science Center, Dallas, Tx. 75216, USA

Among the first CoASH dependent processes to be described was the synthesis of citrate. Novelli and Lipmann (1) showed that CoASH deficient yeast cells were unable to oxidize pyruvate and presumably that CoASH was involved in the synthesis of citrate. Stern and Ochoa (2) showed that a soluble enzyme system from pigeon liver catalyzed the synthesis of citrate and required ATP, acetate, oxalacetate and CoASH. Novelli and Lipmann (3) extended and confirmed these results using extracts from yeast and E. coli. In the ensuing years the enzyme citrate synthase was purified and studied and its catalyzed reaction shown to be (4,5)

$$H_2O + \text{acetyl CoA} + \text{oxalacetate} = \text{citrate} + \text{CoASH} + H^+$$

Esterification of acetate with CoA had, in Lipmann's early terminology, resulted in a "tail" activation of acetate (6). A few years after the description of citrate synthase Srere and Lipmann (7) described another citrate enzyme the ATP citrate lyase (citrate cleavage enzyme) which catalyzed the reaction

$$\text{Citrate} + \text{ATP} + \text{CoA} \underset{}{\overset{Mg^{+2}}{=}} \text{acetyl CoA} + \text{OAA} + \text{ADP} + P_1$$

The similarities between this enzyme and the synthase

were obvious. The utilization of ATP energy allowed the reaction to be driven toward citrate cleavage rather than toward synthesis. Mechanistically the reverse reaction (a very slow one) was still of the "tail" activation type. The chemical basis for the effect of CoA on acetate is said to derive from the increased charge density of a sulfur atom, compared to an oxygen atom, which is responsible for its decreased tendency to participate in resonance. This results in an increased positive charge on the carbonyl of the acetyl moiety and a consequent increased acidity of the protons on the methyl group (8). The enolization of the protons of acetyl CoA catalyzed by citrate synthase cannot be demonstrated unless an "inducer" (S malate or α-ketoglutarate (presumably oxalacetate is the natural inducer)) is present (9,10). Similarly oxalacetate is necessary for the enolization catalyzed by ATP citrate lyase (11).

Another citrate lyase, an apparently simpler enzyme had been isolated from a number of bacterial species (12,13,14)

$$\text{citrate} \overset{M^{+2}}{=} \text{acetate} + \text{oxalacetate}$$

When compared to the other two "citrate lyases" it seemed to be of a completely different sort. No CoASH was involved; what was the mechanism of "tail" activation of acetate? Was this a totally different enzyme mechanism? Was it possible that there was some evolutionary information available in the study of these three enzymes? Citrate lyase has only been found in procaryotes, while the ATP citrate lyase has only been found in eucaryotes. In order to attempt an answer to these questions we started the study of citrate lyase in addition to our ongoing experiments concerning citrate synthase and ATP citrate lyase.

Table 1. Effect of Incubating Various Reagents With Citrate Lyase

Final Concentration (mM) of Reagent	Percent of Control Activity Remaining Upon Incubation at 30° for 20 min
Hydroxylamine	
10	69
50	30
100	7
N Methyl NH_2OH	
2000	5
O Methyl NH_2OH	
2000	100
Sodium borohydride	
68	100
Potassium cyanate	
100	100
Hydrazine	
100	100
Thiosemicarbazine	
56	100
Methyl hydrazine	
100	100
Phenylhydrazine	
100	100

Partially purified citrate lyase (SA 2.9) was incubated with the neutralized reagents in 5 mM KPO_4 pH 7.4. Aliquots were assayed for activity. Control tubes with equivalent concentrations of KCl instead of reagents were always run, and it was also shown in separate experiments that the small amount of reagent in the assay mix had no effect when added directly to an assay mixture.

Our studies on citrate lyase focused at first on its anomalous behavior of the enzyme which we termed a "suicide" or reaction-inactivation process (15). In attempting to influence the "suicide" reaction we found that hydroxylamine was a potent inhibitor of this reaction (Table 1) but borohydride had no effect on the activity. Hydroxylamine has no effect on the activity of citrate synthase or ATP citrate lyase. When the hydroxylamine-inactivated or the reaction-inactivated enzyme was treated with acetic anhydride, the enzyme regained activity. During the course of our studies Eggerer and his coworkers (16) published an elegant study showing that the active citrate lyase was an acetyl enzyme and that the acetyl group turned over during the reaction. They postulated the following

Table 2. Reactivation of Inactivated Enzymes

	Enzyme reaction-inactivated in 0.1 mM dithionitrobenzoate	Enzyme hydroxylamine-inactivated in 0.1 mM dithionitrobenzoate
Activity	0	0
Incubation + Ac_2O 15 min	0.2	0
Incubation + Ac_2O with 0.1 M dithiothreitol	35	44

Citrate lyase was reaction-inactivated with 0.1 M Mg citrate at pH 8.1 in 0.1 M Tris•HCl (pH 8.1) in the presence of 0.1 mM dithionitrobenzoate. Another sample was inactivated in 0.1 M hydroxylamine in 0.1 M Tris•HCl (pH 8.1). The inactivated enzymes were isolated by ammonium sulfate precipitation (about 70%), and were then incubated in the presence or absence of 0.1 M dithiothreitol in 0.1 M Tris•HCl (pH 8.1) for 15 min. 2 μl of Ac_2O were added to each sample, and citrate lyase was assayed immediately. Activity is expressed as enzyme units/ml of original enzyme solution.

enzyme mechanism

ES-acetyl + citrate = ES citryl + acetate

ES citryl = ES acetyl + oxalacetate

---

citrate = acetate + oxalacetate

We had both been able to show that the acetyl group was probably a thioester (Table 2). The insight of Buckel et al. (16) suggested to us that the enzyme might contain pantetheine as the carrier of the acetyl group.

Table 3. Pantothenate Content of Citrate Lyase

Enzyme treatment*	Alkaline phosphatase	mol Pantothenate/mol enzyme+
0.05 M KCl dialysis	+	4.09 ± 0.27
	-	0
8 M Urea-0.1 M dithiothreitol dialysis, followed by 0.05 M KCl dialysis	+	3.94 ± 0.27
	-	0

*Protein was isolated by gel filtration on Bio-gel A 1.5 M in 0.05 M KPO_4 (pH 7.4)-1.6 mM $MgCl_2$. Part was dialyzed against 0.05 M KCl for 60 hr, and part against urea-dithiothreitol for 6 hr, and then against 0.05 M KCl for 60 hr. Each sample was hydrolyzed in sealed tubes in 1 M KOH at 105° for 4 hr. The samples were neutralized with concentrated HCl and buffered with 0.1 M Tris·HCl (pH 8.8). Alkalins phosphatase was added (0.5 mg), and the samples were incubated at 37° for 2 hr. These were then filtered through 0.45- m Millipore filters, and aliquots were taken for pantothenate determinations. Controls were handled identically, but without the addition of alkaline phosphatase.

+Average of triplicate analyses at each of three different concentrations, based on a molecular weight of 560,000. The specific activity of the enzyme is 73 units/mg dry protein.

We analyzed the purified enzyme from K. aerogenes and found that it contained 4 mols of phosphopantetheine per mol of enzyme (Table 3) (17). In addition we grew a pantothenate requiring mutant of the bacteria on ^{14}C-pantothenate and then isolated the citrate lyase using an antibody to the enzyme (18). The isolated complex contained ^{14}C and calculation revealed it contained 4 mols of pantothenate/mol of enzyme in agreement with our earlier studies.

Eggerer and his colleagues were able to confirm these observations but extended them to show that the pantetheine was covalently attached to a small subunit of the enzyme which was similar but not identical to the acyl carrier protein of fatty acid synthesis (19). Both groups were able to show that the acetyl group was also part of the small molecular weight subunit.

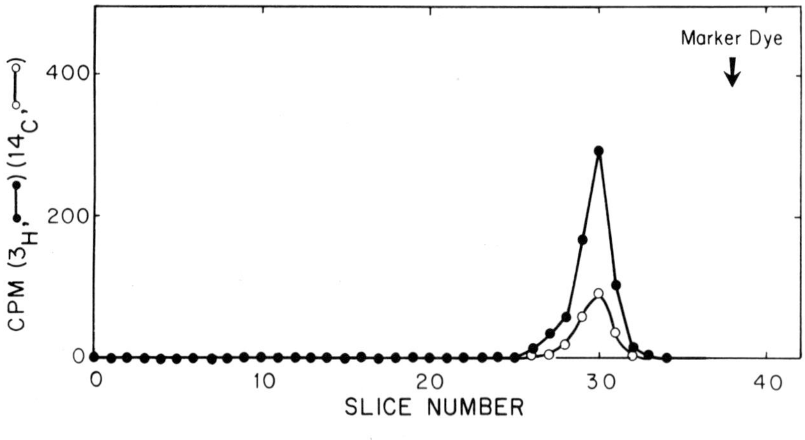

Fig. 1. SDS gel electrophoresis of citrate lyase labelled with ^{14}C-pantothenate and ^3H-acetate. The enzyme containing both ^{14}C-pantothenate and ^3H-acetate was denatured in 1% SDS and was subjected to polyacrylamide gel electrophoresis. It was cut into 2 mm slices and the radioactivity was determined in a liquid scintillation spectrometer.

Fig. 1 shows that when ^{14}C-pantothenate citrate lyase which is subsequently labeled with ^{3}H-acetyl groups, is subjected to SDS gel electrophoresis, both the labels travel with the small subunit (18). Dimroth _et al._ (19) have shown that this "ACP" contains a cysteine residue in addition to the SH moiety of the pantetheine. It remains to be demonstrated on which of these SH groups the acetyl group is located. However, consideration of the other citrate enzymes and of fatty acid synthetase leads us to believe that it is the pantetheine SH which is acetylated.

The structure of citrate lyase had been studied by several groups (19-21). There is consensus that the molecular weight of the native enzyme is about 560,000. Mahadik and SivaRaman (20) indicated the presence of 8 identical subunits but studied the enzyme dissociation in the absence of SH reagents. Bowen and Mortimer (21) showed the presence of 16 identical subunits when mercaptoethanol was added to the denaturating solvents. Dimroth _et al._ (19) agreed with this interpretation except for the addition of the 4 "ACP" subunits.

Our analysis of the protein by gel electrophoresis revealed three nonidentical subunits. One had a molecular weight of about 50,000, one had a molecular weight of 30,000 and the third was the "ACP" subunit with a weight of 10,000 (Fig. 2). When each band was re-electrophoresed, no change in their behavior was noted (Fig. 3). In addition gel chromatography in urea enabled us to separate the subunits from each other, thus indicating that the presence of the different subunits

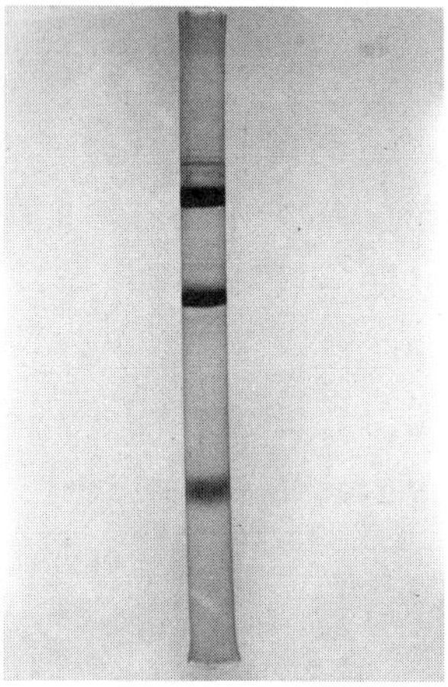

Fig. 2

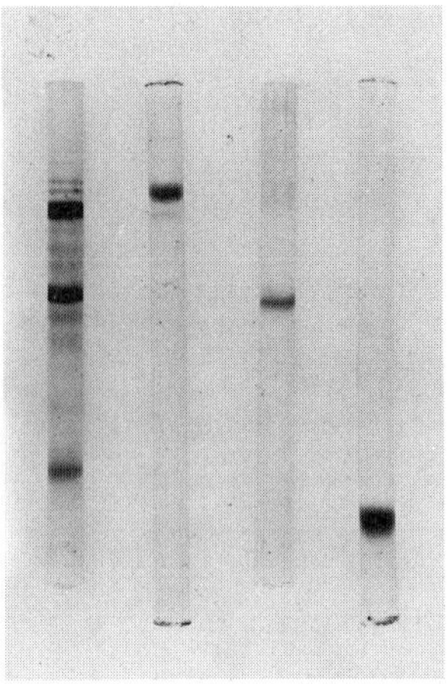

Fig. 3

Fig. 2. Subunits of citrate lyase. Citrate lyase was denatured in 1% SDS containing 0.1 M β-mercaptoethanol and it was subjected to SDS gel electrophoresis and stained with Coomassie blue.

Fig. 3. Re-electrophoresis of the subunits of citrate lyase separated by SDS gel electrophoresis. Citrate lyase was subjected to SDS gel electrophoresis and the 3 major bands were sliced (before staining) and the slices were put on the top of fresh gels and subjected to re-electrophoresis, and stained with Coomassie blue. (Left to right). The reference gel, the slowest moving subunit, the middle protein band, and the fastest moving protein band.

was not an artifact of SDS gel electrophoresis.

Since the proposed enzyme mechanism contains two separate reactions, an acyl transfer and a citrate lyase, and since the structure of the enzyme contains two different subunits in addition to the ACP I cannot resist formulating the enzyme as a multienzyme complex (Fig. 4).

While it is true that the presence of acetyl phosphopantetheine answers the problem of analogy of the three citrate enzymes and the problem of the enolization of the proton on the acetate the new information still leaves several important problems unanswered. First, the mechanism of the acyl transfer is as difficult a mechanistic problem as was the enolization. This re-

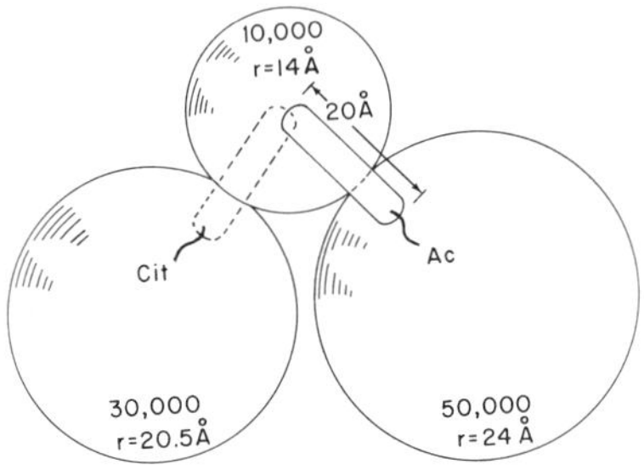

Fig. 4. Model of the multienzyme couples showing relative size of subunits (assuming globular proteins) and of phosphopantetheine.

action is analogous to the CoA transferases discussed by Jencks (22). There is difficulty in understanding how the energy of the thioester is conserved during the transfer. Secondly, there exists the problem of the acetyl enzyme. Schmellenkamp and Eggerer (23) have shown that the acetylation of the enzyme occurs in a way analogous to the activation of acetate shown to occur by Jones *et al.* (24). They postulate (probably correctly) that an acetyl transferase reported by us (25) was due to the ligase and contaminants of acetate and ATP in our acetyl donor compound. The work of Giffhorn *et al.* (26) indicates that in Rhodopseudomonas the deacetylation of the enzyme occurs *in vivo* so that the deacetylation-reacetylation cycle may represent a normal control mechanism akin to the phosphorylation-dephosphorylation control mechanisms described for other enzymes.

On the bases that of the three citrate enzymes, citrate lyase occurs only in procaryotes and that from a stoichiometric view it is the "simplest" reaction I have assumed that it may represent evolutionarily the earliest of the three citrate enzymes. If this is true, we might ask what are the disadvantages of its structural and mechanistic properties to have led to further change. There is an advantage to having a coenzyme covatentzy bound, as is phosphopantetheine in citrate lyase, to its enzyme. As part of the protein fixed spatially in the vicinity of the active site its apparent "microenvironmental" concentration is much higher that an equivalent amount of unbound coenzyme.

On the other hand, a free coenzyme might become, as has coenzyme A, a co-substrate for many different reactions. Its concentration will then be a function of enzyme activity in several different metabolic pathways. An obvious advantage of this situation is that the change in a common substrates' concentration will act as a communication device between the various biochemical activities of the cell. Was this the overriding pressure for the formation of the adenine-containing cosubstrates of phosphate, nicotinamide and pantetheine? Recent metabolic studies have indicated that changes in cell metabolism are regulated in part by changes in the phosphorylation, redox and acylation potentials.

The recently discovered similarity between the activation of pantetheine in citrate lyase and the activation of CoA in eucaryotes supports the idea that bound pantetheine evolved to give rise to free Coenzyme A.

Over half of all proteins have nucleotide binding sites and recent work on the x-ray structure of kinases and dehydrogenases led Rossmann to postulate a basic binding site on proteins for nucleotides (27). The adenylate which is coupled to pantetheine possibly provides the binding site for proteins so that the pantetheine can perform its function as it did in primitive enzymes.

Of course there is no evidence for the evolution of the enzymes as I have described it. One could also consider the possibility that citrate synthase because of its ubiquitous occurrence was the first citrate enzyme to evolve and that the two different citrate lyases, the metal dependent one in procaryotes and the ATP dependent one in eucaryotes, represent the special solutions that each cellular type evolved to solve the problem of reversal of the citrate synthase reaction.

It is also possible that the three citrate enzymes evolved independently of each other from other enzyme precursors. This possibility seems less likely now that we know of the similarity of the reaction mechanisms of the three citrate enzymes.

Citrate lyase may well be one of the simplest and perhaps most primitive multienzyme systems. Further studies on its structure and function will provide information not only on the evolution of the citrate enzymes but also on the development and function of multienzyme systems.

In conclusion I hope I have shown the enormous impact of Lipmann's work in the specific area of the study of the three citrate lyase enzymes.

Acknowledgements
This work was supported in part by a U.S.P.H.S. grant.

References

1. Novelli, G.D., Lipmann, F.: The involvement of Coenzyme A in acetate oxidation in yeast. J. Biol. Chem. 171, 833-834 (1948).
2. Stern, J.R., Ochoa, S.: Enzymatic synthesis of citric acid by condensation of acetate and oxalacetate. J. Biol. Chem. 179, 491-492 (1949).
3. Novelli, G.D., Lipmann, F.: The catalytic function of CoA in citric acid synthesis. J. Biol. Chem. 182, 213-228 (1950).
4. Ochoa, S., Stern, J.R., Schneider, M.C.: Enzymatic synthesis of citric acid II. Crystalline condensing enzyme. J. Biol. Chem. 193, 691-702 (1951).
5. Stern, J.R., Shapiro, B., Stadtman, E.R., Ochoa, S.: Enzymatic synthesis of citric acid III. Reversibility and mechanism. J. Biol. Chem. 193, 703- (1951).
6. Lipmann, F.: Biosynthetic mechanisms, in The Harvey Lectures Series XLIV, 1948-1949. Charles C. Thomas, Illinois (1950).
7. Srere, P.A., Lipmann, F.: An enzymatic reaction between citrate adenosine triphosphate and coenzyme A. J. Amer. Chem. Soc. 75, 4874 (1953).
8. Jaenicke, L., Lynen, F.: Coenzyme A, in The Enzymes (P.D. Boyer, H. Lardy, K. Myrback, eds.), 2nd ed., 3, Part B, 3-103. Academic Press, New York (1960).
9. Eggerer, H.: Zum mechanismus der biologischen unwandlung von citronensaure VI. Citrat-synthase

ist eine acetyl-CoA-enolase. Biochem. Z. 343, 111-138 (1965).

10. Srere, P.A.: A magnetic resonance study of the citrate synthase reaction. Biochem. Biophys. Res. Commun. 26, 609-614 (1967).

11. Das, N., Srere, P.A.: Exchange of methyl protons of acetyl coenzyme A catalyzed by adenosine triphosphate citrate lyase. Biochemistry 11, 1534-1537 (1972).

12. Dagley, S., Dawes, E.A.: Dissimilation of citric ac acid by bacterial extracts. Nature 172, 345-346 (1953).

13. Gillespie, D.C., Gunsalus, I.C.: An adaptive citric acid desmolase in Streptococcus faecalis. Bacteriol. Proc. 80 (1953).

14. Wheat, R.W., Ajl, S.J.: Citritase, the citrate-splitting enzyme from Escherichia coli I. Purification and properties. J. Biol. Chem. 217, 897-907 (1955).

15. Singh, M., Srere, P.A.: The reaction inactivation of citrate lyase from Aerobacter aerogenes. J. Biol. Chem. 246, 3847-3850 (1971).

16. Buckel, W., Buschmeier, V., Eggerer, H.: Der wirkungsmechanismum der citrat-lyase aus Klebsiella aerogenes. Hoppe-Syler's Z. Physiol. Chem. 352, 1195-1205 (1971).

17. Srere, P.A., Bottger, B., Brooks, G.C.: Citrate lyase: A pantothenate-containing enzyme. Proc. Nat. Acad. Sci. USA 69, 1201-1202 (1972).

18. Singh, M., Srere, P.A.: Unpublished results.

19. Dimroth, P., Dittmar, W., Walther, G., Eggerer, H.:

The acyl-carrier protein of citrate lyase. Eur. J. Biochem. 37, 305-315 (1973).

20. Mahadik, S.P., SivaRaman, C.: Citrate lyase: Molecular weight and subunit structure. Biochem. Biophys. Res. Commun. 32, 167-172 (1968).

21. Bowen, T.J., Mortimer, M.G.: Sub-units of citrate oxaloacetate-lyase. Eur. J. Biochem. 23, 262-266 (1971).

22. Jencks, W.P.: Coenzyme A transferases, in The Enzymes (P.D. Boyer, ed.), 3rd ed., IX, Part B, 483-496. Academic Press, New York (1973).

23. Schmellenkamp, H., Eggerer, H.: Mechanism of enzymic acetylation of des-acetyl citrate lyase. Proc. Nat. Acad. Sci. USA 71, 1987-1991 (1974).

24. Jones, M.E., Black, S., Flynn, R.M., Lipmann, F.: Acetyl coenzyme A synthesis for pyrophosphoryl split of adenosine triphosphate. Biochim. Biophys. Acta 12, 141-149 (1953).

25. Singh, M., Bottger, B., Stewart, C., Brooks, G.C., Srere, P.A.: S-Acetyl phosphopantetheine: Deacetyl citrate lyase S-Acetyl transferase from *Klebsiella aerogenes*. Biochem. Biophys. Res. Commun. 53, 1-9 (1973).

26. Giffhorn, F., Beuscher, N., Gottschalk, G.: Regulation of citrate lyase activity in *Rhodopseudomonas gelatinosa*. Biochem. Biophys. Res. Commun. 49, 467-472 (1972).

27. Olsen, K.W., Buehner, M., Ford, G.C., Moras, D., Rossmann, M.G.: The three-dimensional structure of glyceraldehyde-3-phosphate dehydrogenase. Fed. Proc. 33, 1374 (1974).

Cleavage of Small DNAs with Restriction Nucleases

R.E. Streeck, F. Fittler, and H.G. Zachau
Institut für Physiologische Chemie und Physikalische Biochemie der Universität, München, BRD

Restriction endonucleases (1-3) have been widely used for cleaving prokaryotic and eukaryotic DNAs into defined fragments. Some experiments from our laboratory on the use of these enzymes have been summarized recently (4-6). In the following a number of experiments on the cleavage of DNAs from bacterial and animal viruses are described which have not been published elsewhere.

SV40 and PM2 DNA

Recently, the fractionation of Hind DNAases on hydroxy apatite into an early and a late eluted fraction was reported (4). Cleavage of PM2 DNA with an authentic sample of Hind III DNAase suggested that "late eluted" Hind DNAase and Hind III DNAase were identical. The objective of the present experiments with SV40 DNA was to establish a definite correlation of the early and late eluted Hind DNAase fractions (4,5) with the Hind II and III DNAases. The SV40 DNA was obtained from G. Sauer, Heidelberg, and was also isolated in our laboratory according to (7). The cleavage pattern found after digestion with a mixture of early and late eluted Hind DNAases consisted of 11 fragments, as was first shown by Danna and Nathans (8). Early and late eluted Hind DNAase alone cleaved

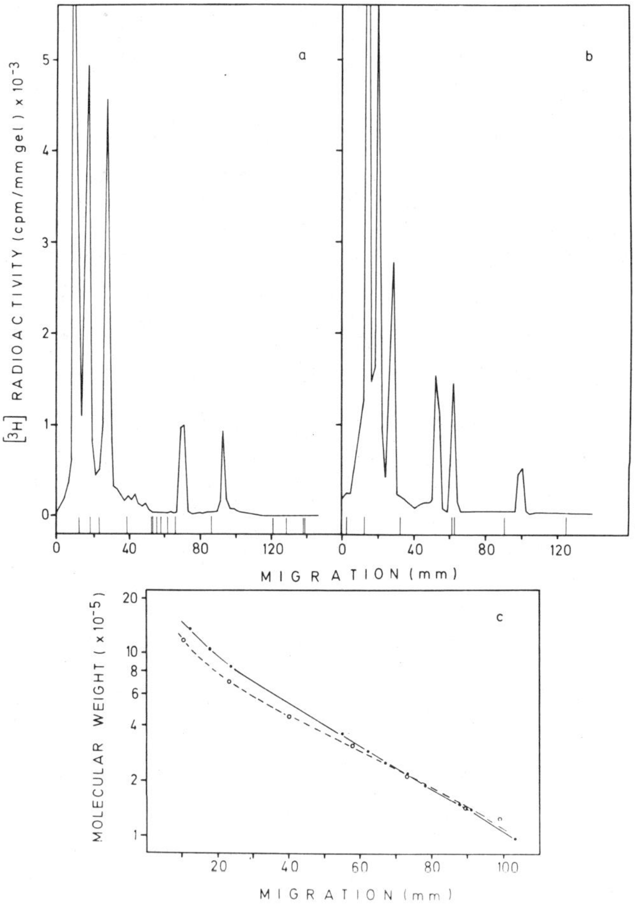

Fig. 1. Cleavage of SV40 DNA by Hind DNAases. 0.34 μg (^{3}H)-labelled SV40 DNA (9.9x10^5 cpm/μg, prepared by G. Sauer, Heidelberg, from strain Rh 911, plaque purified and passaged under low multiplicity of infection) and 10 μg unlabelled PM2 DNA were digested for 6 hours at 30° by an excess of Hind II DNAase (a), Hind III DNAase (b), and by Hind II plus Hind III DNAase (not shown) and submitted to disc electrophoresis (4). The Hind DNAases were prepared by P. Philippsen of this laboratory according to (4). The staining and the determination of radioactivity (Oxymat sample oxidizer) in the gel were carried out as described previously (4,5). The migration of the PM2 DNA fragments (5) is indicated by vertical lines at the bottom of panel (a) and (b). (c) Molecular weight/mobility relations for PM2 DNA fragments (●——●) and SV40 DNA fragments (o----o).

SV40 DNA into 5 and 6 fragments, respectively (Fig. 1a,b). Similar cleavage patterns have been shown before to be characteristic of Hind II and Hind III DNAase (9).

A second objective of the digestion experiments with SV40 DNA was to correlate the molecular weight determination based on SV40 DNA fragments (9) with the one based on PM2 DNA fragments (5). Unlabelled PM2 DNA and a small quantity of highly labelled SV40 DNA were mixed, digested, and submitted to polyacrylamide gel electrophoresis. The calibration curve for the SV40 DNA fragments which was drawn using the molecular weights of (9) had a somewhat smaller slope than the one based on the PM2 DNA fragments (Fig. 1c) the molecular weights of which had been determined adopting a molecular weight of 6.0×10^6 for PM2 DNA (10). The latter curve was consistent with independent data obtained from digestion of λdv, ∅29, mouse and guinea pig satellite DNA (5,6). We cannot exclude the possibility that the deviation of the PM2 and SV40 calibration curves is due to some not understood properties of certain DNA fragments affecting their electrophoretic mobilities in the gel systems or to differences between the Rh 911 strain of SV40 and the 776 strain used in (9).

The fragments obtained from PM2 DNA by Hind II and by Hind II plus Hind III DNAase are useful for molecular weight calibrations of small DNAs since they range from 1.4×10^6 to 0.4×10^5 daltons and are about evenly spaced (5). A digest of PM2 DNA obtained with a restriction nuclease from B. subtilis R (X5) may even be better suited for this purpose.

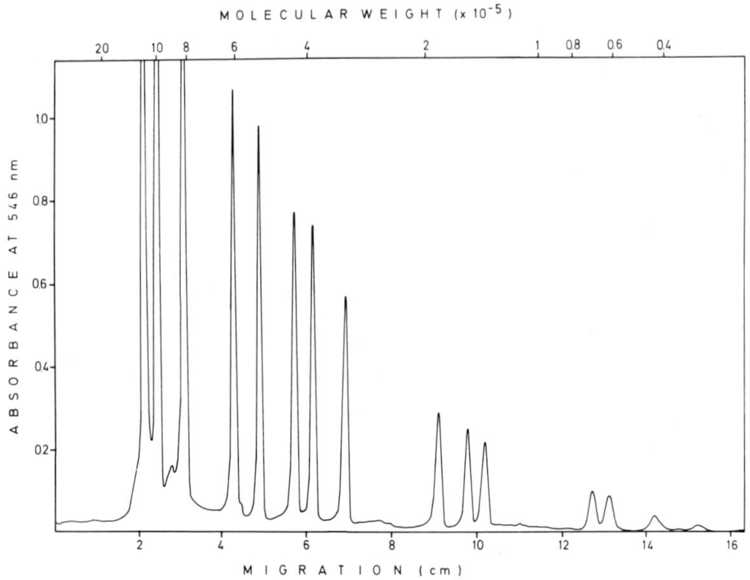

Fig. 2. Cleavage of PM2 DNA by a restriction nuclease from B. subtilis R (X5). The nuclease was isolated by W. Hörz of this laboratory who is grateful to T. Trautner for the strain and for advice on the purification. 10 µg PM2 DNA were cleaved for 5 hours at 37° with an excess of DNAase and submitted to disc electrophoresis.

15 fragments were obtained all of which were separated by gel electrophoresis (Fig. 2).

T7 DNA

A further possibility to characterize the Hind II and III preparations was the digestion of T7 DNA. While Hind II DNAase produced an elaborate pattern of about 50 fragments (Fig. 3, see also ref. 11), Hind III DNAase did not cleave T7 DNA at all. This was to be expected (H.O. Smith, personal communication) and proved the purity of the Hind III preparations.

The integrity of T7 DNA after incubation with Hind III DNAase was shown by electrophoresis in 3% polyacryl-

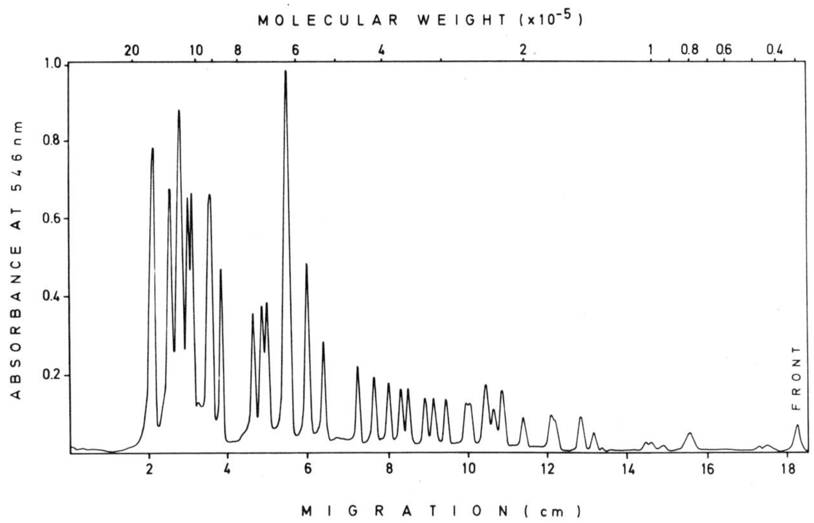

Fig. 3. Cleavage of T7 DNA by Hind II DNAase. 20 μg T7 DNA which was prepared according to (12) was cleaved to completion with Hind II DNAase as described in the legend of Fig. 1 and submitted to disc electrophoresis in 3% acrylamide, 0.1% bisacrylamide. Molecular weights were determined by co-electrophoresis with PM2 DNA fragments.

amide gels and in 1.0% agarose gels, in the absence or presence of added PM2 DNA, which was completely digested into the known pattern of seven fragments (5). The absence of Hind III cleavage sites in T7 DNA was further demonstrated by the fact that the Hind II and the Hind II + III digestion patterns were identical. T7 DNA was also completely resistant towards EcoRI DNAase as was shown by similar experiments. For the Hind III and EcoRI DNAases 8 - 9 cleavage sites each would have been expected in T7 DNA on the basis of statistics but apparently the respective hexanucleotide recognition sequences do not occur within T7 DNA.

ϕ29 DNA

ϕ29 DNA had previously been digested with Hind II DNAase and with a mixture of Hind II and III DNAases

(5). The Hind III DNAase cleavage pattern is shown (Fig. 4) since it yields further information on the attached protein(s) of ϕ29 DNA. These protein(s) are present in a DNA preparation which has undergone a sarcosyl/phenol treatment but they are removed by proteinase K digestion. The Hind III DNAase digest of untreated ϕ29 DNA (Fig. 4a) lacks two fragments which are present in molar amounts in the digest of ϕ29 DNA from which the attached proteins had been removed prior to cleavage (Fig. 4b). One additional fragment each had been found previously in the EcoRI and Hind II digests of proteinase K treated ϕ29 DNA (1.2×10^6 and 3.2×10^5 daltons, respectively). Since ϕ29 DNA is known to have a protein attached which specifically links the two ends of the DNA (13), our findings suggest that the additional fragments are derived from the ends of the DNA.

From the cleavage patterns of the two ϕ29 DNA preparations obtained by simultaneous incubation with different restriction nucleases one can determine the positions of some cleavage sites relative to each other at the ends of the DNA. With Hind II + III DNAase a fragment of 1.6×10^5 daltons was formed (5). A fragment of the same size was seen in Hind III + EcoRI DNAase digests, while Hind II + EcoRI DNAase produced a fragment of 3.2×10^6 daltons. It can be concluded that there is one cleavage site each for Hind II and Hind III DNAase located at distances of 3.2×10^5 and 1.6×10^5 daltons, respectively, from one end. At the other end of ϕ29 DNA there is a Hind III cleavage site at a distance of 1.6×10^6 daltons. It is not clear yet at which end the EcoRI cleavage site is located and where the other terminal

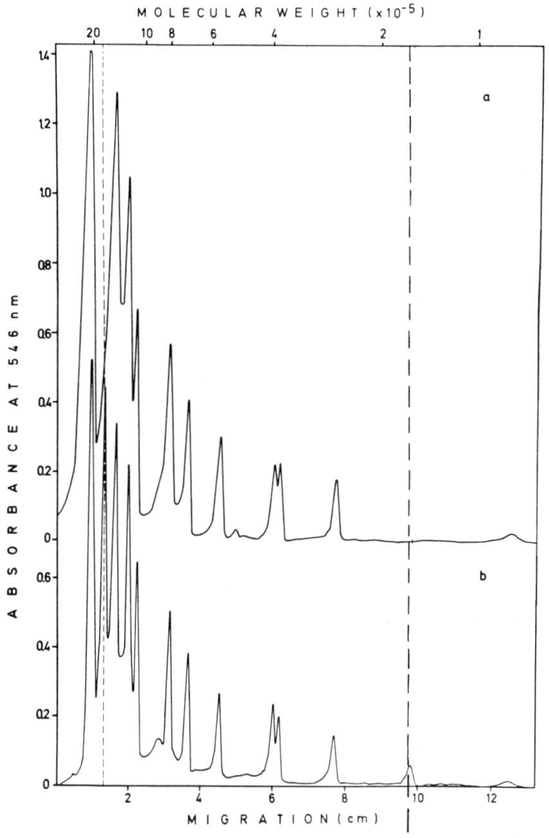

Fig. 4. Cleavage of ϕ29 DNA by Hind III DNAase. The isolation of ϕ29 DNA (14) and the removal of the attached protein(s) by proteinase K digestion have been described (5). 10 µg of either untreated (a) or proteinase K treated (b) ϕ29 DNA were cleaved with Hind III under the conditions of Fig. 1 and submitted to disc electrophoresis in 4% acrylamide, 0.13% bisacrylamide.

fragments in the EcoRI and the Hind II DNAase digests are migrating on the gel. Further experiments with additional restriction nucleases will be helpful in locating more precisely the attachment site of the linker protein and in characterizing the ends of the DNA which are regions of early transcription (Viñuela, personal communication).

Acknowledgment

We are indebted to U. Behrens for expert technical assistance. This study would not have been possible without the gift of restriction nucleases from P. Philippsen and W. Hörz and of SV40 DNA from G. Sauer. The work was supported by Deutsche Forschungsgemeinschaft and Fonds der Chemischen Industrie.

Summary

The DNAs of SV40 and of the T7 and ϕ29 phages were cleaved with two restriction nucleases from Hemophilus influenzae Rd and PM2 DNA was cleaved with a restriction nuclease from B. subtilis R (X5). The fragmentation patterns are discussed with respect to molecular weight/electrophoretic mobility relations. In ϕ29 DNA the effect of specifically bound proteins on the cleavage at terminally located restriction sites is discussed.

References

1 Review: Arber, W.: Restriction and Modification, in Progr. Nucleic Acid Res. and Mol. Biol. 14 (1974), W. Cohn ed., Academic Press.

2 Collection of recent articles: Cold Spring Harbor Symp. Quant. Biol. 38 (1973).

3 Smith, H.O. and Nathans, D.: A Suggested Nomenclature for Bacterial Host Modification and Restriction Systems and their Enzymes, J. Mol. Biol. 81, 419-423 (1973).

4 Philippsen, P., Streeck, R.E., and Zachau, H.G.: Defined Fragments of Calf, Human, and Rat DNA Produced by Restriction Nucleases, Eur. J. Biochem. 45, 479-488 (1974).

5 Streeck, R.E., Philippsen, P., and Zachau, H.G.: Cleavage of Small Bacteriophage and Plasmid DNAs by Restriction Endonucleases, Eur. J. Biochem. 45, 489-499 (1974).

6 Hörz, W., Hess, I., and Zachau, H.G.: Highly Regular Arrangements of a Restriction-Nuclease-Sensitive Site in Rodent Satellite DNA, Eur. J. Biochem. 45, 501-512 (1974).

7 Collins, C. and Sauer, G.: Fate of Infecting Simian Virus 40 DNA in Nonpermissive Cells: Integration into Host DNA, J. Virol. 10, 425-432 (1972).

8 Danna, K. and Nathans, D.: Specific Cleavage of Simian Virus 40 DNA by Restriction Endonuclease from Hemophilus influenzae, Proc. Natl. Acad. Sci. USA 68, 2913-2917 (1971).

9 Danna, K., Sack, G.H. jr., and Nathans, D.: Studies of Simian Virus 40 DNA. A Cleavage Map of the SV40 Genome, J. Mol. Biol. 78, 363-376 (1973).

10 Espejo, R.T., Canelo, E.S., and Sinsheimer, R.L.: DNA of Bacteriophage PM2: A Closed Circular Double-Stranded Molecule, Proc. Natl. Acad. Sci. USA 63, 1164-1168 (1969).

11 Allet, B., Roberts, R., Gesteland, R., and Solem, R.: Class of Promotor Sites for E. coli DNA-Dependent RNA Polymerase, Nature 249, 217-221 (1974).

12 Englund, P.T.: The 3'-Terminal Nucleotide Sequences of T7 DNA, J. Mol. Biol. 66, 209-224 (1972).

13 Ortin, J., Viñuela, E., Salas, M., and Vasquez, C.: DNA-Protein Complex in Circular DNA from Phage ϕ29, Nature New Biol. 234, 275-277 (1971).

14 Hirokawa, H.: Transfecting DNA of Bacillus Bacteriophage ϕ29 that is Protease Sensitive, Proc. Natl. Acad. Sci. USA 69, 1555-1559 (1972).

The Ribosomal and Nonribosomal Synthesis of Guanosine Polyphosphates

Jose Sy
The Rockefeller University, New York, New York 10021, USA

RIBOSOMAL SYNTHESIS OF GUANOSINE POLYPHOSPHATES

Magic Spots I and II (MS I and MS II) were initially found as two unusual nucleotides that were made during amino acid starvation of Escherichia coli auxotrophs (1). The structure of MS I was determined by Cashel and Kalbacher (2) as guanosine 5'-diphosphate-(2' or 3')diphosphate (ppGpp). MS II (pppGpp) was inferred to be the 5'-triphosphate analog of MS I. The accumulation of ppGpp and pppGpp was shown to have an inverse correlation with the rate of stable RNA synthesis in vivo (3-5). But the mechanism by which these compounds inhibit the synthesis of stable RNA has yet to be clarified.

In 1972, Haseltine et al. (6) discovered that these two nucleotides, ppGpp and pppGpp, could be made in vitro by crude ribosomes from stringent strains of E. coli cells, using ATP as donor and GTP or GDP as precursors. They further showed that when crude ribosomes are washed with 0.5 M NH_4Cl, both washed ribosomes and the NH_4Cl wash are needed for guanosine polyphosphate synthesis. They called the NH_4Cl wash factor the stringent factor. Relaxed strains, which failed to accumulate ppGpp and pppGpp during amino acid starvation, were found to be defective in ribosome-bound stringent factor (1).

The fact that Dr. Lipmann has always been interested in energy-rich molecules plus the knowledge that ppGpp and pppGpp are made on the ribosome, made a study of the exact mechanism by which these molecules are synthesized quite irresistible. Questions we wanted to answer were: 1) was the pyrophosphate transferred stepwise or as pyrophosphoryl, and 2) could we definitely localize the position of the pyrophosphate group in either 2'- or 3'-position.

Utilizing the ribosomal system, we have been able to show that ppGpp formation occurs by way of a pyrophosphoryl transfer of the β-γ-pyrophosphate group of ATP to GDP (8), answering the first question. For identification of the position into which the pyrophosphoryl is transfered to be 2' or 3', we used $[\beta\text{-}^{32}P]$ATP as the pyrophosphoryl donor. The position of the transferred pyrophosphate group was then assayed for by preparing $pG\overset{*}{p}$ from $pG\overset{*}{p}p$ (an asterisk over a p, e.g. $\overset{*}{p}$, denotes a ^{32}P-label in that position) with Zn^{++}-activated yeast inorganic pyrophosphatase (9). For determining the position of ^{32}P in the $pG\overset{*}{p}$ obtained by action of the pyrophosphatase, the 3'-specific rye grass nucleotidase was used to assay for liberation of ^{32}P. A nearly complete hydrolysis of the labeled phosphate was found, and indicated that it had been in the 3'-position. When $pG\overset{*}{p}$ was prepared from $ppG\overset{*}{p}p$ by acid instead of enzymatic hydrolysis and this was assayed under similar conditions with 3'-nucleotidase, only about one half of the label was released. This was to be expected since acid hydrolysis of ppGpp should yield an equilibrium mixture of pG3'p and pG2'p. Since two-dimensional thin layer chromatography on polyethyleneimine cellulose (1st dimension: 1.5 M LiCl; 2nd dimension, saturated $(NH_4)_2SO_4$, 1 M NaOAc, isopropanol, 79:19:2) separated pG3'p and pG2'p, by its use the position of ^{32}P could be determined without hydrolysis. In this manner it was confirmed that pG3'p was the product of pyrophosphatase action and that mixtures of pG3'p and pG2'p were produced by acidic hydrolysis causing $3' \rightarrow 2'$ equilibration. The above evidence for insertion of pyrophosphate into 3'-position for experimental reasons was obtained with ppGpp only, defining the structure of MS I to be guanosine 5'-diphosphate-3'-diphosphate:

ppGpp (MS I)

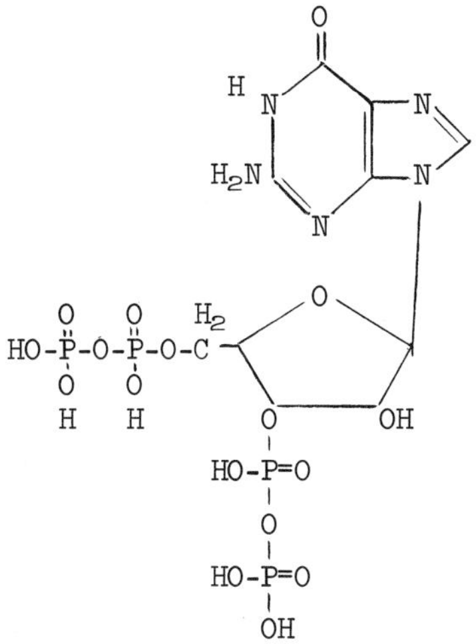

However, because in our experiments pppGpp (MS II) was shown to be a precursor used in the assay of MS I, we conclude the structure of MS II to be guanosine 5'-triphosphate-3'-diphosphate. The structure of ppGpp has recently been confirmed by a ^{13}C-NMR study (10). ppG3'pp and its analog, ppG2'pp, have now been chemically synthesized by Simoncsits and Tomasz (11), and a confirmation using these synthetic compounds is in progress.

Components of the Ribosomal System.

Additional requirements for the ribosomal synthesis of guanosine polyphosphates have recently been defined (12,13): in addition to the stringent factor and ribosomes, mRNA and uncharged tRNA are also required; the latter has to be combined with its specific mRNA to be active. Aminoacyl-tRNA is not inhibitory unless the EF-T factor is added (13). Thiostrepton, fusidic acid, and tetracycline have all been shown to be inhibitory (6,12).

Both the 30S and 50S ribosomal subunits are required for guanosine polyphosphate synthesis (12,13). With regard to ribosomal specificity, we have shown recently (14) that ribosomes from Bacillus brevis and Chlamydomonas reinhardtii chloroplasts are also active in synthesis when supplemented with E. coli stringent factor; however, the cytoplasmic ribosomes of C. reinhardtii are inactive. Richter has reported (15) that with yeast both cytoplasmic and mitochondrial ribosomes do not respond to stringent factor.

The Pyrophosphoryl Donor and Acceptor Specificity.

The nucleotide requirement for the synthetic reaction is rather strict. Only ATP, and to a much lesser extent, dATP, can donate the pyrophosphoryl group (16). GDP, GTP, and to some degree, ITP, have been described as acceptors, and dGTP was found to be inactive (16). Recently, however, we have found (17) that other members of the guanosine 5'-phosphate group will react; both GMP and guanosine 5'-tetraphosphate accepted the pyrophosphate group from ATP in the ribosomal reaction, yielding pGpp and ppppGpp, respectively. The reaction with GMP and guanosine 5'-tetraphosphate was only 10% as efficient as that with GDP or GTP. This specificity is reminiscent of the relative chain-length specificity of acyl-CoA-dehydrogenases. There, high activity of the so-called octanoyl-CoA hydrogenase toward C_8 to C_{12} fatty acyl-CoA, contrasts with low activity with either C_4 or C_{16} fatty acyl-CoA (18).

NONRIBOSOMAL SYNTHESIS OF GUANOSINE POLYPHOSPHATES

While attempting to localize stringent factor on one of the ribosomal subunits, we discovered that a partially purified stringent factor is able to synthesize guanosine polyphosphates without the ribosomal complex (19). The optimal conditions for the nonribosomal reaction were found to be the presence of 20% methanol, low temperatures (20-30°), and in view of its relative slowness, long incubations. With a purified preparation of stringent factor, the presence of 1 mg/ml of bovine serum albumin protects the enzyme against inactivation by denaturation.

For the nonribosomal reaction system, higher concentrations of NH_4^+ (100 mM) and a lower Mg^{++} concentration (4.4 mM) are optimal; it is not inhibited by the antibiotic, thiostrepton. As shown by

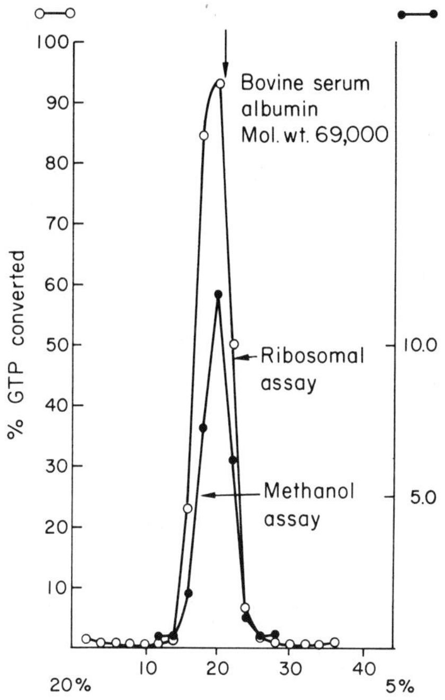

Fig. 1. Sucrose gradient centrifugation of stringent factor. Fraction II stringent factor (60 μg) was centrifuged in a SW 50 rotor at 45,000 rpm for 14 hr on a 5-20% linear sucrose gradient containing 20 mM Tris-Ac, pH 7.8, 20 mM $Mg(OAc)_2$, 100 mM NH_4Cl, and 1 mM dithiothreitol. Two-drop fractions were collected from the bottom and assayed for guanosine polyphosphate synthesis. o——o, ribosomal system: 20-μl aliquots of the gradient fractions were incubated at 30° for 90 min in a final volume of 50 μl that contained 40 mM Tris-Ac, pH 8.0, 4 mM dithiothreitol, 10 mM $Mg(OAc)_2$, 4 mM ATP, 0.4 mM [α-^{32}P]GTP (23 Ci/mole), 27 μg of tRNA, 10 μg of poly AUG, and 80 μg of washed ribosomes. Incubation was stopped with HCOOH and guanosine polyphosphates were assayed as described (8). ●——●, nonribosomal system: 20-μl aliquots of the gradient fractions were incubated at 30° for 18 hr in a final volume of 50 μl that contained 40 mM Tris-Ac, pH 8.0, 4 mM dithiothreitol, 20 mM $Mg(OAc)_2$, 4 mM ATP, 0.4 mM [α-^{32}P]GTP (23 Ci/mole), 50 mM NH_4Cl, 15% methanol, 27 μg of tRNA, and 10 μg of poly AUG. Incubation was stopped with HCOOH and guanosine polyphosphates were assayed as described (8).

sucrose gradient centrifugation, there were no contaminating ribosomal subunits present in the stringent factor preparation used; the enzymatic activity appears only in the lower molecular weight protein fractions.

Fig. 1 shows that on longer centrifugation of the stringent factor preparation in a sucrose gradient there is complete coincidence in the fractions that catalyze the ribosome-dependent synthesis and nonribosomal synthesis of guanosine polyphosphates. The approximate molecular weight of stringent factor, as judged by its sedimentation rate, is 75,000 Daltons.

The major result of finding the reaction in a nonribosomal system is the identification of the stringent factor as being the pyrophosphoryl transferase in the synthesis of ppGpp and pppGpp. This is in contrast to other ribosome-linked reactions where ribosomal proteins have been inferred to be the catalytic sites (20).

REVERSIBILITY OF THE PYROPHOSPHORYL TRANSFER REACTION

The high-energy nature of the 3'-phosphoryl linkage in ppGpp is inferred by analogy with the 3'-energy-rich linkage in 3'-5'-cAMP and the polynucleotide phosphorylase-catalyzed formation of nucleoside diphosphates (21). It is to be expected, then, that the stringent factor should catalyze a pyrophosphoryl transfer and that the reaction should be reversible. I wish here to summarize the results that demonstrate this reversibility. A more detailed report will be communicated elsewhere (22).

The requirements for the reverse reaction are shown in Table 1. As in the synthetic reaction, the reverse reaction requires stringent factor, ribosomes, mRNA, and uncharged tRNA. 5'-AMP is the only nucleotide found to accept the pyrophosphate group from pppGpp. The primary products of the reverse reaction are GTP and ATP. On longer incubation, only GDP and ADP accumulate, due to the presence of GTPase and ATPase in the partially purified stringent factor preparation.

TABLE 1. Requirement for the reverse reaction

	% GDP formed*
Complete	24.6
- 5'-AMP	1.3
- ribosomes	1.7
- stringent factor	0.2
- poly AUG	7.3
- poly AUG - tRNA	3.2
- ribosomes - stringent factor	0.3
No addition	0.3

* % GDP formed represents the amount of input radioactivity converted into GDP. The primary product according to Fig. 2 should be pp$\overset{*}{p}$G which is converted to $\overset{*}{p}$pG by GTPase.

Reverse reactions were assayed by the following method: the standard reaction mixture contained in 20 μl, 20 mM Tris-OAc, pH 8.1, 0.1 mM dithiothreitol, 10 mM $Mg(OAc)_2$, 1.5×10^{-5} M pp$\overset{*}{p}$Gpp (20 Ci/mole), 10 mM 5'-AMP, 29 μg of ammonium chloride-washed ribosomes, 1 μg of fraction II stringent factor, 5 μg of poly AUG, and 5 μg of tRNA. Incubations were at 30° for 90 min and the reaction was stopped by the addition of HCOOH. The resulting precipitates were centrifuged and 10 μl of the supernatant spotted on polyethyleneimine cellulose plates. The thin layer sheets were desalted by immersion in absolute methanol (25) and developed with 1.5 M KH_2PO_4, pH 3.4 (2). The resulting chromatograms were radioautographed overnight and spots corresponding to the various guanosine derivatives were cut out and counted.

Both the synthetic and reverse reactions are not strictly dependent on the presence of the ribosomal acidic proteins L_7 and L_{12}, in contrast to the elongation factor-dependent GTPase activity (23), and both reactions are equally sensitive to inhibition by the antibiotics thiostrepton and tetracycline. Table 2 shows an inhibitory effect found with various nucleotides on the reverse reaction. AMPPCP (β-γ-methylene-adenosine triphosphate) and ATP are very effective inhibitors, whereas ADP and the guanosine

TABLE 2. Inhibition of reverse reaction by nucleotides

Nucleotides added		% GTP + GDP formed
None		42
GDP	10^{-5} M	43
	10^{-4} M	43
	10^{-3} M	32
GTP	10^{-5} M	43
	10^{-4} M	49
	10^{-3} M	26
GMPPCP	10^{-3} M	30
None		46
ADP	10^{-5} M	47
	10^{-4} M	44
	10^{-3} M	33
ATP	10^{-5} M	43
	10^{-4} M	24
	10^{-3} M	3
AMPPCP	10^{-5} M	31
	10^{-4} M	14
	10^{-3} M	5

The reverse reaction was assayed as described in Table 1. Various nucleotides, as indicated, were added as a 1:1 Mg^{++} complex. Incubations were at 30° for 90 min.

nucleotides, GDP, GTP, and GMPPCP (β-γ-methylene-guanosine triphosphate) inhibit only at millimolar concentrations.

Interestingly, pppGpp is a more efficient substrate than ppGpp in the reverse reaction (Fig. 2). This may indicate that pppGpp is the primary product of the stringent factor reaction in the in vivo situation, where it can be rapidly hydrolyzed into ppGpp by the various ribosome-linked reactions that utilize pppGpp, as has

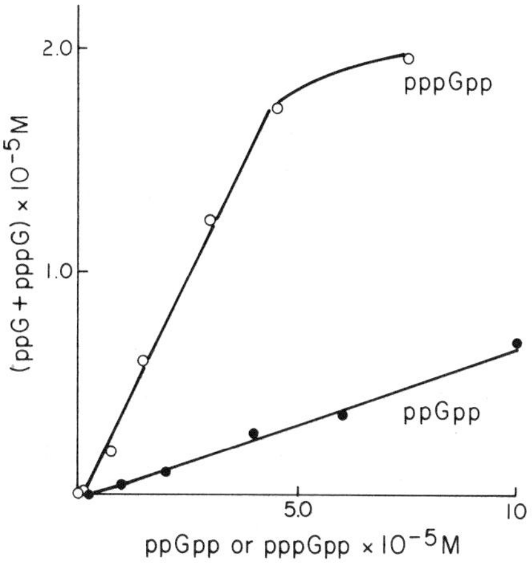

Fig. 2. pppGpp and ppGpp concentration curves. The reverse reactions were performed as described in Table 1. ppGpp was added as a ^{3}H-derivative (130,000 cpm/reaction) at the indicated concentrations. pp$\overset{*}{p}$Gpp was added as the 5'-α-^{32}P-derivative (27,000 cpm/reaction) at the indicated concentrations. Incubations were at 30° for 90 min.

been shown by Hamel and Cashel (24).

The overall reaction catalyzed by the stringent factor can thus be summarized as: (p)ppG + pppA $\rightleftharpoons$ (p)ppGpp + pA. A definite although relatively slow reverse reaction is also catalyzed by the stringent factor without ribosomes. Interestingly, in the reverse direction, addition of methanol has not as great an effect as in the forward reaction.

SUMMARY

In vitro synthesis of guanosine polyphosphates (ppGpp and pppGpp) is catalyzed by a ribosome-linked pyrophosphotransferase, stringent factor. For maximal activity, the enzyme requires ribosomes, mRNA, and codon-specific uncharged tRNA to transfer the β-γ-pyrophosphate moiety of ATP to the 3'-OH position of GDP or GTP. Under appropriate conditions (20% methanol, 20-30°), enzymic activity can be assayed in the absence of the ribosomal complex. The pyrophosphotransferase-catalyzed reaction is now found to be reversible and the overall reaction is: (p)pp$^{5'}$G + ppp$^{5'}$A (p)pp$^{5'}$G$^{3'}$pp + p$^{5'}$A.

******

I should like to dedicate this paper to Dr. Fritz Lipmann on the occasion of his 75th birthday, and to express my sincere gratitude and admiration to him as my mentor and as a founder of modern biochemistry.

Acknowledgment

These studies were supported by a grant to Dr. Lipmann from the United States Public Health Service.

REFERENCES

1. Cashel, M., Gallant, J., Nature 221, 838-841 (1969).
2. Cashel, M., Kalbacher, B., J. Biol. Chem. 245, 2309-2318 (1970).
3. Cashel, M., J. Biol. Chem. 244, 3133-3141 (1969).
4. Lazzarini, R. A., Cashel, M., Gallant, G., J. Biol. Chem. 246 4381-4385 (1971).
5. Fiil, N. P., von Meyerburg, K., Friessen, J. D., J. Mol. Biol. 71, 769-783 (1972).
6. Haseltine, W. A., Block, R., Gilbert, W., Weber, K., Nature 238, 381-384 (1972).
7. Block, R., Haseltine, W. A., J. Mol. Biol. 77, 625-629 (1973).
8. Sy, J., Lipmann, F., Proc. Nat. Acad. Sci. USA 70, 306-309 (1973).
9. Schlesinger, M. J., Coon, M. J., Biochim. Biophys. Acta 41, 30-36 (1960).

10. Que, L., Willie, G. R., Cashel, M., Bodley, J. W., Gray, G. R., Proc. Nat. Acad. Sci. USA 70, 2563-2566 (1973).

11. Simoncsits, A., Tomasz, J., Biochim. Biophys. Acta 340, 509-515 (1974).

12. Pedersen, F. S., Lund, E., Kjeldgaard, N. O., Nature 243, 13-15 (1973).

13. Haseltine, W. A., Block, R., Proc. Nat. Acad. Sci. USA 70, 1564-1568 (1973).

14. Sy, J., Chua, N. H., Ogawa, Y., Lipmann, F., Biochem. Biophys. Res. Commun. 56, 611-616 (1974).

15. Richter, D., FEBS Lett. 34, 291-294 (1973).

16. Cochran, J. W., Byrne, R. W., J. Biol. Chem. 249, 353-360 (1974).

17. Sy, J., Lipmann, F., manuscript in preparation.

18. Crane, F. L., Hauge, J. G., Beinert, H., Biochim. Biophys. Acta 17, 292-294 (1955).

19. Sy, J., Ogawa, Y., Lipmann, F., Proc. Nat. Acad. Sci. USA 70, 2145-2148 (1973).

20. Haselkorn, R., Rothman-Denes, L. B., Ann. Rev. Biochem. 42, 397-438 (1973).

21. Lipmann, F., Advan. Enzyme Regulation 9, 5-16 (1971).

22. Sy, J., Proc. Nat. Acad. Sci. USA, in press.

23. Hamel, E., Kokas, M., Nakamoto, T., J. Biol. Chem. 247, 805-814 (1972).

24. Hamel, E., Cashel, M., Proc. Nat. Acad. Sci. USA 70, 3250-3254 (1973).

25. Randerath, K., Randerath, E., Methods Enzymol. 12, 323-347 (1967).

Phage f2 RNA Structure in Relation to Synthesis of Phage Proteins

P. Szafrański, W. Filipowicz, A. Wodnar-Filipowicz,
L. Zagórska
Institute of Biochemistry and Biophysics, Polish
Academy of Sciences, 02-532 Warsaw, Poland.

INTRODUCTION

The secondary structure of bacteriophage f2 RNA plays an important role in the initiation of the synthesis of three phage proteins: maturation protein, coat protein and RNA-replicase (1,2,3,4). When native f2 RNA is used as template, mainly the initiation site for coat protein is recognized by Escherichia coli ribosomes and initiator tRNA. However, when the ordered structure of f2 RNA has been changed by modification with O-methylhydroxylamine (methoxyamine), which specifically reacts with cytosine ring, a number of ribosomes and fMet-tRNA molecules can be bound to the RNA strand. Such a modification of f2 RNA can be achieved in the presence of 6M guanidine hydrochloride (3). Under these conditions f2 RNA is denatured and all cytosines are exposed for modification. Methoxyamine apears to be a very good tool to study the relationship between the RNA structure and initiation of protein synthesis, since cytosines are not present in the initiation codons.

In this paper further evidence on the role of the ordered structure of f2 RNA in the synthesis of phage pro-

teins is presented. It is shown that in the presence of initiation factors, fMet-tRNA and modified f2 RNA, E. coli ribosomes form polysomes and that under these conditions initiator tRNA occupies the P site on the ribosomes.
The synthesis of phage proteins directed by f2 RNA modified with methoxyamine is strongly inhibited.

METHODS

Preparation of E. coli Q13 protein synthesizing system (S-30), E. coli MRE 600 ribosomes and initiation factors, growth and purification of phage f2 as well as isolation of f2 RNA and preparation of f$[^{3}H]$ Met-tRNA have been described previously (5). Conditions of cell free protein synthesis, polyacrylamide-gel electrophoresis of phage proteins, binding of f$[^{3}H]$ Met-tRNA to ribosomes and estimation of the radioactivity, protein, RNA and phage f2 have also been reported (5).
f2 RNA at a concentration of 1.5-3 mg/ml was modified with 1M methoxyamine, pH 5.5, either in the presence of 10 mM magnesium acetate (non-denaturing conditions) or of 6M guanidine hydrochloride (denaturing conditions) according to Filipowicz et al. (2,3). Details of the modification are given in Legends to Figures.
The procedure of Steitz (6) has been used for $[^{14}C]$-labelling of phage f2 RNA.

RESULTS

f2 RNA-Stimulated Binding of fMet-tRNA to Ribosomes.

This process was followed by studying the binding of f $[^{3}H]$Met-tRNA to E.coli ribosomes at different Mg^{2+} concentrations in a system directed by f2 RNA.

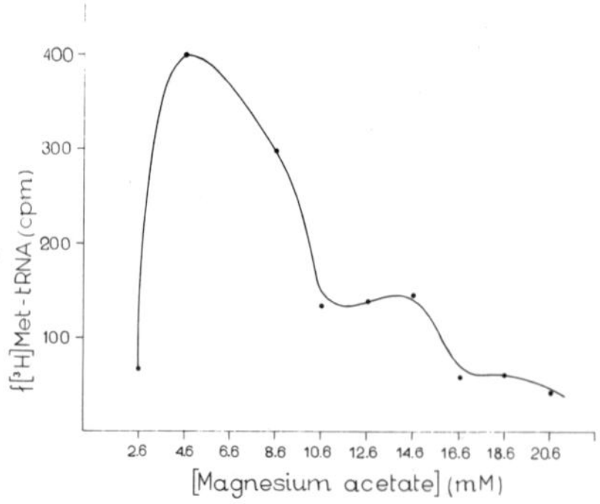

Fig. 1. Binding of f[^{3}H]Met-tRNA to ribosomes at various concentrations of Mg^{2+}.

Binding of f[^{3}H]Met-tRNA to E. coli MRE 600 ribosomes was performed in 5o µl mixtures containing 66 ng of ribosomes, 30 ng of crude initiation factors, 7 pmoles of f[^{3}H] Met-tRNA, 70 ng of f2 RNA and 50 mM Tris-HCl buffer, pH 7.5, containing 80 mM NH_4Cl, 1 mM GTP, 12 mM 2-mercaptoethanol and magnesium acetate, as indicated. After incubation for 20 min at 25°C the amount of f[^{3}H] Met-tRNA bound to ribosomes was determined by millipore filtration (7).

The results presented in Fig. 1 show that the maximum binding of initiator tRNA to ribosomes occurs at about 5 mM Mg^{2+}. This concentration was used in the subsequent experiments on initiation of the synthesis of phage f2 proteins. It can be seen from Fig. 1 that at 10,6 - 14,6 mM Mg^{2+} the amounts of bound f[^{3}H]-Met--tRNA are similar and attain 30 % of the value observed under optimal conditions. Formation of initiation complex (f2 RNA-ribosomes-fMet-tRNA) at all Mg^{2+} concentrations studied depends on the addition of initiation factors.

We have shown previously that in the system directed by native f2 RNA, Ala-tRNA and fMet-tRNA are bound to ribosomes at an equimolar ratio (8). This indicates that ribosomes binds to f2 RNA mainly in the coat protein initiation site, since the initiation codon for coat protein in f2 RNA is followed by a codon specific for alanine (9). Formation of the initiation complex in the presence of native f2 RNA was analysed by sucrose gradient centrifugation and the result is presented in Fig. 2a.

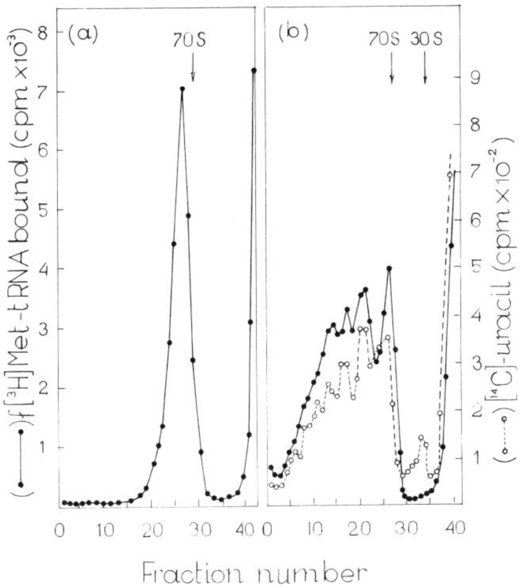

Fig. 2. Sucrose gradient profiles of initiation complexes formed with native (a) and modified (b) f2 RNA.

$[^{14}C]$ Uracil-labelled f2 RNA used in experiment (b) has been modified with 1M methoxyamine in the presence of 6M guanidine hydrochloride for 5 days at 37°C (3). Formation of initiation complexes was performed in incubation mixtures containing 250 ng of E.coli MRE 600 ribosomes, 70 ng of crude initiation factors, 58 pmoles of f$[^3H]$Met-tRNA and (a) 120 ng or (b) 6 ng of f2 RNA in 100 µl of 50 mM Tris-HCl buffer, pH 7.2, containing 5 mM magnesium acetate, 80 mM NH_4Cl, 1 mM GTP and 12 mM 2-mercaptoethanol. Incubation was carried out at 25°C for (a) 20 min and (b) 15 min. Mixtures were layered on the top of 10-40% sucrose gradients and centrifuged for 80 min at 4°C, 40 000 rpm, in a SW 50 rotor of Spinco 50L ultracentrifuge. Fractions were collected on fiber glass filters which were then dried and washed with 5% cold trichloroacetic acid (three times), ethanol-ether and ether. The radioactivity was counted in a Tri-Carb liquid scintillation spectrometer.
Specific activities of f$[^3H]$Met-tRNA was 2215 cpm/pmole in the case of exp. (a) and 1330 cpm/pmole in the case of exp. (b).

It can be seen from Fig. 2a that the initiation complex formed with native f2 RNA migrates in sucrose gradient to the monosome region, as could be expected. However, when f2 RNA modifed with methoxyamine under denaturing conditions was used for initiation complex formation, the presence of heavy polysomal complexes formed of a number of ribosomes distributed along the f2 RNA chain was observed (Fig. 2b). Fractions representing $[^{14}C]$-labelled f2 RNA correspond to the peaks containing f$[^{3}H]$Met-tRNA. This means that the attachment of ribosomes to f2 RNA occurs at the initiation codons accepting initiator tRNA. Calculations performed on the basis of the results presented in Fig. 2 show that the amount of f$[^{3}H]$Met-tRNA bound to the ribosomes is about 60 times greater in the presence of modified f2 RNA, as compared with native f2 RNA. Thus, unfolding of a f2 RNA molecule which results from its modification exposes many normally unfunctional codons recognized by ribosomes and initiatior tRNA.

In order to establish the position of fMet-tRNA on ribosomes in such a polysomal complex, the puromycin reaction was performed. Results are summarized in Table 1.

Table 1. Reaction of puromycin with f$[^{3}H]$Met-tRNA bound to ribosomes in the presence of modified f2 RNA.

Incubation mixture	f$[^{3}H]$ Met-tRNA bound to ribosomes (pmoles)
Complete	8.50
Complete + puromycin	1.43
- f2 RNA	0.83

f2 RNA was modified with methoxyamine in the presence of 6M guanidine hydrochloride for 5 days at 37°C. Binding of f$[^{3}H]$Met-tRNA to the E.coli MRE 600 ribosomes was performed in 50 µl incubation mixtures containing 6 ng of f2 RNA, 120 ng of ribosomes, 35 ng of crude initiation factors and 27 pmoles of f[^{3}H]Met-tRNA (spec. act. 2215 cpm/pmole) in Tris buffer (see legend to Fig. 2). Mixtures were incubated for 20 min at 25°C, and subsequently 40 ng puromycin were added. Incubation was continued for further 15 min. f[^{3}H]Met-tRNA bound to ribosomes after puromycin addition was determined by millipore filtration (7).

As can be seen from Table 1, puromycin removes almost completely fMet-tRNA from the initiation complex formed with modified f2 RNA. These results indicate that fMet-tRNA present in this polysomal complex occupies the P site on ribosomes, similarly as in case of initiation complex formed with native f2 RNA (8).

Synthesis of Phage Proteins Directed by Modified f2 RNA.

f2 RNA modified with methoxyamine under denaturing conditions is deprived of template activity. Therefore, to study the synthesis of phage proteins, f2 RNA was methoxyaminated under non-denaturing conditions. Under these conditions only cytosines not being base-paired can be modified, and f2 RNA retains its ordered structure (2).

This partly modified f2 RNA still exhibits its messenger activity. The synthesis of phage peptides was performed with f2 RNA which has been methoxyaminated for 4 hours under non-denaturing conditions. In this case only few cytosine residues are attacked by methoxyamine. The sucrose gradient profile of polysomes formed in the presence of this RNA is presented in Fig. 3b.

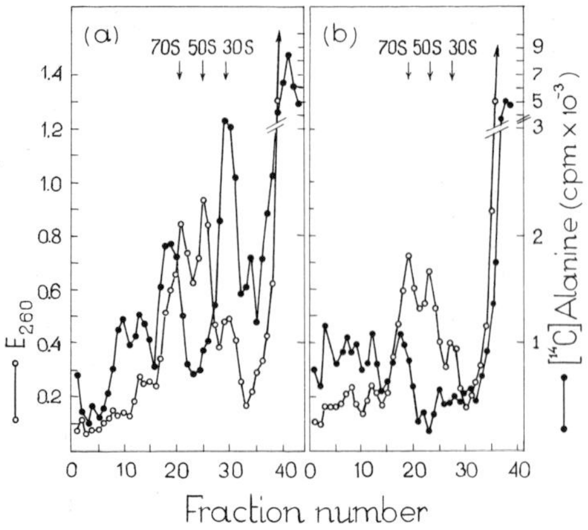

Fig. 3. Sucrose gradient profiles of polysomes formed during polypeptide synthesis in the presence of native (a) and modified (b) f2 RNA.

Synthesis of [^{14}C] alanine-labelled polypeptides was performed in E. coli Q13 S-30 supernatant for 30 min at 37°C, as described previously (5). f2 RNA used in exp. (b) was modified under non-denaturing conditions in the presence of 1M methoxyamine and 10 mM magnesium acetate for 4 hours at 37°C. Mixtures were layered on the top of 10-40% sucrose gradients and centrifuged for 2 hours at 4°C, 39 000 rpm in a SW 50 rotor of Spinco 50L ultracentrifuge. Fractions were collected and radioactivity was estimated as in Fig. 2.

Comparison of the results from Fig. 3b with those obtained with native f2 RNA (Fig. 3a) shows that in the presence of modified f2 RNA, the synthesized [^{14}C] -labelled polypeptides are shifted into the region of heavier polysomes. These results confirm our previous observations 10 that polypeptides formed under the direction of modified f2 RNA cannot be liberated from the ribosomes. This can be explained on the assumption that methoxyaminated cytosines form a block for the translating f2 RNA ribosomes. It can also be seen from Fig. 3b that even after short modification of f2 RNA, the synthesis of phage polypeptides drops markedly.

In order to examine the products synthesized in the presence of modified f2 RNA polyacrylamide-gel electrophoresis was applied.

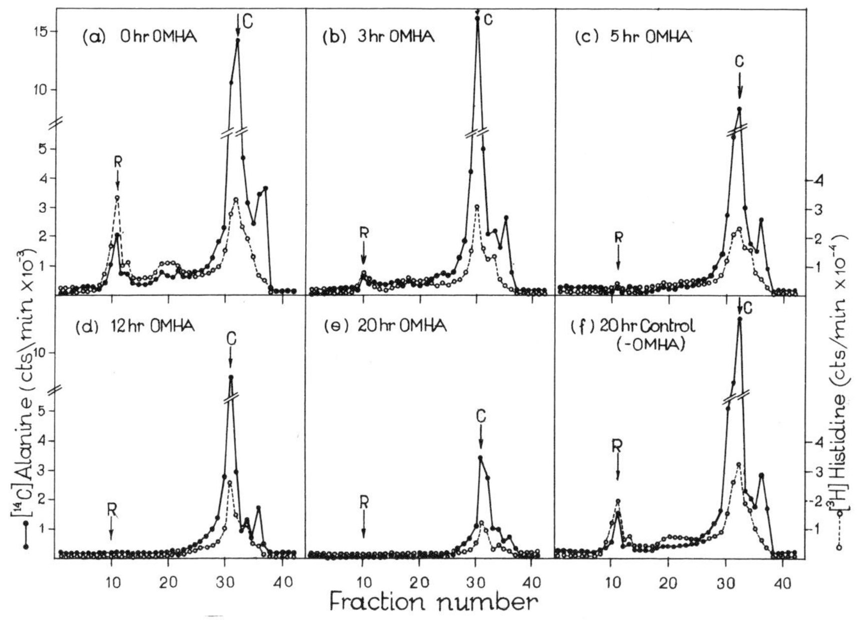

Fig. 4. Polyacrylamide-gel electrophoresis of peptides synthesized in the presence of f2 RNA modified with methoxyamine.

Incorporation of [^{3}H] histidine and [^{14}C] alanine into phage f2 polypeptides was carried out in E. coli Q13 supernatant S-30 (5).
f2 RNA was treated under non-denaturing conditions with 1M methoxyamine in the presence of 10 mM magnesium acetate, pH 5.5, at 21°C for the time indicated in the Figure. Control f2 RNA used in exp. (f) was treated with 1M NaCl, 10mM phosphate, pH 5.5, for 20 hours. Samples were analysed by polyacrylamide-gel electrophoresis, as previously described (5). For estimation of radioactivity distribution, the gels were frozen, cut into 2-mm slices, solubilized and counted in Tri-Carb liquid scintillation spectrometer according to Muto (11). Arrows refer to positions of RNA-replicase (R) and coat protein (C).

Figs. 4 b,c,d and e represented the separation of peptides formed in the presence of f2 RNA modified with methoxyamine during different times. Polyacrylamide-gel electrophoresis of [^{14}C]alanine- and [^{3}H]histidine-labelled polypeptides coded by control f2 RNA treated with methoxyamine for 0 hours (Fig. 4a) shows the synthesis of RNA-replicase and coat protein. The same results are obtained for f2 RNA incubated in 1M NaCl for 20 hours at 21 ^{o}C (Fig. 4f). Histidine counts present in the coat protein region may correspond to some incomplete or degraded RNA-replicase polypeptides. f2 RNA treated with methoxyamine for 3 hours (Fig. 4b) already shows lower ability to synthesize RNA-replicase as compared with native f2 RNA. This ability drops with increasing time of f2 RNA modification. After 12 and 20 hours of methoxyamine action no RNA-replicase synthesis is observed. The coat protein formation is much less inhibited.

DISCUSSION

The unbalanced synthesis of proteins of small bacteriophages is often explained by the polarity of phage gene expression. There is evidence showing that the polar effect results from the secondary structure of phage RNA. It has been shown that when f2 RNA is unfolded the polarity disappears and all three phage proteins can be initiated independently (1). Since unfolded f2 RNA renaturates very qnickly, formaldehyde is often used to destroy the ordered structure of phage RNA. However this regent is not specific and can react with the amino groups of different bases. Methoxyamine used in our experiments specifically reacts with the cytosine rings and changes the ordered structure of f2 RNA leading to the exposure of the normally hidden initiation codons. Under these conditions, in the presence of fMet-tRNA, initiation factors and GTP, ribosomes recognize many initiation codons and form polysomes with f2 RNA. The puromycin reaction shows that in these polysomes, initiator tRNA occupies the P site on ribosomes. This also indicates that under the applied conditions ribosomes can recognize initiation codons without specific arrangement of neighbouring nucleotides. The present results supply direct evidence that the ordered structure of f2 RNA is the major factor preventing initiation at nonfunctional AUG codons. Experiments with translation of f2 RNA modified with methoxyamine show that - with increasing time of modification - f2 RNA loses its template ability. Even after modification of few cytosines the synthesis of RNA-replicase decreases markedly. Formation of coat protein is much less affected under these conditions. This can be explained by the difference in the length of cistrons between coat protein and RNA-replicase. The cistron for the latter protein is four times longer than coat protein cistron and more of its cytosines can be modified. It is suggested that methoxyaminated cytosines can form a hindrance in movement of ribosomes along f2 RNA chain.

SUMMARY

The ordered structure of bacteriophage f2 RNA was modified with O-methylhydroxylamine which specifically reacts with the cytosine ring. The effect of this modification on the initiation of the synthesis of phage f2 proteins was studied by binding f[^{3}H]Met-tRNA to E.coli ribosomes using sucrose gradient analysis. When f2 RNA was modified under denaturing conditions and used for initiation complex formation, the presence of heavy polysomal complexes made up of a number of ribosomes distributed along the f2 RNA chain was observed. Thus, unfolding of a f2 RNA molecule which results from its modification exposed many normally masked initiation codons recognized by ribosomes and initiator tRNA.

The puromycin reaction showed that in the polysomal complexes initiator tRNA occupies the P site on ribosomes. The results supply direct evidence that the ordered structure of f2 RNA is the major factor preventing initiation at the nonfunctional AUG codons.

f2 RNA modified under denaturing conditions is deprived of template activity. Even after modification of few cytosines in f2 RNA, the synthesis of RNA-replicase decreases markedly. The coat protein formation is much less affected under these conditions. Inhibition of the f2 RNA translation can be explained on the assumption that methoxyaminated cytosines form a block for ribosomes during their movement along the RNA chain.

ACKNOWLEDGMENTS

This work was supported in part by the grant 09.3.1. from the Polish Academy of Sciences.

REFERENCES

1. Lodish, H. F.: Secondary structure of bacteriophage f2 Ribonucleic Acid and the Initiation of in vitro protein biosynthesis. J. Mol. Biol. 50, 689 - 702 (1970).

2. Filipowicz, W., Wodnar, A., Szafranski, P.: Reaction of methoxyamine with phage f2 RNA and activity of

modified messenger. FEBS Letters 23, 249 - 253 (1972).

3. Filipowicz, W., Wodnar, A., Zagórska, L., Szafrański, P.: f2 RNA structure and peptide chain initiation: fMet-tRNA binding directed by methoxyamine - modified unfolded or native-like f2 RNAs. Biochem. Biophys. Res. Commun. 49, 1272 - 1279 (1972).
4. Szafrański, P., Filipowicz, W., Wodnar-Filipowicz, A., Zagórska, L.: Messenger activity of phage f2 RNA modified with methoxyamine. In Ribosomes and RNA Metabolism ed. by J. Zelinka and J. Balan, Publishing House of Slovak Academy of Sciences, Bratislava, 1973, p. 321 - 333.
5. Zagórski, W., Filipowicz, W., Wodnar, A., Leonowicz, A., Zagórska, L., Szafrański, P.: The effect of magnesium-ion concentration on the translation of phage-f2 RNA in a cell-free system of Escherichia coli. Eur. J. Biochem. 25, 315 - 322 (1972).
6. Steitz, J. A.: Identification of the A protein as a structure component of bacteriophage R17. J. Mol. Biol. 33, 923 - 936 (1968).
7. Nirenberg, M., Leder, P.: RNA codewords and protein synthesis. The effect of trinucleotides upon the binding of sRNA to ribosomes. Sciences (Washington) 145, 1399 - 1407 (1964).
8. Zagórska, L., Szafrański, P.: Ribosomal binding site of fMet-tRNA in Escherichia coli system directed by phage f2 RNA. Acta Biochim. Polon. 20, 101 - 111 (1973).
9. Webster, R. E., Engelhardt, D. L., Zinder, N. D.: In vitro protein synthesis: chain initiation. Proc. Natl. Acad. Sci. U.S. 55, 155 - 161 (1966).
10. Szafrański, P., Zagórski, W., Filipowicz, W., Wodnar-Filipowicz A., Chroboczek, J.: Regulatory

elements in initiation of phage f2 RNA translation. Symp. on Ribosomes and Biosynthesis of Proteins, Reinhardsbrunn, 13-16.V.1974, Biochemische Gesselschaft der DDR.

11. Muto, A. J.: Messenger activity of nascent ribosomal RNA. J. Mol. Biol. 36, 1 - 14 (1968).

Protein Kinases from Eukaryotic Organisms

Masao Takeda and Yasutomi Nishizuka
Department of Biochemistry, Kobe University School of Medicine
Kobe, Japan

It is our great pleasure to dedicate this article to Prof. Fritz Lipmann on the occasion of his seventyfifth birthday. In the early thirties, an alkali-labile phosphate bond attached to serine hydroxyl in proteins was first demonstrated by Lipmann and Levene in phosvitin (1) and casein (2). The synthesis of the phosphomonoester bond is catalyzed by protein kinase which transfers the terminal phosphate of ATP to the hydroxyl group of either seryl or threonyl residue of proteins. Such protein kinase, which phosphorylates casein and endogenous phosphoproteins, has been found in rat liver by Burnett and Kennedy in 1954 (3). The discovery of 3',5'-cyclic AMP (cAMP)-dependent protein kinase by Walsh, Parkins and Krebs (4) in 1968 has shed light on the role of cAMP in hormonal regulation of cellular metabolism and greatly stimulates the studies on protein kinases. The regulation of biological activities of proteins through phosphorylation and dephosphorylation reactions is now accepted to be one of the most significant mechanisms of controlling cellular metabolism. In 1968, we started to investigate the phosphorylation of chromosomal proteins in comparison with the ADP-ribosylation reaction (5), and four distinctly different types of protein kinase have been thus far characterized which are able to phosphorylate nuclear proteins. It is the purpose of this article to briefly summarize the properties of these protein kinases.

cAMP-dependent protein kinase first described by Walsh *et al.* (4) shows broad substrate specificity and phosphorylates histone (6), glycogen phosphorylase kinase (4), glycogen synthetase (7,8), lipase (9,10), carbonic anhydrase (11), ribosomal proteins (12-14), membrane-associated proteins (15,16) and many other enzymes and proteins (17). In 1968, Yamamura in our laboratory has resolved the liver enzyme into two fractions, cAMP-dependent and independent, both of which phosphorylate, nevertheless, the same specific seryl and threonyl residues of histone and protamine (18). Shortly afterward, these two fractions are shown to be interconvertible forms of a single protein kinase, and this interconversion is mediated by reversible attachment of regulatory unit (R) to catalytic unit (C) of the protein kinase. cAMP is selectively bound to the regulatory unit in an allosteric manner, and dissociates the catalytic unit to exhibit full activity (19). The reaction is reversible (20). The mechanism of activation of the protein kinase is shown in the following equation.

$$\underset{\text{(dependent form)}}{\text{RC}} + \text{cAMP} \rightleftharpoons \text{R-cAMP} + \underset{\text{(independent form)}}{\text{C}}$$

An essentially identical conclusion has been concurrently obtained in other laboratories with protein kinases obtained from bovine adrenal (21), rabbit reticulocytes (22) and rabbit skeletal muscle (23).

Langan has presented evidence that a single intraperitoneal injection of either glucagon or N^6, O^2-dibutyryl cAMP into rat causes a significant increase in the phosphorylation of a specific seryl residue of lysine-rich (F_1) histone (24). The seryl residue is also shown to be phosphorylated by cAMP-dependent protein kinase *in vitro* (25). Using the Langan's system we are able to show that the rate of phosphorylation of F_1 histone *in vivo* is proportional to the intracellular cAMP concentration (26). After glucagon administration, protein kinase activity in rat liver soluble fraction, which is measured *in vitro* without adding cAMP, increases rapidly, reaches the meximum at 5 min and returns to the normal value at 75 min when intracellular cAMP returns almost to the control level. The results seem to indicate that in rat liver cAMP does regulate protein kinase reversibly as demonstrated in *in vitro* systems (26). Further analysis using a cell fractionation technique with nonaqueous solvents has revealed that in rat liver nuclei cAMP-dependent protein kinase is indeed activated by glucagon through enhancement of the cAMP level (27).

In rat liver apparently multiple species of cAMP-dependent protein kinases has been distinguished (28). The multiple kinases are composed of a common catalytic unit and apparently differ from each other in their associated regulatory units. These protein kinases are activated by cAMP in a similar manner, but their apparent affinities for the cyclic nucleotide are slightly different (20,28). However, the multiplicity is partly due to proteolysis during the isolation procedure (20).

The catalytic units of muscle and liver cAMP-dependent protein kinases show essentially identical kinetic and catalytic properties; recombination of regulatory and catalytic units from heterologous sources produces a hybrid cAMP-dependent protein kinase (29,30). cAMP-dependent protein kinase is found also in unicellular eukaryotic organisms. Takai in this laboratory has recently found and purified the enzyme from baker's yeast about 100-fold (31). The mode of activation of this enzyme by cAMP is essentially similar to that described for mammalian enzymes. Table 1 summarizes the properties of yeast cAMP-dependent protein kinase in comparison with rat liver enzyme. These enzymes show closely similar but slightly different properties. Nevertheless, the catalytic and regulatory units from yeast and liver enzymes are crosswise reactive and recombination of these units produces a hybrid cAMP-dependent protein kinase.

Table 1. *Properties of yeast and rat liver cAMP-dependent protein kinases*

Property	Yeast enzyme[a/]	Liver enzyme[b/]
Km value for ATP	1.2×10^{-5} M	0.5×10^{-5} M
Ka value for cAMP	2.0×10^{-8} M	1 to 4×10^{-8} M
Optimum Mg^{2+} ion	5 mM	3 mM
Optimum pH	7.5	7.0
Half-life at 50°	2 min	5 min
Isoelectric point		
Holoenzyme	pH 7.7	pH 4.8 to 5.2
Catalytic unit	pH 6.9	pH 7.4 and 8.2
Regulatory unit	pH 9.1	pH 4.8 to 7.0
Molecular weight		
Holoenzyme	58,000	100,000 to 20,000
Catalytic unit	30,000	35,000
Regulatory unit	28,000	—

a/ The data are taken from Ref. 31.
b/ The data are taken from Refs. 20 and 30.

The fingerprints of radioactive tryptic phosphopeptides of histone phosphorylated separately by protein kinases from either rabbit muscle, rat liver, bovine adrenal, rat brain, baker's yeast, or silkworm pupae are closely similar (30,31,32), suggesting that the enzymes from different sources phosphorylate same specific sites of substrate proteins. In addition, cAMP-dependent protein kinases from rabbit muscle, rat liver, yeast, silkworm pupae are equally active to phosphorylate muscle phosphorylase kinase and glycogen synthetase resulting in the activation and inactivation of the respective enzymes (30,33). An available evidence strongly suggests that cAMP-dependent protein kinases in general show identical spectra for substrate proteins; namely, the enzymes apparently lack tissue- as well as species-specificities at least in their functional activities. A plausible evidence seems to relegate a role of crucial importance of the protein kinase to simultaneous control of several biological reactions in each tissue (32).

CYCLIC GMP-DEPENDENT PROTEIN KINASES

During the course of studies on the distribution of cAMP-dependent protein kinases Kuo and Greengard (34) have found protein kinase fractions in certain arthropods which are preferentially activated by 3',5'-cyclic GMP (cGMP) and designated them as cGMP-dependent protein kinases. This class of enzymes have also been found in some mammalian tissues (35-37). The existence of such protein kinases suggests that the protein phosphorylation may also play a role in the as yet undefined action of cGMP in biological systems. However, no entirely distinct pattern of differences of cGMP-dependent and cAMP-dependent protein kinases has been reported. We have recently purified these two classes of enzymes each about 100-fold from the soluble fraction of silkworm *(Bombyx mori)* pupae, and characterized as briefly summarized in Table 2. Although both enzymes phosphorylate preferentially histone and

Table 2. *Properties of cAMP-dependent and cGMP-dependent protein kinases purified from silkworm*[a/]

Property	cGMP-dependent protein kinase	cAMP-dependent protein kinase
Km value for ATP	3.3×10^{-5} M	5×10^{-6} M
Ka value for cGMP	7.5×10^{-9} M	4×10^{-7} M
for cAMP	2.0×10^{-7} M	1.3×10^{-8} M
Optimum Mg^{2+} ion	3.0 mM	3.0 mM
Optimum pH	pH 6-7	pH 7-8
Isoelectric point[b/]	pH 5.4	pH 5.4
Molecular weight[b/]	140,000	180,000

a/ Nishiyama, K., Katakami, H., Yamamura, H., Takai, Y., Shimomura, R., Nishizuka, Y., unpublished data.
b/ The data represent values for holoenzymes.

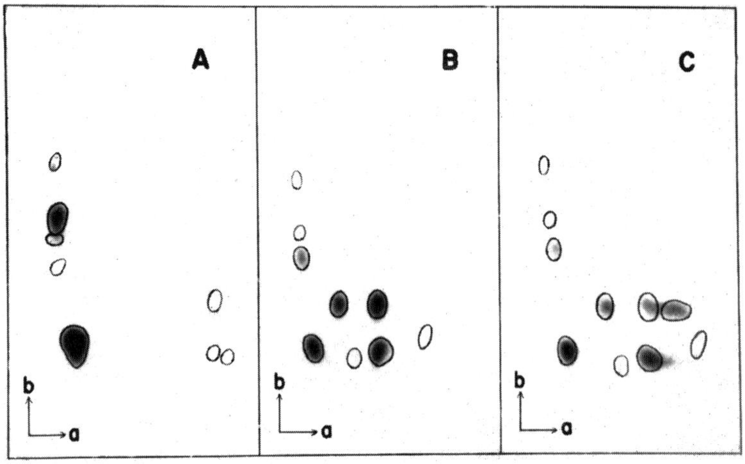

Fig. 1. *Autoradiography of tryptic digests of radioactive histone fully phosphorylated by cyclic nucleotide-dependent protein kinases.* Paper chromatography (Direction a) and paper electrophoresis (Direction b) were carried out as described in Ref. 16. Under these conditions the histone preparations were equally digested to produce identical sets of more than 30 peptide spots visualized by the ninhydrin reaction. A, cGMP-dependent protein kinase (silkworm); B, cAMP-dependent protein kinase (silkworm); C, cAMP-dependent protein kinase (rat liver).

---

protamine, the tryptic phosphopeptides of histone phosphorylated by these protein kinases are apparently different as judged by the fingerprint procedure as shown in Fig. 1. In contrast to cAMP-dependent protein kinase, the cGMP-dependent enzyme is totally inactive in the phosphorylation of muscle phosphorylase kinase and glycogen synthetase. Therefore, cGMP-dependent protein kinase belongs to another entity with entirely different substrate specificity from that of the class of cAMP-dependent protein kinases. Species- and tissue-specificities as well as the natural substrate proteins of this class of enzymes intimately involved in the regulation of biological processes have remained unexplored.

PROTAMINE KINASE

Another type of protein kinase has been obtained from rat brain cytosol which phosphorylates preferentially protamine and to some extent histone (38). The enzyme, referred to as protamine kinase, is independent on cAMP and completely separable from the catalytic unit of cAMP-dependent protein kinase by isoelectrofocusing electrophoresis. Protamine kinase is not controlled by the regulatory unit of cAMP-dependent protein kinase. Protamine kinase and cAMP-dependent protein kinase phosphorylate different seryl and threonyl residues of protamine and histone as judged by the fingerprint procedure. Some properties of rat brain protamine kinase are summarized in Table 3. A similar type of enzymes is found in rat liver, yeast (31) and calf uterus (39). It may be recalled that Langan has suggested the existence of a protein kinase (HK II) in calf liver which phosphorylates a single specific site of F_1 histone (40). Recently, rat brain tubulin has been shown to serve as substrate of protamine kinase (41). Nevertheless, the biological role of this class of enzymes has remained unexplored.

PHOSPHOPROTEIN KINASES

A protein kinase, which phosphorylates seryl and threonyl residues

Table 3. *Properties of rat brain protamine kinase and rat liver nuclear phosphoprotein kinases*

Property	Protamine kinase[a]	Phosphoprotein kinase[b]	
		Kinase A_1	Kinase A_2
Km value for ATP	1.3×10^{-6} M	6×10^{-6} M	1×10^{-5} M
Optimum Mg^{2+} ion	20 mM	20 mM	20 mM
Optimum pH	pH 8	pH 7-8	pH 8
Isoelectric point	pH 5.8	pH 4.3 and 10.5[c]	pH 10.5
Molecular weight	90,000	230,000	40,000

a/ The data are taken from Ref. 38.
b/ The data are taken from Ref. 50.
c/ Each component is indistinguishable from the other in kinetic and catalytic properties.

of bovine casein and egg yolk phosvitin, has been described first by Burnett and Kennedy (3) in rat liver. Subsequently, similar enzymes have been partially purified from various mammalian tissues (42-45), fish roe (46) and brewer's yeast (42). Naturally occurring phosphate acceptors of these enzymes are phosphoproteins, as yet uncategorized, which are almost ubiquitously distributed in biological systems. Such phosphoproteins, containing up to 1.3% phosphorous by weight, have been found to be enriched in rat liver (47) as well as calf thymus (48) nuclei. The phosphate groups bound to these proteins have been shown to turnover rapidly, and a role of the phosphoproteins in regulating some nuclear functions, particularly transcription of genetic activities, has been implied. The phosphorylation of nucelar phosphoproteins is shown in this laboratory to be catalyzed by two enzymes associated with chromatin which may be separated from bulk of phosphoproteins (49, 50). These protein kinases, tentatively referred to as protein kinase A_1 and protein kinase A_2, are partially purified from rat liver chromatin by salt extraction followed by Sephadex G-200 gel filtration and phosphocellulose column chromatography. 3',5'-Cyclic nucleotides do not stimulate or inhibit these kinases. Properties of these kinases are summarized in Table 3. The enzymes show closely similar properties, but are clearly distinguished from each other in substrate specificities. Upon sodium dodecyl sulfate-polyacrylamide gel electrophoresis the nuclear phosphoproteins are shown to be heterogeneous; each protein kinase phosphorylates specific seryl and threonyl residues of different protein molecules. Casein and phosvitin also serve as effective substrates, whereas histone and protamine are less than 7% as active as casein. Similarly to pigeon brain protein kinase described by Rabinowitz and Lipmann (42) both protein kinases act more effectively on native casein than dephosphorylated casein, and seem to differ from the Golgi protein kinase (51) which preferentially phosphorylates dephosphorylated casein. As illustrated in Fig. 2., the tryptic phosphopeptides of casein separately phosphorylated with

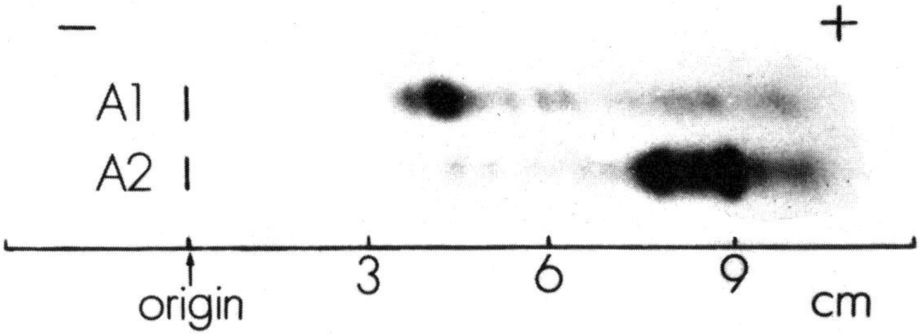

Fig. 2. *Autoradiography of tryptic digests of casein fully phosphorylated by protein kinase* A_1 *and* A_2. Cellogel R.S. electrophoresis and other experimental conditions are described in Ref. 52. $\underline{A_1}$, protein kinase A_1; $\underline{A_2}$, protein kinase A_2.

---

protein kinase A_1 and A_2 are clearly distinguishable each other indicating these kinase phosphorylate different seryl and threonyl residues of the substrate proteins. With nuclear phosphoproteins as substrates both kinases phosphorylate preferentially seryl but also some threonyl residues. With casein as substrate, however, protein kinase A_1 phosphorylates mainly threonyl residues, whereas protein kinase A_2 preferentially seryl residues.

Rat liver cytoplasmic casein kinase of the classical type is also shown to be separable by gel filtration on a Sephadex column into two fractions which phosphorylate specific seryl and threonyl residues of casein and phosvitin (52). The latter kinases react also with nuclear phosphoproteins and are indistinguishable from the nuclear protein kinases in their kinetic and catalytic properties. However, the cytoplasmic enzymes appear not to represent simply a leakage of the nuclear enzymes into the cytoplasm during the isolation procedure. By subcellular fractionation with non-aqueous solvents, these protein kinases are shown to distribute equally in cytoplasm and nuclei although the specific activities of these enzymes are twenty times higher in chromatin than in cytoplasm.

SUMMARY AND DISCUSSION

The proteins containing phosphate in direct covalent linkage was recognized as early as in 1900 by Osborn and Campbell (53) and Levene and Alsberg (54). Since then, a wide variety of *"phosphoproteins"* has been described in the literature. The phosphorylation and dephosphorylation of some of these proteins have been firmly established to be intimately related to the regulation of their functional activities. In most cases, however, the biological roles of *"phosphoproteins"* as well as the enzymes involved have been thus far unknown. In the present article four classes of protein kinases are described which phosphorylate preferentially either histone, protamine or nuclear phosphoproteins. The first class of enzyme is cAMP-dependent, and is widely distributed in eukaryotes including unicellular organisms such as yeast. The enzymes apparently lack tissue- as well as species-specificities at least in their catalytic activities, and phosphorylate many enzymes and proteins including glycogen phosphorylase kinase, glycogen synthetase, lipase, ribosomal proteins and membrane-associated proteins in addition to histone and protamine. A line of evidence obtained in several laboratories including our own indicates that this class of enzymes plays central roles in regulating various biological reactions. The second class of enzymes which are cyclic GMP-dependent is purified and characterized from silkworm pupae. Recently, cGMP-dependent enzymes are also found in brain (35,36), pancreas (37), bladder and uterine tissue (35). However, no substrate other than histone and protamine has been identified for this class of enzymes. Nevertheless, the present studies clearly show that cAMP-dependent and cGMP-dependent enzymes exhibit entirely different kinetic and catalytic properties and serve probably different physiological roles in biological systems. The third class is enzymes which are isolated from the soluble fraction of brain, liver and yeast. The enzymes phosphorylate preferentially protamine and to some extent histone but not casein and phosvitin. The last class of enzymes is associated mainly with chromatin and phosphorylates nuclear phosphoproteins originally found in liver nuclei by Langan (47) and Kleinsmith and Allfrey (48). Casein and phosvitin but not histone and protamine serve as substrates. Two enzymes are shown to belong to this entity: one is rather specific for seryl and the other threonyl residues with casein as substrate. Similar enzymes are also present in the cytoplasmic fraction of various tissues. Cyclic nucleotides have no effect on the latter two classes of enzymes. All protein kinases mentioned above phosphorylate each specific seryl and threonyl residues and are clearly distinguished from each other in their kinetic and catalytic properties. In any case, at least four types of protein kinases described here may be responsible for the reactions to form a serine-phosphate bond in "*phosphoproteins*" in nuclei of eukaryotic organisms. The biological roles of these enzymes are inevitable for further investigations.

Acknowledgments —— Members, past and present, who have contributed to the experiments in the present article include Drs. A. Kumon, H.Yamamura, Y.Ohga, Y.Inoue, H.Higashino, Y.Nakaya, Y.Takai, Messrs. K.Nishiyama, S.Matsumura, K.Sakai, H.Katakami and Miss R. Shimomura. Mrs. S.Nishiyama and Miss M.Kuroda are acknowledged for their secretarial assistance. This investigation has been supported in part by the research grants from the Jane Coffin Childs Memorial Fund for Medical Research and the Scientific Research Fund of the Ministry of Education of Japan.

REFERENCES

(1) Lipmann, F., Levene, P.A., *J. Biol. Chem.* 98, 109-114 (1932).
(2) Lipmann, F., *Biochem. Z.* 262, 3 (1933).

(3) Burnett, G., Kennedy, E.P., *J. Biol. Chem.* 211, 969-980 (1954).
(4) Walsh, D.A., Perkins, J.P. and Krebs, E.G., *J. Biol. Chem.* 243, 3763-3765 (1968).
(5) Nishizuka, Y., Ueda, K., Yoshihara, K., Yamamura, H., Takeda, M., Hayaishi, O., *Cold Spring Hab. Symp. Quant. Biol.* 34, 781-785 (1969).
(6) Langan, T.A., *Science,* 162, 579-581 (1968).
(7) Soderling, T.R., Hickenbottom, J.P., Reimann, E.M., Hunkeler, F.L., Walsh, D.A., Krebs, E.G., *J. Biol. Chem.* 245, 6317-6328 (1970).
(8) Villar-Palasi, C., Larner, J., Shen, L.C., *Ann. N.Y. Acad. Sci.* 185, 74-84 (1971).
(9) Huttunen, J.K., Steinberg, D., Mayer, S.E., *Proc. Nat. Acad. Sci.* 67, 290-295 (1970).
(10) Corbin, J.D., Reimann, E.M., Walsh, D.A., Krebs, E.G., *J. Biol. Chem.* 245, 4849-4851 (1970).
(11) Narumi, S., Miyamoto, E., *Biochim. Biophys. Acta* 350, 215-224 (1974).
(12) Blat, C., Loeb, J.E., *FEBS.Letters* 18, 124-126 (1971).
(13) Eil, C., Wood, I.G., *Biochem. Biophys. Res. Commun.* 43, 1001-1009 (1971).
(14) Walton, G.M., Gill, G.N., Abrass, I.B., Garren, L.D., *Proc. Nat. Acad. Sci.* 68, 880-884 (1971).
(15) Weller, M., Rodnight, R., *Nature,* 225, 187-188 (1970).
(16) Shimomura, R., Matsumura, S., Nishizuka, Y., *J. Biochem.* 75, 1-10 (1974).
(17) Matsumura, S., Nishizuka, Y., *J. Biochem.* 76, in press (1974).
(18) Yamamura, H., Takeda, M., Kumon, A., Nishizuka, Y., *Biochem. Biophys. Res. Commun.* 40, 675-681 (1970).
(19) Kumon, A., Yamamura, H., Nishizuka, Y., *Biochem. Biophys. Res. Commun.* 41, 1290-1297 (1970).
(20) Kumon, A., Nishiyama, K., Yamamura, H., Nishizuka, Y., *J. Biol. Chem.* 247, 3726-3735 (1972).
(21) Gill, G.N., Garren, L.D., *Biochem. Biophys. Res. Commun.* 39, 335-343 (1970).
(22) Tao, M., Salas, M.L., Lipmann, F., *Proc. Nat. Acad. Sci.* 67, 408-414 (1973).
(23) Reimann, E.M., Brostrom, C.O., Carbin, J.D., King, C.A., Krebs, E.G., *Biochem. Biophys. Res. Commun.* 42, 187-194 (1971).
(24) Langan, T.A., *Proc. Nat. Acad. Sci.* 64, 1276-1283 (1969).
(25) Langan, T.A., *J. Biol. Chem.* 244, 5763-5765 (1969).
(26) Takeda, M., Ohga, Y., *J. Biochem.* 73, 621-629 (1973).
(27) Higashino, H., Takeda, M., *J. Biochem.* 75, 189-191 (1974).
(28) Yamamura, H., Kumon, A., Nishiyama, K., Takeda, M., Nishizuka, Y., *Biochem. Biophys. Res. Commun.* 45, 1560-1566 (1971).
(29) Yamamura, H., Kumon, A., Nishizuka, Y., *J. Biol. Chem.* 246, 1544-1547 (1971).
(30) Yamamura, H., Nishiyama, K., Shimomura, R., Nishizuka, Y., *Biochemistry* 12, 856-862 (1973).
(31) Takai, Y., Yamamura, H., Nishizuka, Y., *J. Biol. Chem.* 249, 530-535 (1974).

(32) Yamamura, H., Inoue, Y., Shimomura, R., Nishizuka, Y., *Biochem. Biophys. Res. Commun.* 46, 589-596 (1972).
(33) Takai, Y., Sakai, K., Morishita, Y., Yamamura, H., Nishizuka, Y., *Biochem. Biophys. Res. Commun.* in press (1974).
(34) Kuo, J.F., Greengard, P., *J. Biol. Chem.* 245, 2493-2498 (1970).
(35) Greengard, P., Kuo, J.F., *Advances in Biochemical Psychopharmacology*, vol.3, ed. by P.Greengard and E.Costa, pp.287-306, Raven Press, New York (1970).
(36) Hoffman, F., Sold, G., *Biochem. Biophys. Res. Commun.* 49, 1100-1107 (1972).
(37) Leemput-Coutrez, M.V., Camus, J., Christophe, J., *Biochem. Biophys. Res. Commun.* 54, 182-190 (1973).
(38) Inoue, Y., Yamamura, H., Nishizuka, Y., *Biochem. Biophys. Res. Commun.* 50, 228-236 (1973).
(39) Puca, G.A., Nola, E., Sica, V., Bresciani, F., *Biochem. Biophys. Res. Commun.* 49, 970-976 (1972).
(40) Langan, T.A., *Ann. N.Y. Acad. Sci.* 185, 166-180 (1971).
(41) Eipper, B.A., *J. Biol. Chem.* 249, 1398-1406 (1974).
(42) Rabinowitz, M., Lipmann, F., *J. Biol. Chem.* 235, 1043-1050 (1960).
(43) Sundararajan, T.A., Sampath-Kumar, K.S.V., Sarma, P.S., *Biochim. Biophys. Acta* 29, 449-450 (1958).
(44) Baggie, B., Pinna, L.A., Moret, V., Siliprandi., *Biochim. Biophys. Acta* 207, 515-517 (1970).
(45) Wålinder, O., *Biochim. Biophys. Acta* 293, 140-149 (1973).
(46) Mano, Y., *Biochim. Biophys. Acta* 201, 284-294 (1970).
(47) Langan, T.A., *Regulation of nucleic acid and protein biosynthesis*, vol.10, ed. by V.V. Koningsberger and L.Bosch, pp. 233-242, Elsevier, Amsterdam (1967).
(48) Kleinsmith, L.J., Allfrey V.G., *Biochim. Biophys. Acta* 175, 123-135 (1969).
(49) Takeda, M., Yamamura, H., Ohga, Y., *Biochem. Biophys. Res. Commun.* 42, 103-110 (1971).
(50) Takeda, M., Matsumura, S., Nakaya, Y., *J. Biochem.* 75, 743-751 (1974).
(51) Bingham, E.W., Farrell, Jr., H.M., Basch, J.J., *J. Biol. Chem.* 247, 8193-8194 (1972).
(52) Matsumura, S., Takeda, M., *Biochim. Biophys. Acta* 289, 237-241 (1972).
(53) Osborne, T.B., Campbell, G.F., *J. Am. Chem. Soc.* 22, 413-422 (1900).
(54) Levene, P.A., Alsberg, C.L., *Z. Physiol. Chem.* 31, 543-555 (1901).

The Interaction of Red Blood Cell Protein Factors with Cyclic AMP

Mariano Tao, Khe-Ching Yuh, and M. Marlene Hosey
Department of Biological Chemistry
University of Illinois at the Medical Center
Chicago, Illinois 60612

Introduction

Adenosine cyclic 3':5'-monophosphate (cyclic AMP) is a versatile regulator affecting a number of metabolic processes. The principal action of cyclic AMP is to bind to certain protein factors. Interestingly, all the known cyclic AMP receptors described to date, with the exception of phosphofructokinase, possess no catalytic functions. These receptors act as regulators which can either stimulate or inhibit the activities of enzymes through formation of molecular complexes. The binding of cyclic AMP to the receptors can facilitate or prohibit the formation of these complexes. An example for the inhibitory action of cyclic AMP on molecular complex formation is illustrated by the cyclic AMP-dependent protein kinase (1,2). The inactive enzyme is a complex of two dissimilar functional subunits: a regulatory subunit which binds cyclic AMP and a catalytic subunit. Cyclic AMP causes the dissociation of the complex by binding to the regulatory component. The phosphotransferase activity is contained in the free catalytic moiety. In contrast, the binding of cyclic AMP to a protein factor isolated from *Escherichia coli* promotes the formation of a ternary complex with RNA polymerase and DNA. This results in an increase in the transcription of the genes of inducible enzymes (3).

In considering the regulatory role of cyclic AMP in red blood cells, it is necessary to first identify and study the protein factors capable of interacting with the cyclic nucleotide. Several cyclic AMP-binding proteins have been isolated from rabbit erythrocytes in our laboratory. Three of these factors are associated with the cyclic AMP-dependent protein kinases and represent the regulatory subunits of these enzymes (4). In addition, we have demonstrated the presence of at least two other cyclic AMP-binding protein factors which are not related to any of the cyclic AMP-dependent protein kinases.

Methods

Two methods have been employed for assay of cyclic AMP binding: (a) Millipore filtration. The standard Millipore assay was performed in a total volume of 0.1 ml, containing 1 mM Tris-HCl, pH 7.5; 1 μM cyclic [^{3}H]AMP or cyclic [^{32}P]AMP; 30 μg/ml protamine; 4 mM $MgCl_2$; and binding proteins. All other experimental details were similar to those described previously for the binding assay of cyclic AMP-dependent protein kinases (4). (b) Sephadex G-50 gel filtration. The incubation mixture contained 1 mM Tris-HCl, pH 7.5; 10% sucrose; 1 μM cyclic [^{32}P]AMP; and binding proteins, in a final volume of 0.1 ml. After incubation for 3 min at 37°C, the entire mixture was applied to a Sephadex G-50 column (0.7 x 12 cm) and eluted with 1 mM Tris-HCl, pH 7.5. Three-drop fractions were collected for radioactivity determinations.

Protein kinase activity was assayed as described previously by measuring the amount of ^{32}P incorporated into calf thymus histone by [γ-^{32}P]ATP (4).

Cyclic AMP-dependent protein kinases and non-kinase associated cyclic AMP-binding proteins were purified from rabbit erythrocytes employing conventional enzyme fractionation techniques. Frozen rabbit erythrocytes were purchased from Pel-Freez Biologicals, Inc.

Results

Cyclic AMP-Dependent Protein Kinases

At least three cyclic AMP-dependent protein kinases, I, IIa, and IIb, are found in the soluble fraction of the erythrocyte lysate. Employing conventional enzyme purification techniques, the activities of kinases I, IIa, and IIb are enriched 4,300-fold, 2,000-fold, and 1,500-fold, respectively (4). Table I shows some of the

TABLE I

SOME PHYSICAL PROPERTIES OF CYCLIC AMP-DEPENDENT PROTEIN KINASES

Kinase	Mol Wt[a] Holoenzyme	Sedimentation Coefficient[b] Holoenzyme	Regulatory Subunit	Catalytic Subunit
I	170,000	7.4 S	5 S	4.1 S
IIa	120,000	5.2 S	3.4 S	4.3 S
IIb	240,000	7.2 S	5.8 S	5.8 S (4 S)[c]

[a] Estimated by Sephadex G-200 gel filtration.
[b] Estimated by sucrose density gradient centrifugation in the presence or absence of cyclic AMP.
[c] Obtained from dissociation with protamine.

physical properties of these kinases. In the presence of cyclic AMP, both kinases I and IIa clearly dissociate into two dissimilar functional subunits of different sedimentation coefficients. In contrast, cyclic AMP fails to cause a distinct separation of the two subunits of kinase IIb in a sucrose density gradient although there is an overall reduction in the sedimentation coefficient of the enzyme from 7.2 S to 5.8 S. However, with protamine as the dissociating agent, a 4 S catalytic component of kinase IIb is obtained. Protamine causes the dissociation of the enzyme by complexing with the regulatory subunit (4,5). Studies with kinase I indicate that the interaction of the two dissimilar functional subunits is reversible (4). Furthermore, the regulatory subunit of kinase I can cross-react with the catalytic moiety of either IIa or IIb (4).

In addition to the soluble enzymes, rabbit erythrocytes also contain membrane bound cyclic AMP-dependent protein kinase activity. The rabbit erythrocyte membranes are prepared by the procedure of Dodge _et al_. (6) from freshly withdrawn blood cells. As shown in Table II, erythrocyte membranes catalyze cyclic AMP-stimulated

TABLE II

CYCLIC AMP-STIMULATED PHOSPHORYLATION OF ENDOGENOUS AND EXOGENOUS SUBSTRATES BY ERYTHROCYTE MEMBRANES[a]

	^{32}P Incorporated (pmoles)	
Addition	Control	+ Cyclic AMP
None	30	55
Histone	83	153

[a] The phosphorylation reactions were carried out as described previously (4) in the presence of 0.67 mg/ml of membrane proteins. As indicated, 1.8 mg/ml of histone was added. The amount of ^{32}P incorporated into histone has been corrected for membrane autophosphorylation.

phosphorylation of histone. Furthermore, autophosphorylation of the erythrocyte membranes occurs in the absence of added exogenous substrates.

Cyclic AMP Receptors Not Associated with Cyclic AMP-Dependent Protein Kinases

During the course of our studies on the role of cyclic AMP in red blood cells, we have observed the presence of two additional cyclic AMP receptors (I and II) other than those associated with cyclic AMP-dependent protein kinases. These receptors have been purified to homogeneity. Polyacrylamide gel electrophoresis of these receptors in the presence or absence of either sodium dodecyl sulfate (SDS) or 8 M urea reveals a single protein component in each instance. The molecular weights of these receptors have been ascertained by Sephadex G-200 gel filtration and by polyacrylamide gel electrophoresis employing various gel concentrations (7). Both receptors exhibit a molecular weight of approximately 240,000 daltons. In SDS-polyacrylamide gel, receptors I and II migrate as 48,000 daltons molecular species (Fig. 1).

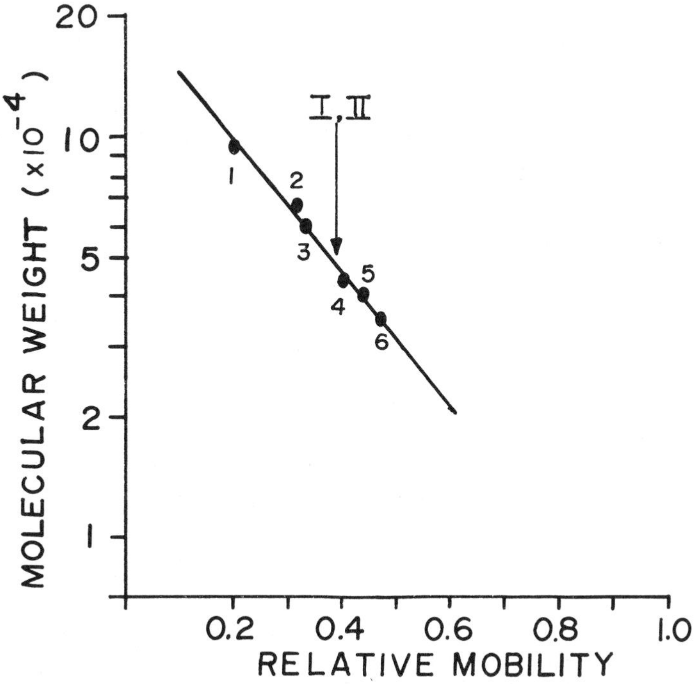

Fig. 1. Estimation of molecular weights of receptors I and II by SDS-polyacrylamide gel electrophoresis. SDS-polyacrylamide gel electrophoresis was carried out as described by Laemmli (8). Measurements and calculations were performed as described by Weber and Osborn (9). (1) phosphorylase _b_, (2) bovine serum albumin, (3) catalase, (4) ovalbumin, (5) creatine kinase, and (6) pepsin.

Neither receptor I nor II is related to or derived from cyclic AMP-dependent protein kinases. As shown in Table III, neither receptor has any effect on the phosphotransferase activity of the catalytic moiety of cyclic AMP-dependent protein kinase I, either in the presence or absence of cyclic AMP. Similar results are obtained with the kinases IIa and IIb. These experiments are performed under the same conditions as those where reversible interactions of the two dissimilar functional subunits of cyclic AMP-dependent protein kinases are demonstrated (4).

TABLE III

LACK OF REGULATION OF THE CATALYTIC SUBUNIT OF CYCLIC AMP-DEPENDENT PROTEIN KINASE I BY RECEPTORS I AND II[a]

Receptor Added	(μg)	^{32}P Incorporated (pmoles)	
		Control	+ Cyclic AMP
I	0	300	330
	12	330	300
	24	350	320
II	0	600	590
	26	500	510
	52	510	570

[a] Details regarding subunit isolation and the reaction conditions have been previously described (4). About 2.5 and 4.0 μg of the catalytic subunit of cyclic AMP-dependent protein kinase I were used when assaying receptor I and II, respectively.

Millipore filtration provides a rapid method for the assay of binding of cyclic AMP to receptors I and II. The efficiency of the binding assay is greatly enhanced by protamine (Fig. 2). A similar effect of protamine on the binding of cyclic AMP by cyclic AMP-dependent protein kinases on Millipore has been reported (4). Protamine does not affect the Km of the interaction of cyclic AMP with receptors I and II. Both receptors bind cyclic AMP with a Km of 3×10^{-7} M - a value obtained from binding studies employing either Sephadex G-50 gel filtration or Millipore filtration. In view of this, a possible interpretation of the effect of protamine on the binding of cyclic AMP is the formation of molecular aggregates between the cyclic AMP-receptor complex and protamine resulting in a greater retention on the membrane. Since the nature of the Millipore binding assay is complex, studies dealing with binding specificities are conducted using Sephadex G-50 gel filtration.

Figure 3 shows the elution profile of an incubation mixture containing cyclic [^{32}P] AMP and receptor I on a Sephadex G-50 column. In the presence of the receptor, a portion of the radioactivity is excluded by the gel and eluted at the void volume. This radioactive material has been identified by thin layer chromatography (10) to be cyclic AMP. A similar experiment with adenosine substituting for cyclic AMP indicates that receptor I may also have an affinity for this nucleoside (Fig. 3). The interaction of adenosine with receptor I has a Km value of 1 x 10^{-7}M. Furthermore, adenosine is found to interfere with the binding of cyclic AMP. As shown in Table IV, the binding of cyclic AMP is reduced by 60% in the presence of equimolar concentration of adenosine. Adenine at substantially higher concentrations is also inhibitory. Other compounds such as AMP, ADP, and ATP have no significant effect on the binding of cyclic AMP. Guanosine, cytidine, uridine, and their phosphorylated derivatives including cyclic GMP, cyclic CMP and cyclic UMP also do not interfere with the formation of the cyclic AMP-receptor complex.

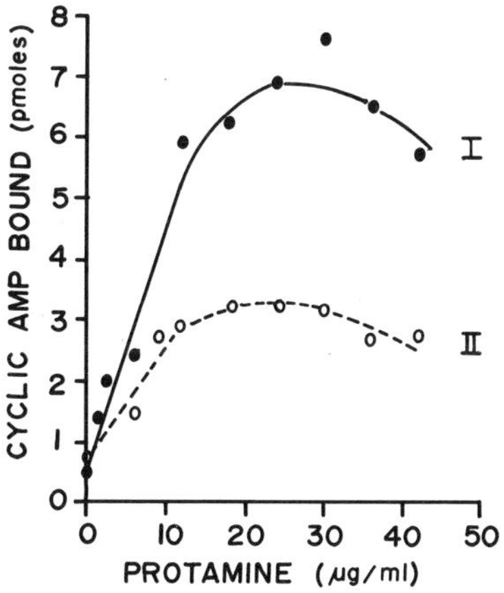

Fig. 2. Effect of protamine on the binding of cyclic AMP by receptors I and II. Cyclic AMP binding assay was carried out by Millipore filtration as described under "Methods" except in the presence of varying concentrations of protamine. The incubation mixture contained either 70 µg/ml of receptor I or 90 µg/ml of receptor II.

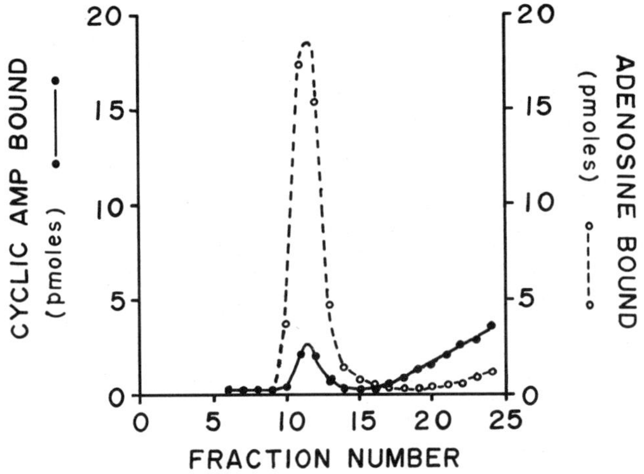

Fig. 3. Binding of cyclic AMP or adenosine to receptor I as determined by Sephadex G-50 gel filtration. Receptor I (0.12 mg/ml) was incubated with 1 µM of either cyclic [^{32}P]AMP or [^{3}H]adenosine, 1 mM of Tris-HCl, pH 7.5, and 10% sucrose. Other experimental details were as described under "Methods".

TABLE IV

EFFECT OF ADENINE AND ITS DERIVATIVES ON THE BINDING OF CYCLIC AMP BY RECEPTOR I[a]

Addition	Cyclic AMP Bound % of Control
None (control)	100
Adenine, 1.0 µM	86
10 µM	66
Adenosine, 1.0 µM	39
10 µM	6
AMP, 10 µM	90
ADP, 10 µM	98
ATP, 10 µM	92

[a] The binding of cyclic [^{32}P]AMP to receptor I (12 µg) was assayed using Sephadex G-50 gel filtration as described under "Methods" with or without the addition of adenine or its derivatives as indicated.

The interaction of receptor I with adenosine has been investigated in greater detail. The results presented in Table V indicate that among several compounds tested, only adenine can inhibit adenosine binding. Interestingly, cyclic AMP has no effect on the binding of adenosine by receptor I. The data suggest that receptor I may contain two types of binding sites: one specific for adenosine while the other binds cyclic AMP and possibly adenosine as well. Further evidence supporting the two binding sites concept is provided by the results in Table VI. The purpose of the experiment presented in Table VI is to determine whether adenosine can bind to receptor I when it is saturated with cyclic AMP. The receptor is incubated with varying amounts of [^{3}H]adenosine in the presence of a saturating concentration of cyclic [^{32}P]AMP. At low concentrations of adenosine, a considerable amount of adenosine is bound to receptor I with no appreciable decrease in cyclic AMP binding. We have established by polyacrylamide gel electrophoresis that both adenosine and cyclic AMP are bound to the same protein factor. This rules out the possibility that the binding of adenosine is due to contaminating proteins in the receptor preparation.

TABLE V

EFFECT OF ADENINE AND ITS DERIVATIVES ON THE BINDING OF ADENOSINE BY RECEPTOR I[a]

Addition	Adenosine Bound % of Control
None (control)	100
Adenine, 1.0 μM	52
10 μM	32
AMP, 10 μM	108
ADP, 10 μM	116
ATP, 10 μM	111
Cyclic AMP, 10 μM	98
0.1 mM	92

[a] The experimental details were as described under Table IV except that cyclic [^{32}P]AMP was replaced by 1.0 μM of [^{3}H]adenosine.

TABLE VI

BINDING OF ADENOSINE TO RECEPTOR I IN THE PRESENCE OF SATURATING AMOUNT OF CYCLIC AMP

$[^3H]$Adenosine (μM)	Cyclic AMP Bound (pmoles)	Adenosine Bound (pmoles)
0	5.3	0
0.10	4.8	6.9
0.25	4.7	21
0.50	3.2	36
0.75	2.8	40
1.00	1.6	43

[a] Receptor I (12 μg) was incubated with a saturating concentration (1 μM) of cyclic $[^{32}P]$AMP and varying amounts of $[^3H]$adenosine. Binding assay was carried out by Sephadex G-50 gel filtration as described under "Methods".

Discussion

The results of our studies clearly establish the presence of multiple cyclic AMP receptors in rabbit erythrocytes. Some of these receptors are associated with the cyclic AMP-dependent protein kinases and serve as regulators of these enzymes. However, at least two receptors (I and II) isolated from the red cells are unrelated to protein kinases. These receptors have unique binding properties. In addition to cyclic AMP, they also bind adenosine. Structural study indicates that both receptors are pentamers composed of monomeric units of 48,000 daltons each. Based on data obtained from polyacrylamide gel electrophoresis under various conditions, these monomeric units are probably identical. Of the two receptors, receptor I is found to constitute the major component and in some preparations of the red cells, receptor II appears to be absent. Whether this is an indication that the two receptors are interrelated remains to be determined.

The interactions of adenosine and cyclic AMP with receptor I have been investigated in greater detail. We have provided evidence to indicate that receptor I may contain two types of binding sites: one specific for adenosine and the other for cyclic AMP. The inhibition of cyclic AMP binding by adenosine further indicates that the nucleoside may also compete for the cyclic AMP-binding site. On the other hand, the inhibition may not be of a competitive type. It is possible that the saturation of the adenosine-binding site(s) by adenosine may alter the conformation of receptor I to a state of low or no affinity for cyclic AMP. Further kinetic analysis is underway to distinguish between these two possibilities.

Under the same conditions, receptor I binds considerably more adenosine than cyclic AMP (Fig. 3). The ratio of cyclic AMP to adenosine bound varies from 1/5 to 1/8. The variation probably reflects the impurity of the cyclic [^{32}P]AMP preparation used. If the cyclic AMP preparation is contaminated with adenosine, this could lead to a lower estimate of the amount of cyclic AMP bound due to inhibition by the nucleoside. However, it is unlikely that the impurity of the cyclic AMP preparation could account for all the difference in the amount of cyclic AMP and adenosine bound. The possibility that receptor I may contain more binding sites for adenosine than for cyclic AMP provides a further explanation for the discrepancy.

The function of these adenosine-cyclic AMP receptors in rabbit red blood cells is not known. It is interesting to note, however, that there is a relationship between adenosine and cyclic AMP metabolism. Recently, several laboratories have reported that adenosine can either inhibit or potentiate the effect of hormones on the adenylate cyclase system (11-14). Whether these receptors have a role in these processes remains to be determined. The availability of these homogeneous receptor preparations is an initial step towards understanding their functions.

Summary

Rabbit red blood cell lysates contain several protein factors capable of interacting with cyclic AMP. Three of these factors found in the soluble fraction are associated with the cyclic AMP-dependent protein kinases and serve as regulators of these enzymes. The sedimentation coefficients of these protein kinase regulators are 5 S, 3.4 S, and 5.8 S, corresponding to the three cyclic AMP-dependent protein kinases of molecular weights, 170,000, 120,000, and 240,000, respectively. Preliminary studies indicate that cyclic AMP-dependent protein kinase activity is also present in the particulate fraction of the lysate in association with the cell membrane.

Two additional cyclic AMP-binding protein factors unrelated to the cyclic AMP-dependent protein kinases have been isolated from rabbit red blood cells and purified to homogeneity. These factors have molecular weight of approximately 240,000 daltons and are pentamers of 48,000 daltons monomeric molecular weight. Preliminary binding studies provide evidence for the existence of at least two distinct binding sites: one specific for adenosine while the other binds cyclic AMP and possibly adenosine as well. The function of these non-kinase associated cyclic AMP-binding protein factors in red blood cell remains to be determined.

Acknowledgments

This work was supported in part by grants from the American Cancer Society (BC-65A) and the National Cancer Institute of the United States Public Health Service (1 R01 CA17036-01). Mariano Tao is an Established Investigator of the American Heart Association.

References

1. Tao, M., Salas, M. L., and Lipmann, F. (1970), Proc. Nat. Acad. Sci. U.S. 67, 408.
2. Gill, G. N., and Garren, L. D. (1970), Biochem. Biophys. Res. Commun. 39, 335.
3. de Crombrugghe, B., Chen, B., Anderson, W., Nissley, P., Gottesman, M., Pastan, I., and Perlman, R. (1971), Nature New Biology 231, 139.
4. Tao, M., and Hackett, P. (1973), J. Biol. Chem. 248, 5324.
5. Tao, M. (1972), Biochem. Biophys. Res. Commun. 46, 56.
6. Dodge, J. T. Mitchell, C., and Hanahan, D. T. (1963) Arch. Biochem. Biophys. 100, 119.
7. Hedrick, J. L., and Smith, A. J. (1968), Arch. Biochem. Biophys. 126, 155.
8. Laemmli, U. K. (1970), Nature 227, 680.
9. Weber, K., and Osborn, M. (1969), J. Biol. Chem. 244, 4406.
10. Tao, M., and Lipmann, F. (1969), Proc. Nat. Acad. Sci. U.S. 63, 86.
11. Sattin, A., and Rall, T. W. (1970), Mol. Pharmacol. 6, 13.
12. Fain, J. N., Pointer, R. H., and Ward, W. F. (1972) J. Biol. Chem. 247, 6866.
13. Schultz, J., and Daly, J. W. (1973), J. Neurochem. 21, 1313.
14. Schwabe, H., Ebert, R., and Erbler, H. C. (1973), Naunyn-Schmiedeberg's Arch. Pharmacol. 276, 133.

On the Linkage and Recombination of Mitochondrial Genes

K. Wakabayashi

Department of Biochemistry, Faculty of Medicine, University of Tokyo, Tokyo, Japan

INTRODUCTION

Extrachromosomal oligomycin resistance in yeast was first reported from this laboratory (1). The plasmagene for oligomycin resistance in mutant, 706R1, was demonstrated to give rise to a change of the sensitivity of the mitochondrial ATPase (2), and it was demonstrated to be a mitochondrial gene. The cross of this mutant with a mitochondrial erythromycin resistant mutant revealed that the recombination between mitochondrial genes took place (3). Bolotin et al. reported that asymmetrical recombination occurred between erythromycin resistance and chloramphencicol resistance (4). Thus the recombination of these three mitochondrial genes was studied (5, 6). It has been reported that the mitochondrial system shares the common aspects with the bacterial system. The genetic study also showed the similar features between mitochondria of yeast and cells of *Escherichia coli*. These are reported and discussed in this paper.

MATERIALS AND METHODS

Media: Yeast extract-peptone-glucose media (YEPD) consisted of 0.35% yeast extract, 0.35% peptone, 0.2% KH_2PO_4, 0.1% $(NH_4)_2SO_4$, 0.1% $MgSO_4$ and 0.1% glucose. Instead of glucose, 2% glycerol was used for a nonfermentable carbon source (medium Y). For the solid media 2.5% agar was used. In the case of minimal media, 0.67% yeast nitrogen base without amino acid (Difco) was replaced with yeast extract and peptone in YEPD media. To assay the oligomycin resistance, 80 μg of oligomycin in 0.2 ml ethanol was spread on the surface of the solid media Y (media YO). To assay the erythromycin or chloramphenicol resistance, erythromycin or chloramphenicol in ethanol solution (60 mg/ml) was added to media Y prior to the solidification to make the final concentration of each drug 3 mg/ml (medium YE or medium YC). To test for the double resistance, media Y containing erythromycin and chloramphenicol each 3 mg/ml (medium YCE) were used. Eighty micrograms of oligomycin was spread on the surface of the solid media YE or YC (medium YEO or medium YCO).

Strains and genetic symbols: Cytoplasmically inherited markers are enclosed in brackets. Oligomycin-resistance, erythromycin-resistance and chloramphenicol-resistance are denoted as (oli), (ery), (cap). These drug resistances can be removed by ethidium bromide treatment. Type I and type II were used for two types of mitochondria according to the previous paper (6). Strains used are listed in Table 1.

Crosses: About 5 x 10^6 cells of opposite mating types, preincubated in YEPD media separately, were mixed and the mixture was incubated for 3.5 hours at 30°C for mating. A portion of the cells was then withdrawn and plated on the minimal media after suitable dilutions. The rest of the cells in the mating mixture were washed with saline, and a portion of the cells was further grown in the minimal media. The cells were transferred to the fresh minimal media 24 hours and again 48 hours after the onset of the mating. The cells were then plated at 72 hours.

Test of drug resistances: Cells in a colony to be tested were suspended in a saline solution and from 10^4 to 10^5 cells were spotted on plates Y, YE, YO, and YEO. The spot method has more advantages than the replica method, especially in detecting the sensitive cells. To test a large number of colonies in three-point crosses, the replica method was also used. Care was taken that not too many cells were plated. The diploids grown on the minimal media were transferred to minimal media in regular pattern and tested for the resistances on plates with antibiotics.

TABLE 1: Genotype and origin of haploid strains

Strain	Genotype		References
706R1 (ATCC)	a his leu thr	(oli_{706})	(2)
102E	α ade	(ery_{102})	(2)
R8	a ade	(oli_{706} ery_{102})	(3)
909A	a ade leu	(oli_{706} ery_{102})	(7)
909B	α thr	(oli_{706} ery_{102})	(7)
903A	α thr	(oli_{706} ery_{102})	
711	α thr	(oli_{706})	(6)
S22	α thr		
1586-2A	a leu met lys his		from S. Nakai
1586-2B	α leu met lys his		from S. Nakai
Y1	a leu thr	(oli_{11})	(6)
1L1A	α ura	(ery_{514} cap_{321})	(6)
1L1D	a thr	(ery_{514} cap_{321})	
D2B	a ura		(7)
55R52C/321	a ura		P.P. Slonimski
3940-1D	α arg his		T. Takahashi

To test the resistances in rho minus cells, the cells were crossed with rho-plus sensitive cells.

Tetrad analysis: Diploids were grown in the presporulation medium (8) and then transferred to sporulation medium (9). Tetrads were dissected by the method of Johnston and Mortimer (10).

RESULTS

Linkage of mitochondrial genes: An oligomycin resistant mutant 706R1 was crossed with an erythromycin resistant mutant 102E and the resulting zygotes were then spread on minimal media after 3.5 hours. Almost all of the zygotes grew well on plates YEO, but each colony possessed heterogenous population of resistant and sensitive cells. This indicated that the original zygotes were heteroplasmic cells, from which the isoplasmic cells segregated. When double resistant diploids were isolated after

TABLE 2: Tetrad analysis of diploids from a cross 102E x 706R1

	oli R : S	ery R : S	a/α	leu	thr
2:2	0	0	12	12	12
3:1	0	0	0	0	0
4:0	12	12	0	0	0

plating double resistant cells on plates YEO, and the progeny was examined, no segregation of single resistant or sensitive cells occurred. Tetrad analysis of the double resistant diploids showed no segregation of single resistant or sensitive cells (Table 2). One of the double resistant haploids, R8, thus obtained was again tested for the segregation of single resistant cells in the progeny. No segregation was observed on 1476 colonies tested. The absence of segregation of two mitochondrial genes during mitosis and meiosis showed the close linkage of these genes.

Double sensitive diploids formed were not caused by the elimination of these mitochondrial genes from the zygotes. The double sensitive diploids were sporulated and the asci were dissected to isolate double sensitive haploids. A haploid S22 was thus obtained and S22ET was obtained from S22 with ethidium bromide treatment. These were crossed with the double resistant haploid R8. Table 3 showed that the oligomycin sensitive diploids and erythromycin sensitive diploids segregated only in the former cross (72 hours cross, 2nd and 3rd lines). These differences

TABLE 3: Segregation of oligomycin resistance and erythromycin resistance

		0	E	0+E	0E	SS
3.5 hr	R8 x 3940-1D	20	0	4	76	0
	R8 x S22	0	0	0	96	3
24 hr	R8 x 3940-1D	20	4	1	29	55
	R8 x S22	14	11	11	64	4
72 hr	R8 x 3940-1D	9	5	0	45	52
	R8 x S22	13	16	0	41	35
	R8 x S22ET	0	0	0	98	0

indicated that the haploid S22 possessed mitochondrial genes of oligomycin-sensitiveness and erythromycin sensitiveness but S22ET was devoid of them. Thus the formation of the double sensitive diploids can be explained by the recombination of the two mitochondrial genes, (ery^R oli^S) and (ery^S oli^R).

Isoplasmic crosses: The cells of type 0+E grew well on the plates with either one of the drugs but showed absence of growth on YEO plates. These cells seemed to harbor a mixture of two resistances that did not complement with each other. The double-resistant R8 was crossed with double-sensitive haploid 3940-1D or S22. The cells of type 0+E were formed at higher rates in the cross R8 x S22 (Table 3)

The number of sensitive deploids was very small in this cross at 24 hours. It was due to the mixture of double resistant and sensitive mitochondria in the diploids, scored as double resistant diploids. Such incomplete segregation was not found after 72 hours. The cross between R8 x 3940-1D gave equal segregation of resistant and sensitive diploids after 24 hours. S22 and R8 were isoplasmic and the isoplasmic crosses gave slow segregation of

TABLE 4: The cross between mitochondria of heterologous types and homologous types

	No. of rec.	Rec. val.	No. of rec.	Rec. val.
	909B x D2B		909B x 55R53C/321	
oli^R cap^R	70		21	
oli^S cap^S	18	35.5	12	13.3
oli^R ery^S	37		21	
oli^S ery^R	25	25.0	14	14.1
ery^R cap^R	43		5	
ery^S cap^S	3	18.5	4	3.6
	706R1 x 1L1A		Y1 x 1L1A	
oli^R cap^R	75		9	
oli^S cap^S	5	26.9	10	12.1
oli^R ery^R	37		13	
oli^S ery^S	16	17.8	7	14.0
ery^S cap^R	50		3	
ery^R cap^S	1	17.1	12	9.5

Total number of colonies examined were: 248 in 909B x D2B, 248 in 909B x 55R53C/321, 297 in 706R1 x 1L1A, 158 in Y1 x 1L1A.

No. of Rec.: Number of recombinants,
Rec. val. : Recombination values

both mitochondria. The segregation of recombinants occurred in a shorter time in heteroplasmic crosses.

Recombination values: Oligomycin-and erythromycin-resistant haploids, which were obtained by tetrad analysis in Table 2 were crossed with chloramphenicol resistant haploids. It was found that the recombination values for (oli) (cap) were larger than the other recombination values, although recombination values deviate depending on the crosses, probably due to the multiple number of mitochondrial genes and also to the effect of the chromosomal genes. Two crosses were shown in which the sum of the recombination values for (oli) (ery) and (ery) (cap) was close to the value for (oli) (cap) (Table 4) and it was suggested that linear arrangement of mitochondrial genes is in the order of (oli) (ery) (cap). Two erythromycin resistances used in these experiments, (ery_{102}) and (ery_{514}), did not seem to make a significant difference in terms of recombination. The cells of 909B and 706R1 were isoplasmic and possessed mitochondrial gene type II. Cells of D2B and 1L1A were also isoplasmic and possessed mitochondrial gene of type I. The cross of the cells of the same types (909B x 55R53C/321, type II xtype II; Y1 x 1L1A, type I x type I) gave different results. These crosses showed that recombination values for (ery)-(cap) were less than those in crosses of type II x type I.

Recombinants: Analysis of the recombinants on three crosses are presented in Table 5. The predominant recombinants were of +-- and ++- type. A sign + is used in this paper for the type of resistance in type II mitochondria. Very few recombinants of type --+ or +-+ were found. Since the arrangement of mitochondri-

TABLE 5: Recombinants of mitochondrial crosses

Types of recombinants			903A x D2B		706R1 x 1L1A		706R1 x 1L1D	
(oli)	(ery)	(cap)	No. of col.	%	No. of col.	%	No. of col.	%
+	+	-	17	10.8	55	18.4	20	15.4
-	+	+	2	1.3	4	1.3	0	0
+	+	-	19	12.4	38	12.8	9	6.9
-	+	-	9	5.7	12	4.0	1	0
+	-	+	1	0.6	0	0	0	0.4
-	-	+	0	0	1	0.3	0	0
+	-	-	27	17.1	37	12.5	10	7.7
-	-	-	83	52.5	150	50.5	82	63.1

No. of col.: Number of colonies

al genes was taken to be in the order of (oli) (ery) (cap), the data indicated that the recombination occurred between the genes of (oli + ery +) and (oli - ery - cap -). (cap +) gene did not seem to be involved in recombination. It was shown in the previous paper that oligomycin resistant recombinants were formed after 3.5. hr in a cross of double resistant cells and double sensitive cells R8 x 1586-2B (3). Thus two crosses were studied on the recombination between (oli) and (ery) after 3.5 hours. A cross 909B x 1586-2A gave oligomycin resistant diploids as in the cross R8 x 3940-1D however, a cross 909A x 1586-2B gave significant numbers of oligomycin resistant diploids (Table 6).

TABLE 6: Transfer of oligomycin resistance in the zygotes

	O	E	O+E	OE	SS
909B x 1586-2A	2	0	1	91	6
909A x 1586-2B	32	3	5	32	28

The decreased number of double resistant diploid was due to the inhibition of the multiplication of mitochondria of type II by the effect of nuclear gene(s) present in 909A (7). The oligomycin resistance, which was transferred to type I mitochondria in the early stage of the conjugation, continued to multiply after the mitochondria of type II disappeared. The recombination occurred in this cross predominantly between (oli +) and (oli - ery -). This polar recombination of oligomycin resistance, under the restriction of mitochondria of type II, indicated that the transmission of oligomycin resistance occurred at first. The transmission of erythromycin resistance occured subsequently in the cross 909B x 1586-2A and it gave rise to the high number of diploids of double resistant type. These data are in agreement with the sequential transfer of mitochondrial genes in the order of (oli) (ery) (cap), from type II mitochondria to type I mitochondria. The most distal gene (cap +) was, however, transferred rarely.

DISCUSSION

The studies of the recombination of mitochondrial genes showed the linear linkage of genes in the order of (oli) (ery) (cap). Further evidence for this order was obtained by kinetic studies on the conjugation of mitochondria of type II and type I. When an oligomycin- and erythromycin-resistant mutant was crossed with a chloramphenicol-resistant mutant, the transfer of the oligomycin resistance started at 2 hours after the onset of the mating of the cells, while erythromycin-resistance was not yet transferred (Wakabayashi, unpublished observation).
The recombination values deviated depending on crosses. These changes are expected in view of the multiple number of mitochondrial genomes. The presence of type O+E indicated that the presence of several copies of mitochondrial genes within a cell. The number of the mitochondrial genes was the subject of controversal data. In spite of multiple number of the mitochondrial genes, the conjugation and recombination of mitochondrial genes occured in a coordinated way. The rapid conjugation of mitochondria might explain the coordinated recombination of mitochondrial genes. It was reported recently that one mitochondrion exists in a yeast cell (11). The multiple number of mitochonrial genes needs to exist in one mitochondrion and the coordinated recombination might be explained in connection with this.
The number of recombinants formed was also effected by the action of nuclear gene(s) (7). The mitochondria of type II disappeared in the presence of nuclear gene(s) and the restriction of type II mitochondria allowed the transfer of mitochondrial genes only at the early stage of mating. In these crosses the transfer of proximal gene, oligomycin resistance occurred most frequently. When type II mitochondria continued to multiply, both recombinants of (oli)-(ery) type can be formed. Absence of (ery - cap +) recombinants was explained by that the transfer of (cap +) gene was

inhibited. The direction of the transfer of mitochondrial genes shown in this paper was opposite to the proposal by Bolotin et al. (4). The formation of recombinants of (oli + ery -) and (oli - ery +) types could not be explained by the transfer of mitochondrial genes in the order of (cap)-(ery)-(oli). Since the mitochondrial genes were transfered from type II mitochondria to type I mitochondria, type II corresponds to the donor type and type I corresponds to the recipient type. In case of *Escherichia coli*, the donor type and the recipient type were denoted as F^+ and F^-. A sign ω^- was adopted by Bolotin et al. (4) for mitochondrial donor type, but it was not used here to avoid the confusion in the usage of + and - sign of F and of the mitochondrial type.
Two erythromycin resistance markers were used in these experiments. It is not certain whether these two erythromycin resistances were located at the same site. The recombination of two resistant mutants did not give sensitive recombinants. These were in contrast to the presence of two distinct loci of oligomycin resistances.

SUMMARY

The mitochondrial drug resistances are linked. Yeast cells of mitochondrial type II and of type I were crossed and the recombination between three mitochondrial genes, oligomycin resistance, erythromycin resistance, and chloramphenicol resistance, was studied. It revealed that the mitochondrial genes were sequentially transfered from the mitochondria of type II to the mitochondria of type I, in the order of (oli)-(ery)-(cap). The transfer of (cap) gene occurred at a low rate.

ACKNOWLEDGEMENTS

The author wishes to thank Dr. P.P. Slonimski for the donation of the strain 55R53C/321, Dr. S. Nakai for the strains 1586-2A and 1586-2B, and Dr. T.takahashi for the strain 3940-1D.

REFERENCES

(1) Wakabayashi, K. and Gunge, N., (1970) FEBS Letters 6, 302.
(2) Wakabayashi, K., (1972) J. Antibiotics 25, 475.
(3) Wakabayashi, K. and Kamei, S., (1973) FEBS Letters 33, 263.
(4) Bolotin, M., Coen D., Deutsh, J., Dujon, B., Netter, P., Petrochilo, E., and Slonismski, P.P., (1971) Ann. Inst. Pasteur 69, 215.
(5) Avner, P.R., Coen, D., Dujon, B., and Slonimski, P.P., (1973) Molec. gen. Genet. 125, 9.
(6) Wakabayashi, K., (1974) J. Antibiotics 27, 373.
(7) Wakabayashi, K., (1974) Proc. Japan Acad. 50, 396.
(8) Miller, G.R., McClary, D.O., and Bowers Jr., W.O., (1962) J.Bacteriol. 85, 725.
(9) McClary, D.O., Nulty, W.L., and Miller, G.R.,(1959) J. Bacteriol. 78, 362.
(10) Johnston, J.R., and Mortimer, R.K., (1959) J. Bacteriol. 78, 292.
(11) Hoffmann, H.P., and Avers, C.J., (1973) Science 181, 749.

Determination of mRNA and 28 S RNA Turnover in Proliferating HeLa Cells

Ulrich Wiegers, Gisela Kramer, Karin Klapproth,
Uta Wiegers and Helmuth Hilz
Institut für Physiologische Chemie der Universität Hamburg,
Germany, 2 Hamburg 20, Martinistraße 52

Introduction

mRNA turnover in mammalian cells has been determined by two principally different methods: a) decay of polysomes in the presence of actinomycin (1, 2), b) analysis of (^{3}H) uridine incorporation into mRNA isolated (3-6) on the basis of its poly(A) content (binding to cellulose or to poly U/ poly dT columns) (7-10).
Both methods have disadvantages. In the case of actinomycin, serious side effects of the antibiotic render these values unvalid (cf. 11-16). The use of specific binding of mRNA via its poly(A) sequences certainly improved the analysis as it allowed determination of mRNA synthesis without the use of inhibitors. However, mRNA separation based on its poly(A) content is not completely specific (10, 15). In none of these approaches, mRNA specific radioactivity has been determined directly, i.e. analysis of cpm and A_{260} of the separated mRNA. Accordingly, mRNA half-lives obtained with this new approach still vary between 0.35 and 2-3 times cell doubling time even for the same cell line (cf. 14, 17).
This paper describes a completely different approach to the determination of mRNA specific radioactivity independent of its poly(A) content, and of 28 S rRNA specific radioactivity without interference by mRNA. Kinetic analysis of the polysomal RNA labeling by exogenous (^{3}H)uridine not only allowed determination of half-lives but also showed the existence of two separate pyrimidine nucleotide precursor pools for mRNA and rRNA, respectively.

Material and Methods

(^{3}H)uridine (uridine-5-T; spec. act. 25 Ci/mmole) was purchased from Buchler-Amersham, Braunschweig;

modified joklik medium (F-13) from Grand Island Biological Comp., USA; cofactors and enzymes from E. Merck (Damrstadt) and from Boehringer (Mannheim). NP 40 was obtained from Shell Comp.; Proteinase K (chromatographically pure) was a generous gift from Dr. H. Lang (E. Merck, Darmstadt).

HeLa S3 cells were grown in suspension cultures as described previously (18). Cell doubling time was 19 - 25 hours.

UTP specific radioactivity was determined in the acid soluble fraction by the method of enzymic displacement (conversion to UDPG) as described previously (19).

Analysis of the base labeling in RNA. Bases of the acid insoluble residue were analyzed for labeling by electrophoresis after alkaline hydrolysis as described previously (20).

(^{3}H)uridine in the medium was analyzed as described previously (20).

Determination of specific radioactivity of polysome-associated mRNA was performed as described previously (21); Polysomes (6 A_{260} x units/ml were incubated at 0° with pancreatic ribonuclease (0.2 µg/ml) for 0, 20, 40 min in duplicates, and precipitated with $HClO_4$. A_{260} and radioactivity of the acid soluble mRNA split products were determined in the supernatant. Values were corrected for base composition and for quenching. Specific radioactivity of intraribosomal mRNA strands was determined as described previously (21) .

Specific radioactivity of 28 S ribosomal RNA was determined as described previously (16): An aliquot of the ribonuclease treated (160 min) polysomes was incubated with proteinase K and 0.1 % SDS in order to inactivate ribonuclease and to isolate the RNA, which was precipitated by the addition of 2.5 vol of ethanol, with standing at -20° overnight. 28 S ribosomal RNA was separated from other RNA species by sucrose gradient centrifugation (5% - 20% w/v in TNE buffer (22); Spinco SW 40 rotor for 18 hours at 25000 rpm and 4°). Fractions of 0.5 ml were collected and specific radioactivity of the isolated fractions determined after alkaline hydrolysis. Values of 28 S RNA were corrected for quenching and for base composition (20).

CsCl density gradient centrifugation Polysomes were fixed with 2.7% formaldehyde and analyzed in CsCl gradients as described previously (21) according to conventional methods (23, 24).

Results and Discussion

1. Analytical Procedures for Selective Determination of mRNA and 28 S RNA

Selective analysis of interribosomal mRNA sections in RNP-free polysomes by controlled ribonuclease digestion. Isolation of HeLa polysomes free of RNP particles of nuclear or cytoplasmic origin was possible when cells were broken in the presence of the non-ionic detergent NP 40, and the polysomes isolated from the post-mitochondrial supernatant by centrifugation through 1.85 M sucrose (21). Analysis of such polysomes from pulse-labeled cultures ((^{3}H)uridine, 75 min) in CsCl density gradients (fig. 1) revealed a single peak (A_{260} and cpm) indicative of HeLa polysomes (21), and the absence of RNP material banding in the region of ϱ = 1.35 - 1.45 (cf. 24).

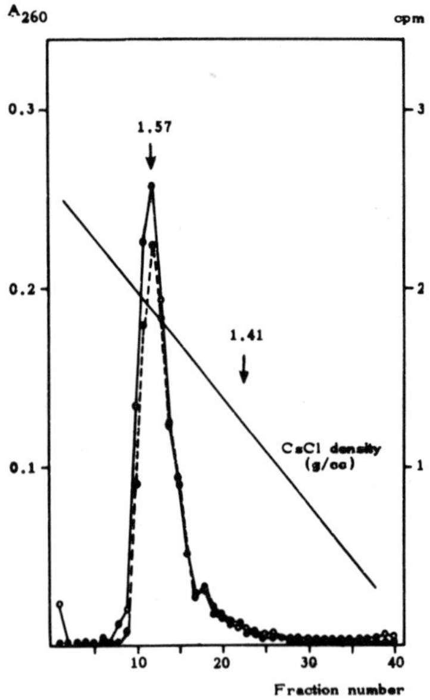

Figure 1: CsCl density gradient analysis of isolated polysomes. - Polysomes from HeLa cells labeled for 75 min with (^{3}H)uridine were isolated and analyzed as described under methods.

o——o A_{260}, •- - - -• cpm x 10^{-2}

Treatment of defined amounts of isolated polysomes with low concentrations of pancreatic ribonuclease (0.2 μg/ml) at 0° selectively hydrolyzed interribosomal mRNA sections to acid-soluble split products (21, 25).

Incubation with ribonuclease	Interribosomal mRNA (dpm/A_{260} of split products)	Intraribosomal mRNA (dpm/A_{260} in gel fractions)
0 - 20 min	115 000	-
20 - 40 min	117 000	114 500

Table 1 : Comparison of specific radioactivities of interribosomal and intraribosomal mRNA produced by ribonuclease treatment at 0°. - Polysomes (6 A_{260} units/ml) prepared from HeLa cells labeled for 75 min with (^{3}H) uridine were incubated with 0.2 μg/ml ribonuclease at 0°. At the times indicated samples were taken for analysis as described under methods and in (21).

The specificity of this procedure could be verified by two observations:
- The kinetics of the reaction showed continuous liberation of acid soluble material the specific radioactivity of which remained constant over the whole reaction period of 80 min ((16) and (table 1)).
- The intraribosomal mRNA sections (50 - 60 nucleotides) protected from the added ribonuclease by the ribosomes had exactly the same specific radioactivity as the interribosomal mRNA split products (table 1). As analysis of intraribosomal mRNA involved completely different purification steps, interference of unrelated RNA species appears very unlikely.

Determination of 28 S rRNA without interference of mRNA. During digestion of interribosomal mRNA with low ribonuclease concentrations at 0°, polysomes were largely degraded to monosomes. Remaining mRNA chains, therefore, were of small size (< 10 S) which could clearly be separated from 28 S RNA by sucrose gradient centrifugation (fig. 2 and (16)).

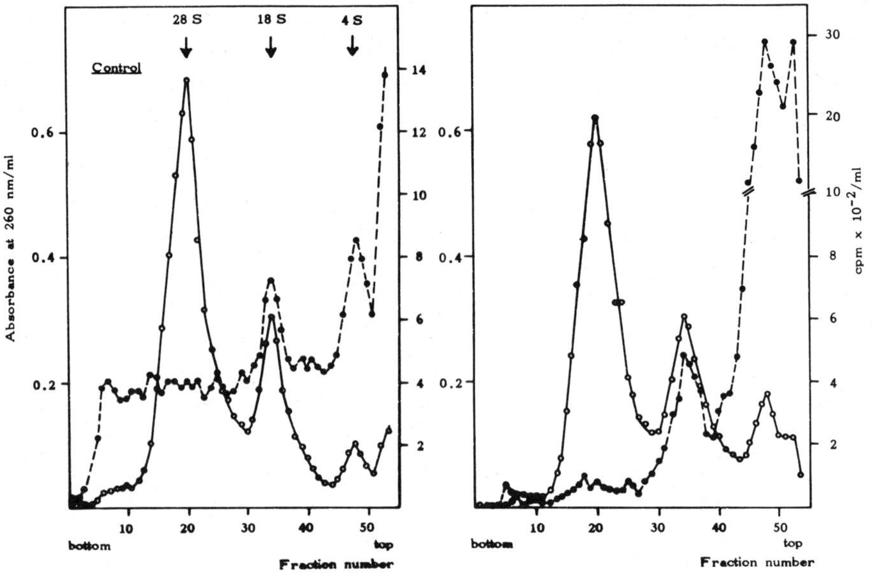

Figure 2: Sucrose gradient analysis of RNA species after ribonuclease treatment of polysomes. - Polysomes (6 A_{260} units/ml) prepared from HeLa cells labeled for 75 min with (^{3}H)uridine were incubated with 0.2 µg ribonuclease/ml at 0° for 160 min. The mixture was then incubated with proteinase K and processed as described under methods.

Use of proteinase K for the isolation of polysomal RNA in the presence of added ribonuclease.

The isolation of the RNA species after controlled degradation of polysomes by added ribonuclease posed special problems. Neither phenol extraction nor diethylpyrocarbonate could prevent degradation of RNA during and after removal of ribosomal proteins (26, 22). Also, SDS did not inactivate ribonuclease irreversibly thus leading to complete degradation of remaining interribosomal mRNA and rRNA subsequent to ethanol precipitation of RNA which also removed SDS (27). The only way to preserve the remaining RNA after controlled degradation of polysomes was subsequent inactivation of ribonuclease by proteinase K (16, 26, 27). This proteolytic enzyme has a high affinity towards ribonuclease and allows instant inactivation of the nuclease. As it is stimulated by SDS, combination of both agents (proteinase K + SDS) is highly effective in destruction of ribonuclease concomitant with the digestion of ribosomal proteins (27).

2. Analysis of mRNA and rRNA Half-Life in Proliferating Cells

(^{3}H)uridine labeling kinetics.

The determination of mRNA half-life by an analysis of the labeling kinetics with (^{3}H)uridine requires a constant supply of the precursor. This is not easily maintained over several cell doubling times when precursor concentration is below 10^{-4} M. Even at this value (^{3}H)uridine had to be replaced continuously or discontinuously in order to retain the concentration within tolerable limits. Under these conditions UTP specific radioactivity approached (^{3}H)uridine specific radioactivity only at exogenous uridine concent-

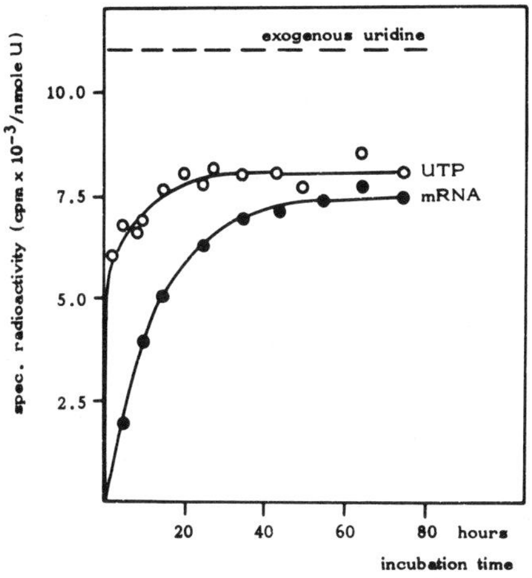

Figure 3: Labeling kinetics of HeLa suspension culture with (^{3}H)uridine (1×10^{-5} M). - HeLa cells (0.4×10^6/ml) were grown in 4 l suspension culture in the presence of (^{3}H)uridine (1×10^{-5} M, 1.1×10^4 cpm/nmole). (^{3}H)uridine was added every 5 hours to maintain uridine concentration within a limited range. In addition, aliquots of the cell suspension taken for analysis were replaced by fresh medium containing (^{3}H)uridine of the same concentration and specific radioactivity as described above. For details of UTP and mRNA determination see under methods.

rations $> 10^{-4}$ M. At lower values, de novo synthesis of UTP was only partially suppressed (20), leading to correspondingly lower steady state specific radioactivities of UTP. While UTP approached steady state values within 4 - 8 hours, mRNA attained this level only after a rather long period (fig. 3). From these data, mRNA half-life could be computed using Greenberg's formula (3), resulting in T/2 = 16.5 hours, which is 0.87 x the cell doubling time.

When the same formula was applied to the kinetics of 28 S RNA, a T/2 > 1000 hours was obtained indicating complete metabolic stability of ribosomal RNA under log growth conditions.

Decay experiments in prelabeled cultures.
In actively proliferating prelabeled cultures (log growth) metabolically stable cell constituents like ribosomal RNA will exhibit theoretical decay rates of their specific radioactivities identical with the rate of cell proliferation: as doubling of cells and doubling of their constituents with synthesis from (non-labeled) precursors will proceed at the same rate, specific radioactivity of a stable component will be decreased to one half after one generation time. All labeled cell components with a true turnover, then, must die away at a faster rate. These theoretical relations will be complicated when reutilization of labeled degradation products takes place. Failure to consider these factors will give rise to erroneous interpretations (e.g. 17).

When HeLa cultures prelabeled for 4 generation times with a constant supply of (^{3}H)uridine were transferred to non-labeled medium, decay rates for 28 S RNA and for UTP[1] indicates reutilization (fig. 4), while mRNA dies away at a rate slightly faster than proliferation indicating true turnover.

To obtain the half life-from decay curves in proliferating cultures, corrections for 'growth' (proliferation) and for 'reutilization' must be introduced. Correction for growth

---

1) Without reutilization of breakdown products, UTP should exhibit a half-life of ~ 1/20 of the cell doubling time as its concentration per cell is sufficient to synthesize about 1/10 of total cellular RNA-U (19)

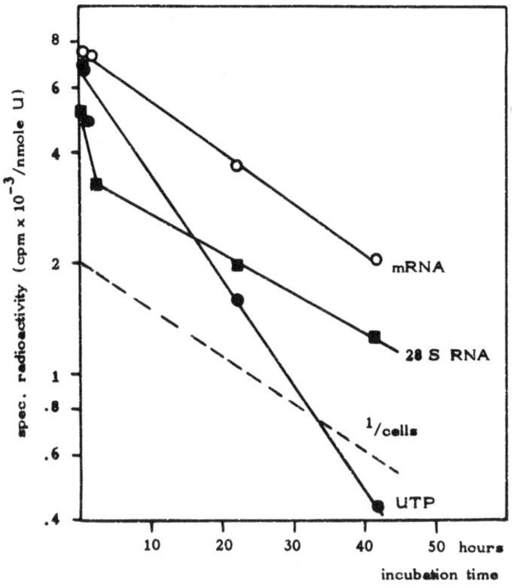

Figure 4: Decay analysis of mRNA, 28 S RNA and UTP in prelabeled cultures. - HeLa suspension cultures were prelabeled for 80 hours with constant supply of (^{3}H)uridine (1×10^{-5} M) as described in Fig. 3, legend. The cells were centrifuged off and resuspended in fresh medium without uridine. At the times indicated aliquots were taken for UTP, mRNA and 28 S RNA analysis, and replaced by fresh medium. Zero time values were taken at the end of the prelabeling period, before centrifugation of the cells. For further details see 'methods'.

can be made graphically by bringing cell doubling rate (= apparent decay rate of metabolically stable constituents) to ∞ (= infinitely slow decay of stable constituents), and correcting the rates of unstable constituents by the same factor. 'Reutilization' can be taken into account by correcting mRNA decay rate by the same factor needed to correct apparent UTP decay to an infinitely rapid rate (= zero cpm in UTP at all times after transfer of culture to non-labeled medium). When these corrections were performed, mRNA half-life was estimated to be 19 hours. At a mean generation time of 24 hours for the culture, the value corresponded to 0.79 x the cell doubling time which is in close agreement with the T/2 = 0.87 x the cell doubling time obtained by the labeling kinetics method.

3. Two Different Precursor Pools for mRNA and rRNA as Revealed by Labeling Kinetics

When HeLa cultures exposed to a constant supply of (^{3}H) uridine (1×10^{-5} M) were analyzed with regard to the approach of steady state specific radioactivities of various RNA fractions, it readily appeared that 28 S RNA-U did not follow the general UTP pool - in contrast to mRNA-U. It leveled off at a distinctly lower value and remained there even after four cell doublings. Furthermore, tRNA-U exhibited labeling kinetics very close to the ribosomal RNA species. This somewhat unexpected behavior was confirmed when labeling of total RNA-U was determined. Again steady state values were approached which were slightly higher than this low level of 28 S RNA. This is not surprising as ribosomal RNA represents the bulk of the cellular RNA. It is interesting to note that the differences in steady state specific radioactivity levels are correlated with the different localizations of transcription: While mRNA is transcribed in nucleus, ribosomal RNA and probably tRNA are synthesized in the nucleolus. There must exist a separate (pyrimidine) nucleotide pool for nucleolar RNA synthesis not in rapid equilibrium with the general nucleotide pool, which is only partially 'fed' by this general pool. Most nucleotides, then, must be provided by local de novo synthesis and used directly for nucleolar RNA formation.

Two observations support this interpretation:

a) (^{3}H)uridine labeled in the 5 position was taken up by the cells and rapidly converted to UTP. However, it also appeared in CTP which is in delayed equilibrium with UTP. As CTP/CDP is also a source of dCTP, label was to be expected in DNA as well. This could indeed be shown: Analysis of DNA bases and their labeling kinetics under exogenous (^{3}H)uridine showed exclusive labeling in DNA-C (29) [2]. Interestingly, DNA-C specific radioactivity approached the same steady state value as mRNA and the general UTP pool again pointing to a function of this pool for nucleic acid synthesis in the extra-nucleolar space.

---

2) The wide-spread opinion that 5-T-uridine provides a specific label for RNA formation because of its inability to give rise to labeled thymidine nucleotides is only valid for short-term ($<$ 60 min labeling) experiments (29).

b) If the provision of the nucleolar RNA synthesis with nucleotides comes about by a system different from the system responsible for the bulk of the cellular UTP (and CTP), if furthermore most of its nucleotides are formed by de novo synthesis as outlined above, this system should respond to alterations in the concentration of exogenous uridine to a lesser degree than the bulk UTP. This prediction could indeed be verified (fig. 5). Total UTP and

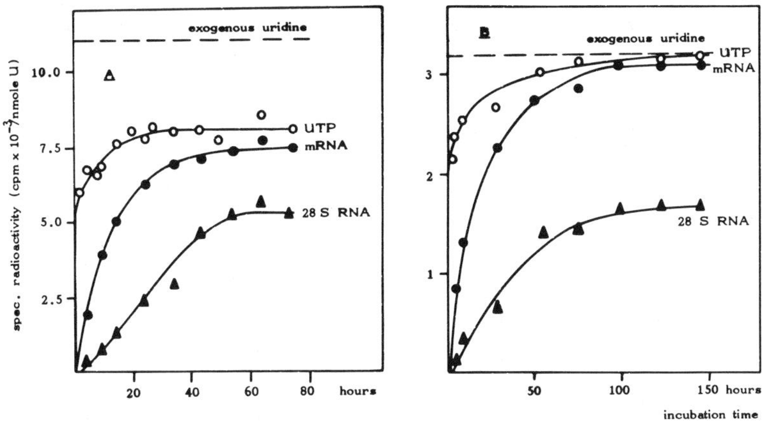

Figure 5: Labeling kinetics of UTP, mRNA and 28 S RNA at 1×10^{-5} M and 2×10^{-4} M (^{3}H)uridine. - HeLa suspension cultures (4×10^5c./ml) were labeled at 1×10^{-5} M and 2×10^{-4} M (^{3}H)uridine. Labeling was performed as described in Figure 3, legend. For experimental details see 'methods'.

mRNA specific radioactivities approached steady state levels 50% below the values of exogenous (^{3}H)uridine, when its concentration was 1×10^{-5} M (fig. 5A). At 2×10^{-4} M, the steady state values became practically identical with the (^{3}H)uridine (fig. 5B). In clear contrast, 28 S RNA specific radioactivities were much less influenced by changes of exogenous uridine concentrations, indicating incomplete equilibration of its precursor pool with the general UTP pool as well as relative independence from the salvage pathways. tRNA and total RNA responded similar to 28 S RNA (not shown).

The finding of a rather slow mRNA turnover in proliferating cultures may not be representative for all mammalian cell systems. The main 'intention' of a proliferating cell is replication of all cell constituents, with no need for adaptations and for rapid changes in mRNA composition. In contrast, resting cells of an adult organism which are subject to hormonal or dietary control should exhibit much faster turnover rates of mRNA. The method of direct RNA and precursor pool analyses presented in this paper will open a rather simple way to the determination of mRNA and rRNA turnover in highly differentiated non-proliferating tissues.

Summary

A new method for the selective and direct determination of mRNA specific radioactivity independent from its poly(A) content, and for 28 S rRNA has been developed. It is based on the selective hydrolysis of interribosomal mRNA sections in RNP-free polysomes by low ribonuclease concentrations under controlled conditions, with the analysis of cpm and A_{260} of the interribosomal split products. Separation by sucrose gradient centrifugation of the residual RNA after digestion of ribonuclease and ribosomal proteins with proteinase K allowed also 28 S RNA analysis without interference by mRNA. Application of the method to the analysis of mRNA and 28 S RNA labeling over several generation times in proliferating HeLa cultures revealed that mRNA half-life was 0.87 times cell doubling time, while 28 S RNA exhibited no turnover ($T/2 > 1000$ h). Similar values were obtained by decay experiments in prelabeled cultures when corrections for growth and reutilization of labeled degradation products were made: Correction for growth could be based on metabolic stability of rRNA, while the degree of reutilization could be extrapolated from the decay of UTP specific radioactivity.

The exact determination of mRNA turnover required not only a constant supply of exogenous (^{3}H)uridine to guarantee a constant specific radioactivity of UTP. It also afforded a correction for RNA-C (which is in delayed equilibrium with RNA-U), and the direct determination of mRNA specific radioactivity (cpm <u>and</u> A_{260} after separation).

Labeling kinetics at different concentrations of exogenous (^{3}H)uridine (constant supply) showed the existence of two different precursor pools for mRNA (and DNA-C) on the one hand, and for rRNA and tRNA on the other.

Acknowledgements

We would like to thank G. Jarmers for propagating cell cultures. This work was supported by grants from the Deutsche Forschungsgemeinschaft.

References

1) Penman, S., Scherrer, K., Becker, Y. and Darnell, J.E., Proc.Natl.Acad.Sci.US 49, 654-662 (1963)
2) Craig, N., Kelley, D., and Perry, R., Biochim. Biophys.Acta 246, 493-498 (1971)
3) Greenberg, J.R., Nature 240, 102-104 (1972)
4) Storb, U., Biochem.Biophys.Res.Commun. 52, 1483-1491 (1973)
5) Schultz, J.A., Can.J.Biochem. 51, 1515-1520 (1973)
6) Perry, R.P., and Kelley, D.E., J.Mol.Biol. 79, 681-696 (1973)
7) Edmonds, M., Vaughan, M.H., and Nagato, H., Proc.Natl.Acad.Sci.US 68, 1336-1340 (1971)
8) Darnell, J.E., Wall, R., and Tushinski, R.J., Proc.Natl.Acad.Sci.US 68, 1321-1325 (1971)
9) Lee, S.Y., Mendecki, J., and Brawermann, J., Proc.Natl.Acad.Sci.US 68, 1331-1335 (1971)
10) Sheldon, R., Jurale, C., and Kates, J., Proc.Natl.Acad.Sci.US 69, 417-421 (1972)
11) Soeiro, R., and Amos, H., Biochim.Biophys.Acta 129, 406-409 (1966)
12) Revel, M., Hiatt, H.H., and Revel, J.P., Science 146, 1311-1313 (1964
13) Sawicki, S., and Goodman, G., J.Cell Biol. 50, 746-761 (1971)
14) Singer, R.H., and Penman, S., Nature 240, 100-102 (1972)
15) Lindberg, U., and Persson, T., Eur.J.Biochem. 31, 246-254 (1972)

16) Wiegers, U., Kramer, G., and Hilz, H., Hoppe-Seyl.Z.Physiol.Chem. 352,843-852(1973)
17) Murphy, W., and Attardi, G., Proc.Natl.Acad.Sci.US 70,115-119(1973)
18) Schlaeger, C., Hoffmann, D., and Hilz, H. Hoppe-Seyl.Z.Physiol.Chem. 350,1017-1022(1969)
19) Kramer, G., Klapproth, K., and Hilz, H., Biochim.Biophys.Acta 262,410-419(1972)
20) Kramer, G., Wiegers, U., and Hilz, H., Biochem. Biophys.Res.Commun. 55,273-281(1973)
21) Wiegers, U., Kramer, G., and Hilz, H., Biochem. Biophys.Res.Commun. 50,1039-1047(1973)
22) Wiegers, U., and Hilz, H., FEBS Letters 23, 77-82(1972)
23) Perry, R.P., and Kelley, D.E., J.Mol.Biol. 35,37-59(1968)
24) Miller, A.O.A., Arch.Biochem.Biophys. 150,282-295 (1972)
25) Kramer, G., and Hilz, H., Hoppe-Seyl.Z.Physiol. Chem. 352,843-852(1971)
26) Wiegers, U., and Hilz, H., Biochem.Biophys.Res. Commun. 44,513-519(1971)
27) Hilz, H., Wiegers U., and Adamietz, P., submitted for publication
28) Wiegers, U., Klapproth, K., and Hilz, H., submitted for publication
29) Wiegers, U., and Klapproth, K., submitted for publication

Structures of Biotin Enzymes

F. Lynen
Max-Planck-Institut für Biochemie, D-8033 Martinsried, West-Germany

I believe it is a splendid idea to come together here - in Berlin - on the occasion of the 75th anniversary of our friend Fritz Lipmann, because he has spent part of his life - some of his very important years - in Berlin. Berlin in these years was not only a prosperous cultural center but also very attractive for the sciences.

David Nachmansohn writes about this period in the prefatory chapter of the 1972 Annual Reviews of Biochemistry: "Berlin became a center of nearly magic attraction to intellectuals in all fields, to artists, writers, philosophers, and scientists. It had 30 to 40 theaters of world fame under leaders such as Max Reinhardt, Barnowski, Erwin Piscator, Leopold Jessner, and many others. The superb quality of the performances by some of the greatest living actors could hardly be surpassed. The musical life was traditionally very strong. There were four superior operas in Berlin playing during the whole year. Painting and sculpture were thriving: Kaete Kollwitz, Barlach, George Grosz, Zille, Liebermann, again to quote a few of the best-known names as illustration."

Fritz Lipmann, with his broad artistic interests, loved this atmosphere. Much later, in 1953, when I came to the United States for the first time, as a fellow of the Rockefeller Foundation, I also spent a month in Lipmann's laboratory at the Massachusetts General Hospital in Boston. At that time - I remember - many most pleasant hours spent together with Freda and Fritz in their "gemütliches" home in Revere Street and I was very impressed because both of them were still very fond of these exciting years in Berlin.

Parallel to the vivid artistic atmosphere in Berlin stood the scientific atmosphere with the center of gravity in Berlin-Dahlem and the institutes of the Kaiser-Wilhelm-Gesellschaft. I should like to cite again David Nachmansohn, when he refers to the famous "Haber-Colloquia" which became a great attraction and played an important role in the activities of all the Kaiser-Wilhelm-Institutes. "In these seminars, which took place every second week, one met not only members of Haber's institute, like Ladenburg, Freundlich, Polanyi or Bonhoefer, but also many others, among them Hahn and Lise Meitner from the chemistry group, von Laue, and members of the three biochemistry groups around Otto Meyerhof, Otto Warburg and Carl Neuberg and of the other biology departments with Mangold, Goldschmidt, Correns, Hartmann and their associates." Berlin-Dahlem is considered as one of the strong routes from which modern biochemistry derived. As an example, Otto Warburg spent practically all his scientific career in Berlin-Dahlem. In this fertile environment he developed the ingenious methods applied today in all biochemical laboratories and there he made his many fundamental discoveries.

One of his discoveries leads me to the topic of my lecture and brings me in close relation to our "Geburtstagskind". I have in mind the elucidation of biochemical functions of the vitamins. The high activity of this group of compounds, so indispensable in the nutrition of human beings and animals, was at first quite mysterious. Eventually, when Warburg studied the yellow enzyme of yeast in connection with Kuhn's purification of vitamin B_2 from whey, this problem was solved. These investigations demonstrated for the first time that vitamins are used for the synthesis of functional groups of enzymes and in this form act catalytically. Later, Fritz Lipmann discovered the vitamin pantothenic acid to be an essential constituent of coenzyme A, the coenzyme standing in the center of the citric acid cycle and lipid metabolism. In 1964 pantothenic acid was found to be also a constituent of the "acyl carrier proteins" (1), which are instrumental in fatty acid synthesis, in the biosynthesis of the polyacetate compounds (2) and of the antibiotic cyclic polypeptides like gramicidine and thyrocidine (1). The studies on the "acyl carrier proteins" gave the explanation why the functional sulphydryl group of cysteamine is bound through pantothenic acid to the adenosine triphosphate in coenzyme A (3).

In our laboratory the biochemical function of the vitamin biotin could be elucidated (4) in studies on ß-methylcrotonyl-CoA carboxylase, an enzyme which participates in the biological degradation of leucine by way of isovaleryl CoA and its dehydrogenation product. This enzyme could easily be isolated from microorganisms grown on isovaleric acid as sole carbon source. The purified enzyme, being about 150-times more active than the crude extract, was found to be homogeneous by sedimentation in the ultracentrifuge and by electrophoresis. By slowly adding ammonium sulfate to the purified enzyme solution Rehn and Apitz-Castro even succeeded in crystallizing the protein (5). By measurements in the analytical ultracentrifuge the sedimentation constant of the pure enzyme was determined to be 20 S which corresponds to a molecular weight of 760 000 daltons (5).

It could be shown that the enzyme contains biotin as prosthetic group covalently bound to the protein and that during the carboxylation a carboxy-biotin-enzyme intermediate is formed (4,6). The carboxylation is thus achieved in two steps, according to the following equations:

$$\text{(1) ATP} + HCO_3^- + \text{biotinenzyme} \xrightleftharpoons{Mg^{++}} CO_2\text{-biotinenzyme} + \text{ADP} + P_i$$

$$\text{(2) } CO_2\text{-biotinenzyme} + CH_3\text{-}\overset{\displaystyle CH_3}{\overset{|}{C}}\text{=CH-CO-SCoA} \rightleftharpoons \text{biotinenzyme} + CH_3\text{-}\overset{\displaystyle CH_2\text{-}COO^-}{\overset{|}{C}}\text{=CH-CO-SCoA}$$

In order to prove the reaction sequence of ß-methylcrotonyl CoA carboxylation we first used exchange experiments with labelled substrates (6). When ^{14}C-ß-methylglutaconyl CoA labelled in the positions 1,3 and 5 is incubated with unlabelled ß-methylcrotonyl CoA and the enzyme, it gives rise to labelled ß-methylcrotonyl CoA by the repeated shuttling of the reaction 2 between CO_2-biotin enzyme and biotin enzyme. It could be demonstrated that the enzyme bound biotin participates because the exchange reaction was strongly inhibited by avidin. The ATP-dependent reaction step was also studied by the exchange technique (6). In this case the exchange reactions between ATP and radioactive ^{32}P-orthophosphate or radioactive ^{14}C-ADP were measured. In agreement with equation 1 it was found that both exchange reactions required the simultaneous addition of four substrates: ATP, ADP, inorganic phosphate, and bicarbonate besides the enzyme.

In the following years the general validity of the sequence of reactions could be proven for the biotin dependent carboxylases of different specificity. In these investigations, in which several laboratories participated, analogous exchange experiments were used (for review see 7). In addition it was possible to isolate the carboxy-biotin enzymes themselves. This important objective was first achieved by Kaziro and Ochoa (8) with propionyl-CoA carboxylase from pig heart.

After having established the intermediate formation of carboxybiotin enzyme by the exchange experiments we turned our efforts to the elucidation of the linkage of carbonic acid with the biotin of the enzyme. We could solve this problem through a fortunate property of ß-methylcrotonyl CoA carboxylase. It was found that free biotin, when present in high concentrations, can replace ß-methylcrotonyl CoA as substrate for the carboxylase, leading to the formation of carboxy-biotin according to the following equation:

$$(3)\ ATP + HCO_3^- + biotin \xrightarrow{Mg^{++}} CO_2\text{-}biotin + ADP + P_i$$

To explain this reaction, we assumed that the carboxylation of added free biotin is achieved by the partial dislocation of the enzyme-bound biotin. This seems to be possible if the biotin of the carboxylase is bound to the protein not only by a covalent linkage through its side-chain carboxyl group, but also by a dissociable linkage between the ring system and a second site on the protein (Figure 1). The second site also

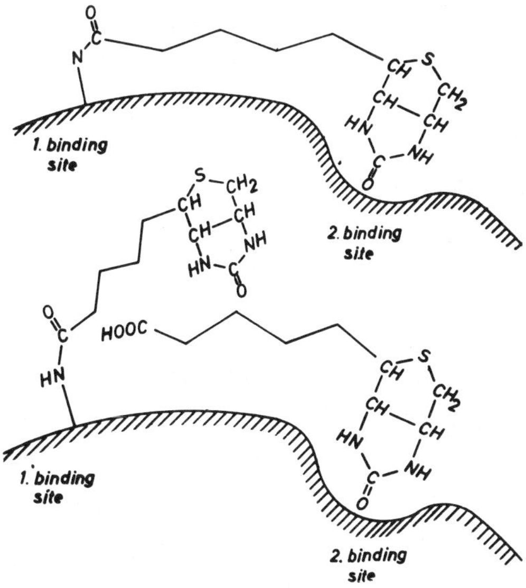

Fig. 1. Proposed mechanism for the binding of free biotin by the enzyme.

might have affinity for free biotin and in the presence of a large excess of the latter, the enzyme-bound biotin is partially displaced and free biotin is carboxylated. If this assumption was correct the carboxy-biotin formed should be a precise model of the carboxy-biotin enzyme.

In experiments with ^{14}C-bicarbonate radioactive $^{14}CO_2$-biotin was produced. The carboxylated biotin was found to be very unstable, especially at acid pH. It was a purely chemical problem then to stabilize the labile radioactive carboxy-biotin by methylation with diazomethane and to identify the methylation product by comparison with authentic compounds as 1'-N-carboxymethyl biotin methylester (6,9). In further studies Knappe, Wenger and Wiegand (10) confirmed the prediction that this same linkage would occur also in the carboxylated enzyme itself. There 1'-N-carboxybiotin is bound to the enzyme protein through amide linkage at the ε-amino group of a lysine residue (Figure 2).

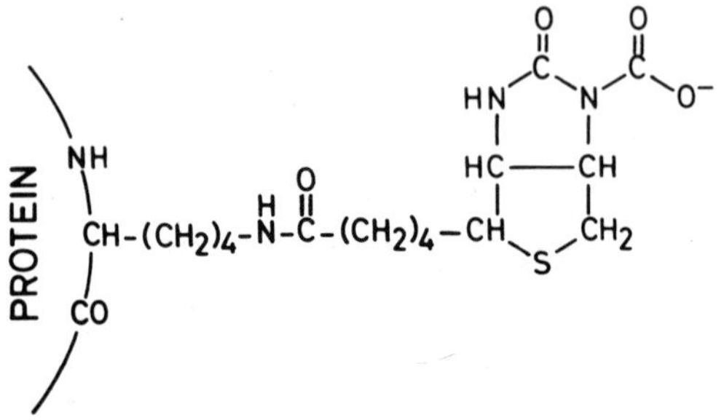

Fig. 2. Chemical structure of carboxybiotin enzyme.

The chemical reactivity of an "active carbonic acid" of this structure derives from the electron attraction of the ureido system, i.e., the cyclic urea moiety of biotin behaves like a weak acid. The carboxy derivatives of such a structure are to some extent comparable to acid anhydrides. The bond between the nitrogen atom and CO_2 is polarized, which augments the electrophilic character of the carboxy group and therefore its ability to enter transcarboxylation reactions(11). Regarding the source of the carboxyl group bound to biotin, the biotin enzymes can be divided into two classes. Enzymes of class I utilize bicarbonate as carboxyl donor and require ATP to drive the formation of the new carbon nitrogen bond. Biotin-containing enzymes, belonging to this group include acetyl-CoA-carboxylase, propionyl-CoA carboxylase, ß-methylcrotonyl-CoA carboxylase, geranoyl-CoA carboxylase and pyruvate carboxylase (7).

Enzymes of class II catalyze the formation of the carboxy-biotinyl intermediate by an ATP-independent transcarboxylation with either a 3-oxo acid or a malonyl-CoA derivative serving as carboxyl donor. Enzymes belonging to this classe include methylmalonyl-CoA: pyruvate transcarboxylase and oxaloacetate decarboxylase (7).

Carboxylation

$$\text{ATP} + HCO_3^- + \text{biotin enzyme} \overset{Mg^{++}}{\rightleftharpoons} CO_2\text{-biotin enzyme} + \text{ADP} + P_i$$

$$CO_2\text{-biotin enzyme} + \text{RH} \rightleftharpoons \text{biotin enzyme} + \text{R-COO}^-$$

Transcarboxylation

$$R_1\text{-COO}^- + \text{biotin enzyme} \rightleftharpoons CO_2\text{-biotin enzyme} + R_1H$$

$$CO_2\text{-biotin enzyme} + R_2H \rightleftharpoons \text{biotin enzyme} + R_2\text{-COO}^-$$

Fig. 3. Classification of biotin enzymes.

It appears that all of the multistep reactions of the biotin enzymes can be accounted for by appropriate combinations of a few basic types of partial reactions, as indicated in Figure 3. In the meantime the beautiful work of Vagelos and his colleagues (12) and of Wood and his coworkers (13) has demonstrated with two bacterial enzymes that biotin enzymes actually represent multienzyme systems. It was demonstrated that the biotinyl prosthetic group resides on a small polypeptide chain, distinct from the catalytic subunits, and appears to function in carboxyl translocation between these subunits. As schematically demonstrated for acetyl-CoA carboxylase of E. coli in Figure 4 the biotin-free carboxylase component (I) catalyzes the carboxylation of the biotinyl prosthetic group on the carrier protein (cross hatched area). Following the translocation of the carboxylated biotinyl group from site I to site II, carboxyl transfer to acetyl CoA is catalyzed by carboxyl transferase. The attachment of CO_2

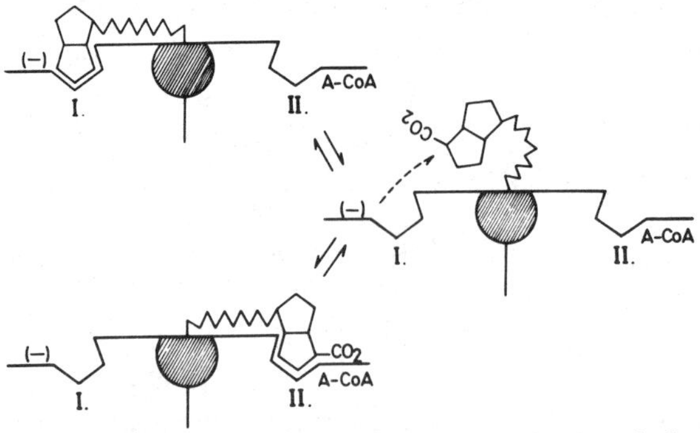

Fig. 4. Scheme of carboxyl translocation in acetyl-CoA carboxylase of E. coli.

to the biocytin structure seems to be very important. This way the system gains a flexible arm of about 14 Å length and this makes the carboxyl translocation between the enzyme components I and II possible.

From these observations it could be concluded that the chemical reaction catalyzed by a given biotin-containing enzyme may result from the specific combination of such subunit enzymes. In realizing that ATP-dependent carboxylations of the biotinyl prosthetic group is a partial reaction common to all biotin enzymes of class I it seemed possible that for a given organism this part structure of class I enzymes is similar or perhaps even identical. If this is true, then also the different part structures catalyzing the second step within class I enzymes should be similar to some extent, irrespective of their different substrate specificity; at least that region of the three-dimensional structure responsible for the association with the first part structure should be similar.

Pursuing this idea, Sumper (14) in our laboratory isolated the two enzymes acetyl CoA carboxylase and pyruvate carboxylase from the same organism, namely yeast cells, and investigated whether a structural relationship between these two enzymes exists. In some experiments ^{14}C-labelled enzymes were used. In order to prepare them yeast cells were cultivated on ^{14}C-biotin.

STRUCTURAL COMPARISON BETWEEN ACETYL-CoA CARBOXYLASE AND PYRUVATE CARBOXYLASE FROM YEAST

The sedimentation coefficient of acetyl-CoA carboxylase was measured by sucrose density gradient centrifugation according to the method of Martin and Ames. The average value of several experiments was 15.5 ± 0.15, when catalase was used as reference. For pyruvate carboxylase from yeast Utter and his colleagues (15) found a sedimentation coefficient of 15.6 S which is in good agreement with the value found for acetyl CoA carboxylase. For a direct comparison of the two enzymes, acetyl-CoA carboxylase and pyruvate carboxylase were centrifuged together in the same tube. As can be seen from Figure 5

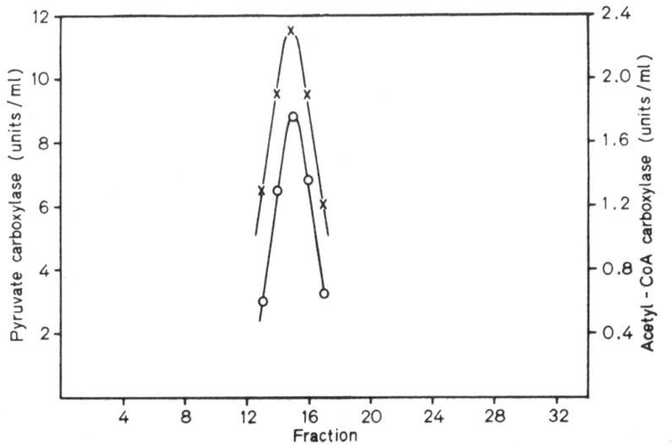

Fig. 5. Comparison of the sedimentation behaviour of acetyl-CoA carboxylase and pyruvate carboxylase (14).

2 mg pyruvate carboxylase and 3 mg acetyl-CoA carboxylase were mixed together and layered over a linear sucrose gradient (5-20 %) in 0.3 M potassium phosphate pH 6.5. Centrifugation was at 39 000 rev./min (SW-40 rotor, L2-ultracentrifuge) for 15 h at 8° C. The gradient was fractionated from the bottom. Pyruvate carboxylase (O); acetyl-CoA carboxylase (X).

no separation of the two enzymes could be observed, what means that the sedimentation coefficients of acetyl CoA carboxylase and pyruvate carboxylase are identical within the experimental error of ± 0,2 S. Assuming a molecular weight of pyruvate carboxylase of 600 000 as measured by Utter and his colleagues (16) it follows that acetyl-CoA carboxylase has a very similar molecular weight.

Acetyl-CoA carboxylase was rapidly inactivated by low ionic strength and pH-values above 7.5. This loss of enzymatic activity was accompanied by a change of sedimentation behaviour. Figure 6 shows in the upper part the dissociation of

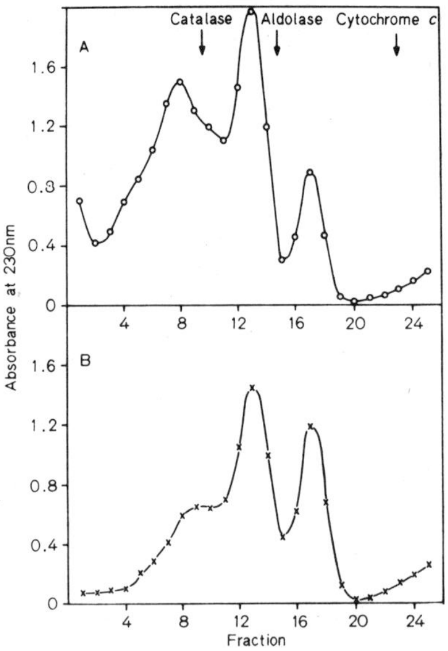

Fig.6. Sucrose density-gradient centrifugation of dissociated acetyl-CoA carboxylase (A) and pyruvate carboxylase (B) (14).

The enzymes were dialyzed against 0.1 M ammonia containing 10 mM 2-mercaptoethanol for 5h at 20°C. Protein concentration during dissociation was 15 mg/ml. The enzyme solutions were layered over a linear sucrose gradient (5-20 %) in 0.1 M Tris-HCl pH 9.0 and centrifuged for 16.5 h at 39 000 rev./min (SW-40 rotor, L2-ultracentrifuge). The temperature was 6° C. Markers were included in separate tubes. The gradients were fractionated from the bottom.

acetyl-CoA carboxylase into subunits with sedimentation coefficients of approximately 12, 9 and 6 S, as estimated by comparison with catalase, aldolase and cytochrome c as reference proteins. Pyruvate carboxylase (lower part), treated in the same way, showed a sedimentation pattern of remarkable similarity to that of acetyl-CoA carboxylase. The different subunits of both enzymes sedimented within the experimental error at identical rates. And in both cases it was found that subunits with sedimentation coefficients less than 15 S could no longer catalyze the carboxylation of acetyl-CoA and pyruvate, respectively.

Both, in the case of acetyl-CoA carboxylase and in that of pyruvate carboxylase, the relative amounts of the 12 S, 9 S and 6 S subunits in the sedimentation patterns were affected by temperature. If the enzymes were dissociated at lower temperatures the amount of the 12 S component decreased in favour of the 9 S and 6 S components. Dissociation of acetyl-CoA carboxylase and pyruvate carboxylase could also be followed by electrophoresis in polyacrylamide gels. Figure 7 presents the electrophoretic behaviour of both enzymes dissociated under identical conditions at alkaline pH. The patterns of both enzymes were again very similar. By comparison of the electrophoretic behaviour of the isolated 12 S, 9 S and 6 S subunits

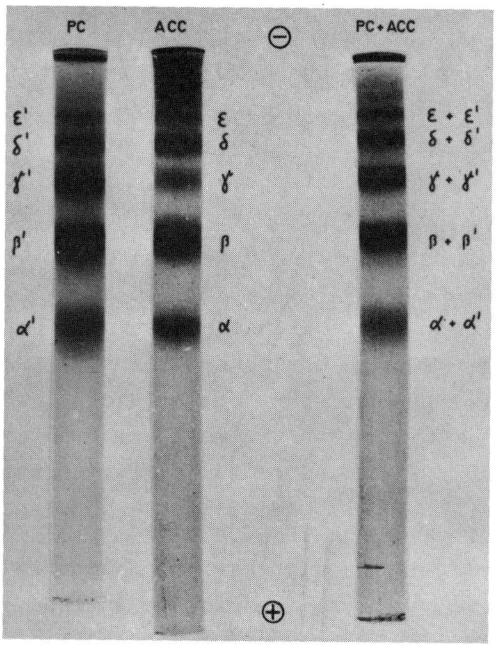

Fig. 7. Polyacrylamide-gel electrophoresis of dissociated acetyl-CoA carboxylase and pyruvate carboxylase (14).

The enzymes were dialyzed against 0.1 M ammonia containing 10 mM 2-mercaptoethanol for 3h at 20° C. (PC) Pyruvate carboxylase; (ACC) acetyl-CoA carboxylase; (PC+ACC) pyruvate carboxylase and acetyl-CoA carboxylase. Approximately 50 μg enzyme was layered onto the 6 % polyacrylamide gel. Buffer system 1 as described by Maurer (17) was used. The negative pole was on the upper end of the figure.

with the electrophoretic pattern it could be shown that the 12 S component is identical with γ, 9 S identical with band ß and the 6 S component is identical with band α.

As already mentioned, the distribution of protein between the various subunits of acetyl CoA carboxylase and pyruvate carboxylase was by no means a constant relation. In contrast, this distribution was highly dependent on the conditions of dissociation used. Especially in the case of pyruvate carboxylase the heavy 12 S and 9 S components were nearly quantitatively convertible into the 6 S subunit. From these observations we may conclude that the 6 S species must be the smallest unit containing the complete primary structure of the native enzyme. This would mean that the heavier components of the dissociated enzymes are aggregates of different numbers of 6 S subunits. To investigate this possibility acetyl CoA carboxylase and pyruvate carboxylase labelled with ^{14}C-biotin were dissociated by alkaline pH and the distribution of radioactivity in the subunits was determined by sucrose density gradient centrifugation. With the assumption of identical 6 S subunits reacting as monomers, dimers, trimers etc. one would predict a constant specific radioactivity across the sedimentation pattern. When Sumper performed the experiment, the patterns of protein and radioactivity corresponded indeed to each other (14).

Further evidence for the aggregation of identical subunits was provided by immunochemical techniques. Antibodies against acetyl-CoA carboxylase and pyruvate carboxylase were prepared by injecting rabbits with the purified enzymes. Using the double diffusion Ouchterlony technique, an immunochemical comparison of the 12 S, 9 S and 6 S subunits of both enzymes was undertaken. The upper part of Figure 8 shows the pattern obtained when antiacetyl CoA carboxylase γ-globulin obtained from rabbits was applied to the centre well and the various subunits, previously separated by centrifugation, and the native enzyme were arranged peripherically. A single connecting band of precipitation was observed both in the case of acetyl-CoA carboxylase (upper part) and in the case of pyruvate carboxylase (lower part). The absence of spurs indicates that there are no antigenic sites present on the natives enzymes that are not present on the 12 S, 9 S and 6 S subunits.

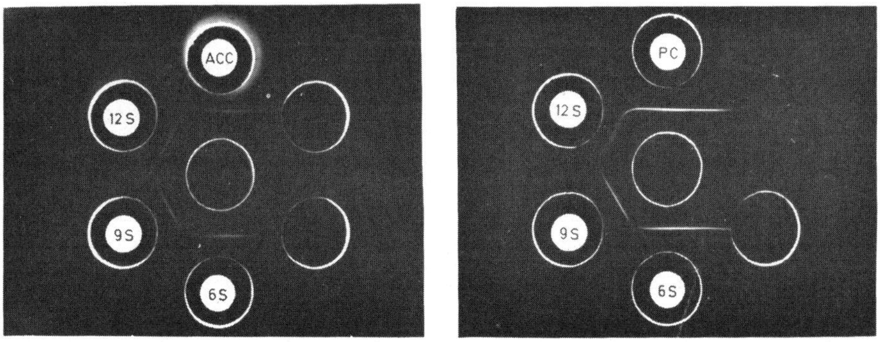

Fig. 8. Immunodiffusion of the separable subunits of acetyl-CoA carboxylase and pyruvate carboxylase (14).

Left part: Agarose gel (0.6 %) contained 100 mM potassium phosphate pH 7.5. The plate was developed at 20°C for 24 h. The center well contained antiacetyl-CoA carboxylase γ-globulin.
Right part: Agarose gel (0.6 %) contained 50 mM Tris-HCl pH 7.5. The plate was developed at 20° C for 48 h. The center well contained antipyruvate carboxylase γ-globulin. The peripheral wells contained native acetyl-CoA carboxylase (AAC) and its subunits as indicated or native pyruvate carboxylase (PC) and its subunits.

Using the relationship between sedimentation coefficients and molecular weights (18) the sedimentation coefficient 9.5 S for a dimer, 12.0 S for a trimer and 15.1 S for a tetramer of the 6 S subunit were obtained. These values are in good agreement with the experimentally found sedimentation coefficients. We thus come to the conclusion that acetyl CoA carboxylase as well as pyruvate carboxylase of molecular weights 600 000 each are composed of four protomers of molecular weight 150 000. The free protomers do not possess enzymic activities. The latter only appear in the tetramers, probably as result of conformational changes due to protein-protein interactions.

Immunochemical studies have also been used in order to prove the hypothesis that for a given organism partial structures of the ATP-dependent enzymes may be similar or perhaps identical, what, as remembered, was the starting point for the comparison of the two carboxylases. It was found that the enzyme activities of both enzymes were inhibited completely by their corresponding antisera. In order to establish insight into which step of the overall reaction was blocked by antibody action, the partial activities of pyruvate carboxylase were assayed. As can be seen from Table 1 the inhibition

Table 1. Inhibition of pyruvate carboxylase by antipyruvate carboxylase γ-globulin (14)

Pyruvate carboxylase preparations, inhibited by antibody to 55, 70 and 90 % of its original activity in the overall reaction, were assayed for partial activities by the exchange techniques.

Inhibition of the overall reaction	Inhibition of ^{14}C-ADP $\rightleftharpoons$ ATP exchange reaction	% Inhibition of ^{14}C-pyruvate $\rightleftharpoons$ oxalacetate exchange reaction
55 %	50 %	10 %
70 %	75 %	15 %
90 %	-	20 %

of ATP-dependent carboxylation of biotin parallels the overall inhibition, whereas the second step, namely the carboxylation of pyruvate, is only slightly influenced.

To check whether the observed similarities of acetyl-CoA carboxylase and pyruvate carboxylase are also reflected in the immunochemical properties, the reaction of antiacetyl-CoA carboxylase γ-globulin with pyruvate carboxylase was studied. It was found that even a 20-fold excess of antibody, necessary to inhibit completely the acetyl-CoA carboxylase, has no effect on the pyruvate carboxylase activity. In a second experiment cross reaction of antiacetyl-CoA carboxylase γ globulin with pyruvate carboxylase was checked by immunoelectrophoresis. No line of precipitation was formed however. We thus come to the conclusion that the first partial reaction, the ATP-dependent carboxylation of biotin, common to both enzymes is catalyzed by protein substructures which are not identical in primary structure, although having identical catalytic functions in both enzymes.

On the basis of these results, we assume the genes which control the sequences of acetyl-CoA carboxylase and pyruvate carboxylase are derived from a common ancestor. By a process of one or more duplications, the ancestral gene gave rise to two or more genes which subsequently evolved independently to code for biotin enzymes with different but still similar functions.

STRUCTURAL STUDIES ON ß-METHYLCROTONYL-CoA CARBOXYLASE OF ACHROMOBACTER

We integrated in our structural studies on Biotin enzymes also experiments with ß-methylcrotonyl-CoA carboxylase, the object, the study of which led us to the elucidation of the chemical mechanism of biotin action. Schiele (19), who continued our work on that enzyme, first investigated the problem whether Achromobacter, when grown on isovaleric acid as sole carbon source produces only ß-methylcrotonyl CoA carboxylase or additionally other biotin enzymes. To answer this question he grew the organism on radioactive ^{14}C-biotin, broke the cells by grinding with alumina or by high speed shaking with glass beads and chromatographed the crude cell extract on a column of Sephadex G 200. Under these conditions two radioactive ^{14}C-biotin containing fractions were eluted from the column (Figure 9). The first one, possessing the higher molecular weight could be identified with ß-methylcrotonyl CoA carboxylase. The molecular weight of the second radioactive peak was much smaller and was estimated to be in the range of 10000.

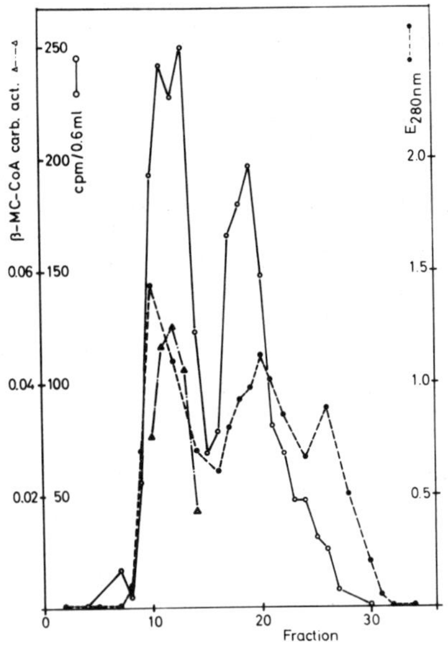

Fig.9. Chromatography of the crude cell extract over Sephadex G 200.

0.7 g freeze dried Achromobacter cells, grown in a medium containing isovaleric acid as carbon source and 0.02 µM ^{14}C-biotin, were suspended in 1 ml of 0.01 M Na_2HPO_4 and broken by grinding with Al_2O_3. After addition of 4 ml of 0.01 M Na_2HPO_4 the suspension was centrifuged for 25 minutes at 100 000 g. 2.2 ml of the supernatant was chromatographed over a Sephadex G 200 column (97.5 x 1 cm) equilibrated with 0.03 M potassium phosphate buffer, pH 7.2. Volume of fractions: 3.25 ml. Radioactivity (o), ß-methylcrotonyl-CoA carboxylase activity (△) and extinction(●).

Separation of the two radioactive proteins could also be achieved by chromatography over DEAE-cellulose (Figure 10). On this column the two labelled proteins appeared in reverse order. As the specific enzyme assay revealed, ß-methylcrotonyl CoA carboxylase activity was associated with the second peak which, dependent on the cultivating conditions contained 50-80 % of the radioactivity.

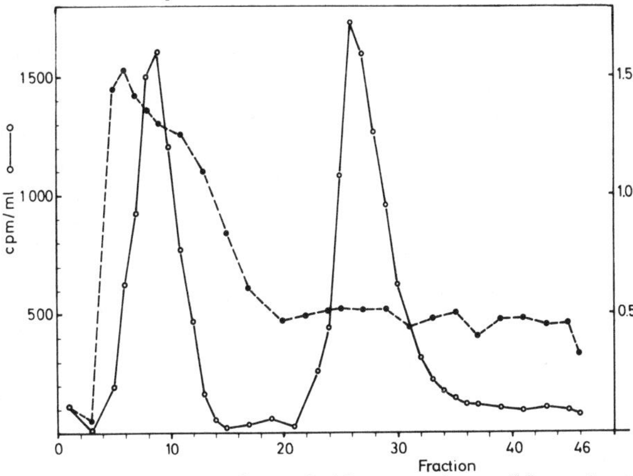

Fig.10. Chromatography of the crude cell extract over DEAE-cellulose.

2 g freeze dried Achromobacter cells, grown as described in the legend of Figure 9, were suspended in 15 ml of 0.01 M Na_2HPO_4 and homogenized in the cell homogenizer of Merkenschlager et al.(20). After centrifugation the supernatant solution was added to a DEAE-cellulose column (16 x 2.4 cm) and the proteins eluted with 400 ml of a linear gradient, containing 0.1 - 0.5 M NaCl in 0.03 M potassium phosphate buffer, pH 7.2. Volume of fractions: 8.7 ml. Radioactivity (O) and extinction at 280 nm (●) of the fractions were measured.

The biological function of the first ^{14}C-biotin protein is not yet known. However, we assume that it may be related to an acetyl CoA carboxylase of Achromobacter. This assumption is based on the observation that the radioactivity of this peak increases at the expense of ß-methylcrotonyl CoA carboxylase if cells grown on isovaleric acid in presence of ^{14}C-biotin are transferred to a new medium which contains only acetic acid as carbon source and no biotin. The results of the chromatographic separation on DEAE-cellulose of extracts from cells, harvested 18 hours or 42 hours after transfer to the new medium revealed that the second peak belonging to ß-methylcrotonyl-CoA carboxylase gradually disappeared, when this enzyme was not any longer required by the organism. After 42 hours this peak was completely gone.

As Achromobacter is known to contain the enzymes of the glyoxalate cycle in rather high activity (21), pyruvate carboxylase otherwise required for gluconeogenesis, appears to be dispensible. On the other hand acetyl-CoA carboxylase must be present, no matter what carbon source has been offered. Fatty acid synthesis, so essential for the construction of membranes, fully depends on the function of this enzyme. We therefore believe that the biotin-carrying protein in the first peak represents a constituent of the acetyl-CoA carboxylase, the structure of which may be similar to the analogous enzyme from E. coli (12). Experiments are now under way, which will prove this assumption.

One other explanation has already been excluded. It might have been possible that the low molecular weight biotin protein represents a constituent of ß-methylcrotonyl-CoA carboxylase. If built like methylmalonyl-CoA pyruvate transcarboxylase from Pr. shermanii (13), where a biotin carboxyl carrier protein is incorporated into the multienzyme complex and assuming a not completely coordinated biosynthesis of the component proteins it seemed possible that some free biotin carboxyl carrier protein besides the one incorporated into the multienzyme complex might exist. If this would be the case, the dissociation of ß-methylcrotonyl-CoA carboxylase by treatment with sodium dodecylsulfate or with urea should lead to the appearance of three proteins, one of which should be identical to the low molecular weight biotin protein. The experimental results obtained, however, were quite different. Schiele used the sodium dodecyl sulfate-polyacrylamide gel electrophoresis according to Weber and Osborn (22) and found under these conditions only two protein bands (Figure 11).

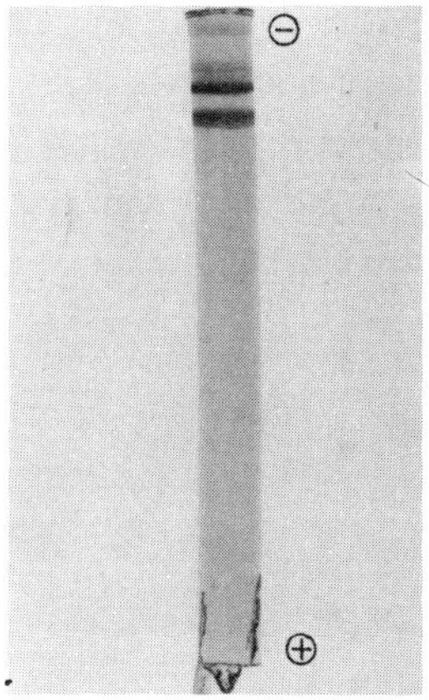

Fig. 11. SDS-disk gel electrophoresis of ß-methylcrotonyl-CoA carboxylase.

The solution containing 0.21 mg ß-methylcrotonyl-CoA carboxylase/ml was dialyzed for 48 hours against 10 mM cysteine in 10 mM glycine-NaOH buffer, pH 9.8. Then sodium dodecyl sulfate was added to a concentration of 0.2 percent and the solution was heated to 80° C for 10 minutes. 4.3 µg protein was added to the 10 percent acrylamide gel (0.27 percent N,N'-methylene bisacrylamide). Electrophoresis for 5 hours at 7 volt/cm. The protein bands were stained with Coomassie Blue.

By comparison with reference proteins the molecular weights were found to be 96000 and 78000 respectively. Assuming each polypeptide to be present in the intact enzyme 4-times gives a calculated molecular weight of 700 000 which compares favourably with the value as measured by Rehn with crystalline ß-methylcrotonyl-CoA carboxylase in the analytical ultracentrifuge (5). When the radioactivity of the two polypeptides was measured, it was found that only the heavier component was radioactive, indicating that only this component carries biotin. The smaller polypeptide was free of radioactivity.

The treatment with sodium dodecyl sulfate leads to an irreversible dissociation of the native enzyme. As Schiele then found, a reversible dissociation can be achieved by dialyzing the purified active enzyme against 10 mM cysteine in 10 mM glycine-NaOH-buffer, pH 9,35, for 2 or 3 days under an atmosphere of nitrogen. This could be demonstrated by sucrose gradient centrifugation as shown in Figure 12. In the upper part the results of centrifugation after such treatment are represented whereas the lower part represents the result of a control experiment, in which the enzyme was dialyzed against 9 mM NaCl in 5 mM K-phosphate buffer, pH 7,0.

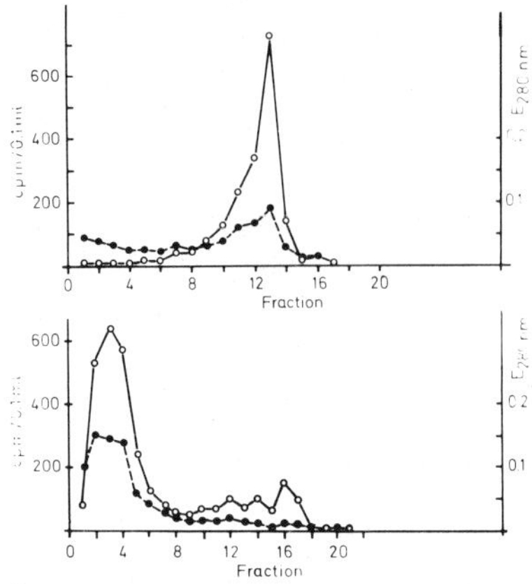

Fig. 12. Sucrose density gradient centrifugation of ^{14}C-biotin carrying ß-methylcrotonyl-CoA carboxylase after dialysis against 5 mM glycine-NaOH buffer, pH 9.35 (upper part) or against 5 mM K-phosphate buffer, pH 7.0 (lower part).

Sucrose gradient 5-20 percent. Centrifugation for 23 hours at 39 000 r.p. m. Open circles: radioactivity (c.p.m./0.1 ml) Closed circles: extinction at 280 nm.

When the solution of the dissociated enzyme was brought to pH 7,5, the subunits partly reassociated to form the high molecular weight species. This reassociation is connected with reactivation. This is illustrated in Figure 13. The figure summarizes the results of an optical assay in which the reaction was started by the addition of the dissociated enzyme, preincubated at pH 8 and room temperature for 20 min. with 10 mM acetyl CoA or 10 mM ß-methylcrotonyl-CoA in presence of 10 mM $MgCl_2$. In the first case a lag phase is seen which is absent in the second case, indicating that the presence of the specific substrate and magnesium ions stimulate the aggregation of the inactive subunits to the active enzyme.

In order to isolate the biotin-containing subunit and to study its enzymic activities, the purified enzyme was dissociated by treatment with urea at pH 7.1 and then passed through a column of DEAE-cellulose. Elution with a NaCl-gradient led to the appearance of a rather sharp fraction which contained all the radioactive ^{14}C-biotin protein, superimposed on a rather broad nonradioactive protein shoulder. The isolated radioac-

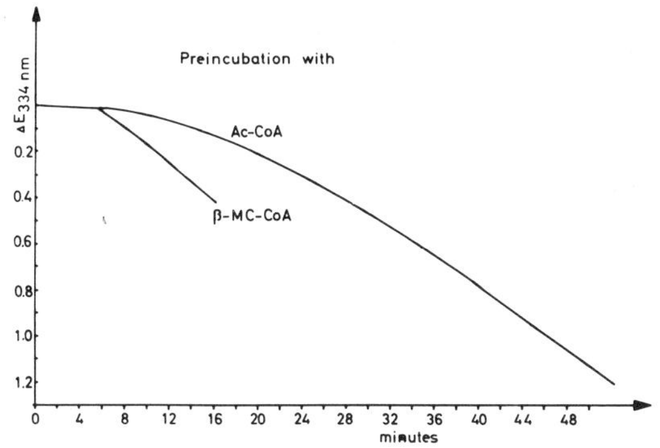

Fig. 13. Kinetics of enzyme reactivation as measured by the optical assay.

ß-methylcrotonyl-CoA carboxylase is measured according to (5) by coupling the ADP-generating carboxylase reaction to pyruvate kinase and lactate dehydrogenase. The cuvette, d = 1 cm, contained in 2 ml: 0.1 M Tris-HCl buffer, pH 8.0; 4 mM $MgCl_2$; 0.5 mM ATP; 0.75 mM phosphoenolpyruvate; 5 mM $KHCO_3$; 1.5 mM NADH; 1 mg serum albumin; 50 µg lactate dehydrogenase, 10 µg pyruvate kinase and 0.1 mM ß-methylcrotonyl-CoA as substrate. After 6 minutes the reaction was started by the addition of 0.08 mg of the dissociated carboxylase preincubated in presence of acetyl-CoA (Ac-CoA) or ß-methylcrotonyl-CoA (ß-MC-CoA) as indicated in the figure (see text). The oxidation of NADH was measured at λ = 334 nm. T = 25° C.

tive protein fraction was found to be unable to catalyze the carboxylation of ß-methylcrotonyl CoA but could still catalyze the carboxylation of added biotin. This is documented by the experiments of Figure 15.

In these experiments the carboxylating activities of the intact enzyme and of the biotin-containing subunit, abbreviated protein B, were measured with our standard optical assay (6) (see legend to Figure 13). The protein fractions were calibrated by their radioactivity due to the covalently bound ^{14}C-biotin. Based on it the capacity of protein B to carboxylate the biotin added reached between 50-60 % of the activity of the intact enzyme. As was known from our previous studies (6) and was confirmed by the present experiments ß-methylcrotonyl-CoA represents a much better substrate than biotin for the intact enzyme. However, protein B was devoid of any activity with ß-methylcrotonyl-CoA. It should be emphasized however that this component of the multienzyme complex is able to carboxylate biotin without any further additions, what indicates that its catalytically active conformation does not depend on changes induced by its association in the complex.

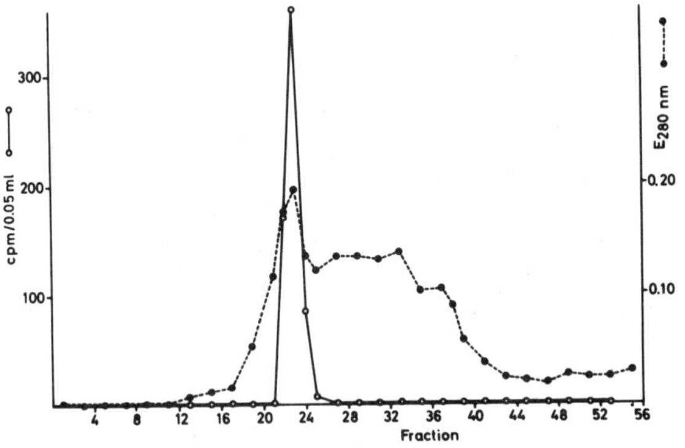

Fig. 14. Isolation of the ^{14}C-biotin containing subunit (protein B) of ß-methylcrotonyl-CoA carboxylase.

3.0 mg of ß-methylcrotonyl-CoA carboxylase (78 000 c.p.m.) after having been incubated with 5 M urea, 0.03 M cysteine and 0.03 M Tris-HCl buffer, pH 7.15 for one hour at 4° C were added to a column of DEAE-cellulose (1.8 x 26 cm) and eluted with 340 ml of a linear NaCl-gradient (0-0.3 M) in 5 M urea, 0.03 M cysteine and 0.03 M Tris-HCl buffer, pH 7.15. Volume of fractions: 6 ml.
Open circles: Radioactivity in c.p.m./0.05 ml
Closed circles: Extinction at 280 nm.

Various trials to isolate the second polypeptide from the dissociated enzyme complex by chromatography over DEAE-cellulose were not successfull. The protein, abbreviated protein A by us, possesses a strong tendency for association and adsorption on glass surfaces. Eventually Schiele succeeded in this task by using an avidin-sepharose column for the separation of both proteins. A solution of ß-methylcrotonyl-CoA carboxylase, dissociated by dialysis against cysteine-glycine-buffer, pH 9,8; was passed over this column prepared according to (23). By complexing with the avidin the radioactive biotin-containing protein B was quantitatively retained by the column whereas protein A appeared in the eluate.

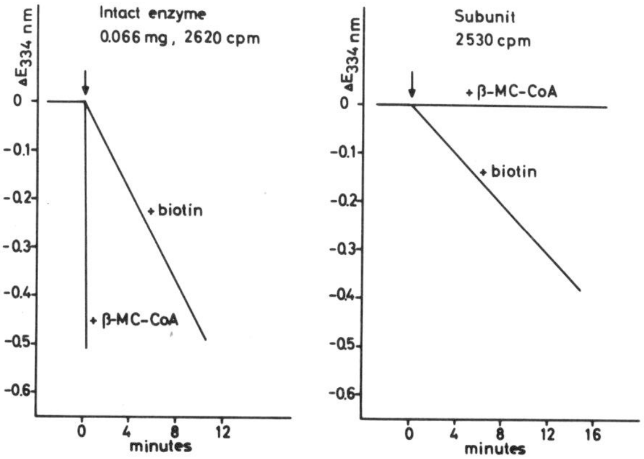

Fig. 15. Enzymic activities of intact enzyme and protein B.
For details of the optical assay see Figure 13. Enzyme or protein B were added as indicated in the figure. The reactions were started with the addition of 0.2 µmole of ß-methylcrotonyl-CoA (ß-MC-CoA) or 60 µmoles of biotin.T = 25° C.

It is characterized by its high extinction at 280 nm and its already mentioned great tendency for aggregation. Checked by sodium dodecyl sulfate disk-gel electrophoresis the isolated protein A proved to be homogeneous and clearly separable from protein B.

When assayed under standard canditions it could be demonstrated that protein B, responsible for the carboxylation of protein-bound biotin or added free biotin, with the addition of

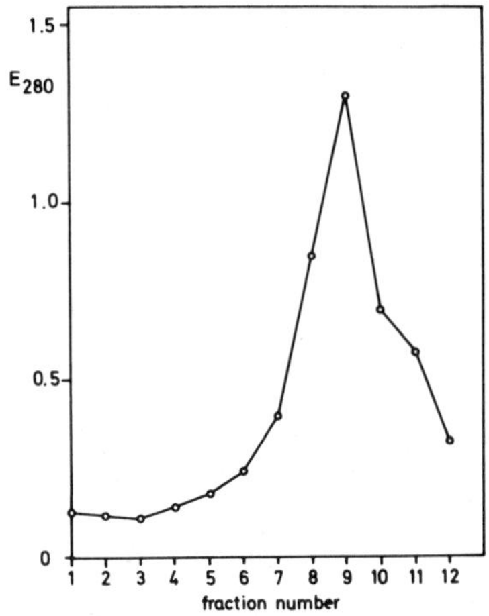

Fig. 16. Chromatographic separation of protein A.

3.8 mg ß-methylcrotonyl-CoA carboxylase dissolved in 1.4 ml were dialyzed for 3 days at 4° C and under an atmosphere of nitrogen against 10 mM glycine buffer, pH 9.8, containing 10 mM cysteine. The solution was filtered over the avidin-Sepharose column (1.1 x 11 cm) prepared according to (23) and possessing a binding capacity of 28.6 nmoles biotin/ml Sepharose. In order to exclude oxygen column and fractions were covered with paraffin oil. Volume of fractions: 1 ml. The curve represents the UV-absorption of each fraction measured at 280 nm. d = 1 cm.

the purified protein A regained the activity to carboxylate ß-methylcrotonyl CoA. This is illustrated in Figure 17.

We thus may conclude that protein A carries the transcarboxylase activity of the multienzyme complex. Preliminary evidence indicates however that this catalytic activity requires its association with protein B as prerequisite. In order to explain this observation it may be assumed that the active conformation of the transcarboxylase requires the specific protein-protein interactions which occur in the association of the separate polypeptide chains within the quaternary structure of the intact ß-methylcrotonyl-CoA carboxylase. From the reported molecular weights of the intact carboxylase

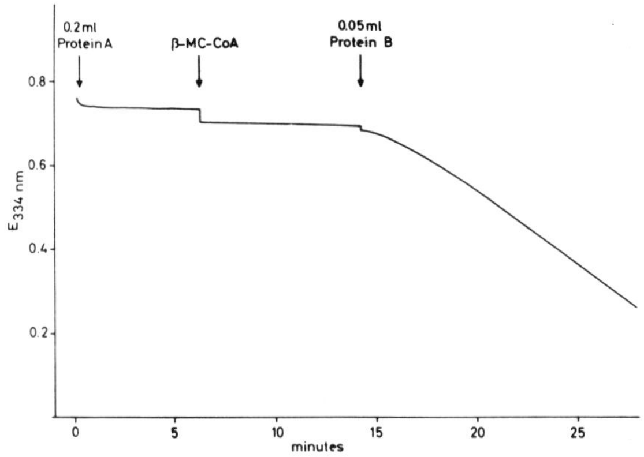

Fig. 17. Restitution of active ß-methylcrotonyl-CoA carboxylase from the separated proteins A and B.

For details of the optical assay see Figure 13. At the arrows the following additions were made: 35 µg of protein A; 0.2 µmole ß-methylcrotonyl-CoA (ß-MC-CoA); 4 µg of protein B. The oxidation of NADH was measured at 334 nm; T = 25° C; d = 1 cm.

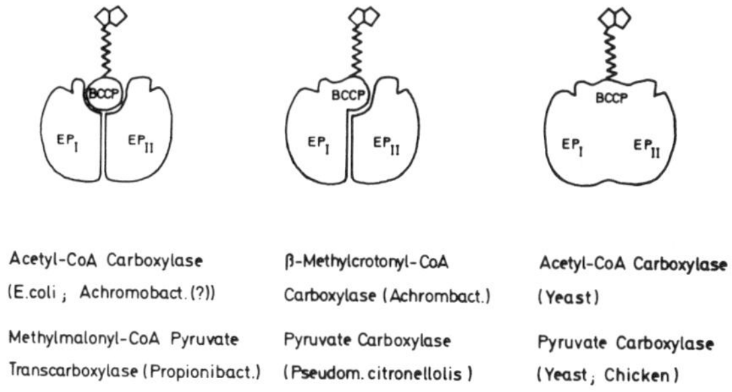

Fig. 18. Classification of biotin enzymes according to their structures.

and of its constituents: Protein A and protein B, the polypeptide structure A_4B_4 for the native enzyme follows.

In summary we thus come to the conclusion that biotin enzymes can be arranged in three groups, as illustrated in Figure 18.

In the first group, represented by acetyl-CoA carboxylase of E. coli and the transcarboxylase of propionibacteria, the active enzyme can be resolved in three types of functional components. 1) The biotin-carboxyl-carrier protein (BCCP) 2) the biotin carboxylase (EP_I) and 3) the carboxyl-transferase (EP_{II}).

In the second group, as represented by ß-methylcrotonyl-CoA carboxylase from Achromobacter only two types of polypeptides are present. One of them carries the biotin carboxylase activity together with the biotin-carboxyl-carrier protein, the other one carries the carboxyl-transferase activity. Each of these polypeptides is present in the native enzyme four times. To this group of carboxylases belongs also pyruvate carboxylase of Pseudomonas citronellolis as recent experiments of Barden and Taylor (24) have shown.

In the third group finally all three functions are incorporated in one multifunctional polypeptide chain, which again occurs several times in the native enzyme. Examples are acetyl-CoA carboxylase and pyruvate carboxylase of yeast and pyruvate carboxylase of chicken, as the studies of Utter and his colleagues (25) have shown.

The various types of carboxylases might represent various stages in the evolution of the enzyme system. According to this concept, the carboxylations may originally have been carried out by separate enzymes with an easily dissociable biotin-carboxyl-carrier protein. In the course of evolution the increasing functional structuring of the cellular interior led to the formation of complexes with multifunctional polypeptide chains, probably as the result of gene fusion. Their biosynthesis is from the kinetic as well as the regular point of view superior to the formation of an aggregate composed of individual proteins. Not only the association process appears to be much simpler, but also the problem of stoichiometry with respect to the component proteins is much easier to overcome.

References

1) Prescott, D.J. and P.R. Vagelos: Acyl carrier protein. Adv. Enzymol. 36, 269-311 (1972).

2) Greull, G.: Das Acyl Carrier Protein aus Penicillium patulum. Thesis, University of Munich, 1973.

3) Lynen, F., E. Oesterhelt, E. Schweizer and K. Willecke: The Biosynthesis of Fatty Acids. In: Cellular Compartmentalization and Control of Fatty Acid Metabolism. Universitetsforlaget, Oslo 1968, 1-24.

4) Lynen, F., J. Knappe, E. Lorch, G. Jütting, and E. Ringelmann: Die biochemische Funktion des Biotins. Angew. Chem. 71, 481-486 (1959).

5) Apitz-Castro, R., K. Rehn and F. Lynen: ß-Methylcrotonyl-CoA-Carboxylase, Kristallisation und einige physikalische Eigenschaften. Eur. J. Biochem. 16, 71-79 (1970).

6) Lynen, F., J. Knappe, E. Lorch, G. Jütting, E. Ringelmann, and J.P. Lachance: Zur biochemischen Funktion des Biotins. II. Reinigung und Wirkungsweise der ß-Methylcrotonyl-Carboxylase. Biochem. Z. 335, 123-167 (1961).

7) Moss, J. and M.D. Lane: The Biotin-dependent enzymes. Adv. Enzymol. 35, 321-4.2 (1971).

8) Kaziro, Y. and S. Ochoa: Mechanism of the Propionyl Carboxylase Reaction I. Carboxylation and Decarboxylation of the Enzyme. J. biol. Chem. 236, 3131-3136 (1961).

9) Knappe, J., E. Ringelmann and F. Lynen: Zur biochemischen Funktion des Biotins III. Die chemische Konstitution des enzymatisch gebildeten Carboxy-biotins. Biochem. Z. 335, 168-176 (1961).

10) Knappe, J., B. Wenger and U. Wiegand: Zur Konstitution der carboxylierten ß-Methyl-crotonyl-Carboxylase (CO_2-Biotin-enzym). Biochem. Z. 337, 232-246 (1963).

11) Knappe, J. and F. Lynen: Kohlensäureübertragung durch Biotinenzyme. 14. Colloq. Ges. physiol. Chem., Mosbach, Baden, 265-275. Berlin, Göttingen und Heidelberg, Springer-Verlag, 1964.

12)Alberts, A.W., A.M. Nervi and P.R. Vagelos: Acetyl-CoA Carboxylase, II. Demonstration of Biotin-Protein and Biotin Carboxylase Subunits. Proc. Nat. Acad. Sci., U.S.A. 63, 1319-1326 (1969).

13)Green, N.M., R.C. Valentine, N.G. Wrigley, F. Ahmad, B. Jacobson and H.G. Wood: Transcarboxylase XI. Electron microscopy and subunit structure. J. biol. Chem. 247, 6284-6298 (1972).

14)Sumper, M. and Ch. Riepertinger: Structural Relationship of Biotin-Containing Enzymes. Acetyl-CoA Carboxylase and Pyruvate Carboxylase from Yeast. Eur. J. Biochem. 29, 237-248 (1972).

15) Young, M.R., B. Tolbert and M.F. Utter: Pyruvate Carboxylase from Saccharomyces cerevisiae. Methods Enzymol. 13, 250-258 (1969).

16) Taylor, B.L., R.E. Barden, and M.F. Utter: Identification of the Reacting Form of Pyruvate Carboxylase. J. Biol. Chem. 247, 7383-7390 (1972).

17) Maurer, H.R.: Disk-Elektrophorese. Walter de Gruyter und Co., Berlin, 1968.

18) Schachmann, H.K.: Ultracentrifugation in Biochemistry. Academic Press, New York, 1959.

19) Schiele, U.: Untersuchungen zur Struktur von Biotinenzymen in Achromobacter. Thesis, University of Munich, 1973.

20) Merkenschlager, M., K. Schlossmann und W. Kurz: Ein mechanischer Zellhomogenisator und seine Anwendbarkeit auf biologische Probleme. Biochem. Z. 329, 332-340 (1957).

21) Jauch, R.: Unpublished results.

22) Weber, K. and M. Osborn: The Reliability of Molecular Weight Determinations by Dodecyl Sulfate-Polyacrylamide Gel Electrophoresis. J. biol. Chem. 244, 4406-4412 (1969).

23) Bodanszky, A. and M. Bodanszky: Sepharose-Avidin Column for the Binding of Biotin or Biotin-Containing Peptides. Experientia 26, 327 (1970).

24) Barden, R.E. and B.L. Taylor: Structure of Pyruvate Carboxylase. Federation Proc.32, 510 (1973).

25) Utter, M.F. and M.C. Scrutton: Pyruvate Carboxylase. Current Topics in Cellular Regulation 1, 253-296 (1969).

Second row: Zachau, Lynen; Krisko, Hirsch-Kauffmann;
Third row: Stadtman, Lipmann; Jones, Eigen;

First row: Lynen, Mrs. Stadtman; The Lipmanns;
Second row: Lipmann, Krebs; Krebs, Mrs. Lipmann, Roskoski, Lipmann;
Third row: Richter, Ureta, Krisko, Lipmann, Wittmann, Zachau, Lynen; Spector, Herrlich;
Fourth row: Chapeville, Hildebrand, Wittmann-Liebold, Bauer; Lipmann, Kleinkauf, Chapeville;

First row: Acs, Krebs, Elliott; Mrs. Lipmann, Krebs;
Second row: Chris Gillespie, Lipmann; Schweiger, Gordon, Roskoski;
Third row: Stadtman, Lipmann, Bennett; Hartmann, Ebashi, Gordon, Boman, Traut;
Fourth row: Mrs. Bauer, Elliott, Mrs. Richter, Srere, Boman; Haenni, Wittmann, Kucan.